The Handbook of Environmental Chemistry

Volume 107

Founding Editor: Otto Hutzinger

Series Editors: Damià Barceló • Andrey G. Kostianoy

In over four decades, *The Handbook of Environmental Chemistry* has established itself as the premier reference source, providing sound and solid knowledge about environmental topics from a chemical perspective. Written by leading experts with practical experience in the field, the series continues to be essential reading for environmental scientists as well as for environmental managers and decision-makers in industry, government, agencies and public-interest groups.

Two distinguished Series Editors, internationally renowned volume editors as well as a prestigious Editorial Board safeguard publication of volumes according to high scientific standards.

Presenting a wide spectrum of viewpoints and approaches in topical volumes, the scope of the series covers topics such as

- local and global changes of natural environment and climate
- anthropogenic impact on the environment
- water, air and soil pollution
- remediation and waste characterization
- environmental contaminants
- biogeochemistry and geoecology
- chemical reactions and processes
- chemical and biological transformations as well as physical transport of chemicals in the environment
- environmental modeling

A particular focus of the series lies on methodological advances in environmental analytical chemistry.

The Handbook of Environmental Chemistry is available both in print and online via https://link.springer.com/bookseries/698. Articles are published online as soon as they have been reviewed and approved for publication.

Meeting the needs of the scientific community, publication of volumes in subseries has been discontinued to achieve a broader scope for the series as a whole.

Nature-Based Solutions for Flood Mitigation

Environmental and Socio-Economic Aspects

Volume Editors: Carla S. S. Ferreira · Zahra Kalantari · Thomas Hartmann · Paulo Pereira

With contributions by

X. Alvarez · D. Barcelo · D. Bekić · R. Bekova · B. Bett · N. Bezak · K. Bogdzevič · O. Bonacci · J. Brabec · O. Cudlín · P. Cudlín · G. Delia · A. Ferreira · C. S. S. Ferreira · A. Figurek · J. M. García-Ruiz · M. Gazdic · E. Gomes · M. Gómez-Rúa · M. Gustafsson · T. Hartmann · M. Hekrle · Z. Hrkić Ilić · P. Hudson · A. Hysa · M. Inácio · J. Jakubínský · A. Janeiro-Otero · G. Johnen · C. Juez · Z. Kalantari · M. Kalinauskas · M. Kapović Solomun · D. Karnauskaitė · M. Kaufmann · M. Khorchani · I. Kotsev · T. Kulić · A. Kuriqi · N. Lana-Renault · T. Lasanta · J. Lindahl · L. Löschner · J. Macháč · Ž. Marjanović-Balaban · P. Matczak · K. Mikša · E. Nadal-Romero · F. C. Pavesi · V. Pechanec · D. Peña-Angulo · P. Pereira · M. Pezzagno · L. V. Pinto · K. Potočki · S. Priest · B. Prodanov · J. Purkyt · O. Rahmati · P. Raška · D. Regüés · R. Ristić · K. Roesel · S. Rusjan · K. Sapač · A. Schindelegger ·

S. Seifollahi-Aghmiuni · L. Slavíková · R. Stričević · N. Šumatić · D. Tumusiime · A. Vacková · J. Vidal-Puga · A. Vidmar · T. Vleesenbeek · H. N. von Platen · W. Warachowska · V. Zupanc

Springer

This publication is based upon work from COST Action LAND4FLOOD: Natural Flood Retention on Private Land (CA16209), supported by COST (European Cooperation in Science and Technology).

COST is a funding agency for research and innovation networks. The Actions help connect research initiatives across Europe and enable scientists to grow their ideas by sharing them with their peers. This boosts their research, career and innovation.

www.cost.eu

COST
EUROPEAN COOPERATION IN SCIENCE & TECHNOLOGY

COST is supported by the Horizon 2020 Framework Programme of the European Union

Editors
Carla S. S. Ferreira
Department of Physical Geography and Bolin Centre for Climate Research
Stockholm University
Stockholm, Sweden

Zahra Kalantari
Department of Sustainable Development, Environmental Science and Engineering (SEED)
KTH Royal Institute of Technology
Stockholm, Sweden

Thomas Hartmann
School of Spatial Planning, Fachgebiet BBV
TU Dortmund University
Dortmund, Germany

Paulo Pereira
Environmental Management Center
Mykolas Romeris University
Vilnius, Lithuania

ISSN 1867-979X ISSN 1616-864X (electronic)
The Handbook of Environmental Chemistry
ISBN 978-3-030-77504-9 ISBN 978-3-030-77505-6 (eBook)
https://doi.org/10.1007/978-3-030-77505-6

This Springer imprint is published by the registered company Springer Nature Switzerland AG
The registered company address is: Gewerbestrasse 11, 6330 Cham, Switzerland

Series Preface

With remarkable vision, Prof. Otto Hutzinger initiated *The Handbook of Environmental Chemistry* in 1980 and became the founding Editor-in-Chief. At that time, environmental chemistry was an emerging field, aiming at a complete description of the Earth's environment, encompassing the physical, chemical, biological, and geological transformations of chemical substances occurring on a local as well as a global scale. Environmental chemistry was intended to provide an account of the impact of man's activities on the natural environment by describing observed changes.

While a considerable amount of knowledge has been accumulated over the last four decades, as reflected in the more than 150 volumes of *The Handbook of Environmental Chemistry*, there are still many scientific and policy challenges ahead due to the complexity and interdisciplinary nature of the field. The series will therefore continue to provide compilations of current knowledge. Contributions are written by leading experts with practical experience in their fields. *The Handbook of Environmental Chemistry* grows with the increases in our scientific understanding, and provides a valuable source not only for scientists but also for environmental managers and decision-makers. Today, the series covers a broad range of environmental topics from a chemical perspective, including methodological advances in environmental analytical chemistry.

In recent years, there has been a growing tendency to include subject matter of societal relevance in the broad view of environmental chemistry. Topics include life cycle analysis, environmental management, sustainable development, and socio-economic, legal and even political problems, among others. While these topics are of great importance for the development and acceptance of *The Handbook of Environmental Chemistry*, the publisher and Editors-in-Chief have decided to keep the handbook essentially a source of information on "hard sciences" with a particular emphasis on chemistry, but also covering biology, geology, hydrology and engineering as applied to environmental sciences.

The volumes of the series are written at an advanced level, addressing the needs of both researchers and graduate students, as well as of people outside the field of "pure" chemistry, including those in industry, business, government, research establishments, and public interest groups. It would be very satisfying to see these volumes used as a basis for graduate courses in environmental chemistry. With its high standards of scientific quality and clarity, *The Handbook of Environmental Chemistry* provides a solid basis from which scientists can share their knowledge on the different aspects of environmental problems, presenting a wide spectrum of viewpoints and approaches.

The Handbook of Environmental Chemistry is available both in print and online via https://link.springer.com/bookseries/698. Articles are published online as soon as they have been approved for publication. Authors, Volume Editors and Editors-in-Chief are rewarded by the broad acceptance of *The Handbook of Environmental Chemistry* by the scientific community, from whom suggestions for new topics to the Editors-in-Chief are always very welcome.

Damià Barceló
Andrey G. Kostianoy
Series Editors

Preface

This book has been developed as an outcome of the LAND4FLOOD network. LAND4FLOOD brings together academics and professionals in flood risk management across Europe and beyond to support the implementation of spatial flood risk management on private land. The initiative is supported by the European COST funding programme (European Cooperation in Science and Technology, www.cost.eu). It builds on a long-term cooperation of a group of researchers around the topic. Numerous publications, collaborations in science and practice, and knowledge and experience-sharing have been produced since its foundation (see land4flood.eu). The main audience of LAND4FLOOD outputs is high-level policy makers, water managers, spatial planners, lawyers, and other stakeholders involved in spatial flood risk management.

Land matters in flood risk management. This is the starting point and key message of LAND4FLOOD. How land matters is explored in numerous publications and outreaches, of which this book is a part of. A special issue in the Journal of Flood Risk Management explored how a catchment-wide and multi-scale perspective on land as a biophysical factor, but also as an important socio-economic institution of interest can potentially contribute to alleviate flood risks [1]. It also points at the questions of scale regarding land and the need to involve multiple disciplines – including hydrology, engineering, economics, and planning. In subsequent years, LAND4FLOOD published further special issues and books focusing on financial schemes for flood recovery [2] and flood resilience of private properties [3] to discuss the role and responsibility of private landowners in dealing with flood risks. The special collection on implementing flood-resilience on the local scale revealed both, the importance of bridging disciplines and the key role private property can play in reducing flood risks [4]. How this can be achieved in different countries is then discussed in another special issue that focused on policies and instruments for mobilizing private land for flood risk management [5]. The book on nature-based flood risk management on private land [6] uncovered the challenges of relying on multiple disciplines to realize measures on private land. LAND4FLOOD shows that land is crucial for realizing nature-based flood mitigation and resilience, how land in private ownership is a potential but also a challenge, and that multiple disciplines need to be involved to make such private land available.

One of the overall lessons is that flood risk management in general and nature-based solutions on private land in particular have not only huge potential for interdisciplinary and also international collaboration, but they also show how difficult it can be to cross disciplinary and national boundaries. This book is an essential contribution as the focus on flood mitigation via nature-based solutions complements the existing knowledge brought together by LAND4FLOOD. In addition, it brings together environmental and socio-economic aspects, enriched with case studies from different countries in a disciplinary book series of Springer, namely on: "The Handbook of Environmental Chemistry". This multidisciplinary perspective allows, not only to understand the different rationales and approaches of disciplines and countries to potentially apply or incorporate some of the merits of other disciplinary approaches, but probably more importantly, it fosters reflections by each discipline on its own rationale and blind spots. This provides a basis for a better implementation of nature-based flood risk management, as it is the major vision of LAND4FLOOD.

References

1. Hartmann T, Jílková J, Schanze J (2018) Land for flood risk management: a catchment-wide and cross-disciplinary perspective. J Flood Risk Manag 11 (1):3–5. https://doi.org/10.1111/jfr3.12344
2. Slavíková L, Hartmann T, Thaler T (2020) Financial schemes for resilient flood recovery. Environ Hazards 19 (3): 223–227. https://doi.org/10.1080/17477891.2019.1703624
3. Hartmann T, Slavíková L, McCarthy S (eds) (2019) Nature-based flood risk management on private land. disciplinary perspectives on a multidisciplinary challenge, 1st edn. Springer, Cham
4. Hartmann T, Jüpner R (2020) Implementing resilience in flood risk management: editorial commentary. WIREs Water. https://doi.org/10.1002/wat2.1465
5. Löschner L, Hartmann T, Priest S, Collentine D (2021) Strategic use of instruments of land policy for mobilising private land for flood risk management. Environ Sci Policy (in press)
6. Hartmann T, van Doorn-Hoekveld W, van Rijswick M, Spit T (2019) Editorial. Water Int 44(5):489–495. https://doi.org/10.1080/02508060.2019.1671464

Ústí nad Labem, Czech Republic
Dortmund, Germany

Lenka Slavikova
Thomas Hartmann

Contents

Part II NBS Effectiveness in Flood Mitigation: Assessment Methods and Case Studies

Part III Socio-Economic Aspects of NBS

Introduction: Nature-Based Solutions for Flood Mitigation

Carla S. S. Ferreira, Zahra Kalantari, Thomas Hartmann, and Paulo Pereira

Abstract Floods are one of the most common natural disasters affecting numerous people worldwide. Over the last years, Nature-Based Solutions (NBS) have gained attention as an emerging approach for flood mitigation that can complement traditional grey infrastructures. NBS provide several ecosystem services, including flood mitigation and improved water quality. Increasing political awareness and interest from the scientific community have led to the implementation of NBS worldwide. This contribution provides an overview of the concept of NBS for flood mitigation, focusing on (1) the environmental impacts of NBS, (2) the effectiveness of NBS in flood mitigation based on several case studies, and (3) the socio-economic aspects of NBS. Compiling the latest research, this book furthers our understanding of the role

C. S. S. Ferreira (✉)
Department of Physical Geography, Stockholm University, and Bolin Centre for Climate Research, Stockholm, Sweden

Navarino Environmental Observatory, Messinia, Greece
e-mail: carla.ferreira@natgeo.su.se

Z. Kalantari
Department of Physical Geography, Stockholm University, and Bolin Centre for Climate Research, Stockholm, Sweden

Navarino Environmental Observatory, Messinia, Greece

KTH, School of Architecture and the Built Environment (ABE), Sustainable Development, Environmental Science and Engineering, Sustainability Assessment and Management, Stockholm, Sweden
e-mail: zahra.kalantari@natgeo.su.se

T. Hartmann
Wageningen University, Wageningen, The Netherlands

J.E.P. University, Ústí nad Labem, Czech Republic
e-mail: thomas.hartmann@wur.nl

P. Pereira
Environmental Management Laboratory, Mykolas Romeris University, Vilnius, Lithuania

Carla S. S. Ferreira, Zahra Kalantari, Thomas Hartmann, and Paulo Pereira (eds.), *Nature-Based Solutions for Flood Mitigation: Environmental and Socio-Economic Aspects*, Hdb Env Chem (2022) 107: 1–8, DOI 10.1007/698_2021_776,

Published online: 3 July 2021

of NBS for flood mitigation and its relation to environmental aspects, to guide students, managers, practitioners, policy-makers, and scientists in future NBS projects.

Keywords Flood mitigation, Nature-based solutions, Socio-economic aspects, Water management, Water quality

Flooding is the most widespread natural disaster, representing 47% of all weather-related disasters, and one of the most expensive in terms of socio-economic damages [1]. About 58 million people worldwide are annually exposed only to river floods, with associated estimated direct costs of €110 billion [2]. The number of deaths, affected people, and economic losses varies significantly by year and continent; however, Asia is where the impacts are higher [3]. In Europe, floods already represent 33% of the natural events recorded between 1900 and 2019 [4]. Dealing with flood events is thus already a major challenge.

The IPCC reports of the past increasingly stress the severity of the impacts and the increase in frequency and intensity that flooding will cause [5]. Flooding is one of the most tangible consequences of climate change and it threatens communities around the globe [6]. The number of flood events and the consequent damages are expected to increase in a climate change context [5]. Several studies suggest that extreme weather-related events such as heavy precipitation, storms, and floods will become more frequent and intense in Europe, although with relevant differences between regions, seasons, and time periods [7]. The average damages of coastal flooding are expected to increase from €3 billion per year to up €38 billion by the end of 2100 [8]. So, the challenges associated with flood risks are very likely to increase substantially in the future.

Floods are defined as the temporary cover of land by water in areas normally dry [9]. They are driven by a combination of meteorological and hydrological extremes, caused by heavy rainfall, snowmelt, or sea level rise. They are also deeply influenced by human factors, which alter natural landscapes and change the water cycle (e.g. urbanization, deforestation) enhancing flood hazard. Increasing number of fluvial and pluvial flood events have been noticed since the second half of the twentieth century [4]. In Europe, the major river floods in 1993 and 1995 were important turning points in the way society responds to floods [10]. In this book, we address floods recorded in rural, urban, and coastal areas, mainly focusing on the most common types of flooding: river floods, pluvial floods, urban floods, and coastal floods.

Traditionally, floods have been managed with a strong focus on engineering infrastructure solutions (called also grey solutions), such as dikes and dams [11]. Nonetheless, these solutions have been increasingly questioned in the past decades. They are planned for floods with a specific return period. The consequence is that most grey solutions protect against often but less severe events, and only a few are designed to protect against extreme events [6]. This provides limited flexibility and adaptability to cope with increasing floods hazard driven by climate change and urbanization [12]. Additionally, possible failure of grey infrastructures can have dramatic social, economic, and environmental consequences, and provide a false sense of security [6], referred to as dike paradox [13]. The ongoing paradigm shift from flood protection to flood risk management in Europe since the 1990s did change the emphasis of grey infrastructure a bit, by taking vulnerabilities into account [14], but it is still largely focused on grey infrastructure.

In recent decades, the need for flexible and multifunctional solutions has led to the emergence of nature-based solutions (NBS) [15]. They are understood as solutions which are inspired by, supported by, or copied from nature [16]. The concept of NBS was first applied in the 2000s as part of integrated climate change mitigation and adaptation, biodiversity protection, and sustainable livelihood actions [17]; however, its use rapidly spreads into other areas. In water management, NBS promise to alter, restore, or use landscape features, for example, to improve soil infiltration, enhance water retention, intercept rainfall, enhance evapotranspiration and therefore reduce surface runoff and flood hazard (flood mitigation) [18], though the effects are often debated in academic literature [19]. NBS also includes the concept of making "space for the rivers", demonstrating a paradigm shift from quickly bypassing the streamflow into downstream areas to cope with floods [20]. Notwithstanding the critique that NBS might be less effective against extreme events or that they need more space [21] is still part of the contemporary debate on flood mitigation.

Over the last years, also the political awareness and interest from the scientific community have led to increasing implementation of NBS worldwide. In Europe, for example, the Action Programme (a framework for policy-making establishing medium- and long-term goals) on flood risk management developed by the European Commission in 2004, defined objectives based on promoting sustainable flood risk management measures and the need to work with natural processes and to deliver multiple benefits from flood risk management [22]. With the publication of the European Flood Directive [9], the paradigm shift towards a more comprehensive flood risk management that embraces flood mitigation was institutionalized [23]. It demands that each member state in the European Union establishes flood risk management plans considering whenever possible "*the maintenance and/or restoration of floodplains, as well as measures to prevent and reduce damage to human health, the environment, cultural heritage and economic activity*" and including the promotion of water retention and mitigation. Thus, NBS started to be considered as relevant technical solutions to complement typical grey infrastructures to mitigate flood hazard.

NBS promise to have – next to flood mitigation – multiple benefits. Besides water regulation and flood risk reduction, NBS can enhance water quality, regulate the climate, improve the quality of life, and support biodiversity [15, 24, 25]. These are just a few examples of the multiple ecosystem services supplied by NBS. The EU recognizes NBS as sustainable solutions to address several environmental, social, and economic challenges [16]. Thus, NBS have been increasingly recognized as a relevant option to meet contemporary and future water resources management challenges and to support achieving several of the United Nations Sustainable Development Goals established for 2030 [3].

Although the multiple ecosystem services provided by NBS and the increasing recognition of their relevance, NBS application is still limited [26]. This is, in part, due to the lack of evidence regarding the impact of NBS on flood mitigation [12]. Although several NBS have been implemented worldwide, most case studies are performed at the small scale [27]. NBS works at large-scale (e.g. on catchment level) are still lacking, as well as comparative studies between the effectiveness of NBS and grey solutions. There is also a lack of evidence on additional environmental, social, and economic benefits from NBS [16]. In addition, NBS in general demand more space than grey infrastructures. This represents a challenge when implementing such measures on private land [28], leading to conflicts of interest, and involving complex property right issues, in particular in urban areas [29]. The implementation of NBS for flood risk reduction is more complex than the implementation of grey solutions, involving different disciplines such as hydrology, ecology, geography, engineering, land management, sociology, and law [21]. Thus, NBS implementation is challenging and requires good communication and coordination between several stakeholders, including policy makers, planners, and engineers.

Bringing together knowledge and experiences from NBS for flood mitigation can help to identify research gaps but also showcase the merits and shortcomings of NBS for flood mitigation. Therefore, this book aims to (1) provide an overview of the typical NBS used for flood mitigation at different scales and in different areas (e.g. from catchment to hillslope scale; from urban to coastal areas); (2) enhance knowledge on the environmental aspects of NBS, particularly in the effectiveness of these solutions for flood mitigation; and (3) discuss socio-economic aspects related with the implementation of NBS, including regulatory aspects, costs of NBS implementation, and the perception of citizens about NBS.

This book, integrated in a book series "The Handbook of Environmental Chemistry", is structured in three main sections. The first section presents a state of the art about different NBS solutions implemented in distinct environments, ranging from floods in coastal to urban areas, and their environmental impacts particularly on water quality and pathogen dispersion. It comprises six chapters based on literature review addressing (1) different structural and non-structural measures typically used in flood mitigation, with particular focus on the main types of NBS used globally, and discusses the need for integrated strategies developed at catchment scale; (2) the main problems of coastal flooding and different types of NBS measures typically implemented to mitigate this problem, with a discussion of the main environmental, social, and economic benefits associated, and the challenges to implement NBS in these kinds of environments; (3) flooding in urban areas, depicting their main causes, the typical NBS measures used in these particular areas with serious space constraints, the current knowledge about their effectiveness on flood mitigation, their main advantages and disadvantages, and their role enhancing urban resilience; (4) environmental problems in urban areas, where most of the world population is concentrated and thus highly susceptible to floods and pollution, discussing how

NBS contribute to improve air, soil, and water quality and thus human well-being; (5) the role of floods on infectious disease epidemics affecting humans and animals, prevalent in developing countries, and how NBS can contribute to prevent and mitigate pathogen dispersion; and (6) the important role of plants as a NBS for water regulation and quality control, presenting and discussing different types of phytoremediation techniques and their relevance particularly in industrial areas.

The second section synthesizes knowledge on the effectiveness of NBS for flood mitigation, based on several case studies including in coastal, rural, and urban areas, and presents different methodologies used to best plan and develop NBS strategies. It includes eight chapters focused on case studies spread all over Europe, addressing NBS at different scales ranging from small and local measures to a network of measures integrated at large scale. Specifically, these chapters address (1) the serious problem of land abandonment in the Mediterranean region, how it affects flood in mountain areas, and the types of NBS measures and strategies that have been implemented to improve water regulation in this area highly prone to land degradation; (2) the role of afforestation on hillslopes and floodplains in flood mitigation in Central Slovenia, based on a combined modelling approach and cost–benefit analysis; (3) the impact of riparian woods and estuaries management on coastal flooding mitigation in Bulgaria; (4) the impact of wetlands on large-scale flood mitigation in Croatia, using long-term discharge data; (5) the impact of different types of habitats on soil water retention and infiltration capacity, and thus on runoff processes and flood mitigation in a medium-sized catchment located in Czech Republic; (6) the effectiveness of a flood mitigation approach combining NBS with grey infrastructures to mitigate small-scale urban floods with distinct return periods; (7) the adaptation of landscape connectivity principles to improve a spatial planning methodology to identify the best locations to place NBS, tested at catchment scale in Portugal and Sweden; and (8) an improved methodology to map the best places to implement NBS, based on the application of the Sponge cities concept at regional scale in Italy, through the identification of rural areas of particular interest for flood retention and landscaping.

The third section explores the socio-economic aspects of NBS, including the perception of people about NBS and several barriers for their implementation, including justice, policy regulatory frameworks, and property right issues. This section comprises six chapters focusing on (1) the combination of game theory model with cost–benefit analysis to support decision-making process regarding different types of NBS, based on its application in four European river basins; (2) a reflection on NBS for flood risk management from a justice perspective, focusing on the social point of view and including the fairness of the decision-making and relevance of public participation, based on case studies from different countries in Europe and Vietnam; (3) lessons from a field study developed in Czechia to assess the perception of people about NBS, including their preferences between different possible types to mitigate pluvial floods and their willingness to pay; (4) legal aspects of NBS for flood mitigation, discussing the coherence of several current laws affecting flood risk management using the example of Lithuania; (5) international experience on implementing NBS on private land, presenting and discussing different possible mechanisms such as expropriation and land-use restrictions, financial incentives and informational measures; and (6) socio-economic aspects of NBS and a review of methodologies and frameworks typically used to perform socio-economic aspects.

This book provides a compilation of the most recent research. By addressing different types of floods and presenting and discussing different types of NBS approaches used in distinct environments and scales, and considering a multidisciplinary overview, this book represents a step further in the knowledge of the role of NBS for flood mitigation, relevant to guide scientists and stakeholders in future NBS projects.

References

1. UNISDR and CRED (2015) The human cost of weather related disasters. 1995-2015. The United Nations Office for Disaster Risk Reduction and Centre for research on the Epidemiology of Disaster. https://www.unisdr.org/2015/docs/climatechange/COP21_WeatherDisastersReport_2015_FINAL.pdf
2. Dottori F, Szewczyk W, Martinez JCC, Zhao F, Alfieri L, Hirabayashi Y, Bianchi A, Mongelli I, Frieler K, Betts R, Feyen L (2018) Increased human and economic losses from river flooding with anthropogenic warming. Nat Clim Chang 8:781–786. https://doi.org/10.1038/s41558-018-0257-z
3. United Nations (2018) Nature-based solutions for water – the United Nations World Water Development Report 2018. Paris, UNESCO. http://www.unesco.org/new/en/natural-sciences/environment/water/wwap/wwdr/2018-nature-based-solutions/
4. Centre for Research on the Epidemiology of Disasters (2019) Natural disasters 2019. The international disaster database. https://www.emdat.be/publications
5. IPCC (2014) Climate change 2014: impacts, adaptation and vulnerability. IPCC WGII AR5, part B: regional aspects, pp. 1267-1327. https://www.ipcc.ch/report/ar5/wg2/
6. Depietri Y, McPhearson T (2017) Integrating the grey, green, and blue in cities: nature-based solutions for climate change adaptation and risk reduction. In: Kabisch N, Korn H, Stadler J, Bonn A (eds) Nature-based solutions to climate change adaptation in urban areas. Theory and practice of urban sustainability transitions. Springer, Cham, pp 91–109. https://doi.org/10.1007/978-3-319-56091-5_6
7. Wijesiri B, Liu A, Goonetilleke A (2020) Impact of global warming on urban stormwater quality: from the perspective of an alternative water resource. J Clean Prod 262:121330. https://doi.org/10.1016/j.jclepro.2020.12133.0
8. Heinzlef C, Robert B, Hémond Y, Serre D (2020) Operating urban resilience strategies to face climate change and associated risks: some advances from theory to application in Canada and France. Cities 104:102762. https://doi.org/10.1016/j.cities.2020.102762
9. Directive 2007/60/EC of the European Parliament and of the Council of 23 October 2007 on the assessment and management of flood risks. https://www.eea.europa.eu/policy-documents/directive-2007-60-ec-of
10. Hartmann T (2010) Reframing polyrational floodplains. Land policy for large areas for temporary emergency retention. Nat Cult 5(1)
11. Patt H, Jüpner R (eds) (2020) Hochwasser-Handbuch. Auswirkungen und Schutz, 3rd edn
12. Kalantari Z, Ferreira CSS, Deal B, Destouni G (2019) Nature-based solutions for meeting environmental and socio-economic challenges in land management and development (editorial). Land Degrad Dev 29(10):3607–3616. https://doi.org/10.1002/ldr.3264

13. Ferdous MR (2019) The levee effect along the Jamuna River in Bangladesh. Water Int 44 (5):496–519
14. van J. Ruiten, L., and Hartmann, T. (2016) The spatial turn and the scenario approach in flood risk management – implementing the European floods directive in the Netherlands. AIMS Environ Sci 3(4):697–713
15. Schanze J (2017) Nature-based solutions in flood risk management – Buzzword or innovation? J Flood Risk Manag 10(3):281–282
16. European Commission (2015) Towards an EU Research and Innovation policy agenda for Nature-Based Solutions & Re-Naturing Cities. Final report of the Horizon 2020 Expert group on "Nature-Based Solutions & Re-Naturing Cities". European Commission, Brussels
17. Escobedo RJ, Giannico V, Jim CY, Sanesi G, Lafortezza R (2019) Urban forests, ecosystem services, green infrastructure and nature-based Ferdous MR et al 2019. The levee effect along the Jamuna River in Bangladesh. Water Int 44(5):496–519
18. European Commission (2021) Future brief: the solution is in nature. Sci Environ Policy. https://ec.europa.eu/environment/integration/research/newsalert/pdf/issue-24-2021-02-the-solution-is-in-nature.pdf
19. Bornschein A, Pohl R (2017) Land use influence on flood routing and retention from the viewpoint of hydromechanics. J Flood Risk Manag 403(2011):103
20. Buuren E, Warner (2012) Space for the river: governance challenges and lessons. In: Warner JF, van Buuren A, Edelenbos J (eds) Making space for the river: governance experiences with multifunctional river flood management in the US and Europe. IWA Publishing, London, pp 187–200. Chapter 14
21. Hartmann T, Slavíková L, McCarthy S (2019) Nature-based solutions in flood risk management. In: Hartmann T, Slavíková L, McCarthy S (eds) Nature-based flood risk management on private land. Disciplinary perspectives on a multidisciplinary challenge. Springer, Berlin. Chapter 1, pp 3–8
22. Commission of the European Communities (2004) Communication from the Commission to the Council, The European Parliament, The European Economic and Social Committee and the Committee of the Regions. Flood risk management: Flood prevention, protection and mitigation. COM(2004)472 final. http://www.waterframeworkdirective.wdd.moa.gov.cy/docs/Floods/communication/com2004_0472en01.pdf
23. Hartmann T, Jüpner R (2014) Editorial. The flood risk management plan between spatial planning and water engineering. In: Hartmann T, Jüpner R (eds) The European flood risk management plan. Between spatial planning and water engineering
24. Collentine D, Futter MN (2016) Realising the potential of natural water retention measures in catchment flood management. Trade-offs and matching interests. J Flood Risk Manag 19 (2):771
25. Almenar JB, Elliot T, Rugani B, Philippe B, Gutierrez TN, Sonnemann G, Geneletti D (2021) Nexus between nature-based solutions, ecosystem services and urban challenges. Land Use Policy 100:104898. https://doi.org/10.1016/j.landusepol.2020.104898
26. Brillinger M et al (2020) Exploring the uptake of nature-based measures in flood risk management: evidence from German federal states. Environ Sci Policy 110:14–23
27. Raška P, Slavíková L, Sheehan J (2019) Scale in nature-based solutions for flood risk management. In: Hartmann T, Slavíková L, McCarthy S (eds) Nature-based flood risk management on private land. Disciplinary perspectives on a multidisciplinary challenge. Springer, Cham, pp 9–20
28. Hartmann T, Jílková J, Schanze J (2018) Land for flood risk management: a catchment-wide and cross-disciplinary perspective. J Flood Risk Manag 11(1):3–5
29. van Straalen F, Hartmann T, Sheehan J (eds) (2018) Property rights and climate change. Land-use under changing environmental conditions. Routledge, New York

Part I
Environmental Impacts of Nature-Based Solutions

Multidimensional Aspects of Floods: Nature-Based Mitigation Measures from Basin to River Reach Scale

Alban Kuriqi and Artan Hysa

Contents

Abstract This chapter aims to deliver a brief overview of flood mitigation and protective measures at different scales within the catchment area and identify the main factors to be considered in flood risk management. It stands on an extensive literature review of the ongoing scholarly discourse on the topic. The main focus is given to novel approaches that are based on Nature-Based Solutions (NBS)

A. Kuriqi
CERIS, Instituto Superior Técnico, Universidade de Lisboa, Lisbon, Portugal
e-mail: alban.kuriqi@tecnico.ulisboa.pt

A. Hysa (✉)
Faculty of Architecture and Engineering, Epoka University, Tirana, Albania
e-mail: ahysa@epoka.edu.al

Carla S. S. Ferreira, Zahra Kalantari, Thomas Hartmann, and Paulo Pereira (eds.),
Nature-Based Solutions for Flood Mitigation: Environmental and Socio-Economic Aspects,
Hdb Env Chem (2022) 107: 11–34, DOI 10.1007/698_2021_773,

Published online: 3 July 2021

principles, which are aligned with emerging public awareness not only towards cost-effective solutions but also towards sustainable proposals which go in harmony with the environment and natural landscapes. First, we provide a synopsis on the main aspects influencing the flood events such as Land-Use/Land Cover (LULC), topography, hydrometeorology, and hydraulics of the river. Later we included novel concepts such as "Sponge Cities", Integrated Water Resource Management (IWRM), and the Sustainable Urban Drainage System (SuDS) as examples of mitigation practices which simultaneously integrate cost-effective non-structural and structural mitigation measures. Further on, we give some insight on the recommendations on the most pressing research questions and conclude the key issues to be considered in the flood risk management concerning NBS approaches. As a result, the study highlights two among many recommendations as to the foremost crucial ones. First, the flood risk reduction agendas must adopt a spatially comprehensive and cross-scale approach while considering flood events. Finally, natural properties of the context such as LULC, topography, hydrometeorology, and hydraulics must be considered simultaneously to utilise the full natural capacity of the context in flood risk mitigation.

Keywords Flood mitigation upscaling, Integrated water resource management, LULC, Nature-based solutions, Sponge city

1 Introduction

1.1 *Climate Change and Uncertainty in Flood Risk Management*

Flooding is one of the most devastating and frequent natural hazards with enormous impacts on the environment, people, and economy over the globe [1]. Flooding occurs due to the overflow of fluvial systems, small streams, or lakes, influenced by heavy and/or intense rainfall. Floods have triggered irreversible tragedies and property damages over decades, where one of the major tragedies is the one of 1931 in Huang He River in China, where more 3.7 million of people had died [2]. Floods frequency, magnitude, duration, timing, as well as severity, differ between regions and depend on several factors such as LULC, meteorological conditions, the geomorphological context of the region [3], and geomorphology of the fluvial system among others [4, 5]. Figure 1 shows where flooding occurs more frequently. Namely, it shows hydrological floodplain defined by bankfull elevation.

In contrast, the topographic floodplain includes the hydrologic floodplain and other lands (flood fringes) up to a defined elevation; usually, it corresponds to 100-year floodplain. Very often, the settlements are placed near or in the zone of the 100-year floodplain. Therefore, they are vulnerable to flood events.

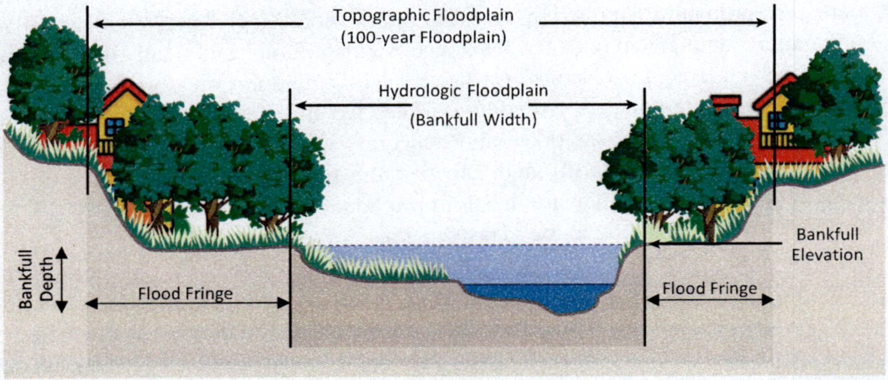

Fig. 1 Cross-section view shows the flood occurrence with regard to water level and floodplain width; it shows hydrologic and topographic floodplains, respectively, modified after Carolyn et al. [6]

In general, fluvial floods in lowland rivers characterised by broad and flat floodplains reveal to be more destructive as people historically tend to reside near to the rivers, which is particularly common in developing countries [1, 4]. Uncontrolled and rapid urbanisation has increased the risk of floods disproportionally; millions of people who live in informal buildings of low standards remain highly vulnerable [7]. Due to precipitation patterns alteration driven mainly by climate changes as a result of the unprecedented increase in greenhouse gas concentrations in the global atmosphere, floods are becoming more frequent and extreme [8, 9]. Climate change has already shifted the timing of fluvial floods. It is also anticipated to intensify their magnitude in the northern hemisphere, particularly in Europe [10]. In such circumstances, the management of an intensified flood risk due to more extensive and frequent floods is becoming more challenging. In a simplified view, flood risk management represents the required strategies of managing existing and potential future flood risk situation, while in a more holistic prospect, it includes several processes such as the planning, design, and implementation of the systems, which intend to reduce the flood risk [11].

Risk management is an intensively discussed and studied topic which regardless of the stakeholder's involvement, consists of three different sets of actions: (1) operation level – defining the actions which are needed to operate an existing system, (2) actions related to project planning level, which is used in case of new projects or regarding re-conceptualisation/revisions of existing projects, and (3) actions that are taken on a project design level, which represents an advanced stage of the second level and provides technical details on the design to achieve an optimal solution for the project implementation [12]. The effectiveness of risk management strategies is also closely connected to the way how people perceive and their willingness to respond to a potential risk that may be induced by floods. In this regard, the so-called protection motivation theory defines the self-preservation behaviour driven by four main factors: (1) the degree of perceived severity of a threat from floods, (2) the perceived probability of the frequency of extreme floods, (3) the effectiveness of any potential recommended response by respective institutions, and finally (4) perceived ability to implement a response (protective and mitigation measures) [13]. Traditional engineering practices referring to protective and mitigation measures include building structures such as dikes and deflectors in the areas vulnerable to floods [14]. Nevertheless, feasibility and effectiveness of such measures in many cases remain questionable as they are meant to provide local protection only.

Furthermore, local protection measures transfer the accumulated risk downstream increasing the burden of the community residing in those areas. Therefore, risk management should be done in a more holistic approach, in larger scale rather than river reach scale by considering many factors triggering the floods and also affected by floods [15]. In this context, the Nature-Based Solutions (NBS) concept has proved to be practical, feasible, and eco-friendly as they offer flood protection without imposing substantial modification in the environment [16].

1.2 Importance of Upscaling in Flood Mitigation

Although floods may affect large areas by causing substantial life losses and damage to the properties, periodic monitoring, planning, and management can reduce the devastating impacts of floods. Integrating NBS as a complementary alternative into the standard civil engineering measures has provided more room for sustainable solutions but in the meantime, defining specific flood protection and/or mitigation strategies is becoming more sophisticated and challenging due to exponential increase of urbanisation [17]. Upscaling prevention or mitigation measures consist of composite measures taken at different spatial scale within the basin, as it considers the enhancement of land cover and implementing mainly structural and/or non-structural measures from basin scale to river reach scale (Fig. 2).

Allocation of an enormous amount of water volume during instantaneous floods is challenging, particularly in urban areas. In this regard, when Flood-Excess Volume (FEV) is reached, the damage risk is higher. In such circumstances, Square-lake mitigation measures proposed by Bokhove et al. [18] (Fig. 3) is a cost-effective solution to manage a high volume of water during extreme floods.

Although square-lake or leaky dam concepts, a flooding prevention measure, moderating the flow of water downstream [19], might be feasible in terms of cost, their effectiveness may not be guaranteed if simultaneously additional measures do not take place at a large scale, i.e., within the entire catchment/basin. Indeed, measures taken within the catchment area can mitigate the risk of extreme floods significantly by ensuring the high effectiveness of the protective measures implemented in the river reach scale [20, 21]. Upscaling of mitigation measures is particularly important because the non-adequate placement of measures can simply transfer the flood burden downstream by increasing the risk, losses, and costs even more [22]. In this context, Fig. 4 shows the importance of upscaling of the mitigation measures and also how local measures alone fail to prevent flooding.

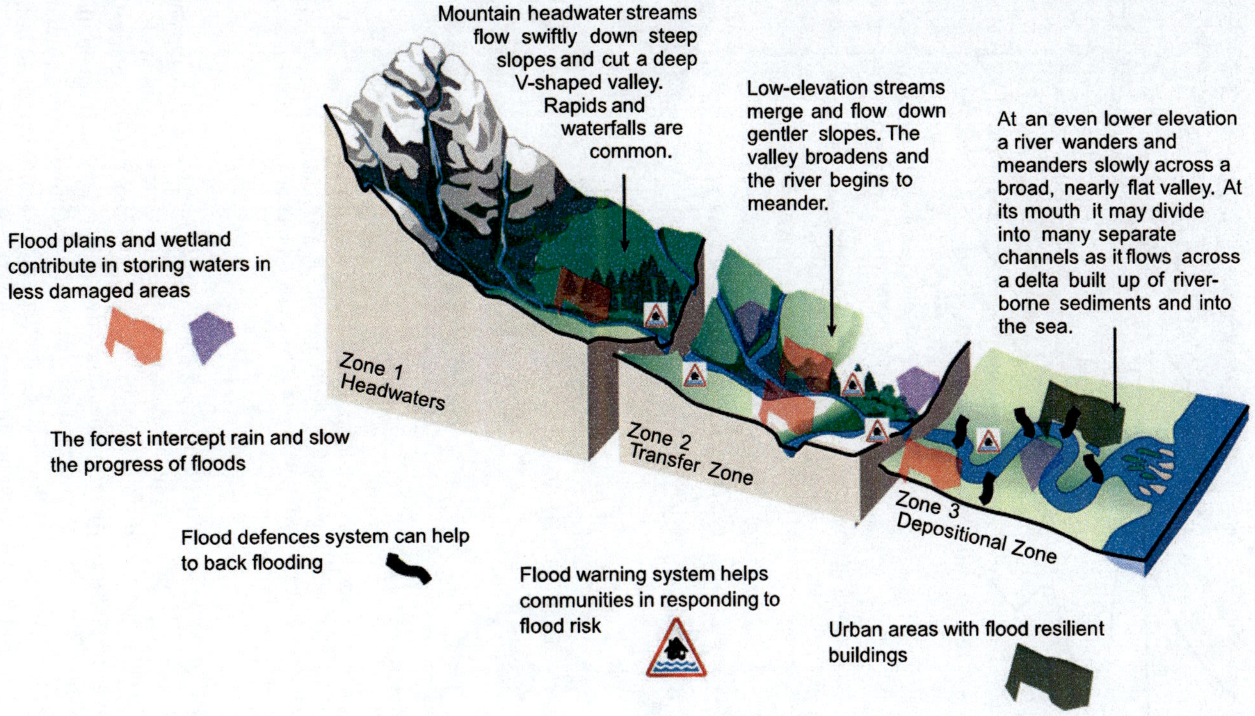

Fig. 2 Characterisation of the rivers Hydrogeomorphological process from basin to river reach scale and actions need to be taken a different spatial scale, modified after Carolyn et al. [6]

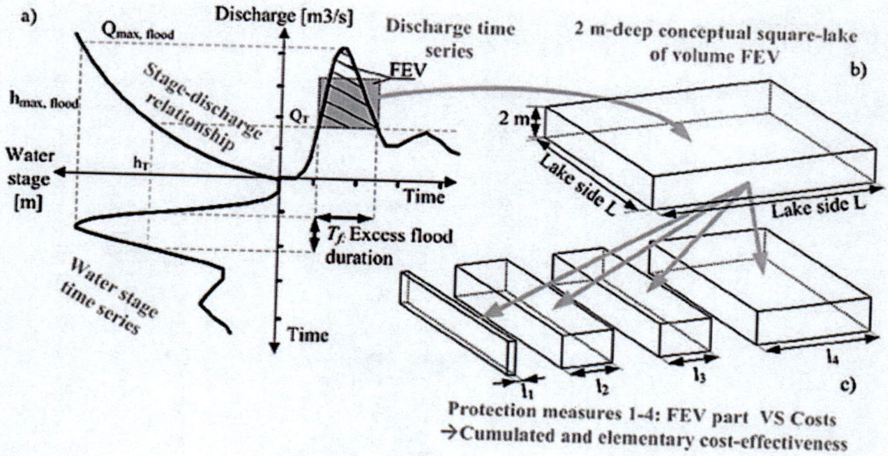

Fig. 3 Schematic representation of square-lake concept for allocation of Flood-Excess Volume (FEV): (**a**) stage–discharge relationship, (**b**) FEV square-lake concept representation, (**c**) FEV-effectiveness assessment considering equivalent measures, afterBokhove et al. [18]

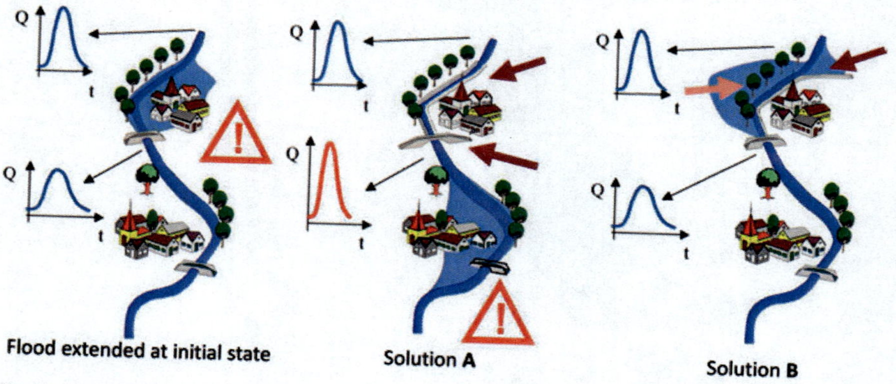

Fig. 4 Flood mitigation upscaling: (**a**) it involves dykes along both sides of the river and channel enlargement, which successfully prevents flooding in one village, (**b**) floods are mitigated in natural flood expansion areas or dry dams/retention basins while local protections and river training are kept to a minimum, after Poulard et al. [14]

As regional flooding tends to be frequent and long-lasting, local planners should adjust the floods mitigation policies based on previous experiences with such devastating events, and this is an opportunity to consider more effective measures at different spatial scales overtime to shield the adverse effects of subsequent floods events [23]. Therefore, a better understanding of the interaction between mitigation and protective measures against flooding at different spatial scales is critical in flood risk management and in implementing the most cost-effective solutions.

This chapter intends to give a brief overview of flood mitigation and protective measures at different scales within the catchment area and also aims to identify the main factors related to different scale to be considered in flood risk management. Namely, the main focus is given to innovative NBS approaches which are not only cost-effective but also a sustainable solution in terms of environment and natural landscape conservation. The chapter is organised as follows: Section 2 presents a synopsis on the main factors influencing flood events; Section 3 describes some of most common and cost-effective non-structural and structural mitigation measures as well as their eco-friendliness; Section 4 gives some insight into recommendations on the most pressing research questions; finally, Section 5 presents a conclusion of the critical issues to be considered in flood risk management concerning NBS approaches.

2 Multidimensional Aspects of Flood Mitigation at the Basin Scale

2.1 Influence of Geomorphological Properties of Earth Surface on Flooding Regimes

Engineered solutions widely control flood risk at the expense of altered flow and sediment regimes, as well as the ecological properties within the riparian zone and beyond [24]. Both elevation profile (i.e. topography and geomorphology) and soil texture (i.e. land cover) features are reported to have a significant impact not only on the physical properties of the basin but also on the functional composition of the riparian lands along the watercourse [25]. For example, floodplain ponds and gravel bars enable inundations. They are reported to have a considerable effect on the enrichment of flora and fauna assemblages along the riparian zone [26]. Therefore, topography and land surface composition are both considered in multi-criteria flood susceptibility assessment procedures and modelling [27].

In this section, we bring a split into two different spatial scales while considering the implications of LULC and topography on flood dynamics (Table 1). Both land surface cover and topographical properties can have different effects on the flood

Table 1 Implications of LULC and topography on hydrodynamics at both basin and riparian scale

Context			Implications	
Zone	Scope	Scale	Topography	LULC
Basin, watershed, catchment area	Regional, national, cross-boundary	>1:5,000	The geomorphology of the watershed contributes to the stream orders, runoff flow, and accumulation	It is affecting the runoff speed and accumulation of water – rainwater carrying capacity of leaves
Riparian zone	Local, site scale	<1:5,000	Define the morphology of the river, riverbanks, and the flood plain, affecting the flood-carrying capacity of the channel	Affecting the riverbank erosion dynamics and land degradation. Slowing down the water flow speed during flooding seasons

risk and preventive capacities at the basin scale as well as in the riparian zone (i.e. it refers to river reach scale). While the spatial scope of the former is expanding to the regional, national, or even international territories, the scope of riparian zone is related to the local and site scale, especially in urbanised lands where the flood risk is highest.

2.2 LULC Data Utility in Flood Risk Mitigation

Generally, LULC is monitored to understand the dynamics of landscape change at a gradient of spatial and temporal scales. The assessment of long-term alterations of the surface cover is useful in understanding landscape dynamics in the territory, especially in areas that are prone to different natural hazards. For example, the vegetation encroachment processes in flood-prone areas along highly modified large rivers are accepted as an issue to be carefully managed [28]. For example, in some cases, this problem is avoided by removing wild vegetation to adjust the width of the channel and to decrease the roughness of the watercourse to increase the discharge rates and avoid flooding [29]. These attempts increase flood prevention capacities as the riparian vegetation is reported to have a considerable impact on the flow friction and flood level [30]. The model developed by Anderson et al. [31] demonstrates that the properties of the riparian zone and the coarseness of the flow channel are determining factors of flow speed, being vital in propagating of flood waves. Thus, the land cover typology like wetlands along the riparian zone have a considerable impact on flood risk reduction [32] and must be comprehensively assessed.

LULC analysis can supply useful spatial information about the surface cover along the waterways. Generally, they are analysed within the riparian zone along the watercourse, which is defined via either fixed or altering width buffer. This approach leads to limitations to the significance that LULC has on the watershed scale as it is unable to reveal the connectivity of natural surfaces starting from the water source to further inland. To cope with this shortcoming, other scholars have highlighted the importance of the transversal (lateral) analysis of LULC (Fig. 5) in relation with the water sources (i.e. ocean, lake, and river) [33, 34]. Analysing LULC beyond the riparian zone helps in defining corridors of natural surfaces in the lateral direction, which can significantly contribute to the water retention capacities and a moderated rainwater discharge into the mainstream.

At the basin scale, the vegetation structure of the land surface on both sides of the watercourse is crucial for defining the roughness of the surfaces of the valley, thus, enabling runoff reduction during the flooding season. The more connected natural surfaces are in the transversal direction, the more moderated the runoff regimes from uplands to the watercourse will be.

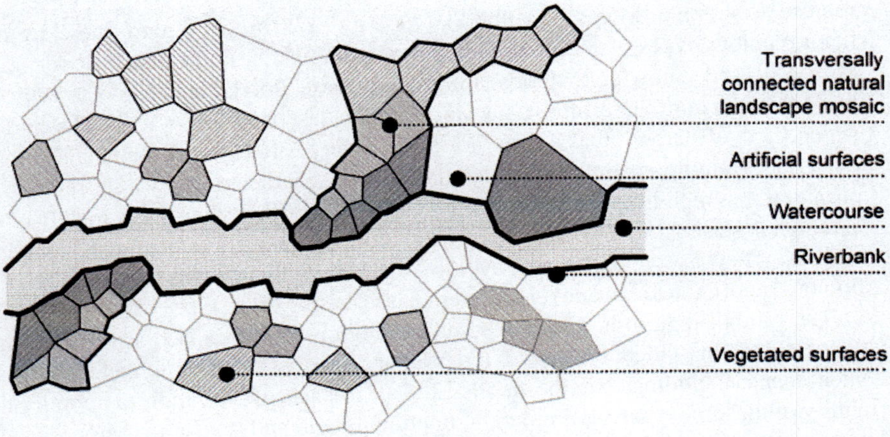

Fig. 5 Conceptual diagram of transversal connectivity of vegetated areas along the watercourse, after Hysa et al. [35]

2.3 Topography as a Static Driver of Water Flow Regime

The implications of topography on flood risk mitigation vary on different spatial scales. For example, while the geomorphology of the watershed (i.e. basin scale) has a direct impact on the runoff amount and speed from the uplands to the main channel, the topography of the riparian zone (i.e. site/river reach scale) holds potential areas suitable for temporarily accommodating water by reducing the flood risk downstream.

The risk-reducing utility of topography is proven by long-run applications of multi-purpose artificial reservoirs constructed on the upstream sub-basins within the watershed. For example, 2.6 million small artificial water bodies in North America contribute to water cycle management by diverting and delaying downstream water flow [36]. While the reservoirs reduce the runoff during high rates of precipitation, they supply restored water for diverse types of usages (i.e. agriculture, recreational, domestic, wildlife, etc.). Moreover, these added water surfaces lead to the flourishing of vegetated surfaces on the upstream lands, thus, enhancing the roughness of the basin surfaces and reducing their runoff capacity. Consequently, the topography and the LULC of the watershed must both be considered when drafting nature-based flooding mitigation.

2.4 Hydrometeorological Aspects of Floods

Flood characteristics change along the seasons and also differ between regions. The hydrometeorological conditions (i.e. precipitation and temperature regimes) represent the main drivers of a flood's typology [37]. In this regard, there are five main typology types, namely: long-rain floods, short-rain floods, flash floods, rain-on-snow, and snowmelt floods [38]. Long-rain floods are characterised by low-intensity rainfall, frontal type storms, they last from days to weeks and in general cover large areas; they are found very often in continental climate [39]. Short-rain floods or the so-called flash floods, characterised by a short duration of rainfall but high intensity, depending on the cloud pattern they occur locally or on a regional scale [38]. Flash floods occur more frequently in arid and semi-arid regions. They are characterised by local high rainfall intensity as well as fast-flowing runoff due to land cover features characterising those type of regions [9, 37].

The rainfall regime in arid and semi-arid regions is a localised structure called convective rain which is one of the primary drivers of several meteorological phenomena, including extreme floods [2, 39]. Because of the high temporal variability of the atmosphere recirculation, flooding in arid and semi-arid regions is very complex, and the occurrence and time duration are hard to predict [8]. Extreme floods occurring during the monsoon season in Asia, mainly in India are main natural hazards threatening millions of people lives [40]. In general, floods driven by high-intensity precipitations are the most unpredictable and destructive ones; in contrast, floods originating from snowmelt are highly predictable and therefore less devastating [2]. However, the latter one depends on several meteorological factors such as short-wave radiation, energy balance, and temperature variability, among others [10]. For instance, rain-on-snow floods originate as a result of a mixture of rainfall and existing snowpack. In this regard, moderate rainfall mixed with snow can generate substantial floods, although not intense [41]. Finally, snowmelt floods occur seasonally, namely during late Spring and early Summer seasons. As mentioned above, this type of flood is not risky in terms of intensity since the snow melting occurs at a low rate [42]. Blöschl et al. [10] proposed several indicators to identify flood typology. Nevertheless, the storm duration is one of the most common indicators that do that. Storm duration depends on factors such as topography and climatology, which may drive substantial spatial differences of the storm type itself [43].

2.5 Hydraulic Aspects of Floods

Indeed, there is a mutual linkage between flood regime and river geomorphology. The typology of floods not only affects humans but also has a substantial impact on several fluvial geomorphological processes; affects both the main channel and the floodplain [40]. Furthermore, it alters sediment transport in longitudinal and horizontal directions, i.e., causing erosion and deposition at a different section of the rivers (Fig. 6).

The type of roughness and river morphology can influence the flood travel time as well as stage – discharge relation [4, 45, 46]. Hydrodynamic modelling is a useful tool in the simulation of flood events to identify the most vulnerable areas. Nevertheless, it requires accurate information about the hydraulic conditions of the river

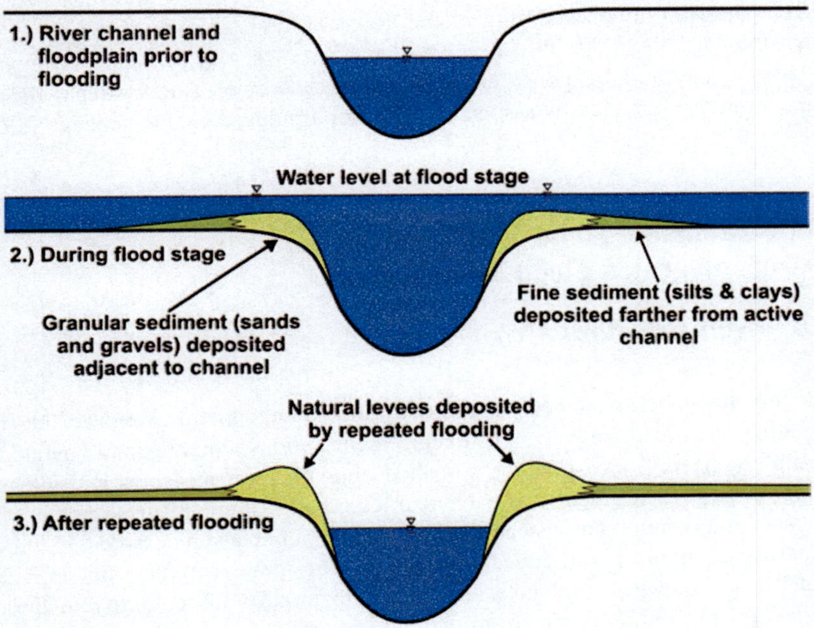

Fig. 6 Floods, river geomorphology precedes interlinkage before and after flooding, after Rogers [44]

channel [4, 47]. In general, 2D hydraulic models are most commonly applied in practice for flood modelling because of their ability in considering both spatial and flow variability in time. Accurate hydraulic information of natural rivers is particularly crucial for generating flood risk maps in high populated areas like urban and peri-urban lands, because it may prevent proper flood management measures or in a worse scenario lead to failure of the implemented measures [47]. Flow resistance is an important parameter that influences the stage–discharge relationship in natural channels [48]. Natural river channels usually have no regular cross-sections. They have variable roughness along the wetter perimeter, which therefore influences the flow resistance.

Moreover, flow resistance is additionally influenced by the longitudinal geomorphology of the river, such as sinuosity and meandering, among others [49]. The hydraulic conditions of a river channel can be improved by periodic cleaning of debris, vegetation, large woody trees, etc., which can enhance the conveyance capacity of the river channel [3]. The unsteady non-uniform hydraulic regime characterises floods. Another important hydraulic factor that influences the shape of the flood hydrograph is the backwater effect created due to water storage or geomorphological irregularities of the river [48]. The backwater effect conveys the secondary flows backwards. It substantially affects the flood routing leading to the formation of a sinuous pattern in the upstream part [50]. The hydraulic regime also influences several physical and biogeochemical processes of the fluvial ecosystem

[51]. Thus, the high variability of the hydraulic regime, such as turbulence could considerably affect the fluvial ecosystem habitat [4, 49, 52]. In this regard, the upstream hydraulic response to the backwater effect of a downstream riffle crest imposes a natural analogy associated with flood-induced channel change [53].

3 Sustainable Flood Mitigation Measures in Support of Integrated Flood Management

3.1 Non-structural Measures

Modern flood risk management practices and strategies aim to reduce the risk of flooding by considering a mix of management options which extend beyond traditional engineering measures the so-called non-structural measures and integrate a wide range of instruments [54]. Non-structural measures are those not involving physical intervention but instead use knowledge, public awareness-raising, previous experiences, training and education, and specific laws, to reduce floods risks. In general, the non-structural measures intend to modify susceptibility to floods to protect people and properties from the flood hazard.

To assure the effectiveness of flood risk management, it is essential to have accurate information on the flood occurrence and typology obtained from non-structural measures, which can significantly decrease the costs of floods for households [55]. Non-structural measures include real-time flood forecasting and warning systems, evacuation systems, land-use planning, flood zoning, preservation of retention ponds, and emergency services, among others [56]. To better manage the risk of floods, a spatial zonation of the flood risk based on Digital Elevation Model (DEM) predictions can provide important information through inundation colour-coded in different flood-vulnerable areas, where different colour may correspond to different levels of flooding [54]. This practical approach would allow people to relate the colour-coded of DEMs to flood warning posts and enable them to take appropriate actions. Other semi-structural or non-structural measures that can reduce the risk of flooding considerably involve wet-proofing approaches such as solidification of walls against water pressure, adapting the flood-prone parts of the settlements with waterproof materials, moving vulnerable instruments to upper floors, risk transfer instruments, flood insurance, evacuation, installing one-way valves on water evacuation pipes to stop the waters from inflowing the house through the pipes and storing paints, and chemicals in the upper parts of the home among others [57]. These kinds of non-structural measures aim to stop the water from inflowing into the house at the highest level as well as they adopt the house to cut the damage in case of flooding. Nevertheless, the efficacy of non-structural measures is sensitive to socio-economic changes and governance provisions policies [54]. Non-structural measures are in better agreement with sustainable development than traditional engineering structural measures, as they are more adjustable, commonly accepted, and environmentally-friendly [58].

Overall, non-structural measures have several benefits such as low implementation cost reducing vulnerability, and easily adaptable; the later on is particularly advantageous considering the uncertainties resulting from climate changes [22].

3.2 Structural Measures

Even though the non-structural measures offer several benefits concerning flood risk management, they would be less effective in many cases without combination with structural measures. Structural river protection and/or mitigation measures such as dams and dikes are among old and traditionally-known measures. They have been constructed for at least four thousand years [11]. Management of an area that is vulnerable to flooding undergoes complex decision-making processes regarding the measures to be implemented in compatibility with land-use related activities and the risk to which environment, human, and their properties are subjected [59]. Structural measures, besides attempts to reduce the water load, contribute to enhancing the resilience of the entire flood defence system, and also promote the readiness to somehow live with floods. Structural flood defences systems may boost urban development at-risk areas. At the same time, the recovery instruments might also provide preventive measures for flood risk management [11]. Before implementing new measures, it is essential to explore if it is possible to create a new space in the existing channel to allocate an extra volume of water to reduce the costs (Fig. 7).

Although structural flood mitigation measures are the most commonly used in practice as proved of being effective in many highly urbanised flood-prone areas, poor implementation and management of these infrastructures may lead to irreversible environmental, geomorphological, and social consequences [60]. Some of the measures may be implemented temporarily to avoid spontaneous risk (Fig. 8).

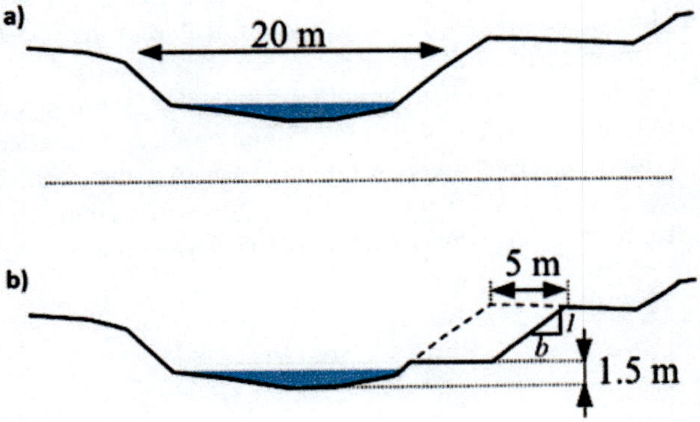

Fig. 7 Schematic representation of side-channel digging to create new room for the extra volume of water (**a**) current profile and (**b**) profile of the giving-room-to-the-river scenario, 5 m wider at 1.5 m high, with the same bank slope *b*, adopted after Bokhove et al. [17]

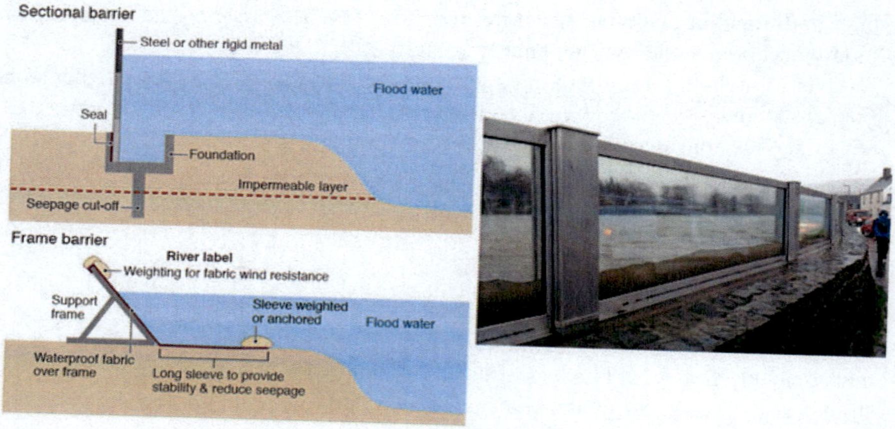

Fig. 8 Schematic representation of two types of temporary flood barrier in left and the right, a picture is showing implementation in practice (Source: https://www.bbc.com/news/uk-25929644)

Fig. 9 Leaky barriers structure applied in the torrential river, after Hankin et al. [19]

Structural measures can be found at different typology, applicable at different scale within the river basin, and target different typology of floods. For instance, leaky barriers (Fig. 9) are mostly applicable in the torrential river and take place in the upland part of the river basin.

Hybrid or combined structural mitigation solution such as tree planting in combination with dykes is also very often applied in practice (Fig. 10).

Such solutions are cost-effective because they can quickly absorb the flooding wave energy and therefore reduce the construction cost of a dyke while in the meantime, increasing the reliability of Dykes [62].

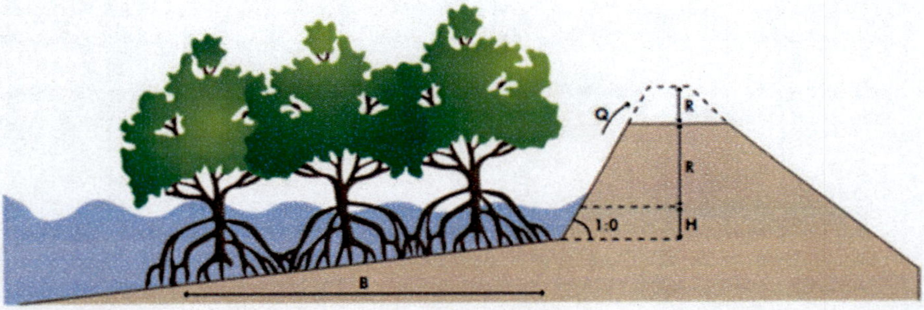

Fig. 10 Schematic representation of a hybrid solution. Mangrove trees can absorb the flood wave energy. As a result, it reduces the dike height that is needed to meet the safety standards after K. van Wesenbeeck et al. [61]

3.3 *The Ecosystem Conservation Aspects of Nature-Based Mitigation Measures*

Particularly in urban areas, large impermeable pavement areas, as well as the roofs of buildings direct rainwater to be collected straight into drainage systems which can quickly become overwhelmed. Therefore, new sustainable approaches such as the "Sponge Cities" [63], Integrated Water Resource Management (IWRM) [64], or the Sustainable Urban Drainage System (SuDS) are receiving considerable interest [65]. According to these new eco-friendly concepts, runoff water coming from impermeable surfaces should seep into the ground or be collected in detention basis and small ponds rather than flushing way through the sewerage system (Fig. 11). So, the collected rainwater can be released in a controlled way, or it can even be used for irrigation or other purposes after validating its quality [66].

Such solutions have huge potential in reducing global warming and in helping people and also ecosystems adapting to a warmer planet. The NBS solutions, namely SuDS and "Sponge Cities", have great potential to sequestering CO_2, and also improve resilience, especially of the urban area, to guarantee sustainable food supplies, to increase biodiversity, and to generate healthier, greener living environments for people and biota [67]. Concerning the biota, particularly the one related to fluvial ecosystems, it is significantly affected by the habitat conditions; degradation of habitat conditions leads to the decline of the aquatic biodiversity [68]. Therefore, flood mitigation measures to take place in the river reach scale must also consider preservation and/or improvement of habitat suitability conditions. In this regard, such measures should guarantee lateral and longitudinal connectivity, spatially and temporally heterogeneous areas with related water bodies, dynamics of water exchanges between surface waters and groundwater, and water quality among others [14, 68]. Figure 12 shows flood mitigation measures in the semi-natural and quasi-natural channel while in the meantime, intending to improve and preserve the habitat conditions.

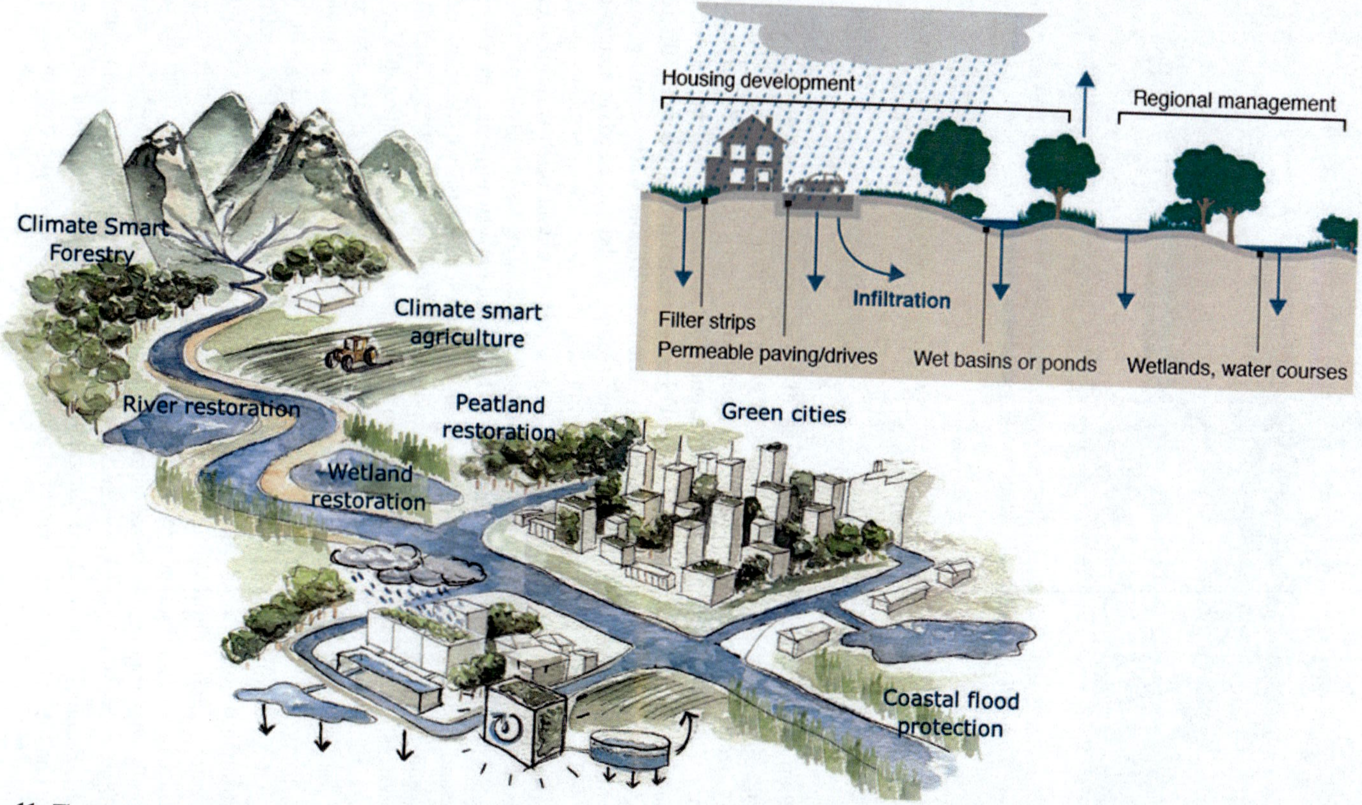

Fig. 11 The upper-right picture shows a cross-section view of SuDS (Source: https://www.susdrain.org), the lower-left picture schematically shows the "Sponge Cities" concept, adapted from Hattum, Blauw [65] (Credits: Natasha de Sena)

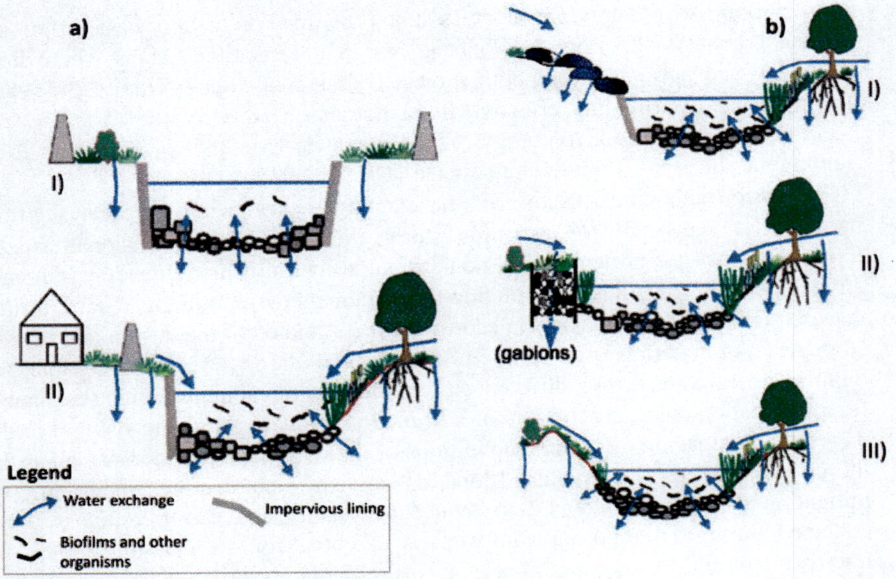

Fig. 12 Schematic representation of the habitat improvement: (**a**) semi-natural channel and (**b**) quasi-natural channel, adapted from Poulard et al. [14]

In the case of the semi-natural channel (Fig. 12a, I), about 50% of the bed is artificial. At the same time, the surface and groundwater connectivity are possible only through the bed. In contrast, in the second scenario (Fig. 12a, II) only about 30% of the bed is artificial; therefore, surface and groundwater connectivity are possible through the bed as well as the riverbank. In the case of the quasi-natural channel (Fig. 12b), almost all types of habitat might be preserved. Nevertheless, the presence of the artificial elements, particularly in the case of Fig. 12b, I), may restrict three-dimensional water connectivity and may also influence the water quality. Overall, the best strategy in such and other types of flood mitigation measures would be the eco-friendly usage design as well as materials [57, 69].

4 Future Research Needs and Recommendations for Improvement of Floods Mitigation Measures

One of the essential aspects in the enhancement of flood risk management strategies is an investment in sciences and communications, which provides a prospect for further improvement and expansion of the context of NBS utility in flood mitigation measures. In this regard, real-time localisation of the most vulnerable areas is an essential task to be considered for further advancement of computer modelling to provide possible accurate information about high-risk flooding areas. Zonation and flood mapping of the flood-prone areas should be done on priority basis [3], in this particular context attention should be given to lower-latitude areas where flood frequency and population are both projected to increase in the future [9]. Many countries fail in providing effective flood mitigation solution due to the lack of information on the flood typology. Therefore, it is essential to define the flood typology at the regional and country scale [10].

Expansion and modernisation of the existing hydrometeorological system are particularly significant for real-time storm tracking. Multidimensional aspects improvement of the structural and non-structural measures is tremendously imperative to enhance the efficacy of the flood management system. In this regard, further research is needed to improve the eco-friendliness related to the design and type of strategies and materials used in flood mitigation solutions [68]. Fostering interdisciplinary research involving different stakeholders is critical in providing sustainable flood mitigation solution with a twofold function; reducing flooding risk but in the meantime, preserving the natural conditions of the riverscape corridor as well as the fluvial ecosystem [14]. Effective flood risk management requires monitoring of mitigation solution with regard to their functionality and eco-friendliness. Data collected through monitoring campaign is essential for decision-makers in the improvement of the existing mitigation measures and developing the new ones [16].

Further research is needed in flood risk perception, to achieve a more inclusive understanding of how risk perceptions affect the vulnerability, capacity, and resilience of individuals and communities facing flooding [13]. The term "resilience" has arisen as the dominant model in flood risk management, mainly related to NBS, which implies the need to plan and design cities that can absorb water during flooding and reproduce natural processes more thoroughly [65]. A better understanding of the flood risk perception is particularly important in urban areas by facilitating the cost-effective and safe expansion of such areas. Finally, further research is also needed to assess the synergistic effects of multiple flood mitigation strategies on protecting community properties [20]. Last but not least, additional research is also needed in terms of policies concerning insurance and jurisdictions aspects related to damage compensation, which might contribute towards enhancement of management practices and governance provisions.

5 Concluding Remarks

This chapter delivered a summary of the existing flood mitigation and protective measures at a cross-scalar context from reach to the basin area. Furthermore, it identified the main natural factors within the context that affects the flooding regimes. It must be considered in flood risk management. This was realised by thoroughly reviewing the ongoing scholarly discourse on the topic. The main focus was given to novel approaches that are based on the principles of NBS. These are advocated to inspire not only cost-effective solutions but also sustainable proposals that are in harmony with the natural environment and native landscapes. We identified four major aspects that have direct implications on the flood regimes, which are LULC, topography, hydrometeorology, and hydraulics of the river. These factors have been discussed in detail to clarify their cross-spatial influence on flood events.

On the other hand, novel concepts such as "Sponge Cities", IWRM, and the SuDS have been reviewed and discussed as examples of mitigation practices that combine cost-effective non-structural and structural mitigation measures. These agendas aim to integrate the social, geophysical, and ecological aspects and provide a comprehensive frame while dealing with urban flooding. Community engagement and activation are crucial dimensions of these approaches, as well as the ecological conservation of existing habitats. Nevertheless, we realised that the importance of the cross-scalar character of flooding phenomena is not considered enough.

In conclusion, this chapter highlights two recommendations. First, the flood risk reduction agendas must adopt a cross-scale approach while considering flood events. The mitigation measures downstream and upstream must complement each other and must be designed in a spatially comprehensive manner. Second, natural properties of the context (i.e. at both basin and reach scale) such as LULC, topography, hydrometeorology, and hydraulics must be considered to utilise the full native capacity in flood risk mitigation. Finally, flood risk reduction agendas must integrate social, geophysical, and ecological properties of the study area while drafting practical nature-based flood risk mitigation measurements.

Acknowledgements Alban Kuriqi was supported by a PhD scholarship granted by Fundação para a Ciência e a Tecnologia, I.P. (FCT), Portugal, under the PhD Programme FLUVIO–River Restoration and Management, grant number: PD/BD/114558/2016. This study was motivated by the synergy generated within the Cost Action CA16209- Land4FLOOD.

References

1. Haque AN, Bithell M, Richards KS (2020) Adaptation to flooding in low-income urban settlements in the least developed countries: a systems approach. Geogr J. https://doi.org/10.1111/geoj.12348
2. Lawford RG, Prowse TD, Hogg WD et al (1995) Hydrometeorological aspects of flood hazards in Canada. Atmosphere-Ocean 33(2):303–328. https://doi.org/10.1080/07055900.1995.9649535
3. Tariq MAUR, van de Giesen N (2012) Floods and flood management in Pakistan. Phys Chem Earth Parts A/B/C 47-48:11–20. https://doi.org/10.1016/j.pce.2011.08.014
4. Ardıçlıoğlu M, Kuriqi A (2019) Calibration of channel roughness in intermittent rivers using HEC-RAS model: case of Sarimsakli creek, Turkey. S.N. Appl Sci 1(9). https://doi.org/10.1007/s42452-019-1141-9
5. Kuriqi A, Koçileri G, Ardiçlioğlu M (2019) Potential of Meyer-Peter and Müller approach for estimation of bed-load sediment transport under different hydraulic regimes. Model Earth Syst Environ 6(1):129–137. https://doi.org/10.1007/s40808-019-00665-0
6. Carolyn A, Hollis A, Leon AB et al (1998) Stream corridors: processes and characteristics. In: Stream corridor restoration: principles, processes, and practices, vol 1. vol A 57.6/2:EN3/PT.653. 15 Federal agencies of the U.S. Gov't, USA, pp 1–89

7. Schismenos S, Stevens GJ, Emmanouloudis D et al (2020) Humanitarian engineering and vulnerable communities: hydropower applications in localised flood response and sustainable development. Int J Sustain Energy. https://doi.org/10.1080/14786451.2020.1779274
8. Dettinger M (2011) Climate change, atmospheric Rivers, and floods in California – a multimodel analysis of storm frequency and magnitude Changes1. JAWRA J Am Water Resour Assoc 47(3):514–523. https://doi.org/10.1111/j.1752-1688.2011.00546.x
9. Hirabayashi Y, Mahendran R, Koirala S et al (2013) Global flood risk under climate change. Nat Clim Chang 3(9):816–821. https://doi.org/10.1038/nclimate1911
10. Blöschl G, Hall J, Parajka J et al (2017) Changing climate shifts timing of European floods. Science 357(6351):588–590. https://doi.org/10.1126/science.aan2506
11. Kundzewicz ZW, Hegger DLT, Matczak P et al (2018) Opinion: flood-risk reduction: structural measures and diverse strategies. Proc Natl Acad Sci U S A 115(49):12321–12325. https://doi.org/10.1073/pnas.1818227115
12. Plate EJ (2002) Flood risk and flood management. J Hydrol 267(1–2):2–11. https://doi.org/10.1016/S0022-1694(02)00135-X
13. Birkholz S, Muro M, Jeffrey P et al (2014) Rethinking the relationship between flood risk perception and flood management. Sci Total Environ 478:12–20. https://doi.org/10.1016/j.scitotenv.2014.01.061
14. Poulard C, Lafont M, Lenar-Matyas A et al (2010) Flood mitigation designs with respect to river ecosystem functions – a problem oriented conceptual approach. Ecol Eng 36(1):69–77. https://doi.org/10.1016/j.ecoleng.2009.09.013
15. Wingfield T, Macdonald N, Peters K et al (2019) Natural flood management: beyond the evidence debate. Area 51(4):743–751. https://doi.org/10.1111/area.12535
16. Short C, Clarke L, Carnelli F et al (2019) Capturing the multiple benefits associated with nature-based solutions: lessons from a natural flood management project in the Cotswolds, U.K. Land Degrad Dev 30(3):241–252. https://doi.org/10.1002/ldr.3205
17. Bokhove O, Kelmanson MA, Kent T et al (2020) Communicating (nature-based) flood-mitigation schemes using flood-excess volume. River Res Appl 35(9):1402–1414
18. Bokhove O, Kelmanson MA, Kent T et al (2019) Communicating (nature-based) flood-mitigation schemes using flood-excess volume. River Res Appl 35(9):1402–1414
19. Hankin B, Hewitt I, Sander G et al (2020) A risk-based, network analysis of distributed in-stream leaky barriers for flood risk management. Nat Hazards Earth Syst Sci. https://doi.org/10.5194/nhess-2019-394
20. Brody SD, Highfield WE (2013) Open space protection and flood mitigation: a national study. Land Use Policy 32:89–95. https://doi.org/10.1016/j.landusepol.2012.10.017
21. Kuriqi A, Ardiçlioglu M, Muceku Y (2016) Investigation of seepage effect on river dike's stability under steady state and transient conditions. Pollack Periodica 11(2):87–104. https://doi.org/10.1556/606.2016.11.2.8
22. Meyer V, Priest S, Kuhlicke C (2011) Economic evaluation of structural and non-structural flood risk management measures: examples from the Mulde River. Nat Hazards 62(2):301–324. https://doi.org/10.1007/s11069-011-9997-z
23. Brody SD, Zahran S, Highfield WE et al (2009) Policy learning for flood mitigation: a longitudinal assessment of the community rating system in Florida. Risk Anal 29 (6):912–929. https://doi.org/10.1111/j.1539-6924.2009.01210.x
24. Poff NL, Zimmerman JKH (2010) Ecological responses to altered flow regimes: a literature review to inform the science and management of environmental flows. Freshw Biol 55 (1):194–205. https://doi.org/10.1111/j.1365-2427.2009.02272.x
25. Janssen P, Piegay H, Pont B et al (2019) How maintenance and restoration measures mediate the response of riparian plant functional composition to environmental gradients on channel margins: insights from a highly degraded large river. Sci Total Environ 656:1312–1325. https://doi.org/10.1016/j.scitotenv.2018.11.434

26. Januschke K, Brunzel S, Haase P et al (2011) Effects of stream restorations on riparian mesohabitats, vegetation and carabid beetles. Biodivers Conserv 20(13):3147–3164. https://doi.org/10.1007/s10531-011-0119-8
27. Khosravi K, Shahabi H, Pham BT et al (2019) A comparative assessment of flood susceptibility modeling using multi-criteria decision-making analysis and machine learning methods. J Hydrol 573:311–323. https://doi.org/10.1016/j.jhydrol.2019.03.073
28. Comiti F, Da Canal M, Surian N et al (2011) Channel adjustments and vegetation cover dynamics in a large gravel bed river over the last 200years. Geomorphology 125(1):147–159. https://doi.org/10.1016/j.geomorph.2010.09.011
29. Shields FD, Nunnally NR (1984) Environmental aspects of clearing and snagging. 110 (1):152–165. https://doi.org/10.1061/(ASCE)0733-9372(1984)110:1(152)
30. Darby SE (1999) Effect of riparian vegetation on flow resistance and flood potential. 125 (5):443–454. https://doi.org/10.1061/(ASCE)0733-9429(1999)125:5(443)
31. Anderson BG, Rutherfurd ID, Western AW (2006) An analysis of the influence of riparian vegetation on the propagation of flood waves. Environ Model Softw 21(9):1290–1296. https://doi.org/10.1016/j.envsoft.2005.04.027
32. Acreman M, Holden J (2013) How wetlands affect floods. Wetlands 33(5):773–786. https://doi.org/10.1007/s13157-013-0473-2
33. Hysa A, Başkaya FAT (2018) Revealing the transversal continuum of natural landscapes in coastal zones – case of the Turkish Mediterranean coast. Ocean Coastal Manag 158:103–115. https://doi.org/10.1016/j.ocecoaman.2018.03.011
34. Hysa A (2021) Introducing transversal connectivity index (TCI) as a method to evaluate the effectiveness of the blue-green infrastructure at metropolitan scale. Ecol Indic 124:107432. https://doi.org/10.1016/j.ecolind.2021.107432
35. Hysa A, Baskaya FAT (2018) A GIS-based method for revealing the transversal continuum of natural landscapes in the coastal zone. MethodsX 5:514–523. https://doi.org/10.1016/j.mex.2018.05.012
36. Smith SV, Renwick WH, Bartley JD et al (2002) Distribution and significance of small, artificial water bodies across the United States landscape. Sci Total Environ 299(1):21–36. https://doi.org/10.1016/S0048-9697(02)00222-X
37. Merz R, Blöschl G (2003) A process typology of regional floods. Water Resour Res 39(12). https://doi.org/10.1029/2002wr001952
38. Yang W, Yang H, Yang D (2020) Classifying floods by quantifying driver contributions in the eastern monsoon region of China. J Hydrol 585. https://doi.org/10.1016/j.jhydrol.2020.124767
39. Cecchini MA, Silva Dias MAF, Machado LAT et al (2020) Macrophysical and microphysical characteristics of convective rain cells observed during SOS-CHUVA. J Geophys Res Atmos 125(13). https://doi.org/10.1029/2019jd031187
40. Kale VS, Ely LL, Enzel Y et al (1994) Geomorphic and hydrologic aspects of monsoon floods on the Narmada and Tapi Rivers in Central India. In: Geomorphology and natural hazards, pp 157–168
41. Li D, Lettenmaier DP, Margulis SA et al (2019) The role of rain-on-snow in flooding over the conterminous United States. Water Resour Res 55(11):8492–8513. https://doi.org/10.1029/2019wr024950
42. Mateo-Lázaro J, Castillo-Mateo J, Sánchez-Navarro J et al (2019) Assessment of the role of snowmelt in a flood event in a gauged catchment. Water 11(3). https://doi.org/10.3390/w11030506
43. De Michele C (2003) A generalised Pareto intensity-duration model of storm rainfall exploiting 2-copulas. J Geophys Res 108(D2). https://doi.org/10.1029/2002jd002534
44. Rogers JD (2008) Development of the New Orleans flood protection system prior to hurricane Katrina. J Geotech Geoenviron 134(5):602–617. https://doi.org/10.1061/(asce)1090-0241(2008)134:5(602)

45. Kuriqi A, Ardiçlioğlu M (2018) Investigation of hydraulic regime at middle part of the Loire River in context of floods and low flow events. Pollack Periodica 13(1):145–156. https://doi.org/10.1556/606.2018.13.1.13
46. Kumar M, Kumari A, Kushwaha DP et al (2020) Estimation of daily stage–discharge relationship by using data-driven techniques of a Perennial River, India. Sustainability 12(19). https://doi.org/10.3390/su12197877
47. Mihu-Pintilie A, Cîmpianu CI, Stoleriu CC et al (2019) Using high-density LiDAR data and 2D streamflow hydraulic modeling to improve urban flood Hazard maps: a HEC-RAS multi-scenario approach. Water 11(9). https://doi.org/10.3390/w11091832
48. Helmiö T, Järvelä J (2004) Hydraulic aspects of environmental flood management in boreal conditions. Boreal Environ Res 9(3):227–241
49. McClain ME, Subalusky AL, Anderson EP et al (2014) Comparing flow regime, channel hydraulics, and biological communities to infer flow–ecology relationships in the Mara River of Kenya and Tanzania. Hydrol Sci J 59(3–4):801–819. https://doi.org/10.1080/02626667.2013.853121
50. Castelltort FX, Bladé E, Balasch JC et al (2020) The backwater effect as a tool to assess formative long-term flood regimes. Quat Int 538:29–43. https://doi.org/10.1016/j.quaint.2020.01.012
51. Christian F, Arturas R, Saulius G et al (2008) Hydraulic regime-based zonation scheme of the Curonian lagoon. Hydrobiologia 611(1):133–146. https://doi.org/10.1007/s10750-008-9454-5
52. Kuriqi A, Pinheiro AN, Sordo-Ward A et al (2020) Water-energy-ecosystem nexus: balancing competing interests at a run-of-river hydropower plant coupling a hydrologic–ecohydraulic approach. Energy Convers Manag 223. https://doi.org/10.1016/j.enconman.2020.113267
53. Pasternack GB, Bounrisavong MK, Parikh KK (2008) Backwater control on riffle–pool hydraulics, fish habitat quality, and sediment transport regime in gravel-bed rivers. J Hydrol 357(1–2):125–139. https://doi.org/10.1016/j.jhydrol.2008.05.014
54. Dawson RJ, Ball T, Werritty J et al (2011) Assessing the effectiveness of non-structural flood management measures in the Thames estuary under conditions of socio-economic and environmental change. Glob Environ Chang 21(2):628–646. https://doi.org/10.1016/j.gloenvcha.2011.01.013
55. Simonovic SP (2002) Two new non-structural measures for sustainable management of floods. Water Int 27(1):38–46. https://doi.org/10.1080/02508060208686976
56. Faisal IMM, Kabir R et al (1999) Non-structural flood mitigation measures for Dhaka City. Urban Water 1(2):145–153
57. Poussin JK, Bubeck P, Aerts JCJH et al (2012) Potential of semi-structural and non-structural adaptation strategies to reduce future flood risk: case study for the Meuse. Nat Hazards Earth Syst Sci 12(11):3455–3471. https://doi.org/10.5194/nhess-12-3455-2012
58. Kundzewicz ZW (2002) Non-structural flood protection and sustainability. Water Int 27(1):3–13. https://doi.org/10.1080/02508060208686972
59. Oliveri E, Mario S (2000) Estimation of urban structural flood damages: the case study of Palermo. Urban Water 2(3):223–234
60. Gilbuena Jr R, Kawamura A, Medina R et al (2013) Environmental impact assessment using a utility-based recursive evidential reasoning approach for structural flood mitigation measures in metro Manila, Philippines. J Environ Manag 131:92–102. https://doi.org/10.1016/j.jenvman.2013.09.020
61. van Wesenbeeck KB, IJff S, Jongman B et al (2017) Implementing naturebased flood protection. World Bank Group, Washington
62. Jongman B (2018) Effective adaptation to rising flood risk. Nat Commun 9(1):1986. https://doi.org/10.1038/s41467-018-04396-1
63. Jiang Y, Zevenbergen C, Ma Y (2018) Urban pluvial flooding and stormwater management: a contemporary review of China's challenges and "sponge cities" strategy. Environ Sci Pol 80:132–143. https://doi.org/10.1016/j.envsci.2017.11.016

64. Gain A, Mondal M, Rahman R (2017) From flood control to water management: a journey of Bangladesh towards integrated water resources management. Water 9(1). https://doi.org/10.3390/w9010055
65. Potter K, Vilcan T (2020) Managing urban flood resilience through the English planning system: insights from the 'SuDS-face'. Philos Trans A Math Phys Eng Sci 378 (2168):20190206. https://doi.org/10.1098/rsta.2019.0206
66. Musz-Pomorska A, Widomski MK, Gołębiowska J (2020) Financial sustainability of selected rain water harvesting systems for single-family house under conditions of eastern Poland. Sustainability 12(12). https://doi.org/10.3390/su12124853
67. Hattum TV, Blauw M, Bergen Jensen M et al (2016) Climate adaptation is a huge opportunity to improve the quality of life in cities. Towards Water Smart Cities Wageningen Environmental Research Gelderland, Belgium
68. Kail J, Guse B, Radinger J et al (2015) A modelling framework to assess the effect of pressures on river abiotic habitat conditions and biota. PLoS One 10(6):e0130228. https://doi.org/10.1371/journal.pone.0130228
69. Masi F, Rizzo A, Regelsberger M (2018) The role of constructed wetlands in a new circular economy, resource oriented, and ecosystem services paradigm. J Environ Manag 216:275–284. https://doi.org/10.1016/j.jenvman.2017.11.086

Nature-Based Solutions to Mitigate Coastal Floods and Associated Socioecological Impacts

Miguel Inácio, Donalda Karnauskaitė, Katažyna Mikša, Eduardo Gomes, Marius Kalinauskas, and Paulo Pereira

Contents

Abstract Despite its importance in supporting the global economy and to accommodate an ever-growing population at the coast, many of the coastal ecosystems such as mangroves, reefs, seagrass meadows, salt marshes and dunes had in the recent years an accentuated decrease in their coverage. The loss of coastal ecosystems, among other problems, leads to the loss of natural capacity for flood mitigation and coastal erosion. Since a considerable share of the coastal population is living in

M. Inácio, K. Mikša, E. Gomes, M. Kalinauskas, and P. Pereira (✉)
Environmental Management Laboratory, Mykolas Romeris University, Vilnius, Lithuania
e-mail: rinacio.miguel@gmail.com; katarzyna.bogdziewicz@gmail.com; eduardojonas@gmail.com; m.kalinauskas@mruni.eu; pereiraub@gmail.com

D. Karnauskaitė
Marine Research Institute, Klaipeda University, Klaipeda, Lithuania
e-mail: donalda.karnauskaite@gmail.com

Carla S. S. Ferreira, Zahra Kalantari, Thomas Hartmann, and Paulo Pereira (eds.),
Nature-Based Solutions for Flood Mitigation: Environmental and Socio-Economic Aspects,
Hdb Env Chem (2022) 107: 35–58, DOI 10.1007/698_2020_675,

Published online: 15 October 2020

flood-prone areas, the loss of capacity of the ecosystems to mitigate the impacts of floods and coastal erosion can increase the vulnerability to natural hazards such as storm surges, hurricanes and typhoons. This is especially relevant in a context of increasing sea-level rise and intensity and frequency of extreme weather events, both increasing the risk to lose lives and assets. Coastal flood mitigation has been done primarily through the use of hard grey infrastructures. However, these types of structures can have long-term impacts on coastal ecosystems, require continuous maintenance and, in the face of extreme events, may represent an inefficient way to prevent coastal degradation. This called the attention of scientists and decision-makers towards the role of nature to mitigate the impacts of coastal floods through nature-based solutions (NBS). NBS, under the framework of ecosystem-based management, are interventions that aim to reduce the impacts of coastal flooding and erosion and simultaneously enhance ecosystems, biodiversity and natural resources. NBS can use (1) natural solutions (e.g. marine protected areas), (2) soft engineering and ecological restoration (e.g. beach nourishment) and (3) hybrid solutions, which integrate natural and grey infrastructures. Recently, NBS are gaining popularity and are part of coastal management strategies in many countries. Despite their efficiency, it is still a new practice, and therefore concerns are raised regarding their environmental and anthropogenic impacts. Also, there are some drawbacks and pitfalls that need to be overcome to increase NBS implementation. In this chapter we make an overview on the need for NBS for coastal flood mitigation, its implementation in a worldwide context, their impacts on the coastal social ecological economics systems, drawbacks and opportunities to improve their acceptance.

Keywords Climate change adaptation, Coastal resilience, Hybrid solutions, Nature-based solutions, Sea-level rise, Storm surge

1 Introduction

Traditional coastal flooding mitigation strategies were implemented using hard engineered measures, commonly referred as “grey” infrastructures [1–3]. Due to their durability, the facility of implementation and effectiveness, grey infrastructures were built as an immediate response to coastal flooding and erosion events [4, 5].

With the aim of protecting human lives and preventing economic losses in highly populated urban areas [4], the construction of grey infrastructures were implemented regardless of their integration in the natural landscape or the long-term effects on coastal ecosystems functions. This led to adverse environmental impacts such as changes in sediment dynamics and local ecology [6–10]. Grey infrastructures require constant maintenance during their lifetime [5]. They are not the best strategies to mitigate climate change impact and other anthropogenic disturbances. Recent effects such as sea-level rise, the increase in urbanization and severity and intensity of storm surges and extreme events like hurricanes exposed their weakness and failed to protecting the coast, with implications in high losses of human lives and assets [11, 12]. After extreme events, the areas where coastal habitats have a high quality are the less affected [13–16]. Coastal habitats (e.g. mangroves, reefs, dunes) have a great capacity to adapt to the new conditions following extreme events. Therefore, when compared to grey measures, they are more sustainable solution to mitigate the impacts of coastal flooding. Healthy coastal habitats are nature-based solutions (NBS) valuable to protect the coast against extreme events.

In the last decades, the use of NBS to mitigate coastal floods has been increasingly used by combining actions to protect, restore and manage ecosystems in a sustainable way, addressing also socio-economic aspects, ensuring human wellbeing and safeguarding biodiversity [5, 17–19]. NBS utilize coastal ecosystems' natural capacity to act as barriers to attenuate and dissipate the effect of storm surges and high-water levels and are seen as a valid and more efficient alternative to traditional coastal protection methods due to its capacity to protect and deliver ecosystem services and other co-benefits [20, 21].

Over time, different NBS for coastal flood mitigation have been developed. The type of NBS to be used is dependent on local socio-economic and environmental factors. Each NBS implementation should be considered as individual. Applying the same measure elsewhere may not be the best practice [22]. However, despite its increasing application cases, there are still challenges and pitfalls regarding the use of NBS such as lack of guidance, political, legal and financial support and monitoring its effectiveness over time [23]. Furthermore, due to the relatively recent adoption, some authors question if there is enough evidence to fully understand the environmental impacts of NBS [24–26].

This chapter aims to synthesize the information on NBS for coastal flood mitigation and erosion protection, including examples of NBS, implemented worldwide, and examples showing the advantages, disadvantages, challenges, opportunities and the associated environmental problems.

2 The Transition from Traditional to Nature-Based Coastal Flood Mitigation

Coastal zones are perhaps one of the most important and complex socioecological systems. These areas host an enormous diversity of terrestrial and marine species, habitats and landscapes [27–29]. This diversity includes some of the most productive and important ecosystems such as mangrove forests, seagrass meadows, coral reefs, intertidal areas, coastal lagoons, coastal dunes, salt marshes and wetlands. Coastal ecosystems provide an enormous array of goods and services, essential to support human development and wellbeing. From basic nutritional and non-nutritional resources, water purification, climate regulation and nursery areas to transportation and cultural identity, coastal zones are important drivers of socio-economic development [30–33].

Coastal zones, defined as the interaction zone between the terrestrial and marine part of Earth [34], cover only 4% of land and 11% of marine waters; however, they host around one-third of the global population [35]. The largest urban agglomerations and megacities are located near the coast [36]. Examples are Tokyo, New York, Newark and Shanghai with 37.2, 20.4 and 20.2 million inhabitants, respectively [37]. In the future, it is expected that population will increase in coastal areas, attaining 625 million in 2030 in low-lying areas [32]. It is estimated that 50% of salt marshes, 35% of mangroves, 30% of coral reef and 29% of seagrass meadows have already been lost [31]. The restoration and destruction of natural habitats are recognized as one of the most important drivers of change in the coastal zone [34, 38].

The capacity of coastal ecosystems to provide flood protection has been degraded as consequence of urban development. Since many of these habitats have been lost or impacted to some degree, these functions are not present or are diminished, leaving coastal communities vulnerable to floods [15, 39, 40]. With around 10% of the global population living in low-lying flood-prone areas much is at stake [41]. In addition to land-use change, the vulnerability of coastal communities is even more exacerbated by climate change [42]. Some of the impacts of climate change in coastal areas are sea-level rise, the increase in severity and occurrence of storm surges and extreme events [43].

The history of coastal protection for flood mitigation is old as ancient civilizations that responded to fluctuations in sea-level rise following the Glacial Age [44]. Early coastal flood protections included the building of dikes in the Netherlands and China, wooden pile groynes in the Baltic coastlines and beach nourishments in the Mediterranean [44–46]. As a consequence of population increase in coastal areas and a chain of impact such as sediment removal, reduction of the sediments transported to the coast by damming rivers, erosion, port and tourism development, stronger and more durable solutions were needed [44, 46, 47]. Technological and scientific developments allowed to engineer coastal protection structures (grey infrastructure) such as breakwaters, ripraps, sea walls, tetrapods, bulkheads and groynes, to address these issues [44]. Grey infrastructures were built all around the world as an immediate response to flooding events or eroding coastlines, where the priority was to save human lives and assets instead of keeping the natural landscape [5, 48].

During the last decades, the concerns about environmental degradation in post-industrial revolution led to a shift in coastal planning and management towards an ecosystem-based approach [49]. Notions of sustainability and sustainable use of natural resources and habitats [50], the ecological contribution of nature to human wellbeing through the deliverance of ecosystem services [51] and green/blue economy [52] brought an environmentally holistic focus. The development of the Integrated Coastal Management (ICM) in the United States and later the Integrated Coastal Zone Management (ICZM) in Europe shifted the coastal protection from a sectorial to an integrative approach [53]. In parallel, concerns of biodiversity and habitat loss developed the implementation of ambitious targets (so-called Aichi

Targets) under the Convention on Biological Diversity (CBD), which aimed at "by 2020, at least 17 % of terrestrial and inland water, and 10 % of coastal and marine areas, especially areas of particular importance for biodiversity and ecosystem services, are conserved through effectively and equitably managed, ecologically representative and well-connected systems of protected areas and other effective area-based conservation measures, and integrated into the wider landscapes and seascapes" [54]. Recently, the United Nations declared the decade of 2021–2030 as the decade for ecosystems restoration. This is one of the greatest challenges for the next years. Approximately 20–50% of blue carbon ecosystems that include seagrass beds, salt marshes and mangroves are degraded. Restoring wetlands can contribute 14% of the capacity to limit the global temperature rise of 2°C [55, 56]. In areas affected by flood, grey infrastructures need to be rebuilt and redesigned, while the habitats described in Box 1 have a natural capacity to restore themselves [57]. In this context, some questions arose regarding the efficiency of grey infrastructures for flood coastal protection and its resilience and adaptability to future climate conditions. This shifted planners and decision-makers towards more nature-friendly alternatives [5].

Box 1 Natural Flood Protection from Coastal Ecosystems

Coastal ecosystems are natural protective barriers against storm surges and sea-level rise, preventing coastal flooding and promoting shoreline stabilization [49]:

- *Coral and other biogenic reefs*: biogenic reefs can be considered the first line of defence for coastal protection for their capacity to attenuate incoming wave energy [58, 59].
- *Seagrasses*: seagrass ecosystems are capable to protect the coast by their capacity to stabilize and maintain sediments in shallow areas and to modify both current flows and wave action [60, 61].
- *Beaches:* highly dynamic changes in beach profile as a response to different wave behaviours and the accretion of sediment creating dissipative surf zones reducing wave energy at the shoreline [62].
- *Dunes:* coastal dunes and dune vegetation act as social, ecological, economic protective buffers against storm surge, wave attack and erosion of the hinterland [63].
- *Mangroves:* mangrove forests contribute directly and indirectly to flood mitigation, by promoting siltation, accretion and stabilization of sediments as well as obstructing the waves energy with its roots and trunks [64].
- *Salt marshes:* salt marsh vegetation buffers the effects of the waves and at the same time prevents erosion [65]. Furthermore, salt marshes also accumulate and store big amounts of water from flooding events.

3 Categorization of NBS for Coastal Flood Mitigation

As many other ecological concepts, several options exist to define and categorize NBS for coastal flood mitigation. However, a common categorization by different authors [66–68] and adopted by Gómez Martín et al. [69] categorizes NBS for coastal flood mitigation into three types: *NBS type 1, low human interventions; NBS type 2, medium human interventions; and NBS type 3, creation of new ecosystems and hybrid solutions.*

3.1 *NBS Type 1: Low Human Interventions*

Also known as natural solutions, these types of NBS include approaches and measures that aim to preserve coastal ecosystems and their natural capacity for flood protection and mitigation, without active physical anthropogenic interventions. Examples of this type of NBS include the creation of coastal and marine protected areas to promote coastal protection by restricting the presence of anthropogenic activities to preserve the well-functioning and resilience of coastal habitats, [70, 71]. Coastal and marine protected areas have been established worldwide, covering 7.44% of all marine areas [72].

3.2 *NBS Type 2: Medium Human Interventions*

Also known as soft engineering approaches, these types of NBS are comprised by extensive and intensive physical approaches. They aim to support the enhancement of ecosystem services provided by coastal habitats in a sustainable way [69]. NBS type 2 are implemented to increase the protection capacity of coastal ecosystems or to complement existing hard engineered structures [73]. This is in general one of the most used type of NBS for coastal flood mitigation. This type of NBS shares the same principles with similar coastal ecosystem-based approaches such as "living shorelines" [74], "soft engineering" [75], "nature-based features or infrastructure" [76], "green/blue infrastructure" [77] and "building with nature" [78]. The most common solutions under type 2 are actions which deal with ecological restoration or rehabilitation of coastal habitats (see Box 2).

Box 2 Examples of NBS Type 2 for Coastal Flood Mitigation Approaches

This type of NBS has been implemented all around the world for many different coastline types and is designed according to existing ecological elements (habitats, functions and processes) within the implementation areas. Some of the strategies have been used by humans since early beginning of coastal protection history. Some of the most popular type 2 NBS for coastal flood mitigation include:

- *Beach nourishment:* also known as beach replenishment is an ancient practice which basically consists of the artificial reallocation of sediment from land-based or offshore sources. Typically, beach nourishments are done as a response to coastal erosion events [79].
- *Mangrove restoration and afforestation:* ecological restoration and engineering actions aiming at returning mangrove forests to a pre-existing condition. Traditional approaches include the plantation of monospecific stands of mangrove seedlings. Mangrove afforestation includes the plantation of mangrove forests for purposes of coastal protection in areas where this ecosystem was not present previously [80].
- *Dune protection and stabilization:* traditional solutions encompass the plantation of dune vegetation for dune stabilization and the use of wooden fences to trap sediment and in some cases limit anthropogenic activities [81]. These actions prevent sediment losses and the re-establishment of sediment dynamics making dunes resilient against storm surges.
- *Salt marsh restoration and recreation:* restoring and re-establishing salt marshes of once reclaimed coastal landscapes require de reopening or de-embankment of existing physical barriers (dikes, levees, etc.) for the re-establishment of tidal hydrodynamics [82]. Other complementary actions, like the removal of invasive species, also contribute to restoring the well-functioning of salt marshes.
- *Coral reef restoration:* reef ecosystems can be done following passive and active restoration strategies. Popular approaches include the direct transplantation and gardening or coral fragments, larval enhancement and the stabilization and creation of substratum for the establishment of coral reefs [83].
- *Seagrass restoration and rehabilitation:* the restoration and rehabilitation of seagrass meadows is mainly done through direct transplantation of plants from donor locations or through mechanical seed dispersal [84].
- *Oyster reefs restoration and establishment:* establishing oyster reefs requires the addition of substratum in which oyster larvae attach. A popular method is the recycling of oyster or clam shells which are designed and agglomerated into "bags" which are then submerged in the location identified as important for the establishment of larvae [85].

(continued)

Box 2 (continued)

- *Barrier island restoration*: barrier island restoration follows a similar approach as beach nourishments. Sediment is pumped to the existing barrier island creating a beach and a dune; furthermore, a second dune is created to allow the development of a marsh and the establishment of vegetation [86].

3.3 NBS Type 3: The Creation of New Ecosystems and Hybrid Solutions

NBS type 3, commonly referred to as hybrid solutions, combine a type 2 NBS with hard grey infrastructures for maximum coastal protection [5]. Hybrid NBS are very popular since they possess the strong protective capacity of hard structures and at the same time contributing to the well-functioning of coastal ecosystems. Examples of hybrid solutions include managed realignments of coastal populations in which hard structures are moved further inland allowing the establishment of natural habitats. The combination of NBS type 2 and hard structures allows for the creation of innovative designs in coastal protections [5, 17].

3.4 Global Implementation of NBS in the Context of Coastal Flooding

As alternatives to hard structures, in the last decades, an increasing number of NBS implementation efforts have been carried out all over the globe to safeguard the coastal population from the damages associated to coastal flooding and make them more resilient to future events (e.g. sea-level rise and extreme weather events) [5, 24, 87]. In some countries NBS implementation is well established in coastal zone management and disaster risk reduction strategies with specific financing schemes, for example, in Europe, the European Commission allocated around 185 million euros between 2014 and 2020 for NBS implementation projects [26], and in the United States, the US Army Corps of Engineers is advised to use NBS to enhance coastal resilience [88]. In recent years, many initiatives have pushed forward the implementation of NBS through projects and pilot studies, compiling case studies databases. Some of these include the "Nature-based Solutions Initiative" by Oxford University (naturebasedsolutionsinitiative.org), "Nature-Based Solutions" (nature-basedsolutions.com), "European Union Repository of Nature-Based Solutions" (oppla.eu), "Think-Nature Platform" (platform.think-nature.eu) and "Natural Hazards Nature-based Solutions" by the World Bank (naturebasedsolutions.org). These databases comprise many case studies and descriptions, which NBS have been applied in coastal areas. All of them contain NBS in coastal areas. Figure 1 illustrates

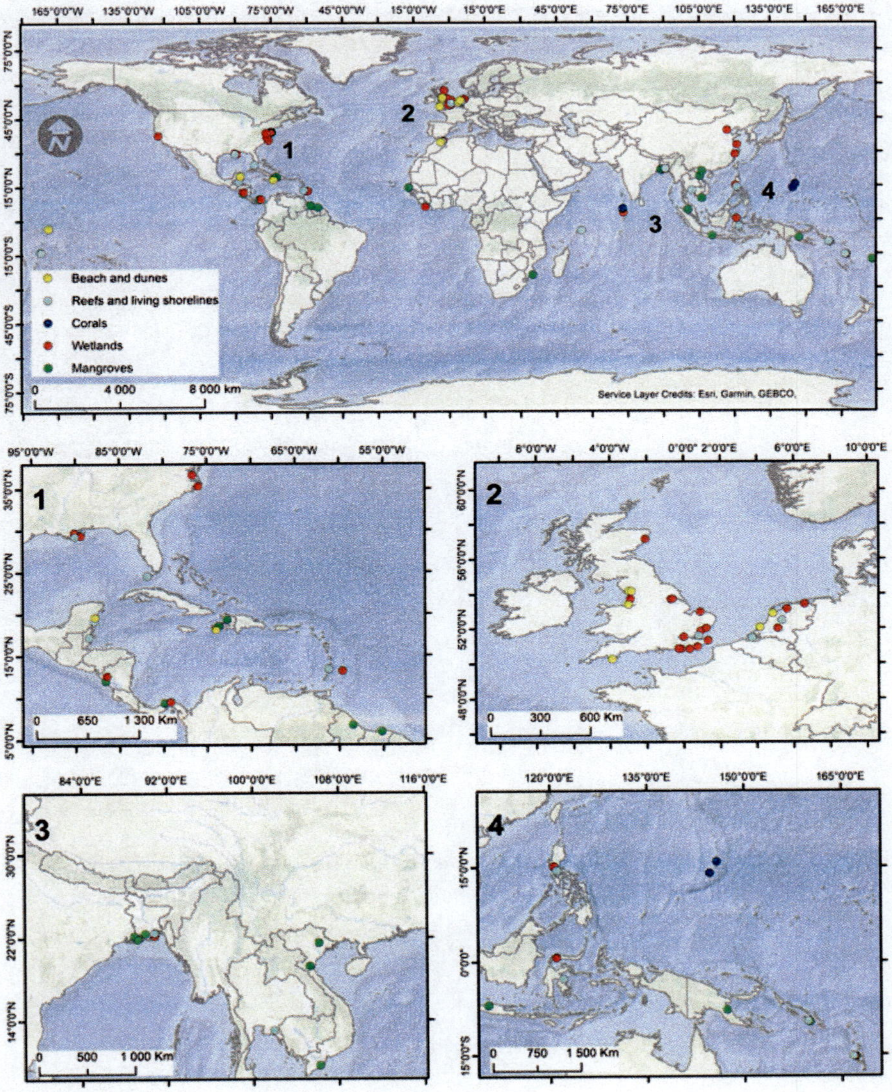

Fig. 1 Global examples of NBS for coastal flood mitigation. The examples were compiled from (1) Natural Hazards Nature-Based Solutions (naturebasedsolutions.org), (2) Naturally Resilient Communities (hnrcsolutions.org) and Narayan et al. [24]

the global distribution of these projects. Beach nourishment in Lithuanian coast is described in Box 3 as an example of a soft engineering solution used to protect coastal assets in detriment of a normal sediment littoral dynamics by built breakwaters to secure port industry.

Box 3 Beach Nourishment in the Lithuanian Coastline

The Lithuanian coastline is composed by sandy beaches and unique coastal dunes. Klaipeda Straight, where the Curonian Lagoon meets the Baltic Sea, transports sediment essential for the littoral dynamics. However, after the development of port industry, two jetties were built at Klaipeda Straight. These structures negatively affect sediment transport along the coast, increasing the deficit in northern part of the coast. The loss of beach width increases the risk of coastal flooding. In some areas, coastal communities are at risk. To ensure sediment dynamics and reduce the risk of erosion and coastal flooding, Lithuania has been using soft engineering solutions: beach nourishment. This method has been applied whenever sediment deficiency is observed.

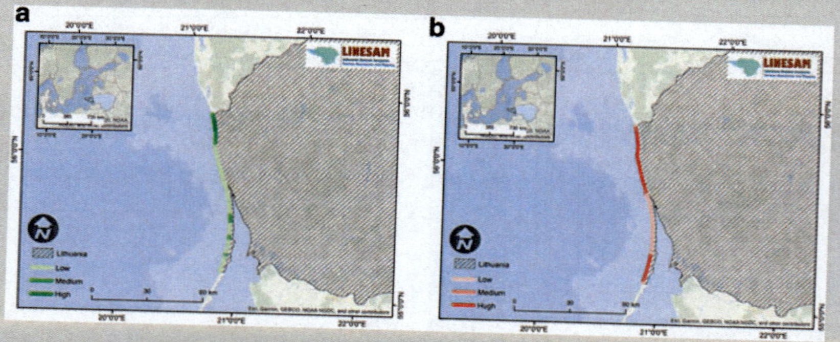

Coastal protection ecosystem service supply (a) and demand (b) in the Lithuanian coastline, mapped within the Lithuanian National Ecosystem Services Assessment and Mapping, following the approach by Liquete et al. [89]. Flood protection supply was assessed using the indicators coastal geomorphology, slope, seabed habitats and land use. Flood protection supply demand was based on indicators of population density, imperviousness and cultural sites. For more information, please consult linesam.mruni.eu.

3.5 *Socioecologic Impacts of NBS*

In the context of coastal flood mitigation, NBS are planned and designed to maximize coastal protection capacity. During the implementation process, an analysis is performed on the potential effects and impacts of the NBS in the socio-economic and ecological setting [73, 90]. This analysis is usually based on the lessons learned from other case studies, which used similar approaches. However, every system is unique, and the design and implementation processes are different. Therefore, it is never possible to predict what will be the impacts of the NBS, unless there is a pilot study before the implementation [91, 92]. Also, monitoring the site before NBS implementation would be an advantage for proper assessment of NBS.

Despite the growing evidence of the efficiency of NBS for coastal flood mitigation [24, 93], monitoring is an aspect that rises concerns among scientists and planners [91, 94–96]. While there are few long-term monitoring results and evidence on the environmental and impacts of NBS, in a short-term perspective, NBS can produce effects beyond providing coastal protection. Some of these effects may be beneficial or problematic [97, 98].

3.5.1 Socioecological Benefits

Depending on the choice of NBS, type 1, 2 or 3, there is either the preservation, restoration or rehabilitation of coastal habitats. Besides enhancing the coastal protection capacity, it also enhances other ecological processes, functions and interactions with surrounding ecosystems. Some implementation driven environmental benefits of NBS include:

- *Increase in habitat availability, biodiversity and species abundance:* the creation or restoration of coastal ecosystems increases the habitat availability for several species. These new habitats may be used as feeding or nursery grounds, nesting and resting areas, or simply hard structures to be colonized. Several case studies showed that reef restoration NBS through artificial structures increased coral establishment, recruitment and colonization by other species [99, 100]. Another environmental impact is the general increase in biodiversity and species abundance. In Mobile Bay (USA), the establishment of an oyster reef for coastal protection purposes attracted a diversity of fish and crab species, some of economic interest for coastal communities [101]. Various salt marsh and wetland restoration actions also reported increased local biodiversity and biomass [102–104].
- *Carbon sequestration:* coastal ecosystems are known for their high capacity to sequester carbon [105, 106]. The establishment of a new or restoration of an existing coastal ecosystems and its associated biodiversity can have a positive effect, increasing carbon sequestration at the local level. The "Sundarbans Mangrove Restoration Project" planted circa 6,000 ha of mangrove trees in India, increasing carbon sequestration almost three times more than expected [107]. A large-scale seagrass restoration of about 1700 ha in Virginia (USA) has shown an increase of carbon sequestration compared to its prior state [108].
- *Support for recreation and tourism industry:* an important human benefit from NBS is the implementation and increase of recreation opportunities, such as birdwatching in salt marshes, snorkelling and diving in coral restored areas or simply hiking along restored wetlands. There is clear evidence that NBS can contribute to their socio-economic wellbeing by supporting the tourism industry. Sauer et al. [109] reported that coastal West Mediterranean areas subjected to restoration were perceived by beachgoers as beneficial for coastal flood protection. Mandić [110] concluded that all dimensions of tourism development can be related to the implementation of NBS. In Pensacola (USA), the effects of re-establishing marsh area breakwaters included an increase in tourism numbers [111].

- *Support for the fishing industry:* restoring and creating coastal habitats through NBS implementation provide an enhancement in the abundance of economically important biological resources, increasing fisheries and coastal livelihoods. Reef restoration in Grenada Islands has increased the abundance of lobsters and octopus, supporting local fisheries [99]. Rezek et al. [112] analysed the long-term performance of seagrass restoration projects in Florida and reported an increase in fishery resources.
- *Improving water purification and quality:* coastal ecosystems, especially mangrove and salt marshes, contribute to the maintenance of the hydrological cycle. Through plant remediation and fixation of pollutants and nutrients, NBS can play an important role in improving water quality, essential for humans and biota [113]. For example, in Tampa Bay (USA), multiple habitat restoration interventions have significantly improved the water quality [114].
- *Self-maintenance and coastal resilience:* different from hard structures, NBS for coastal flood mitigation require much less maintenance [5, 17, 115]. This is because NBS have a natural capacity to cope with the impacts of extreme events and sea-level rise and adapt to new environmental conditions [5]. These characteristics are important for supporting coastal resilience of coastal communities [20]. For example, a study by Rodriguez et al. [116] concluded that oyster reefs would be resilient enough to cope with increasing sea-level rise. Another example, reported by Mo et al. [117] is the self-restoration of salt marshes after being affected by a hurricane in the Gulf of Mexico.
- *Cost-effectiveness:* the above-mentioned examples of co-benefits can only be delivered by NBS solutions, since hard grey infrastructures do not have this capacity. Therefore, when integrating all the benefits of coastal flood mitigation and the environmental and human co-benefits, NBS are more cost-effective than traditional hard structures. Therefore, the investments made for the design and implementation of NBS will be paid off in a short-medium time range. Narayan et al. [24] and Reguero et al. [118] analyse and compare the cost-effectiveness of several NBS for coastal food mitigation. They found that for wave heights up to 0.5 m, the costs of salt marshes and mangrove nature-based defence projects can be two to five times cheaper than a submerged breakwater.

3.5.2 Socioecological Constrains

Even with the best available knowledge and detailed planning, nature process is complex, dynamic and unpredictable, and so is the implementation of NBS. Unforeseen short- and long-term interactions with surrounding natural areas and conflicts of interests with human demands can create constrains for NBS implementation. Some of these constrains include:

- *Introduction or proliferation of alien species:* while the principles for the design and implementation of NBS requires the non-introduction of alien and pest species, this is always a possibility and can have significant negative outcomes on coastal ecosystems [25, 119]. However, in some cases introduced non-native species can help to restore ecosystems, thus their functions and services [120]. This problem can be tackled with an appropriated planning.

- *Continuous human interventions:* despite being still self-sustaining and resilient to coastal flooding hazards, some NBS also require continuous interventions. For example, due to the highly dynamic nature of sediment transport in the coastal zone, beach nourishments need to be replenished regularly increasing the costs of its maintenance [121]. A more cost-effective approach is the establishment of mega-nourishments like the Sand Motor in the Netherlands, which is done to be self-dynamic for decades [122]. Another example, are dune restoration solutions, which would also require maintenance when wooden fences are damaged [123].
- *Land occupation and land-use conflicts:* one of the biggest conflicts is the required area for the implementation of NBS, which may belong to private owners and slow down the implementation process [124–126]. Also, in the NBS implementation areas, there could be other important socio-economic activities. For example, in mangrove areas restoration may be seen as a conflicting solution with other important socio-economic activities like aquaculture (shrimp) which in fact cuts down big portion of mangrove areas [127].
- *Sustainability:* while the process of designing and implementing NBS aims at sustainable use of resources, environmental integrity and socio-economic development, choosing the most sustainable NBS is complicated since sustainability is a long-term metric. Sustainability is not a metric that is measured as an immediate response. These uncertainties may also create resistance amongst the stakeholders [128].

4 Challenges and Opportunities for the Implementation of NBS for Coastal Flood Mitigation

In the context of coastal flood mitigation, the evidence on the effectiveness of NBS is reshaping the world's coastlines from grey to green [17, 129]. Despite its efficiency not only for coastal flood mitigation but for the supporting role in the deliverance of several other ecosystem services and benefits, there are still some challenges and resistances. Most of these challenges and resistances are strongly related to the infancy of the NBS compared to the long-established hard coastal engineering. Recently, to increase the acceptance and implementation of NBS, several authors [23, 25, 124, 130] identified the most pressing challenges to be overcome. It is important to understand that challenges can appear at any step of the process of implementing NBS. Based on those challenges, we have identified some of the most important:

- *Lack of guidance:* despite the recent efforts in creating guidelines for NBS implementation for coastal flood protection by European Commission, The Nature Conservancy, World Wildlife Foundation and the World Bank, the lack

of a clear, standard and stepwise guiding process for the implementation of NBS is one of the most common challenges mentioned. This lack of guidance can hinder, delay or even abandon the implementation of NBS [19, 22]
- *Low and slow acceptance:* despite NBS efficiency, several works [129, 131, 132] revealed that there is still resistance from more conservative stakeholders. Since the traditional coastal flood mitigation was through hard structures, there is still a misconception that these are the most robust against storms and sea-level rise.
- *Stakeholder engagement:* while in countries where NBS for flood mitigation is well established into coastal zone management and ecosystem-based approach, stakeholder engagement aims to include all users and interested parties [133]. The non-inclusion of all interested parties in the decision, design and implementation of NBS can create some resistance problems in the process [109, 134]
- *Lack of monitoring and evidence of effectiveness:* while hundreds of NBS for coastal flood protection have been implemented, only some of them have a monitoring program [17], and very few are long-term. Without a relevant monitoring plan, it is difficult to justify the effectiveness and efficiency of NBS.
- *Lack of financing and insurance:* According to Seddon et al. [25], "less than 5% of climate finance goes towards dealing with climate impacts, and less than 1% goes to coastal protection, infrastructure and disaster risk management including NBS". In general, NBS are financed by public and private funds [135–138]. In countries where NBS are established in their coastal flood mitigation strategies, there are defined financing schemes. However, for developing countries where NBS are slowly gaining importance, financing schemes are lacking. The World Bank has been working in this respect, financing NBS implementation projects around the world (see naturebasedsolutions.org).
- *Time pressure:* usually when a coastal flood mitigation solution is needed, it comes as urgent. While traditional coastal defences exercise their protective capacity since the moment they are implemented, this is not the case for NBS, especially those which require ecological restoration. In the case of NBS, it takes time until the protective capacity is expected to increase as the coastal ecosystem becomes more and more resilient [5, 97].
- *Dilution with other ecological concepts:* the umbrella concept over NBS is the ecosystem-based approach. However, NBS is just one of many concepts. Living shorelines, building with nature, green infrastructure and nature-based infrastructure are examples of concepts that share many principles with NBS [22, 73]. Such "similar" concepts kind of dilute NBS in the perspective of ecosystem-based management. This can hinder the importance of NBS to the frontline as solutions for coastal disaster mitigations.

Challenges can also be looked like opportunities to explore in the future, which would increase the acceptance and implementation of NBS for coastal flood mitigation. Despite that the main aim is coastal flood mitigation, NBS should also be framed to tackle other socio-economic and environmental challenges. This might be an advantage against traditional hard grey structures. Identifying trade-offs and synergies with and among socio-economic and environmental targets from different policies may increase the probability of NBS to be implemented. We have identified some of these frameworks, in which NBS could be framed into:

- *Contributing to the achievement of the United Nations Sustainable Development Goals*: considering the UN 2030 Agenda for Sustainable Development Goals to address global societal challenges, NBS have the potential to substantially contribute to its targets and to assist achieve the full range of SDGs [137]. According to the Nature-Based Solutions for Climate Manifesto developed for the UN Climate Action Summit 2019 "they are a significant complement to decarbonisation, reducing global climate change risks and establishing climate-resilient societies" [139]. NBS provide multiple benefits and are identified as critical for the regeneration and improvement of wellbeing in coastal resilience and ecosystem restoration [22, 140]. They enhance the insurance value of ecosystems and increase carbon sequestration, as well as increase the sustainability in energy use. NBS are increasingly seen as innovative solutions to manage water-related risks while transforming natural capital into a source of green growth and sustainable development [141] . Specifically, NBS are directly relevant to various SDGs: to SDG 1 (tackling poverty), SDG 2 (food security), 3 (health and wellbeing), SDG 4 (quality of education), SDG 6 (sustainable management of water), SDG 7 (climate adaptation strategies may be linked to the goal for clean and sustainable energy), SDG 8 (sustainable economic growth), SDG 10 (reducing inequalities), SDG 11 (sustainable cities and communities), SDG 12 (responsible consumption and production), SDG 13 (climate change), SDG 14 (conservation and sustainable use of oceans, seas and marine resources) and SDG 15 (protection, restoration and promotion of sustainable use of terrestrial ecosystems) [22, 137, 142].

"NBSs are a necessary component of the general global effort to attain the goals of the Paris Agreement on Climate Change" [139]. They improve harmony between people and nature, representing a holistic, people-centred response to climate change. A more robust harmonization of policies across economic, environmental and social agendas is especially important to recognize the multiple dimensions of NBS impacts and co-benefits.

- *Contributing to coastal resilience and climate change adaptation:* Masselink and Lazarus [143] define coastal resilience as "the capacity of the socioeconomic and natural systems in the coastal environment to cope with disturbances, induced by factors such as sea-level rise, extreme events and human impacts, by adapting whilst maintaining their essential functions". A report by the US Army Corps of Engineers [76] and Sutton-Grier et al. [5] are just two examples of a panoply of reports and studies that place NBS for coastal flood mitigation as a central pillar to achieve more resilient ecosystems and communities. Similarly, due to their versatile characteristics of self-adapting and repairing after extreme events and to cope with sea-level rise, NBS will serve as climate change adaptation structures, which coastal communities must rely on to avoid further losses [4].

- *Contributing to the achievement to global targets:* at the global level the CBD, through the Aichi Targets, aims to halt the loss of biodiversity, improve its status and the environmental and socio-economic benefits associated with it, until 2020. Unfortunately, they were not met, but in the future, targets such as the UN decade for restoration can be achieved by applying the NBS principle. Therefore, framing NBS as contributors to CBD and Aichi Targets can potentiate its implementation.
- *Contribution to the enhancement and deliverance of multiple ecosystem services and circular economy (CE):* ecosystem services are the contributions of nature's ecological processes and functions which deliver goods and services for human wellbeing [144]. Usually, ecosystem services are used to show the socio-economic benefits of a preserved environment or an ecological intervention. Several of the above-mentioned case studies identify the multitude of ecosystem services enhanced by the implementation of NBS for flood mitigation. As ecosystem services also include cultural services (tourism), it is easier for the general public to perceive the NBS as a direct input and improvement of their socio-economic and health status. Framing NBS for enhancing ecosystem services could contribute to their implementation [145]. Moreover, integrating NBS in the built environment can contribute to a circular economy through the provision of ecosystem services [146].
- *Contributing to green/blue economy:* the concepts of blue and green economy are built on the premises of economic development with sustainable use of environmental resources for the improvement of human wellbeing, without compromising and jeopardizing natural ecosystems. NBS for coastal flood mitigation can be framed as contributing to green/blue economy since their implementation can, for example, enhance fisheries and tourism industries [147, 148].

5 Concluding Remarks

There is growing evidence on the efficiency of NBS as one of the most cost-effective solutions for their multiplicity of benefits beyond flood mitigation. Furthermore, scientific evidence points out that the integration of coastal ecosystems as NBS increases resilience and the capacity for climate change adaptation, since natural ecosystem cope and adapt easily to new situations. The acceptance of NBS for flood mitigation is increasing with every implementation and because coastal communities experience fewer damages and losses when protected by coastal ecosystems.

Despite the many benefits of NBS, there is still some resistance for its implementation. What we consider to be an impeding factor for NBS implementation is the similarity of terminologies, objectives and aims of similar concepts, like living shorelines, building with nature and green infrastructures. Most of these concepts are under the umbrella of ecosystem-based management, and even if their framework is different, they all aim to the sustainable status of coastal ecosystems. This multitude of concepts in a way dilutes the visibility of NBS as first choice solution for coastal flood mitigation. Another aspect is that, compared to hard structures, NBS are site-specific. Although there are some general frameworks, the lack of concrete guidance and best practices are hardly transferable to global guidelines. Every NBS implementation is unique and the socioecological response is specific. However, it is important to report and monitor its efficiency and increase evidence of its utility. Another hindrance of the NBS is the time needed from implementation until full protection capacity, since grey infrastructures develop coastal protection as soon as they are established. Nevertheless, the versatile nature of NBS allows the inclusion of hard structures in hybrid solutions, which can deliver both short and long-term protection against coastal floods.

To increase the acceptance and implementation of NBS, governments and planners should advocate for NBS as contributors for global environmental and socio-economic targets beyond risk disaster management. NBS must be framed as active contributors to global targets like SDGs, UN decade for restoration, the compliance with climate change adaptation targets, green and blue economy and growth and aiding to reverse global biodiversity and ecosystem loss. Framing NBS in these contexts will also increase its cost-effectiveness and potentiating its implementation.

Overall, the multitude of benefits of NBS implementation to humans and biota allows claiming that these measures can be considered as more effective and sustainable compared to grey infrastructure solutions. However, to fully understand its impacts, it is necessary to promote a long-term monitoring process to support its reliability and credibility.

Acknowledgements Authors were partly funded by the Lithuanian National Ecosystem Services Assessment and Mapping (LINESAM) Project, which has received funding from the European Social Fund project LINESAM no. 09.3.3-LMTK- 712-01-0104 under a grant agreement with the Research Council of Lithuania (LMTLT).

References

1. Schoonees T, Gijón Mancheño A, Scheres B, Bouma TJ, Silva R, Schlurmann T, Schüttrumpf H (2019) Hard structures for coastal protection, towards greener designs. Estuar Coasts 42:1709–1729
2. Arkema KK, Griffin R, Maldonado S, Silver J, Suckale J, Guerry AD (2017) Linking social, ecological, and physical science to advance natural and nature-based protection for coastal communities. Ann N Y Acad Sci 1399:5–26
3. Salgado K, Martinez ML (2017) Is ecosystem-based coastal defense a realistic alternative? Exploring the evidence. J Coast Conserv 21:837–848
4. Depietri Y, McPhearson T (2017) Kabisch N, Korn H, Stadler J, Bonn A (eds) Integrating the Grey, Green, and blue in cities: nature-based solutions for climate change adaptation and risk reduction BT – nature-based solutions to climate change adaptation in urban areas: linkages between science, policy and practice. Springer, Cham, pp 91–109
5. Sutton-Grier AE, Wowk K, Bamford H (2015) Future of our coasts: the potential for natural and hybrid infrastructure to enhance the resilience of our coastal communities, economies and ecosystems. Environ Sci Pol 51:137–148

6. Anthony EJ, Gratiot N (2012) Coastal engineering and large-scale mangrove destruction in Guyana, South America: averting an environmental catastrophe in the making. Ecol Eng 47:268–273
7. Dugan JE, Airoldi L, Chapman MG, Walker SJ, Schlacher T (2011) Wolanski E, McLusky D (eds) 8.02 – estuarine and coastal structures: environmental effects, a focus on shore and nearshore structures. Academic Press, Waltham, pp 17–41
8. Kubowicz-Grajewska A (2015) Morpholithodynamical changes of the beach and the near-shore zone under the impact of submerged breakwaters – a case study (Orłowo Cliff, the Southern Baltic). Oceanologia 57:144–158
9. Rangel-Buitrago N, Williams AT, Anfuso G (2018) Hard protection structures as a principal coastal erosion management strategy along the Caribbean coast of Colombia. A chronicle of pitfalls. Ocean Coast Manag 156:58–75
10. Govarets A, Lauwerts B (2009) Assessment of the impact of coastal defence structures. Biodiversity series: OSPAR Commission
11. Jayaratne R, Mikami T, Esteban M, Shibayama T (2014) Investigation of coastal structure failures due to the 2011 great eastern Japan earthquake tsunami. In: From sea to shore? Meet. Challenges Sea. ICE Publishing, pp 1241–1250
12. Tsai C-H, Tzang S-Y, Hsiao S-S, Cheng C-C, Li H-W (2006) Coastal structure failures and coastal waves on the north coast of Taiwan due to typhoon herb. J Coast Res 2006:393–405
13. Powell EJ, Tyrrell MC, Milliken A, Tirpak JM, Staudinger MD (2019) A review of coastal management approaches to support the integration of ecological and human community planning for climate change. J Coast Conserv 23:1–18
14. Costanza R, Pérez-Maqueo O, Martinez ML, Sutton P, Anderson SJ, Mulder K (2008) The value of coastal wetlands for hurricane protection. AMBIO J Hum Environ 37:241–248. https://doi.org/10.1579/0044-7447(2008)37[241:TVOCWF]20CO;2
15. Arkema KK, Guannel G, Verutes G, Wood SA, Guerry A, Ruckelshaus M, Kareiva P, Lacayo M, Silver JM (2013) Coastal habitats shield people and property from sea-level rise and storms. Nat Clim Chang 3:913–918
16. Das S, Vincent JR (2009) Mangroves protected villages and reduced death toll during Indian super cyclone. Proc Natl Acad Sci U S A 106:7357–7360
17. Morris RL, Konlechner TM, Ghisalberti M, Swearer SE (2018) From grey to green: efficacy of eco-engineering solutions for nature-based coastal defence. Glob Chang Biol 24:1827–1842
18. Morris RL, Strain EMA, Konlechner TM, Fest BJ, Kennedy DM, Arndt SK, Swearer SE (2019) Developing a nature-based coastal defence strategy for Australia. Aust J Civ Eng 17:167–176
19. Sutton-Grier AE, Gittman RK, Arkema KK et al (2018) Investing in natural and nature-based infrastructure: building better along our coasts. Sustain 10:1–11
20. Schueler K (2017) Nature-based solutions to enhance coastal resilience
21. van Wesenbeeck BK, IJff S, Jongman B, Balog-Way SAB, Kaupa SM, Bosche LV, Lange G-M, Holm-Nielsen NB, Nieboer H, Taishi Y (2017) Implementing nature based flood protection: principles and implementation guidance. The World Bank
22. Cohen-Shacham E, Andrade A, Dalton J et al (2019) Core principles for successfully implementing and upscaling nature-based solutions. Environ Sci Pol 98:20–29
23. Brears RC (2020) Nature-based solutions to 21st century challenges. Routledge
24. Narayan S, Beck MW, Reguero BG, Losada IJ, Van Wesenbeeck B, Pontee N, Sanchirico JN, Ingram JC, Lange GM, Burks-Copes KA (2016) The effectiveness, costs and coastal protection benefits of natural and nature-based defences. PLoS One 11:1–17
25. Seddon N, Chausson A, Berry P, Girardin CAJ, Smith A, Turner B (2020) Understanding the value and limits of nature-based solutions to climate change and other global challenges. Philos Trans R Soc B Biol Sci 375:20190120
26. Kumar P, Debele SE, Sahani J et al (2020) Towards an operationalisation of nature-based solutions for natural hazards. Sci Total Environ 731:138855

27. Agardy T, Alder J (2005) Millennium ecosystem assessment: ecosystems and human Well-being: current state and trends, volume 1. In: Millennium Ecosystem Assessment (ed) Millennium ecosystem assessment. Island Press, Washington, pp 513–459
28. Martínez ML, Intralawan A, Vázquez G, Pérez-Maqueo O, Sutton P, Landgrave R (2007) The coasts of our world: ecological, economic and social importance. Ecol Econ 63:254–272
29. Wilson EO, Peter FM (1988) Ecological diversity in coastal zones and oceans
30. Costanza R, Arge R, De Groot R et al (1997) The value of the world's ecosystem services and natural capital. Nature 387:253–260
31. Barbier EB, Hacker SD, Kennedy C, Koch EW, Stiera C, Silliman BR (2011) The value of estuarine and coastal ecosystem services. Ecol Monogr 81:169–193
32. Neumann B, Vafeidis AT, Zimmermann J, Nicholls RJ (2015) Future coastal population growth and exposure to sea-level rise and coastal flooding – a global Assessment. PLoS One 10:e0118571
33. Solé L, Ariza E (2019) A wider view of assessments of ecosystem services in coastal areas: the perspective of social-ecological complexity. Ecol Soc 24. https://doi.org/10.5751/ES-10883-240224
34. Crossland CJ, Kremer HH, Lindeboom H, Crossland JIM, Le Tissier MDA (2005) Coastal fluxes in the Anthropocene: the land-ocean interactions in the coastal zone project of the international geosphere-biosphere Programme. Springer, Berlin
35. Barbier EB (2017) Marine ecosystem services. Curr Biol 27:R507–R510
36. Von Glasow R, Jickells TD, Baklanov A et al (2013) Megacities and large urban agglomerations in the coastal zone: interactions between atmosphere, land, and marine ecosystems. Ambio 42:13–28
37. Blackburn S, Pelling M, Marques C (2019) Megacities and the coast: global context and scope for transformation. In: Wolanski E, Day JW, Elliott M, Ramachandran RBT (eds) Coasts estuaries futur. Elsevier, pp 661–669
38. Pereira P (2020) Ecosystem services in a changing environment. Sci Total Environ 702:13500
39. Rao NS, Ghermandi A, Portela R, Wang X (2015) Global values of coastal ecosystem services: a spatial economic analysis of shoreline protection values. Ecosyst Serv 11:95–105
40. Barbier EB (2015) Valuing the storm protection service of estuarine and coastal ecosystems. Ecosyst Serv 11:32–38
41. Spalding MD, Ruffo S, Lacambra C, Meliane I, Hale LZ, Shepard CC, Beck MW (2014) The role of ecosystems in coastal protection: adapting to climate change and coastal hazards. Ocean Coast Manag 90:50–57
42. Torresan S, Critto A, Dalla Valle M, Harvey N, Marcomini A (2008) Assessing coastal vulnerability to climate change: comparing segmentation at global and regional scales. Sustain Sci 3:45–65
43. Ranasinghe R (2016) Assessing climate change impacts on open sandy coasts: a review. Earth Sci Rev 160:320–332
44. Pranzini E (2018) Coastal erosion and shore protection: a brief historical analysis. J Coast Conserv 22:827–830
45. Charlier RH, Chaineux MCP, Morcos S (2005) Panorama of the history of coastal protection. J Coast Res 211:79–111
46. Pranzini E, Wetzel L, Williams AT (2015) Aspects of coastal erosion and protection in Europe. J Coast Conserv 19:445–459
47. Todd PA, Heery EC, Loke LHL, Thurstan RH, Kotze DJ, Swan C (2019) Towards an urban marine ecology: characterizing the drivers, patterns and processes of marine ecosystems in coastal cities. Oikos 128:1215–1242
48. Nordstrom KF (2014) Living with shore protection structures: a review. Estuar Coast Shelf Sci 150:11–23
49. van Wesenbeeck BK, Griffin JN, van Koningsveld M, Gedan KB, McCoy MW, Silliman BR (2013) Nature-based coastal defenses: can biodiversity help? In: Encycl biodivers, 2nd edn. pp 451–458

50. Neumann B, Ott K, Kenchington R (2017) Strong sustainability in coastal areas: a conceptual interpretation of SDG 14. Sustain Sci 12:1019–1035
51. Lele S, Springate-Baginski O, Lakerveld R, Deb D, Dash P (2013) Ecosystem services: origins, contributions, pitfalls, and alternatives. Conserv Soc 11:343
52. Eikeset AM, Mazzarella AB, Davíðsdóttir B, Klinger DH, Levin SA, Rovenskaya E, Stenseth NC (2018) What is blue growth? The semantics of "sustainable development" of marine environments. Mar Policy 87:177–179
53. Støttrup JG, Dinesen GE, Schumacher J, Gillgren C, Inácio M, Schernewski G (2019) The systems approach framework for collaborative, science-based management of complex systems. J Coast Conserv 23:881. https://doi.org/10.1007/s11852-018-00677-5
54. Rees SE, Foster NL, Langmead O, Pittman S, Johnson DE (2018) Defining the qualitative elements of Aichi biodiversity target 11 with regard to the marine and coastal environment in order to strengthen global efforts for marine biodiversity conservation outlined in the United Nations sustainable development goal 14. Mar Policy 93:241–250
55. UNEP (2020) The UN decade on ecosystem restoration 2021-2030
56. Yu Y, Zhao W, Martinez-Murillo JF, Pereira P (2020) Loess plateau: from degradation to restoration. Sci Total Environ:140206
57. Maly E, Kondo T, Banba M (2017) Experience from the United States: post-Katrina and Sandy. In: Land use management in disaster risk reduction. Springer, Berlin, pp 79–106
58. Elliff CI, Silva IR (2017) Coral reefs as the first line of defense: shoreline protection in face of climate change. Mar Environ Res 127:148–154
59. Martins KA, de Pereira PS, Esteves LS, Williams J (2019) The role of coral reefs in coastal protection: analysis of beach morphology. J Coast Res 92:157
60. Ondiviela B, Losada IJ, Lara JL, Maza M, Galván C, Bouma TJ, van Belzen J (2014) The role of seagrasses in coastal protection in a changing climate. Coast Eng 87:158–168
61. Christianen MJA, van Belzen J, Herman PMJ, van Katwijk MM, Lamers LPM, van Leent PJM, Bouma TJ (2013) Low-canopy Seagrass beds still provide important coastal protection services. PLoS One 8:e62413
62. Hanley ME, Hoggart SPG, Simmonds DJ et al (2014) Shifting sands? Coastal protection by sand banks, beaches and dunes. Coast Eng 87:136–146
63. Sigren JM, Figlus J, Armitage AR (2014) Coastal sand dunes and dune vegetation: restoration, erosion, and storm protection. Shore Beach 82:5–12
64. Othman MA (1994) Value of mangroves in coastal protection. Hydrobiologia 285:277–282
65. Möller I, Kudella M, Rupprecht F et al (2014) Wave attenuation over coastal salt marshes under storm surge conditions. Nat Geosci 7:727–731
66. Cohen-Shacham E, Walters G, Janzen C, Maginnis S (2016) Nature-based solutions to address global societal challenges. IUCN Gland 97
67. Eggermont H, Balian E, Azevedo JMN et al (2015) Nature-based solutions: new influence for environmental management and research in Europe. Gaia 24:243–248
68. Balian E, Eggermont H, Le Roux X (2014) Outputs of the strategic foresight workshop "nature-based solutions in a BiodivERsA context". BiodivERsA, Brussels
69. Gómez Martín E, Máñez Costa M, Schwerdtner Máñez K (2020) An operationalized classification of nature based solutions for water-related hazards: from theory to practice. Ecol Econ 167:106460
70. Roberts CM, O'Leary BC, Mccauley DJ et al (2017) Marine reserves can mitigate and promote adaptation to climate change. Proc Natl Acad Sci U S A 114:6167–6175
71. McLeod E, Salm R, Green A, Almany J (2009) Designing marine protected area networks to address the impacts of climate change. Front Ecol Environ 7:362–370
72. UNEP-WCMC, IUCN (2020) Marine protected planet. UK UNEP-WCMC IUCN, Cambridge
73. Nesshöver C, Assmuth T, Irvine KN et al (2017) The science, policy and practice of nature-based solutions: an interdisciplinary perspective. Sci Total Environ 579:1215–1227
74. Lee Smee D (2019) Coastal ecology: living shorelines reduce coastal Erosion. Curr Biol 29: R411–R413

75. Fiselier JL (2016) Finlayson CM, Everard M, Irvine K, McInnes RJ, Middleton BA, van Dam AA, Davidson NC (eds) Soft engineering for coastal protection: natural Hazard regulation BT – the wetland book: I: structure and function, management and methods. Springer, Dordrecht, pp 1–8
76. Bridges TS, Burks-Copes KA, Bates ME, Collier ZA, Fischenich JC, Piercy CD, Russo EJ, Shafer DJ, Suedel BC, Gailani JZ (2015) Use of natural and nature-based features (NNBF) for coastal resilience. US Army Engineer Research and Development Center, environmental laboratory ...
77. Ruckelshaus MH, Guannel G, Arkema K, Verutes G, Griffin R, Guerry A, Silver J, Faries J, Brenner J, Rosenthal A (2016) Evaluating the benefits of Green infrastructure for coastal areas: location, location, location. Coast Manag 44:504–516
78. de Vriend H, van Koningsveld M, Aarninkhof S (2014) 'Building with nature': the new Dutch approach to coastal and river works. Proc Inst Civ Eng Civ Eng 167:18–24
79. Charlier RH, De Meyer CP (1995) Beach nourishment as efficient coastal protection. Environ Manag Health 6:26–34
80. Spalding M, McIvor A, Tonneijck FH, Tol S, Van Eijk P (2014) Mangroves for coastal defence. Guidelines for coastal managers and policy makers. Wetl Int Nat Conserv:13–34
81. Chapman DM (1984) Dune stabilization. In: Beaches coast Geol. Springer, Boston, pp 379–380
82. Wolters M, Garbutt A, Bakker JP (2005) Salt-marsh restoration: evaluating the success of de-embankments in north-West Europe. Biol Conserv 123:249–268
83. Boström-Einarsson L, Babcock RC, Bayraktarov E et al (2020) Coral restoration – a systematic review of current methods, successes, failures and future directions. PLoS One 15. https://doi.org/10.1371/journal.pone.0226631
84. Calumpong HP, Fonseca MS (2001) Seagrass transplantation and other seagrass restoration methods. In: Global seagrass research methods. Elsevier, pp 425–443
85. Grabowski JH, Peterson CH (2007) Restoring oyster reefs to recover ecosystem services. Ecosyst Eng Plants Protists 4:281–298
86. Oliver B, Ramirez-Avila JJ (2019) Barrier Island restoration: a literature review. In: World environmental water resource congress 2019 Hydraul. Waterw. Water Distrib. Syst. Anal. American Society of Civil Engineers Reston, VA, pp 310–319
87. Raymond CM, Breil M, Nita MR, et al (2017) An impact evaluation framework to support planning and evaluation of nature-based solutions projects. In: Report prepared by the EKLIPSE Expert Working Group on Nature-based Solutions to Promote Climate Resilience in Urban Areas. Centre for Ecology and Hydrology
88. Colgan CS, Beck MW, Narayan S (2017) Financing natural infrastructure for coastal flood damage reduction
89. Liquete C, Zulian G, Delgado I, Stips A, Maes J (2013) Assessment of coastal protection as an ecosystem service in Europe. Ecol Indic 30:205–217
90. Kalantari Z, Ferreira CSS, Deal B, Destouni G (2019) Nature-based solutions for meeting environmental and socio-economic challenges in land management and development. Land degradation and development
91. Frantzeskaki N (2019) Seven lessons for planning nature-based solutions in cities. Environ Sci Pol 93:101–111
92. Nguyen TP (2018) Melaleuca entrapping microsites as a nature based solution to coastal erosion: a pilot study in Kien Giang, Vietnam. Ocean Coast Manag 155:98–103
93. Karamouz M, Heydari Z (2020) Conceptual design framework for coastal flood best management practices. J Water Resour Plan Manag 146:4020041
94. Calliari E, Staccione A, Mysiak J (2019) An assessment framework for climate-proof nature-based solutions. Sci Total Environ 656:691–700
95. Dushkova D, Haase D (2020) Not simply Green: nature-based solutions as a concept and practical approach for sustainability studies and planning agendas in cities. Land 9:19

96. Wamsler C, Wickenberg B, Hanson H, Olsson JA, Stålhammar S, Björn H, Falck H, Gerell D, Oskarsson T, Simonsson E (2020) Environmental and climate policy integration: targeted strategies for overcoming barriers to nature-based solutions and climate change adaptation. J Clean Prod 247:119154
97. van der Nat A, Vellinga P, Leemans R, van Slobbe E (2016) Ranking coastal flood protection designs from engineered to nature-based. Ecol Eng 87:80–90
98. Augusto B, Roebeling P, Rafael S, Ferreira J, Ascenso A, Bodilis C (2020) Short and medium- to long-term impacts of nature-based solutions on urban heat. Sustain Cities Soc 57:102122
99. Reguero BG, Beck MW, Agostini VN, Kramer P, Hancock B (2018) Coral reefs for coastal protection: a new methodological approach and engineering case study in Grenada. J Environ Manag 210:146–161
100. Silva R, Mendoza E, Mariño-Tapia I, Martínez ML, Escalante E (2016) An artificial reef improves coastal protection and provides a base for coral recovery. J Coast Res 75:467–471
101. Scyphers SB, Powers SP, Heck Jr KL, Byron D (2011) Oyster reefs as natural breakwaters mitigate shoreline loss and facilitate fisheries. PLoS One 6:e22396
102. Frisk MG, Miller TJ, Latour RJ, Martell SJD (2011) Assessing biomass gains from marsh restoration in Delaware Bay using Ecopath with Ecosim. Ecol Model 222:190–200
103. van Loon-Steensma JM, Vellinga P (2013) Trade-offs between biodiversity and flood protection services of coastal salt marshes. Curr Opin Environ Sustain 5:320–326
104. Meli P, Benayas JMR, Balvanera P, Ramos MM (2014) Restoration enhances wetland biodiversity and ecosystem service supply, but results are context-dependent: a meta-analysis. PLoS One 9:e93507
105. Macreadie PI, Nielsen DA, Kelleway JJ, Atwood TB, Seymour JR, Petrou K, Connolly RM, Thomson ACG, Trevathan-Tackett SM, Ralph PJ (2017) Can we manage coastal ecosystems to sequester more blue carbon? Front Ecol Environ 15:206–213
106. Thomas S (2014) Blue carbon: knowledge gaps, critical issues, and novel approaches. Ecol Econ 107:22–38
107. Wylie L, Sutton-Grier AE, Moore A (2016) Keys to successful blue carbon projects: lessons learned from global case studies. Mar Policy 65:76–84
108. Greiner JT, McGlathery KJ, Gunnell J, McKee BA (2013) Seagrass restoration enhances "blue carbon" sequestration in coastal waters. PLoS One 8:e72469
109. Sauer I, Roca E, Villares M (2019) Beach users' perceptions of coastal regeneration projects as an adaptation strategy in the Western Mediterranean. J Hosp Tour Res:1096348019889112
110. Mandić A (2019) Nature-based solutions for sustainable tourism development in protected natural areas: a review. Environ Syst Dec:1–20
111. Naturally Resilient Communities Pensacola, Florida. http://nrcsolutions.org/pensacola-florida/. Accessed 8 Jun 2020
112. Rezek RJ, Furman BT, Jung RP, Hall MO, Bell SS (2019) Long-term performance of seagrass restoration projects in Florida, USA. Sci Rep 9:15514
113. Liquete C, Udias A, Conte G, Grizzetti B, Masi F (2016) Integrated valuation of a nature-based solution for water pollution control. Highlighting hidden benefits. Ecosyst Serv 22:392–401
114. Beck MW, Sherwood ET, Henkel JR, Dorans K, Ireland K, Varela P (2019) Assessment of the cumulative effects of restoration activities on water quality in Tampa Bay, Florida. Estuar Coasts 42:1774–1791
115. Gittman RK, Popowich AM, Bruno JF, Peterson CH (2014) Marshes with and without sills protect estuarine shorelines from erosion better than bulkheads during a category 1 hurricane. Ocean Coast Manag 102:94–102
116. Rodriguez AB, Fodrie FJ, Ridge JT, Lindquist NL, Theuerkauf EJ, Coleman SE, Grabowski JH, Brodeur MC, Gittman RK, Keller DA (2014) Oyster reefs can outpace sea-level rise. Nat Clim Chang 4:493–497
117. Mo Y, Kearney MS, Turner RE (2020) The resilience of coastal marshes to hurricanes: the potential impact of excess nutrients. Environ Int 138:105409

118. Reguero BG, Beck MW, Bresch DN, Calil J, Meliane I (2018) Comparing the cost effectiveness of nature-based and coastal adaptation: a case study from the Gulf coast of the United States. PLoS One 13:1–24
119. Gallardo B, Bacher S, Bradley B, Comín FA, Gallien L, Jeschke JM, Sorte CJB, Vilà M (2019) InvasiBES: understanding and managing the impacts of invasive alien species on biodiversity and ecosystem services. NeoBiota 50:109–122
120. Ewel JJ, Putz FE (2004) A place for alien species in ecosystem restoration. Front Ecol Environ 2:354–360
121. Burcharth HF, Hawkins SJ, Zanuttigh B, Lamberti A (2007) Chapter 7 – Conceptual/predesign alternatives. In: Burcharth HF, Hawkins SJ, Zanuttigh B, Lamberti ABT-EDG for LCCS (eds). Elsevier, Oxford, pp 33–45
122. Brown JM, Phelps JJC, Barkwith A, Hurst MD, Ellis MA, Plater AJ (2016) The effectiveness of beach mega-nourishment, assessed over three management epochs. J Environ Manag 184:400–408
123. Pérez-Maqueo O, Martínez ML, Lithgow D, Mendoza-González G, Feagin RA, Gallego-Fernández JB (2013) The coasts and their costs. In: Martínez ML, Gallego-Fernández JB, Hesp PA (eds) Restoration of coastal dunes. Springer, Berlin, pp 289–304
124. Davis M, Krüger I, Hinzmann M (2015) Coastal protection and suds – nature-based solutions
125. Schmidt-Traub G, Locke H, Gao J et al (2020) Integrating climate, biodiversity, and sustainable land use strategies: innovations from China. Natl Sci Rev. https://doi.org/10.1093/nsr/nwaa139
126. Sarabi H, Romme V, Wendling L (2019) Key enablers of and barriers to the uptake and implementation of nature-based solutions in urban settings: a review. Resources 8:121
127. Lovelock CE, Brown BM (2019) Land tenure considerations are key to successful mangrove restoration. Nat Ecol Evol 3:1135
128. Pagano A, Pluchinotta I, Pengal P, Cokan B, Giordano R (2019) Engaging stakeholders in the assessment of NBS effectiveness in flood risk reduction: a participatory system dynamics model for benefits and co-benefits evaluation. Sci Total Environ 690:543–555
129. Denjean B, Altamirano MA, Graveline N et al (2017) Natural assurance scheme: a level playing field framework for Green-Grey infrastructure development. Environ Res 159:24–38
130. Thorslund J, Jarsjo J, Jaramillo F et al (2017) Wetlands as large-scale nature-based solutions: status and challenges for research, engineering and management. Ecol Eng 108:489–497
131. Kabisch N, Korn H, Stadler J, Bonn A (2017) Nature-based solutions to climate change adaptation in urban areas—linkages between science, policy and practice. pp 1–11
132. Kronenberg J, Bergier T, Maliszewska K (2017) The challenge of innovation diffusion: nature-based solutions in Poland BT – nature-based solutions to climate change adaptation in urban areas: linkages between science, policy and practice. In: Kabisch N, Korn H, Stadler J, Bonn A (eds). Springer, Cham, pp 291–305
133. Wamsler C, Alkan-Olsson J, Björn H, Falck H, Hanson H, Oskarsson T, Simonsson E, Zelmerlow F (2020) Beyond participation: when citizen engagement leads to undesirable outcomes for nature-based solutions and climate change adaptation. Clim Chang 158:235–254
134. Roca E, Villares M (2012) Public perceptions of managed realignment strategies: the case study of the Ebro Delta in the Mediterranean basin. Ocean Coast Manag 60:38–47
135. Thiele T, Alleng G, Biermann A et al (2020) Blue infrastructure finance: a new approach, integrating Naturebased solutions for coastal resilience. IUCN, Gland
136. Reguero BG, Beck MW, Schmid D, Stadtmüller D, Raepple J, Schüssele S, Pfliegner K (2020) Financing coastal resilience by combining nature-based risk reduction with insurance. Ecol Econ 169:106487. https://doi.org/10.1016/j.ecolecon.2019.106487
137. Faivre N, Fritz M, Freitas T, de Boissezon B, Vandewoestijne S (2017) Nature-based solutions in the EU: innovating with nature to address social, economic and environmental challenges. Environ Res 159:509–518
138. World Bank (2017) Implementing nature-based flood protection: principles and implementation guidance. World Bank, Washington

139. UNEP (2019) Nature-based solutions for climate manifesto
140. Somarakis G, Stagakis S, Chrysoulakis N (2019) ThinkNature nature-based solutions handbook. https://doi.org/10.26225/jerv-w202
141. Gómez Martín E, Giordano R, Pagano A, van der Keur P, Máñez Costa M (2020) Using a system thinking approach to assess the contribution of nature based solutions to sustainable development goals. Sci Total Environ 738:139693
142. Vasseur L, Horning D, Thornbush M, Cohen-Shacham E, Andrade A, Barrow E, Edwards SR, Wit P, Jones M (2017) Complex problems and unchallenged solutions: bringing ecosystem governance to the forefront of the UN sustainable development goals. Ambio 46:731–742
143. Masselink G, Lazarus ED (2019) Defining coastal resilience. Water 11. https://doi.org/10.3390/w11122587
144. Millennium Ecosystem Assessment (2005) Ecosystems and human well-being: synthesis. Ecosystems 1134:25. https://doi.org/10.1196/annals.1439.003
145. Keesstra S, Nunes J, Novara A, Finger D, Avelar D, Kalantari Z, Cerdà A (2018) The superior effect of nature based solutions in land management for enhancing ecosystem services. Sci Total Environ 610–611:997–1009
146. Pearlmutter D, Theochari D, Nehls T et al (2019) Enhancing the circular economy with nature-based solutions in the built urban environment: green building materials, systems and sites. Blue-Green Syst 2:46–72
147. Maes J, Jacobs S (2017) Nature-based solutions for Europe's sustainable development. Conserv Lett 10:121–124
148. Mustafa S, Estim A, Daning Tuzan A, Cheng Ann C, Leong Seng L, Raehanah Mohd Shaleh S (2019) Nature-based and technology-based solutions for sustainable blue growth and climate change mitigation in marine biodiversity hotspots. Environ Biotechnol 15:1–7

Nature-Based Solutions for Flood Mitigation and Resilience in Urban Areas

Carla Sofia Santos Ferreira, Kristina Potočki, Marijana Kapović-Solomun, and Zahra Kalantari

Contents

C. S. S. Ferreira (✉)
Department of Physical Geography and Bolin Centre for Climate Research, Stockholm University, Stockholm, Sweden

Navarino Environmental Observatory, Costa Navarino, Navarino Dunes, Messinia, Greece

Research Centre of Natural Resources, Environment and Society (CERNAS), Polytechnic Institute of Coimbra, Coimbra Agriculture School, Coimbra, Portugal

K. Potočki
Department of Hydroscience and Engineering, Faculty of Civil Engineering, University of Zagreb, Zagreb, Croatia

M. Kapović-Solomun
Faculty of Forestry, University of Banja Luka, Banja Luka, Bosnia and Herzegovina

Z. Kalantari
Department of Physical Geography and Bolin Centre for Climate Research, Stockholm University, Stockholm, Sweden

Navarino Environmental Observatory, Messinia, Greece

School of Architecture and the Built Environment, KTH Royal Institute of Technology, Stockholm, Sweden

Carla S. S. Ferreira, Zahra Kalantari, Thomas Hartmann, and Paulo Pereira (eds.),
Nature-Based Solutions for Flood Mitigation: Environmental and Socio-Economic Aspects,
Hdb Env Chem (2022) 107: 59–78, DOI 10.1007/698_2021_758,

Abstract Urban areas face several environmental problems and risks related to water management, such as floods and degradation of water quality, enhancing population vulnerability and threatening urban sustainability. These problems are expected to be exacerbated with increasing urbanization and climate change, which leads to higher frequency and intensity of hydrometeorological extremes. Moving towards more flood resilient cities has proven a major challenge, particularly considering the high concentration of population and economic activities and, thus, high pressure on limited available space. Nature-based solutions (NBS) in urban areas favour stormwater retention, infiltration, and filtration, contributing to flood mitigation and enhancement of water quality. The effectiveness of different NBS on stormwater management, however, is influenced by design and placement aspects, but a network of connected NBS elements can improve flood mitigation and enhance urban resilience. Stronger evidence of the advantages of NBS, however, is still required to overcome the current challenges and barriers impairing their wider implementation in urban areas.

Keywords Flood mitigation, Nature-based solutions, Urban areas, Urban resilience, Water pollution

1 Introduction

Urbanization has increased considerably over the last century, driven by the increasing urban population [1]. Today, 55% of the world's population lives in cities and the United Nations projection indicates that by 2050 this share will increase to 68%, and urban population will reach 6.7 billion people [2]. Physical expansion of urban areas is even faster than expansion of urban populations, due to the occupation of nearby land (peri-urban areas) motivated by the lower living costs than recorded in urban centres, easy mobility/transport, and the demand for improved quality of life [3]. The occupation of nearby rural areas involves an increasing consumption of natural resources, habitat loss, and environmental degradation and consequent decrease in ecosystem services supply, including water and climate regulations [4–6].

Number of floods in the world is rising since 1950s and this is associated with the changes in hydrological cycle and more frequent occurrence of hydrometeorological extremes [7]. Combined effects of more frequent occurrence of extreme events together with the development of urban settlements result in increasing occurrence of urban floods [8]. Removal of vegetation and expansion of sealed surfaces in urban areas additionally disrupts the hydrological cycle, i.e. reduces rainfall interception, evapotranspiration, and infiltration and thus increases runoff [9]. Urban water management practices based on nature-based solutions (NBS) are promising strategies to maintain the urban hydrological cycle as close as possible to the natural state. While reducing floods, NBS improve mitigation and adaptation to global changes (including both land-use and climate change) and provides services for maintaining and restoring ecological functions [10, 11].

Urban green infrastructure is a broad concept that supports NBS through integration of green and blue spaces in urban areas, thus sustaining water resources together with maintenance of biodiversity and ecological functions [12]. In the context of urban water management and mitigation of urban floods, similar concepts and solutions based on elements of green infrastructure can be found in the literature and differing in terminology depending on the part of world where they are developed. Some examples are Sustainable Urban Drainage Systems (SUDS), Runoff Best Management Practices (BMP), Low-Impact Development (LID), Water-Sensitive Urban Design (WSUD), Integrated Urban Water Management (IUWM), and Sponge city (SC) [13, 14]. NBS therefore contain solutions from specific techniques in urban drainage to the broad principles, such as sustainable development of urban areas [13]. Ecosystem restoration and climate change adaptation achieved by multiple functions of NBS contribute to the implementation of UN 2030 Agenda for Sustainable Development Goals and lead to enhanced development of circular economy [15]. However, the increasing extent and complexity of the urban systems pose major challenges for water management, and particularly to the implementation of NBS and to foster urban resilience [16, 17].

This chapter aims to present and discuss the main approaches used in urban flood risk management, and the most widely NSB measures implemented for flood mitigation in urban areas, based on literature review. Additionally, this chapter discusses the role of NBS to improve urban resilience and the main advantages and barriers to implement NBS in urban environments.

2 Urban Flood Risk Management Approaches

The urban water cycle is disrupted due to the extensive impervious surfaces, and their associated impacts on increasing flood hazard have been recognized for decades [18]. The traditional paradigm of *flood protection* founded on structural measures has been abandoned due to the high costs and inherent uncertainties regarding their effectiveness. Thus, a new approach based on flood risk management was slowly introduced in water management legislation at the turn of the century. For example, European Union has adopted the Water Framework Directive [19] and subsequent Floods Directive [20]. Water Framework Directive introduced an approach to integrated river basin management through development of River Basin Management Plans and commits EU member states to achieve good qualitative and quantitative status of all water bodies. Steps for assessment and management of flood risks are prescribed in Floods Directive. Measures focused on prevention, protection, and preparedness are proposed in Flood Directive through the development of Flood Risk Management Plans. Flood Risk Management Plans need to be coordinated with the River Basin Management Plans, together with the implementation of all the relevant environmental objectives from the Water

Directive. Flood risk management is therefore an integral part of integrated river basin management and incorporates the concept of *living with flood risk* [21].

Flood risk is defined as "a 'product' of the probability of flood and their consequences", or, as the "product of flooding hazard and society's vulnerability to flood hazard" [22]. Quantification of flood risk in urban areas presents a challenge due to the complex interrelationships of different flood sources and the effectiveness of management measures, so an integrated approach for flood risk management is needed [23]. The main phases included in the design and implementation of an integrated Flood Risk Management plan include (1) flood risk assessment, including risk perception and risk tolerance; (2) risk reduction through implementation of adaptative strategies and measures [22]; (3) emergency management; and (4) short- and long-term recovery [24, 25]. Development and implementation of the Flood Risk Management approach requires trans-sectoral governance, cross-sectoral cooperation and planning, interdisciplinarity, and inclusion of different stakeholders. Although this adds complexity to the Flood Risk Management, this wide integral approach enables coordination between social, hydrological, and ecological systems providing framework for better adaptation to climate change and sustainable development of urban areas [21, 25].

In urban areas, flood mitigation is performed through a series of structural and non-structural measures. Typically, structural measures rely on "grey" solutions, i.e. hard-engineering structures for flood defence such as channels, pipelines, and storage tanks included in urban stormwater drainage systems, which provide quick conveyance and drainage of stormwater runoff. The application and maintenance of these conventional methods have proved costly and insufficient to cope with challenges of more frequent precipitation extremes and consequent floods in urban areas, driven by climate changes [26]. Urban drainage systems are designed so that they can accept runoff caused by design rain, i.e. rain of a certain duration and recurrence period (usually 1–5 years). Design rain is determined by statistical analysis of historical rain events and does not consider changes caused by climate change recorded after the construction of the system. Therefore, the drainage system in circumstances of higher frequency and intensity of rain events, although designed and dimensioned according to the rules of the profession, can no longer successfully care excess water resulting from more frequent flooding [27]. Land-use changes during urbanization process are characterized by an increased share of impervious surfaces, resulting in reduced infiltration which ultimately leads to an accelerated and increased volume of surface runoff to be managed. Previous studies have shown that an increase in impermeable surfaces by 30% compared to the state before urbanization results in a twofold increase in flooding over a 100-year return period [28]. Also, complex interactions between urban and natural system present challenges to modelling urban flood processes, since hydrological models are usually based on simplified surface runoff processes and hydraulic models on simplified piped systems [29].

The transition from traditional urban water management system towards nature-based urban flood management intends to reestablish hydrological conditions before urbanization, i.e. reducing and delaying runoff, through the incorporation of green

Fig. 1 Integration of green infrastructure in Singapore (Photo by: Ana Sović Kržić)

elements that increase infiltration, evaporation, and retain water [30]. An increasing number of cities all around the world have been implementing green solutions, regulations, programmes, and incentives enabling flood protection based on NBS. Singapore (Fig. 1), Berlin, and several cities in China present good examples of NBS for stormwater management [31–33].

NBS applications for Flood Risk Management in urban areas, however, must consider specific local conditions and a multidisciplinary approach, in order to implement economically, environmentally, technologically, and socially sustainable solutions. Operationalization of NBS for floods and other hydrometeorological hazards can be established through a set of principles that describe co-design, co-development, co-deployment, and demonstration of the NBS effectiveness. Research should be conducted with impact/scenario modelling together with the incorporation of related policy frameworks. Achieving these steps is possible through shared knowledge and skills of stakeholders, researchers, experts, and end-users from different fields, including engineering, hydrology, urban planning, landscape architecture, ecology, economics, law, and other professions [34].

3 Nature-Based Solutions for Urban Flood Mitigation

Water management in urban areas is established to mitigate the impacts of development on water cycling by means of NBS, namely through the implementation of Green Infrastructures. The Green Infrastructure concept appeared in the last decade

[35] as a result of the urbanization pressure and the shortage of green and blue spaces within urban areas. Green Infrastructure can be described as a system of natural areas, features, and green spaces in rural and urban, terrestrial, freshwater, coastal, and marine areas [36]. It includes a network of natural and designed landscape components with important role on water regulation and flood risk mitigation and management [37], as well as reduction of water pollution [2]. In the context of urban water management, Sustainable Urban Drainage Systems have been also presented as water management elements, based on natural hydrological processes. However, whereas Sustainable Urban Drainage Systems are devoted to more specific techniques on smaller spatial and functional scales, Green Infrastructure is used on larger scales and involves a multitude of stakeholders, such as local authorities and private landowners [13].

There are many permeable vegetated surfaces integrating Green Infrastructure in urban areas, such as green corridors, urban parks, urban gardens, urban forests, urban grasslands, and other recreation zones [38, 39]. The water elements integrating urban Green Infrastructure include rivers, lakes, canals, ponds, and floodplains, which provide an additional capacity to cope water during rainfall events [40]. In turn, Sustainable Urban Drainage Systems, which are incorporated into urban drainage systems, include green roofs, retention and detention ponds, wetlands, infiltration tranches, bioretention basins, rain gardens or swales, and impervious pavements (Fig. 2). These structures are mainly implemented as source control techniques, to reduce the amount but also improve the quality of stormwater at or near its source [41].

Sustainable Urban Drainage Systems and Green Infrastructure can be designed for temporary water storage and runoff reduction, but also to provide additional ecosystem services such as regulation of water quality and cultural services for citizens (e.g. aesthetics and recreation). Regarding water regulation, they provide infiltration, detention attenuation, conveyance, and water harvesting as the main management options for runoff quantity control and peak flow reduction. The use of vegetation in the NBS measures (e.g. green roof, infiltration gardens, and urban forests) additionally provides rainfall interception and evapotranspiration, enabling water to return to the atmosphere [42].

Some of the most widely used NBS to mitigate runoff and address issues of poor surface water quality include wetlands and runoff ponds (e.g. retention ponds, flood storage reservoirs, shallow impoundments), which contain water during dry weather and are designed to hold extra water when it rains [43]. While wetlands restoration has been performed to renew their natural functions (e.g. by removing underground drainage tiles), constructed wetlands have been also implemented to improve food mitigation and surface water quality. Typically, constructed wetlands are created through excavation of upland soils to elevations that will support the growth of wetland species, but they can involve also dyke installations [2]. Constructed wetlands establish a hydrological regime which mimic the functionality of natural wetlands and facilitate filtration of polluted stormwater runoff and pollutant absorption [43].

Fig. 2 Examples of NBS in urban areas: (**a**) green facade and (**b**) green roof in Riga, Latvia, (**c**) bioretention basin and (**d**) infiltration trench in Riga, Latvia, (**e**) detention basin in Pula, Croatia, and (**f**) wetland in Ghent, Belgium

Detention basins, which are grassed depressions or basins created by excavation into which runoff generated during rainfall events is channelled, are designed to temporarily detain and facilitate the slow filtration of runoff. They play important roles in regulating water flows and maintaining water quality by retaining sediments and reducing nutrient and metals, as a result of settling of particulate pollutants and uptake by vegetation [43, 44]. Additional bioretention structures, such as pits backfilled with soil, mulch, and/or vegetation used to retain and infiltrate runoff, also rely on biophysical processes within the soil matrix to reduce the volume of stormwater and pollutant characteristics [2].

The performance of NBS on flood mitigation, however, is highly dependent on rainfall return periods [45]. Based on field and laboratory studies, for example, porous pavements showed better performance in respect of peak runoff reduction than green roofs and bioretention cells under different storms [46]. Bioretention cells, in turn, revealed more effective in the reduction of runoff volume [42]. In general, filter trenches, soakaways, and green roofs are typically designed to cope with moderate rainfall events, whereas elements such as retention ponds, swales, and detention basins can cope with heavier rainfalls [45].

As presented above, several studies have shown the positive impacts of NBS on water infiltration, retention, interception, transpiration, evaporation, and mitigation of surface runoff, and thus their role in managing flood risk [39, 47]. However, the performance of different NBS in flood protection is strongly linked to different spatial allocations and to different patterns of their installation. Small-scale examples of several NBS showed better performance in surface runoff reduction than single NBS [48] and that was also the case with NBS that were spatially distributed but with good hydrological linkages [49]. Spatial allocation tools are therefore used to estimate optimal hydrological functioning and perform spatial analysis [50]. Mapping of Green Infrastructure is considered a prerequisite to improve its functionality, but consensus is still lacking about using appropriate typology, mapping methods, and tools for specific applications [39].

Some researchers argue better effectiveness of NBS over the grey infrastructures [51, 52]. Others, however, have found that although NBS provide flood reduction gains, under intensive rainfalls and in cases when only green measures are applied, its performance is questionable [11]. Grey infrastructures provide rapid conveyance and transport of runoff into downslope areas [53], particularly relevant in dense urban areas, and can be designed and constructed to manage large volumes of stormwater (e.g. driven by 50-year floods [54]). Nowadays, this grey approach is considered to offer low sustainability, while NBS provide numerous complementary benefits, such as climate regulation and supporting biodiversity [55]. Thus, a combination of NBS and grey measures has been advocated as the best option for stormwater management and urban flood mitigation [54, 56].

4 The Role of NBS to Improve Urban Resilience

4.1 Sustainability and Urban Resilience Principles

Urban resilience can be defined as the ability of a system to develop the resources, skills, and capacities needed to maintain or rapidly return to desired functions in the face of a disturbance (e.g. flood and climate change) and limit its negative impacts [57]. It enables the urban system to prepare and plan for (pre-disaster actions to mitigate hazards by reducing their frequency, intensity, and duration), absorb (minimize potential damages and losses), rapidly recover from, and adapt to stressors and adverse events (Fig. 3) [58].

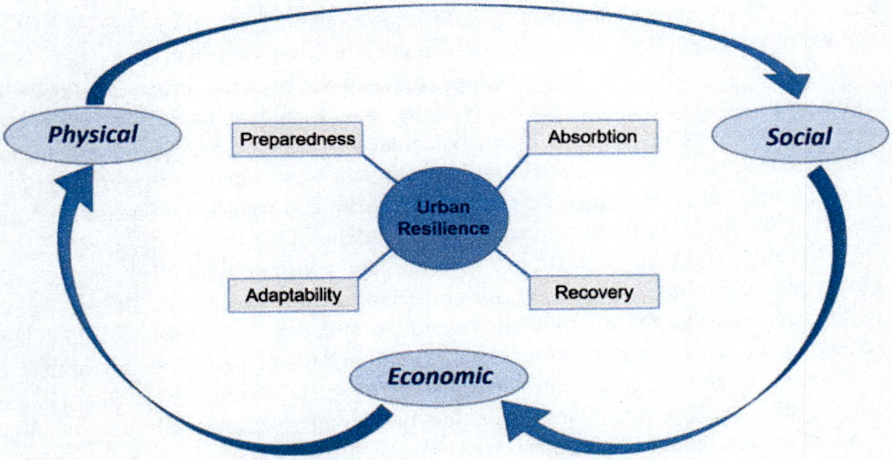

Fig. 3 Three dimensions and four principles of urban resilience

In order to plan and prepare for disturbance (e.g. flood event), it is important to assess the consequences based on past experiences, namely by building knowledge about previous disturbance, exposure, vulnerability, and monitoring of critical slow variables. Absorbing disturbances requires robust infrastructures, and creation of buffer capacity to dynamically cope with disturbance and maintain the desired functions of the urban system. Urban systems must provide diverse responses, ensured by different spatial diversity, to recover from disturbances. Adaptability involves changes driven by institutional learning capacity and reflectivity, which requires innovative and transdisciplinary practices, and flexibility in spatial planning, to quickly modify and transform the urban system and maintain the desired functions into the future [59]. Besides the physical component of urban resilience, the social networks that connect resources to vulnerable social groups (social resilience) and the economic recovery of the urban areas (economic resilience) are important aspects to include in resilience thinking [60]. The need for urban resilience has been reinforced by the European Commission, the World Bank, and the United Nations and has become a major focus for guiding planners and decision-makers [58] and to support the achievement of Sustainable Development Goals [15].

4.2 NBS Contribution for Urban Resilience

Governments and local authorities are increasingly involved in resilience-building strategies, seeking to design and implement sustainable solutions, which combine the maximization of tradeoffs between positive and negative effects of, for example, urbanization and climate change, with sustainable development and environmental concerns, to guarantee liveable conditions in urban areas [58]. Towards this end, there is a perspective change in the conception, planning, and development of the built, infrastructural, operational, and functional forms of urban areas [61]. NBS have been identified as a promising approach to enhance urban resilience by providing flexible and adaptable solutions, based on the delivery of ecosystem services [43, 60]. In terms of flood risk management, for example, "re-naturing" urban areas reestablish some natural hydrological processes (e.g. water infiltration and purification), providing clear departing from the traditional resistance-based approach focused on flood-safe solutions (e.g. dams) [17].

Depending on their type, function, design, and configuration, NBS may contribute to urban resilience by integrating properties such as diversity, efficiency, flexibility, modularity, multifunctionality, and redundancy into urban planning and design [58]. Protecting, restoring, and enhancing green and blue infrastructures across spatial and temporal scales in urban areas enhance resilience of urban systems to disturbances. For example, vegetation buffers in riparian zones provide flood protection and reduce the occurrence of extreme urban heat events, identified as two key socio-natural disasters requiring preparation, recovery, and resilience [62]. Wetlands are known for their provisioning of ecosystem services, and thus constructed wetlands have great potential for use as NBS to address a variety of environmental, social, and economic challenges. Common multi-beneficial ecosystem services derived from wetlands include water quality protection [63], groundwater level and soil moisture regulation [64, 65], flood regulation and sediment retention [66], and biodiversity support [76]. As the frequency of natural extreme events increases, it is becoming increasingly important to deploy NBS such as wetlands, both locally and at larger scales, in flood risk mitigation measures that strengthen the resilience of the urban landscape [2]. Wetlands are often described as natural sponges, due to their long hydraulic residence time combined with their vegetative features, which play an important role in reducing downstream peak flows, erosion rates, and nutrient retention [67, 68]. However, despite their importance, there has been a rapid and sustained decline in wetland areas globally. The absolute scope of global wetland losses is uncertain, and the rate of loss has slowed substantially in some regions of the world, such as the USA and Europe, in recent decades [69]. Nevertheless, many regions worldwide are still experiencing rapid wetland loss [70, 71].

According to the domino effect concept, based on a chain reaction causing changes in a territory, some urban areas may be affected by floods even if they are not located directly in the risk area. Indeed, in the interconnected space of urban areas, risks have impacts beyond spatial municipality boundaries [57]. Since most urban catchments begin prior to and continue beyond municipal boundaries, different approaches to impervious cover regulation and water management strategies may marginalize the benefits of a municipality's effort to implement NBS. Because of the spatial and institutional mismatch, NBS strategies requires collaborative or polycentric governance approaches. As a result, a growing emphasis on NBS as a significant contributor to urban resilience necessitates a more thorough understanding of the institutional fit between the social infrastructure for governance [62].

The links between NBS and urban resilience, however, should also consider the resilience (or vulnerability) of ecosystems themselves. For example, climate change impacts ecosystems and may affect their capacity to function and provide services. The extent to which urban ecosystems, as isolated green spaces within the urban areas, can themselves be resilient may be limited, but could be supported with active management, by selection of temperature-adapted species, creation of connected networks, and control of habitat disturbance and destruction processes [72].

5 Challenges and Barriers to Implement NBS in Urban Areas

5.1 The Role of Urban Planning in NBS Implementation

Urban planning can play a substantial role to support the implementation of NBS, in response to the challenges of attainting urban resilience and environmental sustainability [73]. Urban resilience from NBS applications, however, must consider the interconnectivity of the urban green spaces at local, regional, or even national scales, to better assess the mitigation of floods. Connectivity refers to the physical connection between green elements (structural connectivity), but also the connection between natural and ecological processes, such as water and geochemical cycles (functional connectivity) [74]. Thus, although one small NBS may (partially) lose functionality during a rainfall event, a larger connected network of NBS can have the potential to function as a decentralized stormwater management infrastructure and thus ameliorate flood risks [2]. As a decentralized approach to stormwater management, NBS are usually inherently more resilient than large, centralized grey infrastructures [60]. According to WWAP/UN Water [2], climate change adaptation will not be possible without a range of NBS that deal with increasing water variability and extremes induced by changing climate. Furthermore, open spaces provided by NBS have a potential for disaster management, since they can be utilized for emergency evacuation and as shelters [58].

The planning and implementation of NBS can be supported by policy approaches. For example, regulation of impervious cover within a city and mandates that new buildings or developments must include green spaces [62]. Some urban areas, in turn, are incentivizing NBS through subsidies for rainwater harvesting or relied on grant programs for the adoption of green roofs [60]. Cities such as New York and the already mentioned Singapore have adopted an NBS approach based on urban green infrastructure to combat climate change and associated problems such as urban floods and to achieve overall socio-economic resilience by delivering ecosystem services [73].

Successful implementation of sustainable NBS to cope with a range of current and future challenges requires the involvement of all relevant sources of expertise and interests in the planning and decision-making process, due to the multi-dimensionality and complexity of NBS [75, 76]. The involvement of a wide range of stakeholders and actors in turn requires deployment of different communication tools and methods. Successful implementation of NBS in urban planning relies on a proactive approach where implementation early in the planning process is key [77, 78].

5.2 Effectiveness of NBS

In order to compare the effectiveness of NBS with that of technology-based grey solutions in urban areas, further research and onsite monitoring are needed to capture the diverse co-benefits that NBS can provide [79, 80]. Multiple social, environmental, and economic co-benefits can be associated with NBS, in addition to their direct benefits, and the challenge is to link and capture these co-benefits in evaluations [77]. The current evidence on NBS performance is largely imbalanced and mainly focuses on a few ecosystem services. Most previous studies addressing the ecosystem services provided by NBS have focused on local climate regulation (40%) and recreation (20%), while only 8% have focused on water regulation [81]. There is thus a major knowledge gap in the evidence based on NBS performance. Consequently, there is an urgent need to investigate a wider range of aspects and to develop assessment models that can be applied at different locations, thus helping to reduce the geographical bias in the literature [81].

To evaluate the economic effectiveness of NBS, Potschin et al. [82] suggest validation methods such as "avoided costs" from, e.g. damage or problems that would arise if NBS were not implemented. Cost–benefit analysis can also be used to help decision-makers choose between different NBS [83]. It should be stressed that additional methods may be required to assess the full economic effectiveness of NBS. For instance, Raymond et al. [84] argue that cost–benefit analysis can be insufficient for evaluating the economic effectiveness of NBS, since it cannot account for the long-term cumulative benefits provided by NBS, and suggest combining it with methods such as participatory assessments, group modelling, and integrated sustainability assessment.

Data availability is currently one of the main factors preventing full-scale implementation of NBS [79]. This lack of data can be overcome by widespread onsite monitoring [77]. Future monitoring efforts need to cover both the process of implementing NBS and the outcomes, including the final benefits of a particular NBS, how it is perceived and how it responds to the challenge for which it was implemented [84]. In order to enable effective monitoring of these aspects, indicators of NBS performance covering a range of social, economic, and technical aspects must be developed [77, 83]. Raymond et al. [83] suggest working with measurable indicators to assess, monitor, and communicate the effectiveness of different NBS. However, it remains unclear the effectiveness of NBS over a longer temporal scale, which NBS would be most effective in the long run and which would produce effective results immediately after implementation. Therefore, when assessing the effectiveness of a given NBS, it is important to consider the possible time lapse between its initial effect and the point when it reaches full effectiveness.

5.3 Advantages and Disadvantages of NBS

There are four main advantages of NBS: (1) *sustainable systematic and integrative approach,* (2) *resource efficiency,* (3) *long-term cost-efficiency,* and (4) *co-benefits.* The systematic and integrative approach is a strong advantage of NBS [75]. NBS applied in a suitable manner can, in an innovative way, use natural elements to achieve environmental and societal goals [10]. More specifically, NBS can provide energy- and resource-efficient measures that combat climate change and, at the same time, support and protect natural capital [85]. For example, green roofs and walls provide thermal insulation of buildings [86], and pervious pavements can reduce surface temperatures up to 4°C, due to lower reflection and evaporation [87].

In many cases NBS have been proven to be more cost-effective and multifunctional over the long term than grey solutions [88]. This is a consequence of their often-low maintenance costs and flexibility of application [89]. In addition, NBS provide a variety of multiple benefits, often including socio-cultural values such as recreation, increased biodiversity, and cultural heritage [90]. Pollution control and opportunities for enhancement of human well-being are other co-benefits provided by NBS [89, 91]. Green roofs and walls, for example, provide air pollution reduction and carbon sequestration [92], and habitat for different species [36].

There are four main disadvantages of NBS: (1) *longer time frame compared with grey solutions,* (2) *space-consuming,* (3) *ecosystem disservices,* and (4) *segregation and environmental injustice.* A particular disadvantage of NBS is the generally longer time frame before reaching full potential and effects compared with grey solutions [77]. Solutions based on ecosystem services require a significant time frame to create or restore a habitat, which can be an obstacle in fast-growing urban areas and a reason for choosing conventional grey solutions [10]. In addition, local conditions have to be well understood in transdisciplinary ways, in order to choose the most beneficial NBS to exploit the full potential at a specific site. This requires expertise and experience in relevant areas, which may be costly [75]. Finally, NBS in urban planning and policy development processes can be time-consuming, unless clear strategies are established. The multidisciplinary process related to NBS involves different stakeholders with multiple different interests and assets [75, 84]. Many NBS projects in an urban context, e.g. open stormwater management, require more space than grey solutions such as underground systems. Therefore, a potential conflict between NBS and the global goal of increased urban compactness can be regarded as a drawback [90]. Apart from the multiple benefits provided by NBS, they can also supply "ecosystem disservices" (EDS) [93]. For instance, NBS involving open water surfaces, such as wetlands and stormwater handling systems, in combination with increased temperatures, could enhance the risk of infection by vector-borne infectious diseases, including malaria and dengue fever [94]. Therefore, it is important to use modelling tools to evaluate multiple benefits of SUDS [95, 96]. Implementation of NBS in urban areas may also not be beneficial for all citizens. It can even lead to segregation, through displacement of population groups that cannot afford the higher rents and land prices resulting from the higher reputation and living standards brought about by NBS [77].

6 Final Considerations

Most people in the world live in urban areas; therefore, it is important to develop resilient cities and ensure adequate proportion of green and blue urban spaces for human well-being. NBS provide sustainable water management, since it relays on natural processes to manage stormwater quantity and quality. Based on vegetated surfaces, NBS provide opportunities for water interception, evapotranspiration, infiltration, and filtration, and thus, reduced surface runoff and water pollution. Besides the relevant contribution for flood risk mitigation and to support water quality within urban areas, NBS comprise multifunctional spaces able to deliver a wide array of ecosystem services beyond water management, such as climate regulation, improving air quality, provision of habitat and support to biodiversity, and contribution to human satisfaction. However, development of a stronger evidence based on NBS is a key aspect for successful NBS implementation, particularly empirical evidence demonstrating the effectiveness of NBS [77]. Since NBS for water management depend on many factors, improving the knowledge in different hydrological, environmental, socio-economic, and management conditions, and providing well-established historic evidence of their positive impacts, will be relevant to support increasing NBS applications.

Several NBS have been implemented at different scales within the urban areas, such as urban forests, gardens, wetlands, infiltration trenches, and green roofs. The effectiveness of NBS for water management varies with their design, size, and local conditions. Nevertheless, it is rather a network of connected NBS than small isolated elements that can effectively mitigate urban floods and thus contribute to enhance urban resilience, namely through adaptation to climate changes. Nevertheless, relatively limited knowledge is available to compare the effectiveness of NBS with conventional alternatives. Filling this information gap is key to better assess the advantages of combining NBS and grey infrastructures in water management plans and to enhance the urban resilience. This will be useful to promote private sector investment in NBS and to advocate for policy changes supporting NBS and promoting NBS to political leaders [2].

Implementing NBS in urban areas, however, is inherently complex, due to increasing environmental, social, and economic challenges and the limited space to fulfil a wide range of needs. Urban planning can play a substantial role to support the implementation of NBS and to manage tradeoffs and conflicts while assuring social equity. Governance systems must improve and legitimize the delivery of ecosystem services by reinforcing the means to prioritize and implement NBS and thus enhance sustainable development and urban resilience.

Acknowledgements This research was supported by the Portuguese Foundation for Science and Technology, through the post-doctoral grant SFRH/BPD/120093/2016, the ERA-LEARN 2020 UrbanGaia project (grant agreement No. 645782), a project funded from The Swedish Research Council (Svenska Forskningsrådet) Formas (grant 2017-00608), and by the Ministry of Scientific-Technological Development, Higher Education and Information Society of the Republic of Srpska (grant no. 19/6-020/961-43/18). The open access of this chapter was funded by LAND4FLOOD COST Action (No. 16209) supported by COST (European Cooperation in Science and Technology, www.cost.eu).

References

1. Seto KC, Sánchez-Rodríguez R, Fragkias M (2010) The new geography of contemporary urbanization and the environment. Annu Rev Env Resour 35
2. WWAP (United Nations World Water Assessment Programme)/UN-Water (2018) The United Nations World Water Development Report 2018: nature-based solutions for water. UNESCO, Paris
3. Ferreira CSS, Kalantari Z, Salvati L, Canfora L, Zambon I, Walsh RPD (2019) Urban areas. In: Pereira P (ed) Advances in chemical pollution, environmental management and protection, vol 4, Chapter 6, ISBN: 9780128164150. Elsevier, Amsterdam, pp 207–249
4. Seto KC, Parnell S, Elmqvist T (2013) A global outlook on urbanization. In: Urbanization, biodiversity and ecosystem services: challenges and opportunities. Springer, Dordrecht, pp 1–12
5. Wang J, Zhou W, Pickett ST, Yu W, Li W (2019) A multiscale analysis of urbanization effects on ecosystem services supply in an urban megaregion. Sci Total Environ 662:824–833
6. Leitão IA, Ferreira CSS, Ferreira AJD (2019) Assessing long-term changes in potential ecosystem services of a peri-urbanizing Mediterranean catchment. Sci Total Environ 660:993–1003
7. Adikari Y, Yoshitani J (2009) Global trends in water-related disasters: an insight for policymakers. World Water Assessment Programme Side Publication Series, Insights. The United Nations, UNESCO. International Centre for Water Hazard and Risk Management (ICHARM)
8. Jha AK, Bloch R, Lamond J (2012) Cities and flooding: a guide to integrated urban flood risk management for the 21st century. The World Bank
9. Shuster WD, Bonta J, Thurston H, Warnemuende E, Smith DR (2005) Impacts of impervious surface on watershed hydrology: a review. Urban Water J 2(4):263–275
10. Eggermont H, Balian E, Azevedo JMN, Beumer V, Brodin T, Claudet J, Reuter K (2015) Nature-based solutions: new influence for environmental management and research in Europe. GAIA – Ecol Perspect Sci Soc 24(4):243–248
11. Zölch T, Henze L, Keilholz P, Pauleit S (2017) Regulating urban surface runoff through nature-based solutions – an assessment at the micro-scale. Environ Res 157:135–144
12. Benedict MA, McMahon ET (2002) Green infrastructure: smart conservation for the 21st century. Renew Resour J 20(3):12–17
13. Casal-Campos A, Fu G, Butler D, Moore A (2015) An integrated environmental assessment of green and gray infrastructure strategies for robust decision making. Environ Sci Technol 49 (14):8307–8314
14. Zevenbergen C, Fu D, Pathirana A (2018) Transitioning to sponge cities: challenges and opportunities to address urban water problems in China. Water 10(9):1230
15. United Nations (2015) Transforming our World: The 2030 Agenda for Sustainable Development https://sustainabledevelopment.un.org/post2015/transformingourworld/publication Accessed 28 July 2020

16. Aguilar-Barajas I, Sisto NP, Ramirez AI, Magaña-Rueda V (2019) Building urban resilience and knowledge co-production in the face of weather hazards: flash floods in the Monterrey Metropolitan Area (Mexico). Environ Sci Policy 99:37–47
17. Leandro J, Chen K-F, Wood RR, Ludwig R (2020) A scalable flood-resilience-index for measuring climate change adaptation: Munich city. Water Res 173:115502
18. Leopold LB (1968) Hydrology for urban land planning: a guidebook on the hydrologic effects of urban land use (vol 554). US Department of the Interior, Geological Survey
19. European Community (2000) Directive 2000/60/EC of the European parliament and of the council establishing a framework for community action in the field of water policy. Off J Eur Commun L327
20. European Community (2007) Directive 2007/60/EC of the European Parliament and of the Council of 23 October 2007 on the assessment and management of flood risks. Official Journal of the European Community L288/27
21. Ashley RM, Blanksby J, Chapman J, Zhou J (2007) Towards integrated approaches to reduce flood risk in urban areas. Advances in urban flood management, pp 415–432
22. Hall JW, Meadowcroft IC, Sayers PB, Bramley ME (2003) Integrated flood risk management in England and Wales. Nat Hazards Rev 4(3):126–135
23. Dawson RJ, Speight L, Hall JW, Djordjevic S, Savic D, Leandro J (2008) Attribution of flood risk in urban areas. J Hydroinf 10(4):275–288
24. ten Brinke WB, Saeijs GE, Helsloot I, van Alphen J (2008) Safety chain approach in flood risk management. In: Proceedings of the Institution of Civil Engineers-Municipal Engineer, vol 161 (2), Thomas Telford Ltd., pp 93–102
25. Serra-Llobet A, Conrad E, Schaefer K (2016) Governing for integrated water and flood risk management: comparing top-down and bottom-up approaches in Spain and California. Water 8 (10):445
26. Hoyer J, Dickhaut W, Kronwitter L, Weber B (2011) Water sensitive urban design: principles and inspiration for sustainable stormwater management in the city of the future. Jovis, Berlin
27. Potočki K, Vouk D, Kuspilić N (2018) Reduction of flood risk in urban areas with integral green solutions. In: VII. Croatian National Platform for disaster risk reduction conference. The National Protection and Rescue Directorate, Zagreb, pp 70–75, (in Croatian)
28. Hollis GE (1975) The effect of urbanization on floods of different recurrence interval. Water Resour Res 11(3):431–435
29. Salvadore E, Bronders J, Batelaan O (2015) Hydrological modelling of urbanized catchments: a review and future directions. J Hydrol 529:62–81
30. Brown RR, Keath N, Wong TH (2009) Urban water management in cities: historical, current and future regimes. Water Sci Technol 59(5):847–855
31. Tan YS, Lee TJ, Tan K (2009) Clean, green and blue: Singapore's journey towards environmental and water sustainability. Institute of Southeast Asian Studies
32. Liu L, Jensen MB (2018) Green infrastructure for sustainable urban water management: practices of five forerunner cities. Cities 74:126–133
33. Lyu HM, Xu YS, Cheng WC, Arulrajah A (2018) Flooding hazards across southern China and prospective sustainability measures. Sustainability 10(5):1682
34. Kumar P, Debele SE, Sahani J, Aragão L, Barisani F, Basu B, Edo AS (2020) Towards an operationalisation of nature-based solutions for natural hazards. Sci Total Environ:138855
35. Selman P (2008) What do we mean by sustainable landscape? Sustain Sci Pract Pol 4(2):23–28
36. Naumann S, McKenna D, Gerdes H, Herbert S, Landgrebe-Trinkunaite R, Kaphengst T (2011) Design, implementation and cost elements of green infrastructure projects. Final report. European Commission, Brussels
37. Elmhagen B, Destouni G, Angerbjörn A, Borgström S, Boyd E, Cousins SAO, Dalén L, Ehrlén J, Ermold M, Hambäck PA, Hedlund J, Hylander K, Jaramillo F, Lagerholm VK, Lyon SW, Moor H, Nykvist B, Pasanen-Mortensen M, Plue J, Prieto C, Van der Velde Y, Lindborg R (2015) Interacting effects of change in climate, human population, land use, and water use on biodiversity and ecosystem services. Ecol Soc 20(1):23

38. Gill SE, Handley JF, Ennos AR, Pauleit S (2007) Adapting cities for climate change: the role of the green infrastructure. Built Environ 33(1):115–133
39. Wang J, Banzhaf E (2018) Towards a better understanding of green infrastructure: a critical review. Ecol Indic 85:758–772
40. Abbott J, Davies P, Simkins P, Morgan C, Levin D, Robinson P (2013) Creating water sensitive places – scoping the potential for water sensitive urban design in the UK. CIRIA, London
41. European Environment Agency (EEA) (2012) Urban adaptation to climate change in Europe. Challenges and opportunities for cities together with supportive national and European policies. EEA Report No. 2/2012. European Environment Agency, Luxembourg. Retrieved from: https://www.eea.europa.eu/publications/urban-adaptation-to-climate-change
42. Davis AP (2008) Field performance of bioretention: hydrology impacts. J Hydrol Eng 13 (2):90–95
43. Kampa E, Lago M, Anzaldúa G, Vidaurre R, Tarpey J (2019) Investing in nature for European water security. The Nature Conservancy, ICLEI and Ecologic Institute. Retrieved from https://www.ecologic.eu/17059
44. Kapović SM (2019) Small retention in polish forests from a Forest management perspective – copying of existing could be right path. Chapter in book: nature-based flood risk management on private land. Springer, chapter, https://doi.org/10.1007/978-3-030-23842-1_5, Book ISBN 978-3-030-23842-1
45. Woods-Ballard B, Kellagher R, Martin P, Jefferies C, Bray R, Shaffer P (2007) The SUDS manual. CIRIA C697. Construction Industry Research and Information Association, London
46. Chui TFM, Liu X, Zhan W (2016) Assessing cost-effectiveness of specific LID practice designs in response to large storm events. J Hydrol 533:353–364
47. De Vleeschauwer K, Weustenraad J, Nolf C, Wolfs V, De Meulder B, Shannon K, Willems P (2014) Green–blue water in the city: quantification of impact of source control versus end-of-pipe solutions on sewer and river floods. Water Sci Technol 70(11):1825–1837
48. Jia H, Wang X, Ti C, Zhai Y, Field R, Tafuri AN, Shaw LY (2015) Field monitoring of a LID-BMP treatment train system in China. Environ Monit Assess 187(6):373
49. Kim HW, Park Y (2016) Urban green infrastructure and local flooding: the impact of landscape patterns on peak runoff in four Texas MSAs. Appl Geogr 77:72–81
50. Zhang K, Chui TFM (2018) A comprehensive review of spatial allocation of LID-BMP-GI practices: strategies and optimization tools. Sci Total Environ 621:915–929
51. Thorne C, Lawson E, Ozawa C, Hamlin S, Smith L (2015) Overcoming uncertainty and barriers to adoption of blue-green infrastructure for urban flood risk management. J Flood Risk Manag. https://doi.org/10.1111/jfr3.12218
52. Moreno J, Smith KM, Mijic A (2017) Economic analysis of wider benefits to facilitate SuDS uptake in London, UK. Sustain Cities Soc 28:411–419
53. Ferreira CSS, Walsh RPD, Nunes JPC, Steenhuis TS, Nunes M, de Lima JLMP, Coelho COA, Ferreira AJD (2016) Impact of urban development on streamflow regime of a Portuguese peri-urban Mediterranean catchment. J Soil Sediment 16:2580–2593
54. Alves A, Vojinovic Z, Kapelan Z, Sanchez A, Gersonius B (2020) Exploring trade-offs among the multiple benefits of green-blue-grey infrastructure for urban flood mitigation. Sci Total Environ 703:134980
55. Almenar NB, Elliot T, Rugani B, Philippe B, Gutierrez TN, Sonnemann G, Geneletti D (2021) Nexus between nature-based solutions, ecosystem services and urban challenges. Land Use Policy 100:104898
56. Jiang Y, Zevenbergen C, Ma Y (2018) Urban pluvial flooding and stormwater management: a contemporary review of China's challenges and "sponge cities" strategy. Environ Sci Policy 80:132–143
57. Heinzlef C, Robert B, Hémond Y, Serre D (2020) Operating urban resilience strategies to face climate change and associated risks: some advances from theory to application in Canada and France. Cities 104:102762
58. Sharifi A (2019) Urban form resilience: a meso-scale analysis. Cities 93:238–252

59. Wardekker A, Wilk B, Trown V, Uittenbroek C, Mees H, Driessen P, Wassen M, Molenaar A, Walda J, Runhaar H (2020) A diagnostic tool for supporting policymaking on urban resilience. Cities 101:102691
60. Zuniga-Teran AA, Gerlak AK, Mayer B, Evans TP, Lansey KE (2020) Urban resilience and green infrastructure systems: towards a multidimensional evaluation. Curr Opin Environ Sustain 44:42–47
61. D'Alpaos C, Andreolli F (2020) Urban quality in the city of the future: a bibliometric multicriteria assessment model. Ecol Indic 117:106575
62. Bixler RP, Lieberknecht K, Atshan S, Zutz CP, Richter SM, Belaire JA (2020) Reframing urban governance for resilience implementation: the role of network closure and other insights from a network approach. Cities 103:102726
63. Verhoeven JTA, Arheimer B, Yin C, Hefting MM (2006) Regional and global concerns over wetlands and water quality. Trends Ecol Evol 21:96–103
64. Xiong S, Johansson ME, Hughes FMR, Hayes A, Richards KS, Nilsson C (2003) Interactive effects of soil moisture, vegetation canopy, plant litter and seed addition on plant diversity in a wetland community. J Ecol 91:976–986
65. Hefting M, Clément JC, Dowrick D, Cosandey AC, Bernal S, Cimpian C, Tatur A, Burt TP, Pinay G (2004) Water table elevation controls on soil nitrogen cycling in riparian wetlands along a European climatic gradient. Biogeochemistry 67:113–134
66. Acreman M, Holden J (2013) How wetlands affect floods. Wetlands 33(5):773–786
67. Gibbs JP (2000) Wetland loss and biodiversity conservation. Conserv Biol 14:314–317
68. Seifollahi-Aghmiuni S, Nockrach M, Kalantari Z (2019) The potential of wetlands in achieving the sustainable development goals of the 2030 Agenda. Water 11:609
69. Borja S, Kalantari Z, Destouni G (2020) Global wetting by seasonal surface water over the last decades. Earth's Future 8:e2019EF001449
70. Davidson NC (2014) How much wetland has the world lost? Long-term and recent trends in global wetland area. Mar Freshw Res 65:934–941
71. Mitsch WJ, Gosselink JG (2015) Wetlands, 5th edn. Wiley, Hoboken, 744 p
72. Bush J, Doyon A (2019) Building urban resilience with nature-based solutions: how can urban planning contribute? Cities 95:102483
73. Gulsrud NM, Hertzog K, Shears I (2018) Innovative urban forestry governance in Melbourne?: Investigating "green placemaking" as a nature-based solution. Environ Res 161:158–167
74. Mukherjee M, Takara K (2018) Urban green space as a countermeasure to increasing urban risk and the UGS-3CC resilience framework. Int J Disaster Risk Reduction 28:854–861
75. Nesshöver C, Assmuth T, Irvine KN, Rusch GM, Waylen KA, Delbaere B, Krauze K (2017) The science, policy and practice of nature-based solutions: an interdisciplinary perspective. Sci Total Environ 579:1215–1227
76. Kalantari Z, Ferreira CSS, Page J, Goldenberg R, Olsson J, Destouni G (2019) Meeting sustainable development challenges in growing cities: coupled social-ecological systems modeling of land use and water changes. J Environ Manage 245:471–480
77. Kabisch N, Frantzeskaki N, Pauleit S, Naumann S, Davis M, Artmann M, Haase D, Knapp S, Korn H, Stadler J, Zaunberger K, Bonn A (2016) Nature-based solutions to climate change mitigation and adaptation in urban areas: perspectives on indicators, knowledge gaps, barriers, and opportunities for action. Ecol Soc 21(2):39
78. Pan H, Page J, Zhang L, Cong C, Ferreira C, Jonsson E, Näsström H, Destouni G, Deal B, Kalantari Z (2019) Understanding interactions between urban development policies and GHG emissions: a case study in Stockholm region. Ambio. https://link.springer.com/article/10.1007/s13280-019-01290-y
79. Calliari E, Staccione A, Mysiak J (2018) An assessment framework for climate-proof nature-based solutions. Sci Total Environ 656:691–700

80. Frantzeskaki N, McPhearson T, Collier MJ, Kendal D, Bulkeley H, Dumitru A, Walsh C, Noble K, van Wyk E, Ordóñez C, Oke C, Pintér L (2019) Nature-based solutions for urban climate change adaptation: linking the science, policy, and practice community for evidence-based decision-making. Bioscience. https://doi.org/10.1093/biosci/biz042
81. Naturvation (2019) Assessment of biophysical and ecological services provided by urban nature-based solutions: a review. [Online]. Retrieved from: https://naturvation.eu/sites/default/files/result/files/naturvation_briefing_note_assessment_of_biophysical_and_ecological_services_provided_by_urban_nature-based_solutions.pdfOssa-
82. Potschin M, Kretsch C, Haines-Young R, Furman E, Berry P Baró F (2015) Nature-based solutions. OpenNESS ecosystem service reference book. OpenNESS synthesis paper. Retrieved from: http://www.openness-project.eu/library/reference-book/sp-NBS
83. Raymond M, Frantzeskaki N, Kabisch N, Berryd P, Breile M, Razvan Nitaf M, Genelettig D, Calfapietrah C (2017) A framework for assessing and implementing the co-benefits of nature-based solutions in urban areas. Environ Sci Policy 77:15–24
84. Raymond CM, Berry P, Breil M, Nita MR, Kabisch N, de Bel M, Enzi V, Frantzeskaki N, Geneletti D, Cardinaletti M, Lovinger L, Basnou C, Monteiro A, Robrecht H, Sgrigna G, Munari L, Calfapietra C (2017) An impact evaluation framework to support planning and evaluation of nature-based solutions projects. Report prepared by the EKLIPSE expert working group on nature-based solutions to promote climate resilience in urban areas. Centre for Ecology & Hydrology, Wallingford
85. Kalantari Z, Santos Ferreira CS, Keestra S, Destouni G (2018) Nature-based solutions for flood-drought risk mitigation in vulnerable urbanizing parts of East-Africa. Curr Opin Environ Sci Health 5:73–78
86. Santamouris M (2014) On the energy impact of urban heat island and global warming on buildings. Energ Buildings 82:100–113
87. Foster J, Lowe A, Winkelman S (2011) The value of green infrastructure for urban climate adaptation. Center Clean Air Policy 750(1):1–52
88. Bauduceau N, Berry P, Cecchi C, Elmqvist T, Fernandez M, Hartig T, Raskin-Delisle K (2015) Towards an EU research and innovation policy agenda for nature-based solutions & re-naturing cities: final report of the horizon 2020 expert group on 'Nature-based Solutions and Re-naturing Cities'
89. Depietri Y, McPhearson T (2017) Integrating the grey, green, and blue in cities: nature-based solutions for climate change adaptation and risk reduction. In: Nature-based solutions to climate change adaptation in urban areas. Springer, Cham, pp 91–109
90. Kati V, Jari N (2016) Bottom-up thinking – identifying socio-cultural values of ecosystem services in local blue–green infrastructure planning in Helsinki, Finland. Land Use Policy 50:537–547
91. Goldenberg R, Kalantari Z, Destouni G (2018) Increased access to nearby green–blue areas associated with greater metropolitan population well-being. Land Degrad Dev 29 (10):3607–3616
92. Bianchini F, Hewage K (2012) Probabilistic social cost-bene fi t analysis for green roofs a lifecycle approach. Build Environ 58:152–162
93. von Döhren P, Haase D (2015) Ecosystem disservices research: a review of the state of the art with a focus on cities. Ecol Indic 52:490–497
94. Finlayson CM, Horwitz P, Weinstein P (2015) Wetlands and human health, vol 5. Springer, Dordrecht
95. Jayasooriya VM, Ng AWM (2014) Tools for modeling of stormwater management and economics of green infrastructure practices: a review. Water Air Soil Pollut 225(8):2055

96. Halkijević I, Bekić D, Lončar G, Potočki K, Gilja G, Carević D (2019) Latest developments in the scope of sustainable water management. In: Lakušić S, Mandić Ivanković A (eds) Future trends in civil engineering 2019. University of Zagreb faculty of Civil Engineering, Zagreb, pp 146–173

Nature-Based Solutions Impact on Urban Environment Chemistry: Air, Soil, and Water

Paulo Pereira, Miguel Inácio, Donalda Karnauskaitė, Katažyna Bogdzevič, Eduardo Gomes, Marius Kalinauskas, and Damia Barcelo

Contents

Abstract Urban areas are the largest source of pollutants that directly or indirectly will end up in the air, soil, and water. It is paramount to find solutions to reduce the impact of pollution on climate change, ecosystem services, biodiversity loss, and human health. Nature-based solutions (NBS) can mitigate the effects of anthropogenic activities significantly and act as a buffer to immobilize filtrate and uptake pollutants. There is a wide range of advantages of implementing NBS to reduce air, water, and soil pollution since they increase ecosystem services supply. This is key to make cities more liveable and sustainable, especially in areas where there are the largest agglomerations of people. This chapter will review the impacts of the air, soil, and water pollution on ecosystems, biodiversity, human health, and the NBS that can be used to minimize this.

P. Pereira (✉), M. Inácio, D. Karnauskaitė, K. Bogdzevič, E. Gomes, and M. Kalinauskas
Environmental Management Laboratory, Mykolas Romeris University, Vilnius, Lithuania

D. Barcelo
Catalan Institution for Research and Advanced Studies (ICRA), Barcelona, Spain

Carla S. S. Ferreira, Zahra Kalantari, Thomas Hartmann, and Paulo Pereira (eds.),
Nature-Based Solutions for Flood Mitigation: Environmental and Socio-Economic Aspects,
Hdb Env Chem (2022) 107: 79–138, DOI 10.1007/698_2021_760,

Published online: 6 May 2021

Keywords Ecosystem services, Nature-based solutions, Pollutants, Sustainable development, Urban areas

1 Introduction

Urban areas are associated with high pollution levels related to air, soil, and water degradation [1, 2]. We are living in an urbanized world. According to the World Bank[1] more than 50% of the population lives in cities, and it is expected that by 2045, the number of inhabitants in urban areas will double. Approximately 7 of 10 persons will live in cities. The increasing pressure in these areas due to population growth is the cause of deregulated urban development, also known as urban sprawl [3]. Urban sprawl is a global phenomenon but is especially evident in developing countries where planning is inexistent or inefficient, as reported in several works [4–6]. The urban areas' growth is imposing tremendous land consumption and bringing problems such as pollution and greenhouse gases (GHG) emission [7, 8]. Apart from the negative impacts on the ecosystems and the services provided [9], pollution also affects human health [10]. Air, soil, and water quality in urban areas are key to a healthy environment. Due to the environmental degradation in urban areas, several efforts have been carried out to reduce the impacts of cities on air, soil, and water. Measures such as compact cities [7], limiting or forbidden vehicle circulation in urban centres, removal of pollutant industries from the cities, increasing the network of bicycle and walking pathways, increasing the parking prices, removing parking lots from city centres, cheap and clean (e.g., electric) public transport, diesel taxes [11], air-filtration infrastructures [12], improving sewage and water treatment facilities [13], and greening the city [14] have been established to reduce environmental degradation. The green cities concept is a reaction to the problems provoked by sprawled urban areas and to support cities to be more liveable and sustainable and less dispersed [15]. Water is also understood as an essential factor for urban population wellbeing; therefore, the blue infrastructure is also a relevant aspect in urban areas sustainability [16, 17]. Both green and blue infrastructures (GBI) provide a wide range of ecosystem services (e.g., climate and water regulation, air purification), mitigating the impacts of grey infrastructure [18, 19]. GBI term is based on using natural solutions or processes [20] to increase wellbeing in urban areas. Nature-Based Solution (NBS) are "actions to protect, sustainably manage and restore natural or modified ecosystems that address societal challenges effectively and adaptively, simultaneously providing human wellbeing and biodiversity benefits".[2] These actions can be carried out using GBI or a mix between green-blue-grey infrastructure, also known as hybrid approaches [21]. In this work, NBS is

[1]https://www.worldbank.org/en/topic/urbandevelopment/overview (Accessed 30-08-2020).

[2]https://www.iucn.org/theme/ecosystem-management/our-work/iucn-global-standard-nature-based-solutions (Accessed 30-08-2020).

Fig. 1 Impacts of anthropogenic activities on biodiversity and ecosystem services

understood as a natural (GBI) or hybrid (green-blue-grey) approach. NBS can contribute importantly to increase the environmental quality in urban areas and remove pollutants from air [22], soil [23], and water [24]. This chapter provides an overview of the urban areas' impacts on environmental chemistry and how NBS can mitigate the effects of anthropogenic activities.

2 Urban Impacts on Environmental Chemistry

Anthropogenic activities have a multitude of negative impacts on the environment. Factories, vehicle circulation, wastewater, and agricultural waste management occupy an important part of urban and peri-urban areas [25]. Also, human practices, e.g., bonfires [26] and litter [27], contribute to the accumulation of pollutants in the environment (Fig. 1). Pollution is a global problem that has a very pervasive impact on society. Since the impacts of pollution are transversal to all aspects of our daily life, it is extremely connected to all Sustainable Development Goals (SDGs).[3] The chemical emitted to the atmosphere, deposited on soil, and transported to the water will significantly impact the quantity and quality of the ecosystem services provided, biodiversity, human health, and economy (Fig. 1) [28].

[3]https://www.unenvironment.org/beatpollution/global-response-pollution (Accessed 30-08-2020).

2.1 Air

Air quality is a major concern in urban areas, and there is an increasing number of days with a high concentration of pollutants in world metropoles. Practically, air pollution affects all countries and is considered one of the critical aspects to achieve United Nations (UN) SDGs, especially in Goal 3 (Good Health and Wellbeing), 7 (Affordable and Clean Energy), and 11 (Sustainable Cities and Communities) [29]. In 2019, particulate matter ($PM_{2.5}$) concentrations were a grave concern in southeastern Asia, especially in Bangladesh, Pakistan, Mongolia, Afghanistan, India, Mongolia. In Europe, Bosnia and Herzegovina are the countries where particulate matter ($PM_{2.5}$) levels were the highest. Also, 30 out of 31 of the most polluted ($PM_{2.5}$) urban areas are located in Asia [30]. Concerning PM_{10}, the urban areas with the highest concentrations are located in Africa (Cairo) and southeast Asia (Delhi, Dhaka, Mumbai, and Kolkata) [29]. In Europe (2017), the highest concentration of this pollutant is observed in Po Valley (Italy), the Balkans region, and Poland (Fig. 2a).

Urban areas are responsible for the emission of a large amount of GHG [31]. For instance, cities release more than 70% of carbon dioxide (CO_2) [32]. The increase of CO_2 is strongly linked to urban and economic development, energy consumption, population dynamics, and energy utilization [8]. CO_2 emissions have been increasing in many regions of China, the USA [32], and European cities such as Rome [33] and London [34]. Nevertheless, the CO_2 concentration is increasing at the global level since the second half of the twentieth century [35]. Traffic is also an important contributor to CO_2 emissions. In Europe, automobile circulation and the consequent CO_2 release into the atmosphere showed a positive trend between 1990 and 2015 in Austria, Denmark, France, Greece, Ireland, The Netherlands, Portugal, and Spain [36]. However, if we considered the CO_2 emissions from all the sources, it is observed a decrease in all the 27 countries of the European Union (1990–2017) [37]. Carbon monoxide (CO) is also emitted into the atmosphere due to the combustion of low-quality fossil fuels, transportation, and industry [38]. This tasteless, odourless, and colourless gas is one of the pollutants responsible for smog. In 2017, the highest CO concentrations were identified in India and China [39]. From 2000 to 2014, the global concentration of CO is decreasing[4], and this was also identified in the USA (1980–2019)[5] and in other major urban areas such as Beijing and Moscow (1998 and 2017) [40].

Nitrogen oxide (NO_x) is a gas strongly influenced by traffic circulation. In 2016, the countries that emitted the highest quantities were China, Russia, India, Brazil, and the USA. Except for the USA, between 1990 and 2016, it was observed in the remaining countries an increasing trend of NO_x. In other areas of the globe such as Europe, the opposite trend was identified, mainly in the United Kingdom, Germany, France, Scandinavian countries,[6] and the Czech Republic [41]. This was also observed in other European Union countries [37].

[4]https://climate.nasa.gov/news/2291/fourteen-years-of-carbon-monoxide-from-mopitt/ (Consulted in 15/10/2020).

[5]https://www.epa.gov/air-trends/carbon-monoxide-trends (Consulted in 15/10/2020).

[6]https://cait.wri.org/ (Consulted in 15/10/2020).

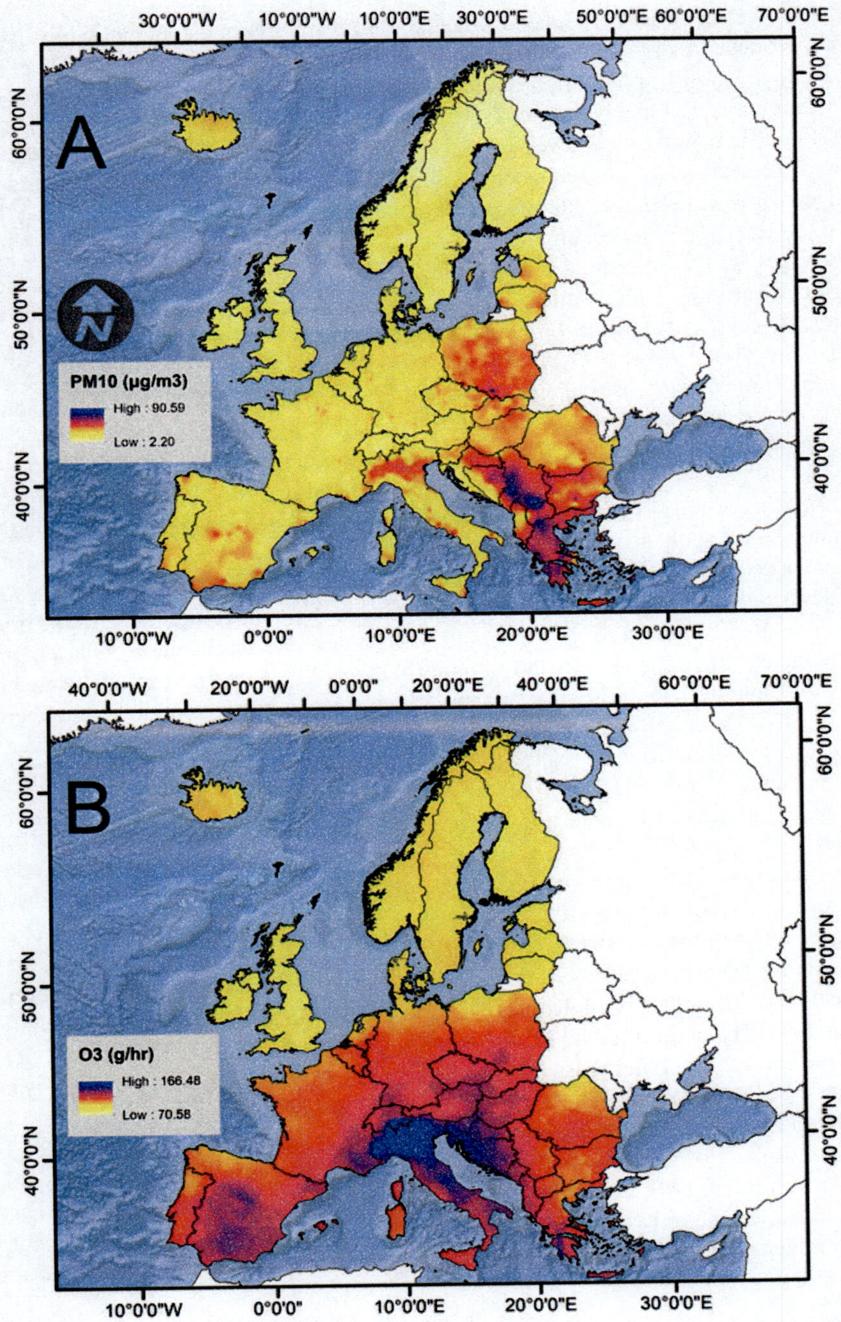

Fig. 2 (**a**) PM_{10} annual average values in 2017 at Urban and Suburban Background stations and (**b**) Ozone 93.2 percentile of maximum daily 8-h running mean values in 2017 at Urban and Suburban Background stations. Data source: https://www.eea.europa.eu/data-and-maps/data/interpolated-air-quality-data-2. Data were interpolated using Ordinary Kriging method (349 stations)

Methane (CH_4) is produced from agricultural and livestock practices, but also as a consequence of practices related to urban environments such as the organic waste decay in landfills, production of oil, coal, natural gas, and coal.[7] It has increased due to human activities [42]. After CO_2, CH_4 is the most significant contributor to the GHG effect. At a global level, this GHG concentration is increasing [35], especially in East Asia as a consequence of industrialization [43]. In Europe, there is a decreasing trend due to the restrictive policies regarding GHG emissions imposed in the EU [44]. In cities, an increasing trend has been observed (e.g., Beijing, [45]), whereas in others, no trend was identified (e.g., Los Angeles, [46]). CH_4 has a higher global warming potential than CO_2. Therefore, this reduction is relevant to decrease greenhouse gas emissions on climate change [47].

Tropospheric ozone (O_3) is an important greenhouse gas that has been increasing globally due to land-use change, industrialization, and traffic [48]. O_3 is formed due to the interaction between CO, NO_x, CH_4, volatile organic compounds (VOC), and sunlight. O_3 concentrations have been decreasing at a global level, especially after 2000 [49]. However, in several urban areas located in China [50, 51], South Korea [52], Malaysia [53], Taiwan [54], Portugal [55], and Spain [56] there is an increasing trend as a consequence of the increasing temperatures and vehicle traffic. On the other hand, a decrease due to the measures imposed in traffic circulation is identified in some cities located in the USA [57], Canada [58], Czech Republic [59], France [60], Greece [61], United Kingdom [62], and Iran [63]. O_3 concentration increases with temperature increasing [64], therefore the amounts of this pollutant in the atmosphere are high in low latitudes. In 2017, O_3 concentrations were especially high in the Iberian Peninsula, Greece, Balkans, and Italy. The highest concentrations were observed in Po Valley (Italy), which is related to this area's high industrialization (Fig. 2b).

Sulphur oxides (SO_x) are composed especially of sulphur dioxide (SO_2) and sulphur trioxide (SO_3), and they are produced by the combustion of sulphur-rich fuels such as petroleum, coal, or crude [65]. SO_2 is a gas with an unpleasant odour and colourless.[8] The global trends in the emissions of this gas are heterogeneous. In Europe and North America, there is a marked decrease between 1990 and 2015 as a consequence of the industrialization reduction. On the other hand, in east Asia, there was an increase between 1990 and 2005, followed by a reduction [66]. In several urban areas located in South Korea [67], China [68], Malaysia [69], Iran [70], United Kingdom [71], and Portugal [72] there is a decreasing trend in the emission of SO_2. On the contrary, in other areas there is observed an increasing trend such as in cities located in Saudi Arabia [73], India [74], and Brazil [75].

Anthropogenic activities in urban areas are also responsible for the emission to the atmosphere of several other pollutants such as metals and metalloids (e.g., Iron (Fe), Aluminium (Al), Lead (Pb), Arsenic (AS), Cadmium (Cd), Mercury (Hg)) and polycyclic aromatic hydrocarbons (PAHs, e.g., anthracene, fluoranthene, benzo[j] fluoranthene, benzo[a]anthracene, benzo[a]pyrene). Metals and metalloids are

[7] https://www.epa.gov/ghgemissions/overview-greenhouse-gases (Consulted in 15/10/2020).

[8] https://www.eea.europa.eu/help/glossary/eper-chemicals-glossary/sulphur-oxides-sox (Consulted on 17/10/2020).

associated with the presence of heavy industry and traffic. In Europe, the emission of these elements in urban areas has been decreasing from 2000 to 2017 due to the reduction of industrial activities and the imposition of limits to these elements' emission [76]. However, in other areas such as China [77], an increasing trend in these pollutants' emissions has been identified.

Microplastics (< 5 mm particles) are an emerging pollutant, which primary sources are fibber fragments in clothes, erosion of rubber tires, house furniture, building materials, industrial emissions, landfills, waste incineration, plastics used in agriculture, particle resuspension from traffic, cosmetics, plastic litter, and debris. They are rich in chemical elements such as polyethylene terephthalate (PET), polystyrene (PS), polypropylene (PP), polyvinyl chloride (PVC), polyurethane (PUR), epoxy resin (EP), alkyd resin (ALK), rayon (RY), and polyethylene (PE) [78–81]. As a consequence of wind transport, several works identified their presence in suspension [82, 83], contributing to air quality reduction.

Urban air pollution is increasing in some areas of the globe, while it is decreasing in others. The emission of pollutants into the atmosphere has detrimental impacts on human health [35]. In 2013, air pollution was the fourth world cause of death [84]. The exposition to high levels of pollutants increases several diseases, especially in children and the elderly population.[9] For example, $PM_{2.5}$ and PM_{10} are responsible for lung cancer, cardiovascular disease, respiratory disease, myocardial infarctions, increasing the risk of premature deaths and years of life lost [37, 85]. People's exposition to urban pollutants is decreasing in the USA [86] and Europe [87] as a consequence of more strict air quality legislation [88]. On the other hand, they are increasing in other parts of the world, such as China [89]. It is estimated that 4.2 million deaths are attributed to air pollution [90]. Globally, the deaths attributed to $PM_{2.5}$ between 1990 and 2015 increased [91]. In 2016, the European Union countries where the premature deaths related to $PM_{2.5}$ were the highest were Germany, Italy, and Poland (Fig. 3). Carugno et al. [92] observed that in Lombardy (Italy), the estimated annual deaths between 2003–2006, 2007–2010, 2011–2014 due to PM_{10} were 343, 254, and 208, respectively. This affects older people, especially [93]. In China, a study carried out by Maji et al. [94] found that in the years of 2014 and 2015, the annual premature death is estimated at 722.370 persons. Also, several works highlighted the positive relationship between mortality and $PM_{2.5}$ and PM_{10} in several cities such as Seoul [95], Wuhan [96], Guangzhou [97], and Paris [98]. The PM_{10} impacts on mortality are also related to specific events. For instance, in the Netherlands, between 1995 and 2012, the increase of PM_{10} after the New Year's fireworks increased the mortality in the following days [99]. On the other hand, areas with a high concentration of PM_{10} and $PM_{2.5}$ are vulnerable to disseminating diseases such as COVID-19 [100]. The lockdown imposed by COVID-19 reduced these pollutants drastically in urban areas [100–103].

High levels of CO, NO_x, O_3, SO_x, metals, and metalloids in the atmosphere have dramatic impacts on human health [104–108]. They increase cancer risks,

[9] https://www.who.int/news/item/15-11-2019-what-are-health-consequences-of-air-pollution-on-populations (Consulted in 17/10/2020).

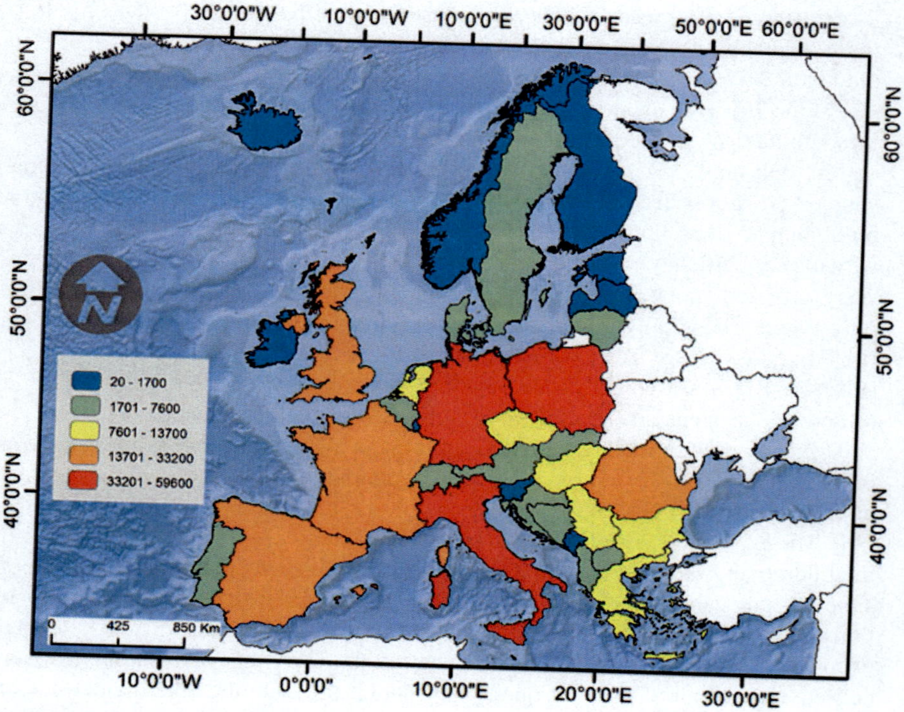

Fig. 3 Premature deaths caused by $PM_{2.5}$ in Europe in 2016. Classes according to Natural Breaks (Jenks) method. Data source [37]

respiratory problems, circulatory diseases, heart failure, mental illness, and anaemia [109–112]. As PM_{10} and $PM_{2.5}$, these gases are responsible for an important number of deaths and a decrease in life years in urban areas. For instance, between 1995 and 2015, 2,102 persons died due to CO poisoning [113]. The increase of NO_x emissions (2008–2015) due to Volkswagen Group "defeat devices" was responsible for 13,000 life years lost [114]. In Chinese urban areas during 2014, 89,391 deaths were attributed to ambient O_3 [115], and in Beijing (2016), an exposition to an SO_2 up to 122.08 $\mu g/m^3$ resulted in 27,854 hospital patients and 884 deaths [116]. Like PM_{10} and $PM_{2.5}$, specific events related to COVID-19 (e.g., lockdown) decreased the concentration of CO, NO_x, O_3, and SO_x in urban areas [117, 118] substantially.

Microplastics' presence in the air is likely inhaled and accumulated in the human body, negatively affecting health [119]. The studies of microplastics' impact on human health are not so developed as the elements mentioned above. On average, it is estimated that humans inhale approximately 26–130 airborne microplastics per day. This inhalation depends on the microplastic particle and size and can be deposited in the lungs and subsequently migrate to the lymphatic system. In high concentrations, microplastics likely increase lesions in the respiratory system. It can be a source of other kinds of diseases such as asthma, bronchitis, or pneumonia [80, 120].

The health impacts of air pollution are massive. Table 1 synthesizes the studies focused on the economic costs associated with urban air pollution. For instance, the majority of the works focused on $PM_{2.5}$ and PM_{10} and in China. The great majority of the published works studied the economic impacts of one year or a short term. It is clear that a single event [121, 122] can result in very high healthcare system expenses.

Air pollution also has negative impacts on agriculture and the environment. High concentrations of $PM_{2.5}$ [123], O_3 [124], SO_2, and NO_2 [125] can contribute to crop productivity decrease. In urban areas, the development of agriculture is also a matter of concern due to the high exposition to air pollutants [126]. The results of the impact of urbanization on urban agriculture are inconclusive. Some highlight that there is no threat to urban agriculture carried out in urban areas [127–129], while others show that the accumulation of pollutants in soils and plant leaves might have a negative impact in human health [130, 131]. In any case, it is vital to minimize the impact of air pollution accumulation in soils and plants and, therefore, reduce the risk of entering the food chain [126].

2.2 *Soil*

Soils in urban areas are affected negatively in multiple ways (e.g., sealing, pollutants accumulation). Urban soils are considered, for example, the ones constructed for gardens, sealed by asphalt or concrete, garbage heaps, and mine spoil. Commonly, they are classified as Technosols [148]. Soil pollution is a significant problem in urban areas and is linked with all SDGs directly or indirectly. However, in the urban context, it is more related with goals 2 (Zero Hunger), 3 (Good Health and Wellbeing), 6 (Clean Water and Sanitation), 11 (Sustainable Cities and Communities), 14 (Life Below Water), and 15 (Life on Land). Soil pollution in urban areas is a consequence of the deposition of pollutants from anthropogenic activities (e.g., industry, traffic). The deposition of pollutants can be due to point source and diffuse soil pollution. Point source pollution is caused by the release of pollutants in the soil, for example, in landfills, industrialized areas, waste and wastewater disposal, high use of agrochemicals, mining, and oil leakage. Diffuse soil pollution is a consequence of the spread of pollutants in the soil without knowing the source. These pollutants were usually released in other places and subsequently transformed or diluted and deposited on soils [149]. Pollutant accumulation is one of the most severe threats affecting soils since it increases their toxicity and substantially reduces soil functions and the capacity to provide ecosystem services in quality and quantity [150]. For instance, urban soils have a significantly reduced capacity to provide food but can play an important role in supporting human infrastructures [148]. In the European Union, approximately 4.5 million sites are estimated to be contaminated [151]. In Fig. 4 it can be observed that the areas with the highest contamination of Lead (Pb) and Mercury (Hg) are located in the industrialized/urban areas of Ruhr (Germany) and Manchester/Liverpool/London (United Kingdom). Also, previous

Table 1 Urban pollution and implications for economic health costs/The search was carried out in Google scholar (2000–2020)

Pollutants	Studied period	Country	Urban area	Economical costs	Reference
$PM_{2.5}$	2012–2016	China	Hong Kong	US$1.5–1.8 billion	Nam et al. [132]
PM_{10}	2000–2004		Beijing	US$1670 and $3,655 million annually	Zhang et al. [133]
$PM_{2.5}$	January 2013			~ US$180 million Yuan	Du and Li [134]
				US$253.8 million	Gao et al. [121]
PM_{10}	2012			583.02 million Yuan	Zhao et al. [122]
$PM_{2.5}$ and PM_{10}	2001		Shanghai	US$625.40 million	Kan and Chen [135]
SO_2	2000–2007		Taiyuan	0.8–1.7 billion Yuan	Zhang et al. [136]
PM_{10}, NO_2, O_3, and SO_2	2010–2013		Pearl River Delta region	Between US $14,768 and US $25,305 million	Lu et al. [137]
PM_{10}	2006		113 Chinese cities	341.403 billion Yuan	Renjie et al. [138]
$PM_{2.5}$	–	South Korea	Seoul	US$10029 million	Lee et al. [139]
$PM_{2.5}$	2013	India	Nagpur	US$2.2 billion	Etchie et al. [140]
$PM_{2.5}$	2017	Iran	Tehran	US$3 billion	Bayat et al. [141]
PM_{10} and O_3	2010	Greece	Thessaloniki	531.225 million euros	Vlachokostas et al. [142]
PM_{10}, SO_x, NO_x, VOC, CH_4, CO, CO_2, N_2O and NH_3	1994–1998	Spain	Madrid	620.8 million euros	Monzon and Guerrero [143]
$PM_{2.5}$	2012	North Macedonia	Skopje	Between 570 and 1,470 million euros	Sanchez-Martinez et al. [144]
$PM_{2.5}$, PM_{10}, CO, SO_2, NO_2, and O_3	2008–2017	Brazil	Sao Paulo	US$111 million	Curvelo Santana et al. [145]
PM_{10} and CO	1991–1994			US$ 3,222,676 million	Miraglia et al. [146]
$PM_{2.5}$	2000	USA	83 urban areas	US$31 billion	Levy et al. [147]

As criteria, only peer-reviewed works were considered. Keywords: Air AND Urban AND pollution AND health AND economic AND costs works observed high concentrations of metals and metalloids [152], PAHs [153], radionuclides [154] in urban parks, schools, and residential areas.

Soil pollutants accumulation alters soil functions, and this has implications on natural biogeochemical cycles. These changes can be driven by acidity, nutrient status, the bioavailability of toxic elements, disruption of soil biodiversity, plant removal, and litter deposition. The soil vulnerability to pollutants depends essentially on the type of soil, base saturation, cation exchange capacity, base cations supply by weathering process, organic matter content, and water table position [148, 155]. For example, the presence of high amounts of metals and metalloids in soil reduces microbiological abundance, diversity [156], and community structure [157], disrupts the regulation of biological activity, C transformations, nutrient cycling [158], and increases CH_4 emissions [159]. The high presence of PAHs [160], radionuclides [161], hydrocarbons [162], pharmaceuticals and personal care products [163], microglasses, microplastics [164], herbicides [165], and pesticides [166] in soils also affects negatively microbiological processes. Several studies highlighted the intensive usage of pesticides and herbicides in urban gardens for weed management [167].

A soil with high toxicity imposes very high stress on plant development [168]. The plants' metals and metalloid uptake affect their growth, development, water plant relation, ionic imbalance, alteration in elemental composition, degradation of photosynthetic pigments and chloroplast, and reduced photosynthetic rate. The interaction between the ions is complex since the high concentration of one element can affect the uptake and transport of other elements. When plants are exposed to high amounts of metals or metalloids, they try to prevent their uptake by roots and be transported to the plant's aerial parts [169]. For example, Zn′s high contents inhibit plant growth [170]. Cd toxicity affects mineral nutrient uptake and translocation, plant growth, development, and metabolism [171]. In high concentrations, Nickel (Ni) can induce competition with essential nutrients, oxidative stress, retardation of germination, changes in enzymatic activities, disruption of cell structure and dehydration, shoot and root production, leaf spots, and foliar necrosis, biomass production, abnormal flower shape, and branching system and mitotic root tip disturbance [172]. Also, plant Pb uptake decreases plant shoot growth, photosynthesis alteration, and lipid composition change [173]. Even in small concentrations, As can cause significant biochemical, physiological, and morphological changes in plants [174].

PAHs also are uptaked by plants [175]. However, this is different according to the different species, leaf morphology, and the presence of metals and metalloids [176, 177]. The accumulation of PAHs in plants affects plant morphology, wax content, and adsorption [177], increases root deformations [178], affects root symbiosis, plant growth, and C allocation [179]. Also, several works highlighted that plants have the capacity to uptake pharmaceutical products [180, 181]. As in PAHs case, the pharmaceuticals soil uptake depends on plant species [182]. Some works observed that pharmaceuticals and care products do not affect plant growth, while others observed a decrease [183]. High concentrations of pharmaceuticals in soil reduce the number of leaf photosynthetic elements and increase the burn edges and white spots. Produce changes in several plant hormones (jasmonates, auxins, abscisic acid, cytokinins) [184] increase toxicity [185] and oxidative stress

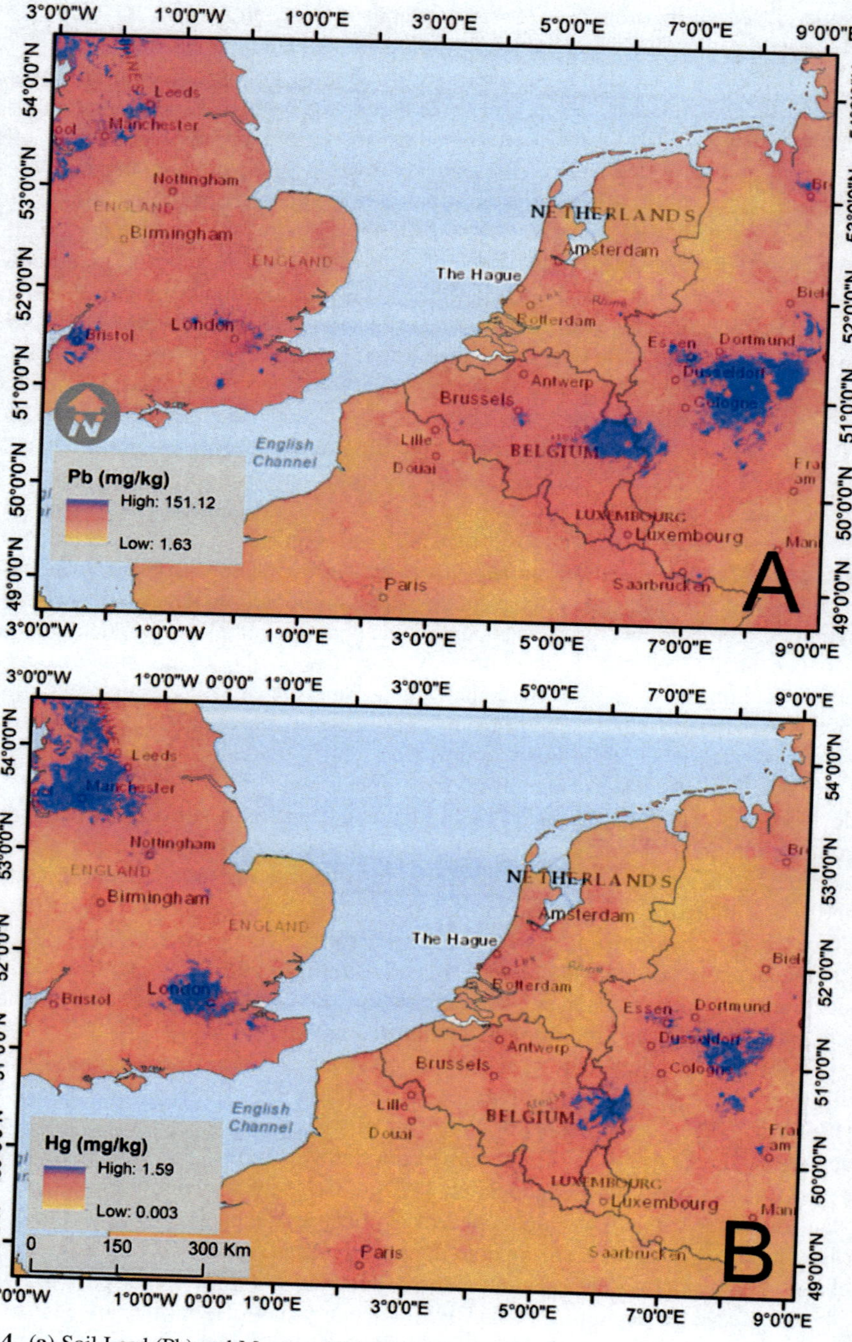

Fig. 4 (**a**) Soil Lead (Pb) and Mercury (Hg) in Northwest of Europe. Source: https://esdac.jrc.ec.europa.eu/resource-type/soil-threats-data

[186]. In the same line, plants can uptake pesticides and herbicides affect plant development, morphology, cellular alterations, chloroplast, leaf thickness [187], size, form [188], root nutrient composition [189], and growth [190].

Soils in urban areas have high concentrations of pollutants, which have a negative implication on human health. Humans are exposed to soil pollutants in different ways, such as ingestion, skin adsorption, penetration, and respiration [126]. A recent revision carried out by Brevik et al. [126] showed that the majority of the works were focused on metals and metalloids. A high and long-term exposition of these elements affects blood, lungs, kidneys, liver, brain and induces neurological disorders (e.g., Alzheimer and Parkinson) and cancer [126, 191, 192]. The high concentration of polychlorinated biphenyls and persistent organic pollutants in soil impacts childbirth, child weight, and foetal growth. Pesticides and herbicides are responsible for short-term health problems such as headaches, skin eye irritation, nausea, dizziness, or more severe problems such as asthma, cancer, diabetes, fertility, and immune system problems [126, 193, 194]. Other elements present in urban soils, such as radionuclides, impact human health [195]. Radionuclides are natural elements; however, they can be deposited on soil via nuclear or medical waste. The high concentrations of these elements can increase leukaemia or cancer [126, 196]. The presence of pollutants in urban soils is extremely concerning, especially when urban agriculture practices are increasing [197]. These elements can be easily uptake by plants and, if in high concentrations, can be a concern regarding human health [198, 199]. Urban agriculture provides a wide range of ecosystem services (e.g., C sequestration, flood regulation, food provision, recreation). However, the consumption of the food produced in these areas may have some risk regarding food contamination and the associated effects on human health [200].

2.3 *Water*

The urbanization process is a cause of water degradation [201]. As in the case of air and soil, water pollution is linked directly or indirectly with all SDGs. In the urban environment, it is especially connected with goal 3 (Good Health and Wellbeing), 6 (Clean Water and Sanitation), 9 (Industry Innovation and Infrastructure), 11 (Sustainable Cities and Communities), 13 (Climate Action), 14 (Life Below Water), and 15 (Life on Land). The high levels of pollutants emission in urban areas and the consequent deposition onto soil surface are responsible for water pollution (leaching or surface runoff transport). Also, urban areas have other water pollution sources such as inefficient water treatment systems, waste disposal, industrial activities, and construction. Freshwater pollution is increasing in many parts of the globe, especially in developing countries [202]. For instance, approximately every day 2 million tons of human waste is released into water bodies. In developing countries, 90% of the sewage sludge is released into water bodies without treatment. Industrial activities release into water bodies approximately 300–400 megatons of waste per year.[10] In Europe, 60% of the water bodies are highly polluted. The areas where the situation is critical are located in Central Europe and the Baltic region. After hydromorphological pressures (e.g., land reclamation, dams, weirs), diffuse pollution sources are the highest pressure in surface water bodies. Both diffuse and point source pollution are responsible for 38% and 18% of surface water contamination, respectively. For instance, urban wastewater, industrial discharges, and storm runoff are the major causes of point source pollution. Hg is responsible for the majority of surface water bodies do not achieve a good quality status. The increase of this element is attributed to urban sewage water and atmospheric deposition. The wastewaters discharged in the rivers are also rich in other hazardous elements such as PAHs, Hg, Cd, Pb, and Ni [203]. Surface water and sediment transported in runoff are responsible for water pollution in urban areas. In Box 1, we identified the areas where polluted surface waters can be accumulated in Vilnius (Lithuania).

Box 1 Potential Surface Water and Sediment Pollution Accumulation Index in Vilnius (Lithuania)

After intense rainfall and snowmelt periods, waterlogging is a recurrent phenomenon in urban areas [204]. The water and sediments that are transported and accumulated on roads have a high level of pollutants [205], representing a critical threat to environmental degradation (e.g., soil pollution, water bodies pollution) and human health [206]. Here we developed a simple model to analyse Vilnius's (Lithuania) potential surface water and sediment pollution accumulation index. This will be helpful to understand the areas where pollutants can be accumulated in the city. The framework applied is described below. To construct this model, we used imperviousness (20 m resolution) data from 2018[11] to identify the sealed areas. A topographic wetness index (20 m resolution) was used to assess high water accumulation areas. This index was calculated from the digital elevation model using Saga QGIS (3.14). Areas with an index higher than 10 correspond to high water accumulation areas [207]. The point sources of pollution (gas stations, power plants, and factories) were vectorized from Google maps. Subsequently, a kernel density was applied using ArcMap 10.2. Water lines and road density were calculated with the ArcMap 10.2-line density tool. Water lines were used to identify the areas where the water and sediments were transported and roads/railways as a source of pollutants (e.g., traffic). These data were obtained from Lithuanian Cadastre.[12]

(continued)

[10]https://en.unesco.org/waterquality-iiwq/wq-challenge (Consulted in 25/10/2020).

[11]https://land.copernicus.eu/pan-european/high-resolution-layers/imperviousness/status-maps/imperviousness-density-2018.

[12]https://www.registrucentras.lt/.

Box 1 (continued)

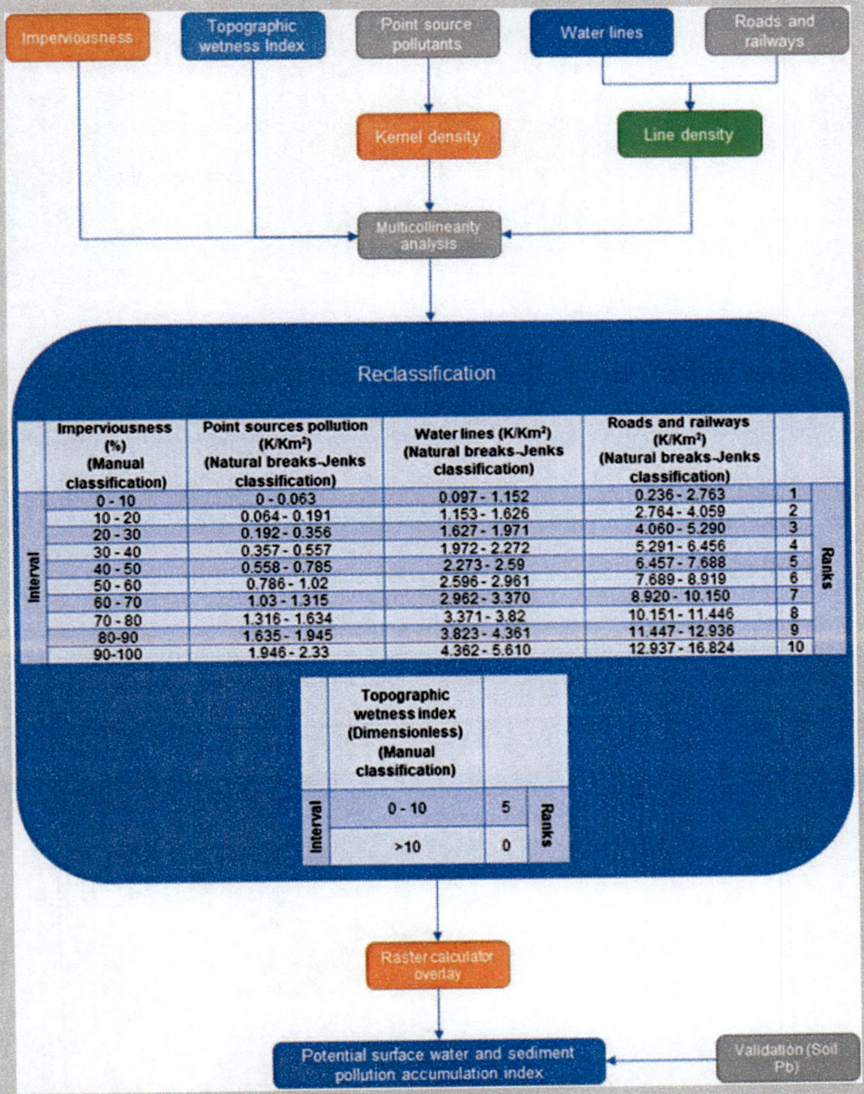

Interval	Imperviousness (%) (Manual classification)	Point sources pollution (K/Km²) (Natural breaks-Jenks classification)	Water lines (K/Km²) (Natural breaks-Jenks classification)	Roads and railways (K/Km²) (Natural breaks-Jenks classification)	Ranks
	0 - 10	0 - 0.063	0.097 - 1.152	0.236 - 2.763	1
	10 - 20	0.064 - 0.191	1.153 - 1.626	2.764 - 4.059	2
	20 - 30	0.192 - 0.356	1.627 - 1.971	4.060 - 5.290	3
	30 - 40	0.357 - 0.557	1.972 - 2.272	5.291 - 6.456	4
	40 - 50	0.558 - 0.785	2.273 - 2.59	6.457 - 7.688	5
	50 - 60	0.786 - 1.02	2.596 - 2.961	7.689 - 8.919	6
	60 - 70	1.03 - 1.315	2.962 - 3.370	8.920 - 10.150	7
	70 - 80	1.316 - 1.634	3.371 - 3.82	10.151 - 11.446	8
	80-90	1.635 - 1.945	3.823 - 4.361	11.447 - 12.936	9
	90-100	1.946 - 2.33	4.362 - 5.610	12.937 - 16.824	10

Interval	Topographic wetness index (Dimensionless) (Manual classification)	Ranks
	0 - 10	5
	>10	0

Previous to modelling the index, a multicollinearity analysis was applied to assess the correlation between the variables. In this case, none of the variables had a correlation higher than 0.70, or the variance inflation factor higher than 10. Therefore, all the variables were used in the model. Data were reclassified to be assessed at a similar scale range. After data reclassification, data were overlaid using ArcMap 10.2 raster calculator tool to assess the potential surface water and

(continued)

Box 1 (continued)

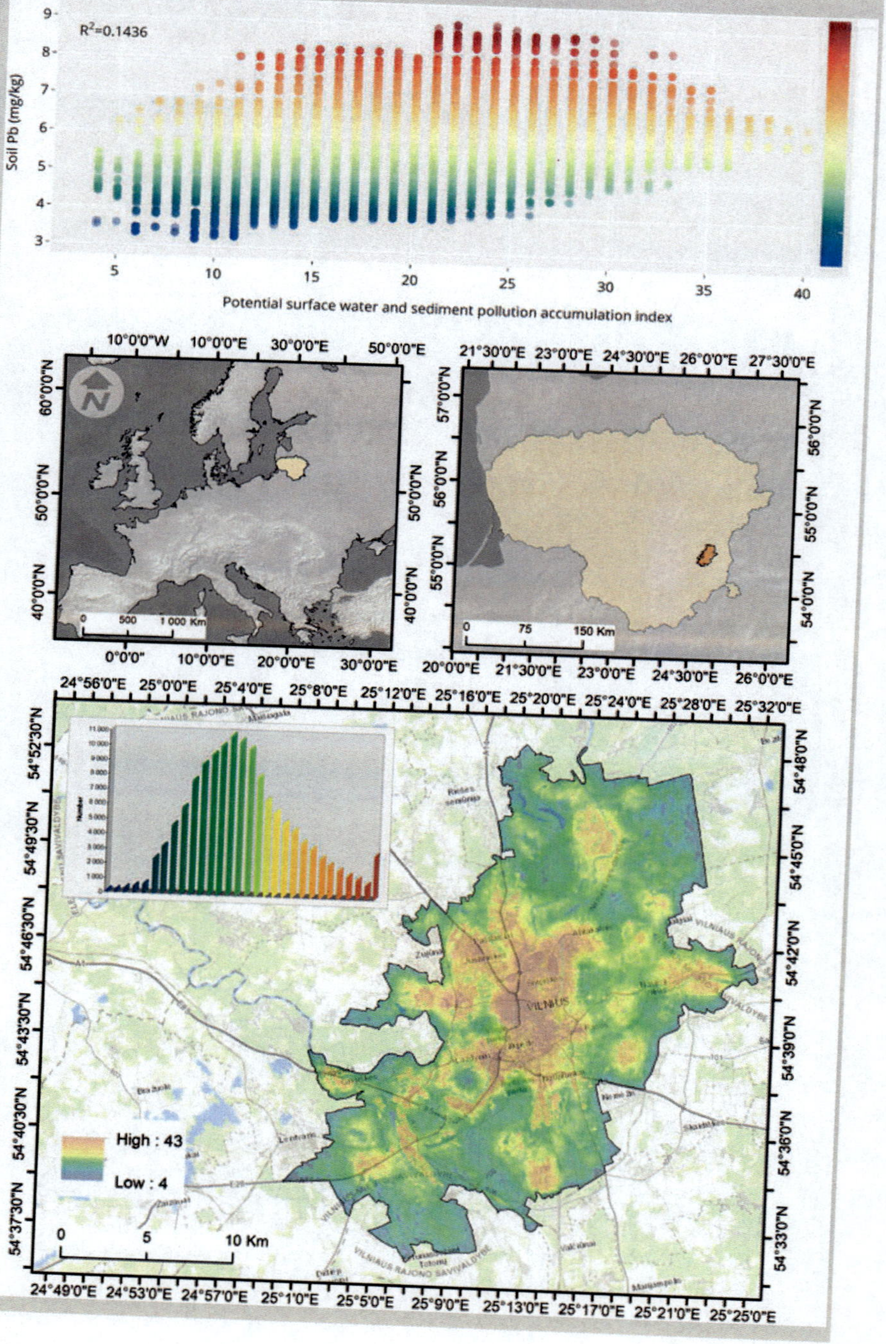

(continued)

Box 1 (continued)

sediment pollution accumulation index in Vilnius. The model validation was not so accurate ($R^2 = 0.1436$; $r = 0.37, p < 0.001$), and this is likely attributed to the large differences between data resolution (20 and 1,000 m). Nevertheless, it was not possible to find data at a better resolution. Despite this limitation, it is possible to identify some relationship between soil Pb and the potential surface water and sediment pollution accumulation index. The results are shown below, and the areas located in the centre of the city are the ones where there is the highest probability of polluted water accumulation. However, some small scattered centres are observed as well, and this is attributed to urban sprawl that increases the impervious area and the expansion of pollution sources (e.g., gas stations, roads).

Urban surface water pollution is a global problem. It is especially evident in developing countries, where an increasing trend is observed due to population growth, urban expansion with control, consumption increase, and the lack of sewage systems [208]. On the other hand, in developed countries, the trend is inverse. For instance, European Union member states made a substantial effort to reduce surface water contamination. From 1992 to 2016, there was a substantial decrease in orthophosphate in European rivers [203]. Although there are good reasons to have a positive perspective regarding urban surface water pollution, several problems are related to emerging pollutants. Metals and metalloids [209, 210] and PAHs [211], pH, EC, turbidity, dissolved solids, suspended solids, biological oxygen demand, chemical oxygen demand, total organic carbon, dissolved organic carbon, nitrate, ammonia [209, 212–214] have been traditionally studied in urban catchments as a consequence of the impact of urban activities (e.g., traffic, industry, wastewater). Recently, other types of pollutants have been identified in high contents such as microplastics, pesticides, herbicides [215], pharmaceutical and care products [216], and drugs such as heroin, cocaine, methadone in urban lakes, wetlands, rivers, coastal lagoons, and estuaries [217, 218]. These pollutants are also found in drinking water, i.e. microplastics [219], pesticides [220], herbicides [221], pharmaceutical and care products [222].

The increase of N and phosphorus (P) is a cause for urban and peri-urban water bodies' eutrophication and water quality degradation [223]. Eutrophication is responsible for biodiversity loss [224], hypoxic "dead zones" that decrease shellfish and fish production, light, the increase of harmful algal blooms that threaten drinking water safety and GHG emission [225, 226]. The change in precipitation patterns due to climate change is expected to increase eutrophication [227]. Metals and metalloids increase water toxicity and decrease water bodies' biodiversity [228]. The high content of these elements in water affects flora and fauna dramatically. For instance, it is widely known that fish [229] and shellfish [230] accumulate high amounts of metals and metalloids. A similar situation is observed in other pollutants such as microplastics [231], pesticides [232], herbicides [233], and pharmaceutical and care products [234]. Overall, the accumulation of these pollutants in water bodies results in ecosystem services, biodiversity, and human health degradation since fish and shellfish are consumed, and these pollutants enter the food chain [235, 236]. The impacts on fauna are dramatic and affect all the trophic levels. High contents of metals and metalloids affect fish behaviour [237], organs (e.g., kidney, liver, gills, muscle) [238], nervous system [239], metabolism [240]. Microplastic ingestion causes freshwater fauna intestinal lesions [241], larval growth [242], consumption, reproduction, and survival [243]. Also, high concentrations of PAHs in water harm aquatic fauna development, membrane damage, embryo development, oedema, changes in cardiac rhythm, swimming capacity [244]. Pesticides damage fish tissues [245], brain, muscles [246], organs (e.g., liver and kidney), and nervous system [247, 248]. Finally, pharmaceutical and care products affect aquatic organisms' reproductive systems, hormones [249], and behaviour [250]. Overall, these elements in aquatic ecosystems are of great environmental concern [251].

Groundwater is the world largest freshwater reserve, therefore is a key resource.[13] In urban areas, the contaminated surface water accumulates in different areas of the city (Box 1), infiltrates and increases the transport of pollutants to groundwater reserves. It is well known that the urbanization process (e.g., residential, industrial, commercial) and peri-urban agriculture aggravate groundwater pollution [252, 253] (Fig. 5). Urban sprawl and the expansion of human activities are increasing the potential sources (e.g., traffic, residential areas) of groundwater pollution [254, 255]. Recently, Burri et al. [256] revised the impacts of urban activities on groundwater pollution. It is a complex process, and since groundwater moves sluggishly, there is a lag process between the beginning of the pollution process and groundwater pollution. Contamination is an irreversible process or difficult to restore[13]. Shallow unconfined aquifers are the most vulnerable to pollution because the proximity to the surface is reduced, and the soil's capacity to filtrate the pollutants is low. In deeper aquifers, the vulnerability to pollution is reduced [253]. Several works observed that a large number of pollutants (e.g., metals and metalloids, nitrites, nitrates, pharmaceuticals, and care products) are present in groundwater reserves located close to urbanized and industrial areas located in America – USA [257], Brazil [258], Asia – China [259], India [260], Bangladesh [261], Jordan [262], Europe – Spain [263], France [264], Italy [265], Germany [266], Africa – Cameroon [267], South Africa, Mozambique [268], Nigeria [269] and Oceania – Australia [270], and New Zealand [271]. Overall, urban groundwater pollution is a global problem that hampers the availability and quality of the largest freshwater resource. Climate change is expected to negatively impact groundwater

[13] https://www.un.org/waterforlifedecade/waterforlifevoices/groundwater.shtml (Assessed in 21-11-2020).

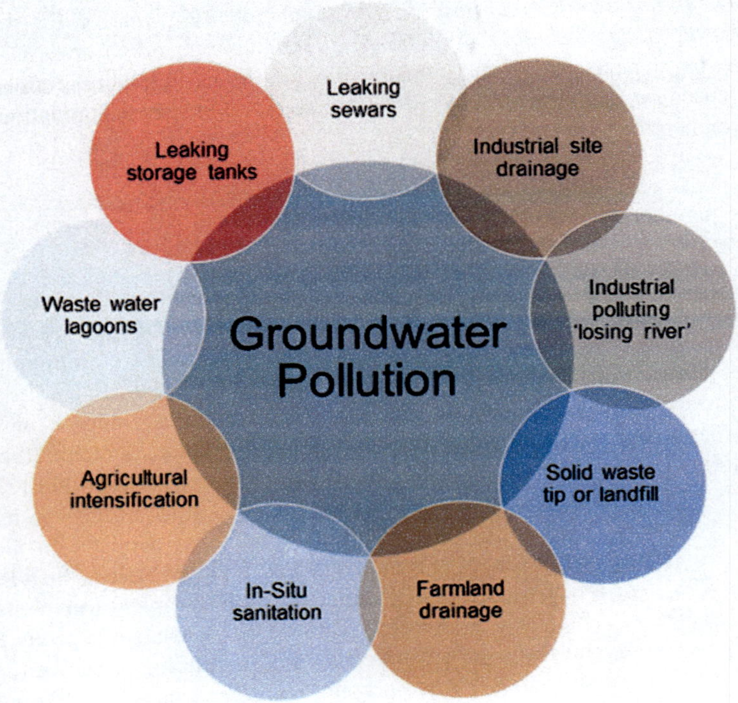

Fig. 5 Causes of groundwater pollution

quality since precipitation will be low and more irregular, leading to aquifer salinization and accumulation of pollutants [272, 273].

3 Nature-Based Solutions and Urban Environmental Chemistry

The previous sections clearly demonstrated that urban and industrial pollution is an important threat to natural resources and human health. Therefore, it is key to minimize the effects of anthropogenic activities. NBS is a sustainable and valuable option to increase biodiversity, ecosystem services, life quality, generate green jobs, and create new urban areas. It can also contribute importantly to climate change adaptation and mitigation (e.g., microclimate regulation, carbon sequestration, and flood retention) [274]. Raymond et al. [275] proposed a framework to implement NBS in urban areas and evaluate their co-benefits: *"1) identify problem or opportunity; 2) select and assess NBS and related actions; 3) design NBS implementation processes; 4) implement NBS; 5) frequently engage stakeholders and communicate co-benefits; 6) transfer and upscale NBS, and 7) monitor and evaluate co-benefits across all stages"*. Reduce the amount of pollutants in urban areas is critical to improving the ecosystems and human wellbeing. In the following sections, we will be focused on the NBS used to reduce air, soil, and water contamination in urban environments.

3.1 Air

Air pollutants removal is critical to decreasing environmental degradation and health problems. Urban forests are crucial to reduce the concentration of air pollutants. The studies focused on PM dry deposition are divided into (1) stand scale (relation between scale and environmental variables), (2) vertical direction (pollutant removal by dry deposition in branch, stem, or leaf surfaces), and lateral direction (the impacts of building design, distance to pollutant sources and vegetation type) [276]. The capacity of trees to capture pollutants can be understood as an NBS to reduce air pollution. The NBS effectiveness to reduce air pollution depends on plant type [277]. Usually, coniferous trees have a higher PM capture capacity than broadleaf species [278]. The articular needle shape, finer and more complex structure increases coniferous leaves' efficiency to capture PM. However, the leaves capacity to retain PM depend on several factors such as tree canopy morphology (e.g., canopy type, branch, and leaf density and leaf micromorphology: wax, trichomes, and roughness), deposition velocity (amount of particulates deposited), magnetic deposition velocity (amount of particulates deposited in clean leaves or trees), particle number, deposition amount, particle cover, the concentration of other pollutants, and meteorological conditions such as wind and precipitation [279, 280]. Leaves can capture PM by absorbing pollutants via leaf stomata or intercept particles in the leaf surface [276]. Also, the capacity of the leaves to retain PM depends on the development stage. Leaves that are fully developed have a high capacity to retain pollutants. Forest land use retains a higher number of particulates than grasslands. Nevertheless, grass cover contributes significantly to reduce wind erosion [281, 282]. Recently a review carried out by Han et al. [276] summarized the capacity of urban forests to remove PM. The studies available show that urban trees have a high capacity to decrease air pollutants. In the USA, Nowak et al. [283] estimated that urban trees could remove 711,000 metric tons (US$3.8 billion value) of O_3, PM_{10}, NO_2, SO_2, CO. Nyelele et al. [284] observed that trees in Bronx (New York, USA) in 2010 removed approximately 5.1 tons of $PM_{2.5}$. Tallis et al. [285] found that urban canopy in the Greater London Authority (UK) could capture annually between 852 and 2,121 tons of PM_{10}. Jim and Chen [286] observed in Guangzhou (China) that the annual removal of PM SO_2, NO_2 was 312.03 Mg. The effectiveness of pollutants captures increased with increasing tree density. Wu et al. [287] reported that in 2015 (Shenzhen City, China), urban vegetation removed a total of 1,000.1 tons of $PM_{2.5}$. In the central area of Beijing (China), Yang et al. [288] found that 2.4 million trees captured 1,261.4 tons of pollutants (PM_{10}, O_3, SO_2, and NO_2) from the air in 2002.

In Canada, Nowak et al. [289] observed that in 2010 urban trees removed 16,500 tons of air pollution (CO, NO_2, O_3, $PM_{2.5}$, and SO_2) that had a human health benefit of 227.2 million Canadian dollars. Forests located near urban areas can mitigate the effects of extreme events. For instance, they can reduce PM transport during dust storms, as observed in Israel [290].

Urban forests (part of GI) can store and sequestrate a large amount of C. This is especially relevant in urban areas where anthropogenic activities release significant quantities of CO_2. The capacity to sequestrate C varies according to the species and is influenced by temperature, precipitation, groundwater level, and surface water level (in riparian forests). For instance, forests in tropical climates absorb C faster than in temperate areas. Also, mature trees, forests with a high diversity have a high capacity to sequestrate C [290–294]. Nevertheless, this capability is strongly related to local conditions. The urban heat island effect decreases the tree's capacity to sequester C [295]. Urban forests can sequester high amounts of C, as shown in Table 2, and the capacity to uptake CH_4 and N_2O [296]. In this context, urban forests play an essential role in climate change mitigation in urban areas, mainly because cities are responsible for 40–70% of worldwide GHG emissions [297]. For instance, De la Sota et al. [298] found that the GI management implemented in Lugo (Spain) contributed to a carbon uptake of 0.26 tons ha^{-1}. Doukalianou et al. [299] observed that in peri-urban forests of Xanthi (Greece), the coniferous forests thinned with high intensity have a high capacity to mitigate climate change.

Green roofs provide many ecosystem services in urban areas (e.g., food supply, microclimate regulation, flood regulation, water purification, and recreation) and can reduce atmosphere pollutants and climate change [300]. Their economic value is considerable. In Lisbon, the green roof's net value is €320 million [301]. The benefits of green roofs to mitigate climate change can be direct (green roof layers) or indirect (building energy consumptions) [302]. For instance, green roofs indirectly decrease air pollution by decreasing the urban heat island and building energy demand [302]. They can be classified as intensive or extensive depending on their purposes, and their efficiency varies according to the building materials [303]. Green roofs' capacity to capture pollutants is higher than conventional roofs [304]. In Chicago (USA), Yang et al. [305] observed that a green roof area of 19.8 ha removed a total of 1,675 kg of air pollutants (O_3, NO_2, PM_{10}, and SO_2). Gourdji [306] found that in Montreal (Canada), the green roof vegetation type affected air pollutant removal capacity. Pinus species were more efficient in removing PM, O_3, and NO_2 than other species. The capacity of green roofs to sequestrate C was observed in various urban areas such as Chiba (Japan) [307], Phitsanulok (Thailand) [308], Haifa (Israel) [309], Tartu (Estonia) [310], Thessaloniki (Greece) [311], Palma de Mallorca (Spain) [312], and Mexico City [313].

Other types of GI such as green walls [314], gardens [315], cemeteries [316], street trees [317], lawns [318] are known to have a positive impact on air quality and C sequestration. Despite the positive impact of GI in air pollutant removal and C sequestration, there are some trade-offs. Vegetation releases into the atmosphere by volatile biogenic compounds that can increase the amount of O_3 in the atmosphere and amplify the impact of O_3 pollutant events [319].

Table 2 Carbon sequestration by urban forests

Country	City	Type of forest	Period	Carbon sequestrated	Value	Reference
China	Shenyang	Ecological and public welfare forest; attached forest; landscape and relaxation forest; production and management forest and road forest	2006	337,000 tons	$US 13.88 million	Liu and Li [320]
	Hangzhou	*Pinus massoniana* and other tropical pines; *Cunninghamia lanceolata*; Evergreen, broad-leaved trees	1991–1996	1,328, 166.55 tons	–	Zhao et al., [321]
	[a]35 urban areas	–	2010	1.90 million tons	–	Chen [322]
India	Vadodara	–	–	73.59 tons		Kiran and Kinnary [323]
	Bilaspur	*Delonix Tamarindus indica, Ficus religiosa, Albizia lebbeck, Ficus benghalensis, Azadirachta indica, Peltophorum pterocarpum, Samanea saman,* and *Senna siamea*	2019	176.64 tons	–	Ragula and Chandra [324]
Thailand	Bangkok	*Polyalthia longifolia Sonn., Mangifera indica L.*, and *Pithecellobium dulce (Roxb.) Benth.* – Most important	2006	16,271 tons	–	Intasen et al. [325]
Turkey	Isparta	For several trees, please see Tuğluer et al. [326]	2015	1.9388 tons	–	Tuğluer et al. [326]
USA	Gainesville	Pine Rockland, mangrove, and Melaleuca quinquenervia	2005–2006	564,490 tons/year	–	Escobedo et al. [327]
	Miami-Dade		2008	54,566 tons/year	–	
	Several urban areas (29)[b]	–	Several years	25.6 million tons/year	$US2.0 billion	Nowak et al. [328]
	Auburn university campus	–	2009–2010	2,970 kg/year	–	Martin et al. [329]
	Scotlandville (Louisiana)	*Salix nigra, Quercus nigra,* and *Ulmus americana*	2014	3.880 tons	$ US 52.595	Ning et al. [330]
Canada	Several urban areas	–	1990	660.2 kt	–	Pasher et al. [331]
		–	2012	662.8 kt	–	McGovern and Pasher [332]

Argentina	Mendoza	*Morus alba*	2015	6117.4 tons	–	Martinez-Carretero et al. [333]
Spain	Barcelona	For different types of species, see Chaparro and Terradas [334]	2004	2.14 tons ha^{-1} $year^{-1}$	–	Chaparro and Terradas [334]
Italy	Rome	–	2010	3,197 $MgCO_2$ ha^{-1} $year^{-1}$	€23.537/ha.	Gratani et al. [335]
Korea	Chuncheon	*Pinus densiflora*, *Pinus koraiensis*, and *Quercus spp.* (*Q. mongolica and Q. aliena*)	–	1.25 ha^{-1} $year^{-1}$	–	Jo [336]
	Kangleung		–	1.47 ha^{-1} $year^{-1}$	–	
	Seoul-Jungnang		–	1.69 ha^{-1} $year^{-1}$	–	
	Daejeon	*Zelkova serrata, Platanus occidentalis, Ginkgo biloba, Chionanthus retusa*, and *Acer pseudo-sieboldianum*	–	216.31 tons/year	–	Park et al. [337]
South Africa	Tshwane	*Combretum erythrophyllum* (Burch.) *Sond., Searsia lancea*, and *Searsia pendulina (Jacq.)*	2002–2032	54,630 tons	$US 3,000,000	Stoffberg et al. [338]
Nigeria	Port Harcourt	–	–	67,979.08 tons	–	Agbelade et al. [339]
	Ilorin	–	–	91,512.49 tons	–	

The search was carried out in Google scholar (2000–2020). As criteria, only peer-reviewed works were considered. Keywords: Urban AND Forests AND Carbon AND sequestration

[a]Harbin; Changchun; Shenyang; Dalian; Urumqi; Xining; Lanzhou; Yinchuan; Xi'an; Kunming; Guiyang; Chongqing; Chengdu; Shijiazhuang; Taiyuan; Hohhot; Beijing; Tianjin; Ji'nan; Zhengzhou; Qingdao; Nanjing; Hefei; Shanghai; Hangzhou; Ningbo; Nanchang; Wuhan; Changsha; Fuzhou; Xiamen; Guangzhou; Nanning; Haikou; Shenzhen

[b]Arlington, TX (2009); Atlanta, GA (1997); Baltimore, MD (2009); Boston, MA (1996); Casper, WY (2006); Chicago, IL (2007); Freehold, NJ (1998); Gainesville, FL (2007); Golden, CO (2007); Hartford, CT (2007); Jersey City, NJ (1998); Lincoln, NE (2008–09); Los Angeles, CA (2007–08); Milwaukee, WI (2008); Minneapolis, MN (2004); Moorestown, NJ (2000); Morgantown, WV (2004); New York, NY (1996); Oakland, CA (1989); Omaha, NE (2008–09); Philadelphia, PA (1996); Roanoke, VA (2010); Sacramento, CA (2007); San Francisco, CA (CA); Scranton, PA (2006); Syracuse, NY (2009); Washington, DC (2004); Woodbridge, NJ (2000)

Soils can be a sink or a source of GHGs, depending on the management carried out. Urban soils usually are subjected to a high level of degradation [339]. Therefore, they can be a source of GHGs. For instance, the soil's capacity to sequester C decreases with the increasing sealing, age of the urbanization, compaction, reduced organic matter, and microbiological activity [340–343]. Lebed-Sharlevich et al. [344] found that urbic technosoils located in Moscow (Russia) released 2 times more CO_2 than natural soils. Similar results were identified in another Russian city (Kursk) by Sarzhanov et al. [345]. Liu et al. [342] observed that 70% of Beijing's soils (China) have a limited C sequestration capacity. Lu et al. [346] found that in Lahti (Finland), sealed soils emitted 15 times more C than park soils. In the USA, Milner and Ramaswami [347] observed that in the cities where urban sprawl is fast, the C sequestration is low. Riches et al. [348] reported that Melbourne's turf herbicide treatment (Australia) was associated with increased N_2O emissions. Also, Townsend-Small and Czimczik [349] found that irrigation practices and frequent fertilization in a turfgrass located in Irvine (USA) could not reduce the GHG emission.

Despite the reduced capacity of urban soils to GHG's sequestration, hotspots inside the urban areas (e.g., urban forests, gardens, green roofs) contribute to climate change mitigation [350]. Urban forests are key to increase soil C sequestration. Lv et al. [351] identified that the forest area in the city of Harbin (China) was key to retain soil C. The forest type and age have implications on soil GHG's sequestration. Setälä et al. [352] found that in urban parks in Helsinki and Lahti (Finland) the oldest parks with evergreen species can sequestrate more C and N. Urban agriculture carried out in rooftops and households contributes directly to increase C sequestration and reduce GHG emission by composting (e.g., solid waste, biochar) and indirectly by reducing the supply chain [315]. Llorach-Massana et al. [353] observed in a rooftop greenhouse located in Barcelona (Spain) that soilless crops reduced by half the GHG (C and N_2O) compared to conventional farming. In London (Sutton region), Kulak et al. [354] observed that if the unused areas in the urban fringe were used for food production, a reduction of 34 t CO2e ha − 1 a − 1 would be estimated. In Ouagadougou (Burkina Faso) and Tamale (Ghana), Häring et al. [355] found that the addition of rice husk biochar in an urban vegetable farm doubled soil C and increased N. Chen et al. [356] observed that the addition of biochar in a green roof soil located in Najing (China) increased the soil capacity to sequester C.

Wetlands can act as NBS to sequester C. However, this important benefit has a significant trade-off. N_2O and CH_4 emission, two very powerful GHG [357, 358]. In urban areas, several wetlands are constructed, and they provide several ecosystem services such as microclimate and flood regulation, C sequestration and recreation. Some are built to use a natural process in wastewater treatment. There are several types of constructed wetlands: (1) horizontal subsurface flow, (2) vertical subsurface flow, (3) free water surface, (4) surface flow, and (5) or a combination of several systems (hybrid). The emissions of CH_4 are high in free water surface and low in vertical subsurface flow. Concerning N_2O, the emissions are high in horizontal subsurface flow and low in vertical subsurface flow. Finally, the CO_2 emissions are high in subsurface flow and low free water surface. The best options to reduce

GHG emissions are hybrid systems [359]. The problem of constructed wetlands establishment is greenhouse gas emissions, especially CH_4 and N_2O [360, 361]. Constructed wetlands can act as a source or sink of GHG depending on the plant species used to vegetate (e.g., phenology and density of vegetation), wastewater flow and composition (chemical oxygen demand (COD)/nitrogen ratio), environmental conditions (e.g., temperature and solar radiation) and management [362]. Also, stormwater basins that are constructed to retain floods are a source of GHG. The problem is similar to the described in the constructed wetlands. For example, Gorsky et al. [363] found that stormwater ponds in southeastern Virginia (USA) were an important source of CH_4. Similar results were observed by McPhillips and Walter [364] in Ithaca (New York, USA). However, there are differences in the GHG emitted by constructed wetlands and stormwater retention basins. Badiou et al. [365] found that stormwater retention basins emitted more GHG than constructed wetlands. Some studies highlighted that these infrastructures are important contributors to GHG emission in urban areas [366, 367]. Although GHG emissions' negative aspects are a matter of concern, constructed wetlands and stormwater retention basins can sequester important amounts of C due to the sedimentation in their bottom. For instance, Stumpner et al. [368] observed a C burial in a constructed wetland located in Sacramento-San Joaquin Delta (California, USA) of 0.63 kg C m^{-2} $year^{-1}$. Merriman et al. [369] found that stormwater retention basins sequestered significant amounts of C in some areas located in the USA (78.4 g C m^{-2} $year^{-1}$), Singapore (135 g C m^{-2} $year^{-1}$), and Sweden (75.8 g C m^{-2} $year^{-1}$). There are important trade-offs in the construction of wetlands and stormwater retention areas that need to be minimized. The best approach to increase C sequestration and mitigate GHG remains an enigma [370].

3.2 Soil

As highlighted in previous sections, soils are an important sink of pollutants. However, the accumulation of toxic elements can represent a problem for human health. Therefore, it is vital to find NBS to reduce the number of pollutants in the soil. For example, several well-known techniques to remediate soil metals and metalloid pollution include electrokinetic extraction, capping, encapsulation, soil flushing, soil washing, stabilization, solidification, and landfilling vitrification, phytoremediation, and bioremediation [370]. These methods are divided into containment, extraction/removal, and solidification/stabilization (Fig. 6).

In this chapter, we are focused on NBS techniques. Therefore, the attention will be directed to extraction/removal techniques. Phytoremediation techniques were analysed in another chapter published in this book (chapter "The Role of Plants in Water Regulation and Pollution Control"). Therefore, here we will revise the other bioremediation approaches. Soil bioremediation can be divided into ex situ (biopile, bioreactor, land farming, and windrow) and in situ (biosparging, bioventing, bioslurping, biostimulation, bioaugmentation, composting, biochar, and

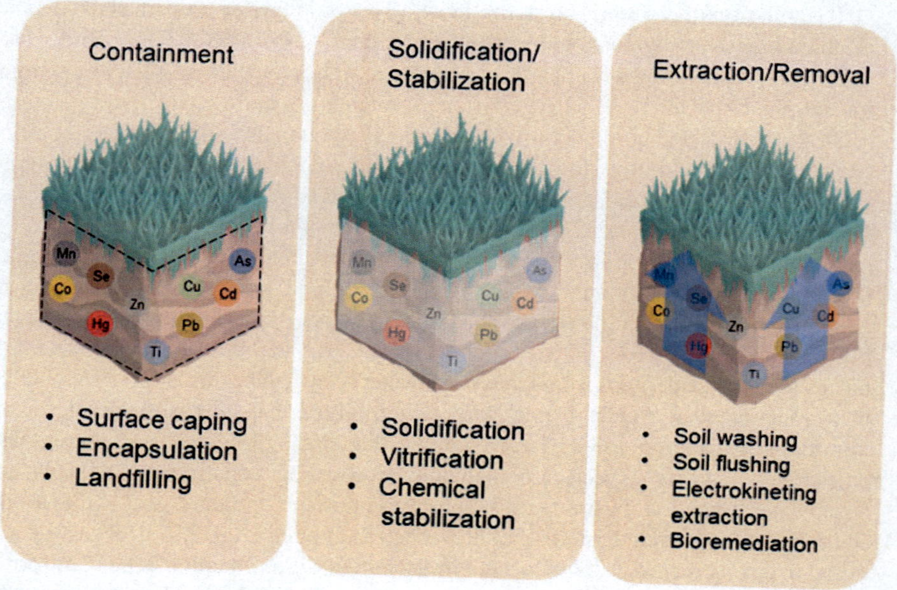

Fig. 6 Methodologies (NBS and non-NBS) applied for soil remediation. Based on: Liu et al. [371]

phytoremediation). Ex-situ techniques consist of excavating and removing the soil to be treated in another place, while in-situ, the remediation is carried out in the polluted site. The application of the different techniques has to consider the parent material of the site, location, cost of the treatment, and the type, degree, and depth of the polluted layer [372]. A detailed review of the different soil remediation methods was carried out by Ossai et al. [373]. Since this chapter focuses on NBS, we will be mainly focused on in-situ techniques. Nevertheless, although ex-situ methods are very destructive to soils, they showed high efficiency in removing hydrocarbons [374, 375], metals and metalloids [376], PAHs [377], emergent pollutants and emergent pollutants, i.e. pharmaceuticals [378] and pesticides [379].

In-situ biological techniques to remove pollutants induce a minimum disturbance in soil structure and are less costly than ex-situ measures. However, these techniques need a long time to effectively reduce soil contamination to acceptable levels (e.g., below recommendation levels). They are also not effective in all soil types, such as saline soils [380]. If soils have a high-level of contamination, the microbiological activity is reduced, decreasing biologic methods' effectiveness [373]. These limitations can be tackled by combining different methodologies (physical, chemical, and biological). This can increase the remediation efficiency. Thus, several works applied in soil remediation have different approaches [381–383]. Many works reviewed the effectiveness of the different bioremediation techniques in soil pollutant removal [372, 373, 380].

Biosparging stimulates microbial activities by injecting oxygen (and nutrients if needed) into soil subsurface. The oxygen is applied in the saturated zone, promoting the upward movement of pollutants to the unsaturated zone to be biodegraded. This method is applied to remove the pollutants adsorbed to the soil particulates located above the water table. It is widely applied in the removal of groundwater pollutants as well. This technique's success depends on two aspects: (1) pollutant biodegradability and (2) soil permeability [372, 384, 385]. According to EPA [386] and Gaur et al. [387], this method can offer some advantages such as (1) cheap technique; (2) under favourable conditions (e.g., reduced soil pollutants) the treatment can last from 6 months to 2 years; (3) soil disturbance is reduced; (4) easy to install; (5) promotes biodegradation at a high airflow rate, and 6) treats a large number of hydrocarbons. However, there are some shortcomings in the application of this method, such as (1) is only totally reliable in environments where air sparging is possible; (2) the relations between physical, chemical, and biological spheres are not totally understood, and (3) is difficult to predict the airflow direction. Biosparging is extensively applied in removing hydrocarbons [387].

Bioventing is a method similar to biosparging and was one of the first methods to be applied at a larger scale. Consists of delivering oxygen, moisture, and nutrients (e.g., N) through direct injection into the vadose zone and stimulating microbiological activity [372, 388, 389]. As biosparging, bioventing is very popular to remove petroleum products (e.g., kerosene, diesel fuel, gasoline). However, their effectiveness depends on the soil type and environmental conditions [390]. This technique has also been applied in soil contaminated by metals and metalloids [384] and pesticides [391]. Bioventing offers several advantages such as (1) easy to establish; (2) cheap technique; (3) easy to combine with other techniques; (4) reduced soil disturbance; (5) low air injection enhances biological degradation, and (6) can be applied under anaerobic conditions. The disadvantages of applying bioventing are (1) reduced efficiency in removing pollutants to an acceptable level (meet legislative requirements); (2) not suitable in certain soil conditions (e.g., high clay content and low permeability), and (3) requires substantial time to be effective [386, 387].

Bioslurping combines different approaches, such as bioventing, soil vapour extraction, and vacuum enhanced pumping. As in biosparging and bioventing, oxygen is provided to stimulate the microbes to degrade the pollutants. It is a method mainly applied to remediate soils polluted with light non-aqueous phase liquids (e.g., benzene, ethyl benzene, styrene) in unsaturated and saturated zones (mainly from the capillary fringe). However, it can also be applied in soils polluted with VOC [372]. The main advantage of applying this method is because it is cheap. However, there are several disadvantages, such as (1) it is very dependent on soil water content since reduced soil moisture decreases bioremediation effectiveness; (2) high soil moisture limits oxygen penetration; (3) not efficient in impermeable soils; (4) at reduced temperatures, the remediation is slow, and (5) can extract a large amount of water that may need post-treatment [392].[14] Despite the shortcomings, Kim et al.

[14]https://frtr.gov/matrix2/section4/4-35.html (Consulted in 10-12-2020).

[393] found that bioslurping successfully reduced soil contaminates (e.g., petroleum hydrocarbons) in a period of 2 years.

Biostimulation technique is based on the release of several materials (e.g., nutrients, amendments, biopolymers, bio-surfactants, slow-release fertilizers, oxygen) and water to increase microbiological biodegradation. It increases soil limiting elements (e.g., N, P, C, Potassium (K)) in reduced concentrations in contaminated soils. It is considered one of the most successful methods for removing hydrocarbons. Nevertheless, it needs the presence of the correct microorganisms to deliver the precise dose of nutrients (C:N:P-30:5:1) and be efficient [373, 394]. The lack of adequate nutrients to stimulate microbiological activity can delay biostimulation's capacity to degrade soil pollutants [395]. There are several advantages of applying this technique: (1) it is carried out by native microorganisms, (2) it is cheap and can be applied in all polluted areas, and (3) it does not have negative implications on the environment. However, it has several shortcomings: (1) it depends very much on the environmental factors (e.g., soil pH, temperature, moisture); (2) if the pollutants are highly attached to soil particles and the pollutants are not biodegradable, this technique cannot be applied, and (3) the results depend very much on the site where it is applied and it requires a high investment in monitoring [394]. Despite the limitations, biostimulation has been successful in the removal of metals and metalloids [396], petroleum products [397], herbicides [395], PAHs, and pharmaceuticals [398].

Bioaugmentation involves the addition of different microbiological communities (genetically modified, allochthonous, or autochthonous), bacteria, or fungi with the capacity to degrade pollutants [373, 387, 399]. The increasing of soil microbes increases the degradation of pollutants in the area where it is applied. These methods' success depends on the microbes' capacity to adapt to the polluted site and soil physical and chemical properties (e.g., type, aeration, pH, temperature, organic matter, nutrients content) and compete with the native biota. The selection of the microbial consortia is key to the efficiency of this technique. Usually, they have to be adaptable to a medium with high pollutants, easy cultivating, and fast-growing. For instance, one way to increase the remediation's success is to isolate several bacteria from the polluted soil, culturing them in a laboratory environment to increase their pre-adaptation and return these bacteria to the polluted soil. This process is known as soil reinoculation with native bacteria. Bioaugmentation is an advantageous method. However, it has some limitations in soils with reduced moisture [394]. This method was applied in different environments, and it was effective in the removal of petroleum products [400], PAHs [401], pharmaceuticals [402], and pesticides [399] .

Composting can be considered an NBS to remove soil pollutants. This method involves the addition of nutrients, tilling, watering, and microbes in organic waste. The composting process requires temperatures between 50 and 65°C to compost soils with petroleum products. The remediation of polluted soils using this approach is achieved with frequent aeration, tillage, and watering [373]. For the process to be efficient, the initial pH, C/N ratio, aeration rate, moisture content, temperature, soil/compost ratio, and soil type need to be monitored [403]. The three most important designs of composting are (1) aerobic static pile; (2) produces the compost in a bioreactor, and (3) windrow compost. From all the methods, windrow compost is the most cost-effective. In this technique, different organic wastes and agricultural materials are added (e.g., straw, manure).[15] Soil composting increases organic matter, improves soil properties, and enhances microbiological activity. The high organic matter content can also increase the pollutants' adsorption and reduce their solubility [404]. The composting process converts, binds, and degrades pollutants and transforms them into unhazardous components [373]. However, the method has several shortcomings such as (1) imposes a high soil disturbance (e.g., excavation); (2) high demand for compost space and labour; (3) in windrow compost, there is a risk of dust emissions; (4) in the presence of volatile or semi-volatile organic compounds air emissions need to be monitored; (5) residual that is not degraded needs careful disposal; (6) during the degradation process odours, greenhouse gases, and toxins can be emitted, and (7) some types of compost may have metals and metalloids or dioxins [405][15]. Composting has been used in the removal of several soil pollutants such as petroleum hydrocarbons [403], PAHs [405], metals and metalloids [406], and pharmaceuticals [407].

Biochar application in soils is recognized to have several environmental benefits such as mitigating GHG emissions, sequestering C, soil amendment, and pollutants immobilization [408, 409]. Biochar is created at temperatures between 400 and 700°C with reduced oxygen. In these combustion conditions, solid residue or biochar, gases, and liquids are produced. Biochar is rich in C, volatile matter, mineral matter, moisture, and very recalcitrant. Biochar's physical structure is rich in pores with different dimensions that favour microbiological colonization. It has a reduced bulk density, high pH, and base nutrients (e.g., Ca, Mg, N). The application of biochar in soils increases aeration, water content, and increases nutrients [410–412]. Overall, the application of biochar on soils has several advantages, such as soil fertility and soil water content improvement, reduced GHG emissions, and waste recycling. Nevertheless, there are several shortcomings: (1) increase of soil alkalinity, harmful to some plants; (2) inhibition of plant germination; (3) the increase of soil nutrients depends on the soil where biochar is applied. Normally the effectiveness of biochar is reduced in soils with high clay content; (4) the high capacity of biochar to retain nutrients decreases the amount of these elements to plants (e.g., N); (5) some types of biochars can be rich in PAHs, VOCs, or dissolved organic C; (6) aged biochar in the soil has negative impacts on earthworms, fungi and reduces root biomass; (7) may not be suitable to all soils; (8) pollutants adsorption is selective, and (9) biochar is composed by fine ash that can impose problems related to health (e.g., respiratory problems) [413–416]. Despite the mentioned drawbacks, biochar has been applied as a solution to soil remediation, and a large number of works confirmed its efficiency to remove/immobilize metals and metalloids [417–419]. This capacity to remove/immobilize was also observed in PAHs [420], petroleum products [421], pesticides [422], herbicides [423], and pharmaceuticals [424].

[15] http://www.cpeo.org/techtree/ttdescript/compost.htm (Accessed in 10-12-2020).

Fig. 7 Constructed wetlands located in Vilnius (Lithuania)

Microplastics became a global problem, and there is a need to develop NBS that tackles these problems. Some methodologies have been developed using microbes, enzymes [425, 426], and fungi [427]. Nevertheless, most of these works were carried out in the laboratory, and field experiments are needed to see these methods' effectiveness.

The revision of soil remediation methods was not intensive. Previous works did it in a detailed manner [372, 373, 387]. However, the techniques described applied individually or combined can be a very good NBS to reduce soil pollutants in urban areas and reduce the probability of water bodies contamination.

3.3 Water

Water bodies in urban environments provide a vast number of ecosystem services. One of the crucial regulating ecosystem services provided by wetlands is water chemicals regulation. Urban runoff usually has a high concentration of various toxic pollutants that normally end in water bodies, contributing to biodiversity loss and water quality [428, 429]. Therefore, there is a need to decrease the impact of urban runoff on water quality. Natural or constructed wetlands (Fig. 7) are important to pollutant retention and recycling, decreasing water treatment costs [430]. They prevent pollutants' transport to other water bodies such as lakes or rivers [431, 432]. However, despite the importance of constructed wetlands to provide an important number of ecosystem services, if the number of pollutants transported is very high, they can become toxic and threaten the fauna and flora. This is especially serious in industrial constructed wetlands [433].

Constructed wetlands are very beneficial to stormwater treatment and can provide many ecosystem services such as the breakdown of organic compounds, nutrients sequestration, sediment retention, water storage, and peak-flow reduction [434]. These infrastructures present several advantages such as (1) are easy to maintain; (2) have a long lifetime; (3) have reduced operational costs, and (4) can treat a wide range of pollutants. However, they have some disadvantages: (1) have, to some extent, a stochastic behaviour and (2) are land-intensive infrastructures [435]. The different types of constructed wetlands and their importance for C sequestration were described in Air sub-section. These infrastructures have been widely used with success to treat wastewater and are considered a sustainable option [436]. Constructed wetlands are intricate systems where there is a complex interaction between a substrate, water, plants, microorganisms, and water. The different processes (physical, chemical, and biological) such as microbial degradation, photodegradation, plant uptake, sedimentation, sorption, and volatilization can coincide. Their effectiveness depends on the climate where they are installed. Constructed wetlands established in warmer zones have higher efficiency in removing pollutants than in the cold ones. The reduced temperature harms their efficiency. Also, pollutant removal performance depends on the age, macrophytes presence and type, microorganisms, subtract that support the macrophytes, pH, dissolved oxygen, design, artificial aeration, effluent recirculation, operational mode. Overall, the constructed wetland design depends on the pollutants' characteristics that need to be removed [437–441]. Previous works did a deep revision about the importance of the constructed wetlands for urban stormwater runoff [434], environmental pollution control [442], metals and metalloids accumulation in plant tissues [443], oxygen supply [444], greywater recycle and reuse [445], microbial nitrogen removal [446], landfill leachate treatment [447] and the fate and removal of (1) organic matter and nitrogen [448]; (2) P [449]; (3) petroleum products [450]; (4) boron [451]; (5) Fe and sulphur cycling [452]; (6) pharmaceuticals [453–455] and (7) pesticides [456]. Therefore, much information is available about the impacts of constructed wetlands in pollutants removal.

Constructed wetlands have high efficiency in retaining suspended solids, biological oxygen demand (BOD), and COD. For example, Vyzmal [457] found that a gravel-based horizontal flow constructed wetlands are efficient to remove suspended solids. If maintained and loaded correctly, these infrastructures can maintain good performance for at least 20 years. Also, Koskiaho and Puustinen [440] observed that older constructed wetlands have a high capacity of retaining solids than younger ones due to the more abundant vegetation. In Brazil, Benvenuti et al. [458] reported that a constructed floating wetland removed with efficiency 78, 56, and 55% of suspended solids, BOD and COD, respectively. For instance, the type of material has an important impact on the constructed wetland's capacity to remove pollutants. Khalifa et al. [459] found that the use of polystyrene foam in constructed wetlands increased the removal efficiency of suspended solids (from 83 to 88.5%), BOD (72 to 88%), and COD (71% to 88%) when compared with the materials used (rubber, plastic, and gravel) without polystyrene foam. Hybrid constructed wetlands (subsurface, vertical flow, and horizontal flow beds) have a high capacity to remove pollutants. In Poland, Gizińska-Górna et al. [460] observed that within 3 years, a hybrid, constructed wetland could remove a high number of organic compounds. On average, 92.7, 96.6, and 95% of suspended solids, BOD, and COD, respectively. Also, in China, Zheng et al. [461] reported that a constructed wetland reduced the amount of COD (74.5%) and BOD (94.4%) that reach an urban river.

Constructed wetlands retain high amounts of N, P, and K, decreasing the impact of urban runoff and wastewater in water bodies eutrophication. Bai et al. [462] reported that a floating-bed constructed wetland reduced from April to October 2016 substantial amounts of ammonia (80.90%), total N (71.12%), and total P (78.44%) to reach an urban river. Plants in constructed wetlands have a high capacity to uptake chemical elements. Schwammberger et al. [463] found that in a constructed floating wetland built for stormwater, plants uptake large amounts of nutrients from the water. In the 16-month study, plants removed 20.20 ± 2.88 and 15.00 ± 2.07 kg of total N and 12.59 ± 1.64 and 7.20 ± 1.56 of total K. In Australia, a constructed wetland decreased 2 to 4 times the amount of N and P reaching downward streams [464]. Zheng et al. [465] reported that a hybrid constructed wetland (e.g., constructed with local sand, gravel and slag) built in China could remove on average 69.2% of total P, 56.3% of ammonia, and 57.5% of total N, decreasing the number of nutrients that reached an urban river. The capacity of removal was high in the autumn and low in the winter. Sedimentation processes play an important role in nutrient retention in constructed wetlands. For instance, Griffiths and Mitsch [466] observed in a constructed wetland for urban stormwater treatment in Naples, Florida, that P's sedimentation rates were 7.8 g m^{-2} $year^{-1}$ and N were 81.7 g m^{-2} $year^{-1}$. In this context, constructed wetlands must be designed to minimize sediment resuspension. In Uganda (Africa), Kabenge et al. [467] observed in a constructed wetland built for stormwater runoff and colonized by cattail (*Typha latifolia*) and bulrush (*Scirpus lacustris*) thats important amounts of total N (72.8%) and total P (62.8%) were removed 8 days after the hydraulic retention.

As mentioned previously, stormwater runoff contains many metals and metalloids, and constructed wetlands act as a sink, preventing them from reaching water bodies. Metals and metalloids immobilization is specially made by adsorption onto sediments and plant uptake [468]. Walaszek et al. [469] observed that in a stormwater-constructed wetland located in Strasbourg, eastern France, during 3-year monitoring (2015–2017) there was a reduction above 97% of Cr, Co, Cu, Pb, and Zn. Also, the capacity of the constructed wetland to retain metals and metalloids depends on the storm dimension. Maniquiz-Redillas and Kim [470] reported that the capacity of constructed wetlands (Cheonan, Korea) to retain metals and metalloids is higher in larger storms than in the small storms. However, the highest number of pollutants was transported during small storms. Cr, Ni, Cu, and Cd were difficult to retain because they were mainly dissolved. Pb, Fe, and Zn were easier to immobilize because they have a high capacity to bind onto sediments. Plants also have a high capacity to retain metals and metalloids. In Egypt, *Typha latifolia* and *Cyperus papyrus* substrate with zeolite planted in a constructed wetland could remove 72 and 84% of Cu and Zn, respectively [471]. Also, Vymazal and Březinová [456] found that *Phragmites australis* can sequester many metals and metalloids.

Petroleum products (e.g., diesel, oil, gasoline) are present in soils and roads due to traffic and are easily transported in the runoff. Constructed wetlands can remove these organic compounds at a low cost [435] up to 80–90% [450]. For instance, Tromp et al. [472] found that a constructed wetland that drained a motorway near

Amsterdam (The Netherlands) monitored during 18 months retained a very high amount of PAHs (90–95%). Schmitt et al. [473] reported that all the PAHs in urban runoff were retained in a sedimentation pond located in Strasbourg (France). In Greece, two subsurface flow and two free water surface constructed wetlands that drained a highway's runoff retained 59% of PAHs in a two-year experiment [474]. Constructed wetlands vegetation has a very good capacity to degrade PAHs. Qin et al. [475] observed that submerged macrophytes have an important role in PAH's degradation.

Constructed wetlands have a high potential to remove pharmaceuticals. This capacity depends on the season, hydraulic mode, pH, temperature, oxygen, and redox potential and design [476]. A comprehensive review of the mechanisms involved in pharmaceutical removal was carried out recently by Vo et al. [477]. Zhang et al. [478] found that constructed wetlands (Jinan, China) could remove up to 80–90% of pharmaceuticals. Subsurface flow constructed wetland type had the highest capacity to remove naproxen, gemfibrozil, and ibuprofen, while surface flow removed high amounts of diclofenac and ketoprofen. In Ukraine, Vystavna et al. [479] reported a decrease in paracetamol (5%), diclofenac (97%), caffeine (80%), and triclosan (88%) reaching an urban river during an experiment carried out between 2012 and 2015. Also, Park et al. [480] found that soil organic matter has a high capacity to retain several pharmaceuticals such as atenolol, ibuprofen, and carbamazepine. As in previous pollutants, vegetation proved a high capacity to immobilize pollutants. Hijosa-Valsero et al. [481] reported that Typha angustifolia and Phragmites australis can retain a high amount of pharmaceutical and personal care products in their root system. Finally, as in the previous pollutants, several works highlighted the high effectiveness of constructed wetlands to remove pesticides [482] and herbicides [483].

4 Conclusions

Urban activities have important impacts on air, soil, and water quality. These effects are extremely detrimental to human health and biodiversity, increasing the degradation of urban areas. GHG emission and the pollution imposed by the release of PM, CO, O_3, SO_2 and NO_x, metals and metalloids and microplastics to the atmosphere are associated with increased health economic losses, respiratory diseases, mortality, and crop losses. The pollutants released from industrial and traffic activities are deposited in soil, increasing their toxicity. High levels of pollutants reduce the soil capacity to provide ecosystem services in quality and quantity and impose high stress on plants development and affect directly (e.g., skin contact) or indirectly (e.g., food consumption) health. Since soils in urban areas are sealed or highly compacted, all the pollutants are easily transported in the runoff, degrading surface and groundwater reserves' quality.

It is urgent to find solutions to reduce our footprint, and NBS is key to reducing the impact of anthropogenic activities. Also, the establishment of these measures will increase the ecosystem services quality and quantity. Despite the stress imposed by pollutants in plant development, they have a high capacity to remove pollutants from the atmosphere, especially urban forests and green roofs that store and sequester a large amount of C. Soils also have a high capacity to sequester C. However, depending on the management, they can be a source as well. In urban forests, the potential for urban soils sequesters C is high compared to other urban soils. Wetlands can positively affect C sequestration; however, they are important sources of potent GHG, such as CH_4 and N_2O. Bioremediation techniques (including biosparging, bioventing, bioslurping, biostimulation, bioaugmentation, composting, biochar, and phytoremediation) can serve as NBS to reduce soil contamination. There are a vast number of methods used in industrial and urban areas. Most of them are based on microbiological processes. When different methods are combined (physical, chemical, and biological), the effectiveness of removing soil pollution is high. Urban areas produce a large amount of waste and stormwater. Constructed wetlands are one of the most widespread NBS used to retain pollutants and treat water. In many cases, they is a very efficient strategy to reduce the number of pollutants that reach rivers and other water bodies.

Pollution negatively affects plants, soils, and water by decreasing their functions. However, they are very efficient in reducing human activities' impact by capturing, transforming, and consuming the wide range of pollutants produced. It is a paradox that we are degrading nature with urban growth and can still minimize our effects through nature. Future studies should be focused on how the GBI used as NBS is negatively affected by acting as a pollutant buffer.

Acknowledgments Authors were partly funded by the Lithuanian National Ecosystem Services Assessment and Mapping (LINESAM) Project, which has received funding from the European Social Fund project LINESAM no. 09.3.3-LMTK- 712-01-0104, under a grant agreement with the Research Council of Lithuania (LMTLT).

References

1. Lewis AC (2018) The changing face of urban air pollution. Science 359:744–745
2. Pereira P, Barcelo D, Panagos P (2020) Soil and water threats in a changing environment. Environ Res 186:109501
3. EEA (2016) Urban sprawl in Europe. Joint EEA-FOEN report. Luxembourg Publications Office of the European Union
4. Wei YD, Ewing R (2018) Urban expansion, sprawl and inequality. Landsc Urban Plann 177:259–265
5. Hatab AA, Cavinato MER, Lindemer A, Lagerkvist CJ (2019) Urban sprawl, food security and agricultural systems in developing countries: a systematic review of the literature. Cities 94:129–142
6. Li G, Li F (2019) Urban sprawl in China: differences and socioeconomic drivers. Sci Total Environ 673:367–377
7. Cadenas-Rodriguez M, Dupont-Courtade L, Oueslati W (2016) Air pollution and urban structure linkages: evidence from European cities. Renew Sust Energ Rev 53:1–9

8. Sarkodie S, Owusu PA, Leirvik T (2020) Global effect of urban sprawl, industrialization, trade and economic development on carbon dioxide emissions. Environ Res Lett 15:034049
9. Yuan Y, Chen D, Wu S, Mo L, Tong G, Yan D (2019) Urban sprawl decreases the value of ecosystem services and intensifies the supply scarcity of ecosystem services in China. Sci Total Environ 697:134170
10. Hankey S, Marshall JD (2017) Urban form, air pollution, and health. Curr Environ Health Rep 4:491–503
11. Viana M, de Leeuw F, Bartonova A, Castell N, Ozturk E, Gonzalez Ortiz A (2020) Air quality mitigation in European cities: status and challenges ahead. Eviron Int 143:105907
12. Liu G, Xiao M, Zhang X, Gal C, Chen X, Liu L, Pan S, Wu J, Tang L Clements-Croome D (2017) A review of air filtration technologies for sustainable and healthy building ventilation. Sustain Cities Soc 32:375–396
13. Sadr SMK, Saroj DP, Mierzwa JC, McGrane SJ, Skouteris G, Farmani R, Kazos X, Aumeier B, Kouchaki B, Ouki SK (2018) A multi expert decision support tool for the evaluation of advanced wastewater treatment trains: a novel approach to improve urban sustainability. Environ Sci Pol 90:1–10
14. Tan PY, Wang J, Sia A (2013) Perspectives on five decades of the urban greening of Singapore. Cities 32:24–32
15. Brilhante O, Klaas J (2018) Green city concept and a method to measure green city performance over time applied to fifty cities globally: influence of GDP, population size and energy efficiency. Sustainability 10:2031
16. Haase D (2015) Reflections about blue ecosystem services in cities. Sustain Water Q Ecol 5:77–83
17. Andreucci MB, Russo A, Olszewska-Guizzo A (2019) Designing urban green blue infrastructure for mental health and elderly wellbeing. Sustainability 11:6425
18. He BJ, Zhu J, Zhao DX, Gou ZH, Qi JD, Wang J (2019) Co-benefits approach: opportunities for implementing sponge city and urban heat island mitigation. Land Use Policy 86:147–157
19. Zuniga-Teran AA, Gerlak AK, Mayer B, Evans TP, Lansey KE (2020) Urban resilience and green infrastructure systems: towards a multidimensional evaluation. Curr Opin Environ Sustain 44:42–47
20. Alves A, Vojinovic Z, Kapelan Z, Sanchez A, Gersonius B (2020) Exploring trade-offs among the multiple benefits of green-blue-grey infrastructure for urban flood mitigation. Sci Total Environ 703:134980
21. Mulligan J, Buckachi V, Clause JC, Jewell R, Kirimi F, Odbert C (2020) Hybrid infrastructures, hybrid governance: new evidence from Nairobi (Kenya) on green-blue-grey infrastructure in informal settlements. Anthropocene 29:100227
22. Gopalakrishnan V, Ziv G, Hirabayashi S, Bakshi BR (2019) Nature-based solutions can compete with technology for mitigating air emissions across the United States. Environ Sci Technol 53:13228–13237
23. Xing Y, Jones P, Donnison I (2017) Characterisation of nature-based solutions for the built environment. Sustainability 9:149. https://doi.org/10.3390/su9010149
24. Liquete C, Udias A, Conte G, Grizzetti B, Masi F (2016) Integrated valuation of a nature-based solution for water pollution control. Highlighting hidden benefits. Ecosyst Serv 22:392–401
25. Okta Pribadi D, Pauleit S (2016) Peri-urban agriculture in Jabodetabek metropolitan area and its relationship with the urban socioeconomic system. Land Use Policy 55:265–274
26. Francos M, Ubeda X, Pereira P (2020) Impact of bonfires on soil properties in an urban park in Vilnius (Lithuania). Environ Res 181:108895
27. Tresch S, Frey D, Le Bayon RC, Zanetta A, Rasche F, Fliessbach A, Moretti M (2019) Litter decomposition driven by soil fauna, plant diversity and soil management in urban gardens. Sci Total Environ 658:1614–1629
28. Pereira P (2020) Ecosystem services in a changing environment. Sci Total Environ 702:135008

29. WHO (2016) Ambient air pollution: a global assessment of a burden disease. WHO, Geneve, p 131
30. IQAIR (2020) 2019 world air quality report. Region & City PM2.5 ranking. https://www.iqair.com/
31. Byrne B, Strong K, Colebatch O, You Y, Wunch D, Ars S, Jones DBA, Fogal P, Mittermeier RL, Worthy D, Griffith T (2020) Monitoring urban greenhouse gases using open-path Fourier transform spectroscopy. Atmos Ocean 58:25–45
32. Mitchell LE, Lin JC, Bowling DR, Pataki DE, Strong C, Schauer AJ, Bares R, Bush SE, Stephans BB, Mendoza D, Malilla D, Holland L, Gurney KR, Ehleringer JR (2018) Long-term urban carbon dioxide observations reveal spatial and temporal dynamics related to urban characteristics and growth. Proc Natl Acad Sci U S A 20:2912–2917
33. Gratani L, Varone L (2005) Daily and seasonal variation of CO2 in the city of Rome in relationship with the traffic volume. Atmos Environ 39:2619–2624
34. Zanni AM, Al B (2010) Emissions of CO2 from road freight transport in London: trends and policies for long run reductions. Energy Policy 38:1774–1786
35. WMO (2018) Greenhouse gas bulletin. The state of greenhouse gases in the atmosphere based on global observations through 2017 14, 8 p
36. Gonzalez RM, Marrero GA, Rodrigiez-Lopez J, Marrero AS (2019) Analyzing CO2 emissions from passenger cars in Europe: a dynamic panel data approach. Energy Policy 129:1271–1281
37. EEA (2019) Annual European Union greenhouse gas inventory 1990–2017 and inventory report 2019. European Commission DG Climate Action BU 5 2/158 B-1049 Brussels
38. Gregorczyk-Maga I, Maga M, Wachsmann A, Janik MK, Chrzastek-Janik O, Bajkowski M, Partyka P, Koziej M (2019) Air pollution may affect the assessment of smoking habits by exhaled carbon monoxide measurements. Environ Res 172:258–265
39. Borsdorf T, De Brugh JA, Hu H, Aben I, Hasekamp O, Landgraf J (2018) Measuring carbon monoxide with TROPOMI: first results and a comparison with ECMWF-IFS analysis data. Geophys Res Lett 45:2826–2832
40. Wang P, Elansky NF, Timofeev YM, Wang G, Golitsyn GS, Makarova MV, Rakitin VS, Shtabkin Y, Skorokhod AI, Grechko EI, Fokeeva EV, Safronov AN, Ran L, Wang T (2018) Long-term trends of carbon monoxide Total columnar amount in urban areas and background regions: ground- and satellite-based spectroscopic measurements. Adv Atmos Sci 35:785–795
41. Hůnová I, Bäumelt V, Modlík M (2020) Long-term trends in nitrogen oxides at different types of monitoring stations in the Czech Republic. Sci Total Environ 699:134378
42. Forster P, Ramaswamy V, Artaxo P, Berntsen T, Betts R, Fahey DW, Haywood J, Lean J, Lowe DC, Myhre G, Nganga J, Prinn R, Raga G, Schulz M, Van Dorland R (2007) Changes in atmospheric constituents and in radiative forcing. In: Solomon S, Qin D, Manning M, Chen Z, Marquis M, Averyt KB, Tignor M, Miller HL (eds) Climate change 2007: the physical science basis. Contribution of working group i to the fourth assessment report of the intergovernmental panel on climate change. Cambridge University Press, Cambridge
43. Ito A, Tohjama Y, Saito T, Umezawa T, Hajaima T, Hirata R, Saito M, Terao Y (2019) Methane budget of East Asia, 1990–2015: a bottom-up evaluation. Sci Total Environ 676:40–42
44. Wunch D, Jones DBA, Toon GC, Deutscher NM, Hase F, Notholt J, Sussmann R, Warneke T, Kuenen J, van der Gon HD, Fisher JA, Maasakkers JD (2019) Emissions of methane in Europe inferred by total column measurements. Atmos Chem Phys 19:3963–3980
45. Wang YS, Zhou L, Wang MX, Zheng XH (2001) Trends of atmospheric methane in Beijing. Chemosph Global Change Sci 3:65–71
46. Wong CK, Pongetti TJ, Oda T, Rao P, Gurney KR, Newman S, Duren RM, Miller CE, Yung YL, Sander SP (2016) Methane budget of East Asia, 1990–2015: a bottom-up evaluation. Atmos Chem Phys 16:13121–13130
47. Jing X, Mira D, Cluff DL (2018) The combustion mitigation of methane as a non-CO_2 greenhouse gas. Prog Energy Combust Sci 66:176–199

48. Yeung LY, Murray LT, Martinerie P, Witrant E, Hu H, Banerjee A, Orsi A, Chappellaz J (2019) Isotopic constraint on the twentieth-century increase in tropospheric ozone. Nature 570:224–227
49. van Dingenen R, Crippa M, Maenhout G, Guizzardi D, Dentener F (2018) Global trends of methane emissions and their impacts on ozone concentrations. EUR 29394 EN, Publications Office of the European Union, Luxembourg
50. Gao W, Tie X, Xu J, Huang R, Mao X, Zhou G, Chang L (2017) Long-term trend of O3 in a mega City (Shanghai), China: characteristics, causes, and interactions with precursors. Sci Total Environ 603–604:425–433
51. Liu Y, Wang T (2020) Worsening urban ozone pollution in China from 2013 to 2017 – part 2: the effects of emission changes and implications for multi-pollutant control. Atmos Chem Phys 20:6323–6337
52. Chang LK, Choi JY, Lee S, Lee D, Jo YL, Kim CH (2020) Interpretation of decadal-scale ozone production efficiency in the Seoul metropolitan area: implication for ozone abatement. Atmos Environ 243:117846
53. Ahamad F, Griffths PT, Latif MT, Juneng L, Xiang CJ (2020) Ozone trends from two decades of ground level observation in Malaysia. Atmos 11:755
54. Chou CKK, Liu SC, Lin CY, Shiu CJ, Chang KH (2006) The trend of surface ozone in Taipei, Taiwan, and its causes: implications for ozone control strategies. Atmos Environ 21:3898–3908
55. Fernández-Guisuraga F, Castro A, Alves C, Calvo A, Alonso-Blanco E, Blanco-Alegre C, Rocha A, Fraile R (2016) Nitrogen oxides and ozone in Portugal: trends and ozone estimation in an urban and a rural site. Environ Sci Pollut Res 23:17171–17182
56. Querol X, Alastuey A, Reche C, Orio A, Pallares M, Reina F, Dieguez JJ, Mantilla E, Escudero M, Alonso L, Gangoiti G, Millan M (2016) On the origin of the highest ozone episodes in Spain. Sci Total Environ 572:379–389
57. Venkata VSV, Kommalapati RR, Huque Z (2018) Long-term meteorologically independent trend analysis of ozone air quality at an urban site in the greater Houston area. J Air Waste Manag Assoc 68:1051–1064
58. Vingarzan R, Taylor B (2003) Trend analysis of ground level ozone in the greater Vancouver/Fraser Valley area of British Columbia. Atmos Environ 37:2159–2171
59. Hůnová I, Bäumelt V (2018) Observation-based trends in ambient ozone in the Czech Republic over the past two decades. Atmos Environ 172:157–167
60. Sicard P, Serra R, Rossello P (2016) Spatiotemporal trends in ground-level ozone concentrations and metrics in France over the time period 1999–2012. Environ Res 149:122–144
61. Mavroidis I, Ilia M (2012) Trends of NOx, NO2 and O3 concentrations at three different types of air quality monitoring stations in Athens, Greece. Atmos Environ 63:135–147
62. Munir S, Chen H, Ropkins K (2013) Quantifying temporal trends in ground level ozone concentration in the UK. Sci Total Environ 458–460:217–227
63. Faridi S, Shamsipour M, Krzyzanowski M, Künzli N, Amini H, Malkawi M, Momeniha F, Gholampour A, Sadegh Hassanvand M, Naddafi K (2018) Long-term trends and health impact of PM2.5 and O3 in Tehran, Iran, 2006–2015. Environ Int 114:37–49
64. Bloomer BJ, Stehr JW, Piety CA, Salawitch RJ, Dickerson RR (2009) Observed relationships of ozone air pollution with temperature and emissions. Geophys Res Lett:36. https://doi.org/10.1029/2009GL037308
65. Balmes JR, Eisner MD (2016) 74 – indoor and outdoor air pollution. In: Broaddus VC, Mason RJ, Ernst JD, King Jr TD, Lazarus SC, Murray JF, Nadel JA, Slutsky AS, Gotway MB (eds) Textbook of respiratory mediine, vol 2. 6th edn. Elsevier, Amsterdam, pp 1331–1342.e5
66. Aas W, Mortier A, Bowersox V, Cherian R, Faluvegi G, Fagerli H, Hand J, Klimont Z, Galy-Lacaux C, Lehmann CMB, Lund Myhre C, Myhre G, Olivié D, Sato K, Quaas J, Rao PSP, Schulz M, Shindell D, Skeie RB, Stein A, Takemura T, Tsyro S, Vet S, Xu X (2019) Global and regional trends of atmospheric sulfur. Sci Rep 9:953

67. Ray S, Kim KH (2014) The pollution status of sulfur dioxide in major urban areas of Korea between 1989 and 2010. Atmos Res 147–148:101–110
68. Sun S, Zhao G, Wang T, Jin J, Wang P, Lin Y, Li H, Ying Q, Mao H (2019) Past and future trends of vehicle emissions in Tianjin, China, from 2000 to 2030. Atmos Environ 209:182–191
69. Binyehmed FM, Abdullah AM, Zainal Z, Zawawi RM, Elhadi Elawad RE (2016) Trend and status of SO2 pollution as a corrosive agent at four different monitoring stations in the Klang Valley, Malaysia. Int J Adv Sci Tech Res 6:302–317
70. Jamaati H, Attarchi M, Hassani S, Farid E, Seyedmehdi SM, Pormehr PS (2018) Investigating air quality status and air pollutant trends over the metropolitan area of Tehran, Iran over the past decade between 2005 and 2014. Environ Health Toxicol 33:e2018010
71. Gil-Alana LA, Yaya OS, Carmona-Gonzalez N (2020) Air quality in London: evidence of persistence, seasonality and trends. Theor Appl Climatol 142:103–115
72. Brito J, Bernardo A, Zagalo C, Goncalves LL (2021) Quantitative analysis of air pollution and mortality in Portugal: current trends and links following proposed biological pathways. Sci Total Environ 755:142473
73. Munir S, Habeebullah TM, Seroji AR, Gabr SS, Mohammed AMF, Morsy EA (2013) Quantifying temporal trends of atmospheric pollutants in Makkah (1997–2012). Atmos Environ 77:647–655
74. Chaudhiri S, Dutta D (2014) Mann–Kendall trend of pollutants, temperature and humidity over an urban station of India with forecast verification using different ARIMA models. Environ Monit Assess 186:4719–4742
75. Ladim AA, Teixeira EC, Agudelo-Castaneda D, Schneider I, Silva LFO, Wiegant F, Kumar P (2018) Spatio-temporal variations of sulfur dioxide concentrations in industrial and urban area via a new statistical approach. Air Qual Atmos Health 11:801–813
76. EEA (2019) Air quality in Europe – 2019 report. EEA report. Publications Office of the European Union No 10/2019, Luxembourg
77. Wang K, Tian H, Hua S, Zhu C, Gao C, Xue Y, Hao J, Wang Y, Zhou J (2016) A comprehensive emission inventory of multiple air pollutants from iron and steel industry in China: temporal trends and spatial variation characteristics. Sci Total Environ 559:7–14
78. GESAMP (2015) Sources, fate and effects of microplastics in the marine environment: a global assessment. In: Kershaw PJ (ed) IMO/FAO/UNESCO-IOC/UNIDO/WMO/IAEA/UN/UNEP/UNDP Joint Group of Experts on the Scientific Aspects of Marine Environmental Protection. Rep. Stud. GESAMP No. 90, 96 p
79. Wu WM, Yang J, Criddle CS (2017) Microplastics pollution and reduction strategies. Front Environ Sci Eng 11:6
80. Prata JC (2018) Airborne microplastics: consequences to human health? Environ Pollut 234:115–126
81. Liu K, Wang X, Wei N, Song Z, Li D (2019) Accurate quantification and transport estimation of suspended atmospheric microplastics in megacities: implications for human health. Environ Int 132:105127
82. Abbasi S, Keshavarzi B, Moore F, Turner A, Kelly FJ, Dominguez AO, Jaafarzadeh N (2019) Distribution and potential health impacts of microplastics and microrubbers in air and street dusts from Asaluyeh County, Iran. Environ Pollut 244:153–164
83. Akhbarizadeh R, Dobaradaran S, Amouei Torkmahalleh M, Saeedi R, Aibaghi R, Faraji Ghasemi F (2021) Suspended fine particulate matter (PM2.5), microplastics (MPs), and polycyclic aromatic hydrocarbons (PAHs) in air: their possible relationships and health implications. Environ Res 192:110339
84. World Bank (2016) The cost of air pollution strengthening the economic case for action. World Bank, Washington
85. Yin H, Xu L (2018) Comparative study of PM10/PM2.5-bound PAHs in downtown Beijing, China: concentrations, sources, and health risks. J Clean Prod 177:674–683

86. Rosofsky A, Levy JI, Zanobetti A, Janulewicz P, Fabian MP (2018) Temporal trends in air pollution exposure inequality in Massachusetts. Environ Res 161:76–86
87. Aleksandropoulou V, Lazaridis M (2017) Trends in population exposure to particulate matter in urban areas of Greece during the last decade. Sci Total Environ 581–582:399–412
88. Union E (2018) Air pollution: our health still insufficiently protected. European Court of Auditors, Luxembourg
89. Liu M, Huang Y, Ma Z, Jin Z, Liu X, Wang H, liu Y, Wang J, Jantunen m BJ, Kinney PL (2017) Spatial and temporal trends in the mortality burden of air pollution in China: 2004–2012. Environ Int 98:75–81
90. Shaddick G, Thomas ML, Mudu P, Ruggeri G, Gumy S (2020) Half the world's population are exposed to increasing air pollution. NPJ Clim Atmos Sci 3:23
91. Cohen AJ, Brauer M, Burnet R, Anderson HR, Frostad J, Estep K, Balakrishnan K, Brunekreef B, Dandona L, Dandona R, Feigin V, Freedman G, Hubbell B, Jobling A, Kan H, Knibbs L, Liu Y, Martin R, Forouzanfar MH (2017) Estimates and 25-year trends of the global burden of disease attributable to ambient air pollution: an analysis of data from the global burden of diseases study 2015. Lancet 389:1907–1918
92. Carugno M, Consonni D, Alberto Bertazzi P, Biggeri A, Baccini M (2017) Temporal trends of PM10 and its impact on mortality in Lombardy, Italy. Environ Pollut 227:280–286
93. Zhang C, Ding R, Xiao C, Xu Y, Cheng H, Zhu F, Lei R, Di D, Zhao Q, Cao J (2017) Association between air pollution and cardiovascular mortality in Hefei, China: a time-series analysis. Environ Pollut 229:790–797
94. Maji KJ, Arora M, Dikshit AK (2017) Burden of disease attributed to ambient $PM_{2.5}$ and PM_{10} exposure in 190 cities in China. Environ Sci Pollut Res 24:11559–11572
95. Chooi Y, Kim H, Lee JT (2018) Temporal variability of short term effects of PM10 on mortality in Seoul, Korea. Sci Total Environ 644:122–128
96. Zhang Y, Peng M, Yu C, Zhang L (2017) Burden of mortality and years of life lost due to ambient PM_{10} pollution in Wuhan, China. Environ Pollut 230:1073–1080
97. Gu Y, Lin H, Liu H, Xiao J, Zeng W, Li Z, Lv X, Ma W (2017) The interaction between ambient PM_{10} and NO_2 on mortality in Guangzhou, China. Int J Environ Res Public Health 14:1381
98. Maesano CN, Morel G, Matynia A, Ratsombath N, Bonnety J, Legros G, Da Costa P, Prud'homme J, Annesi-Maesano I (2020) Impacts on human mortality due to reductions in PM10 concentrations through different traffic scenarios in Paris, France. Sci Total Environ 698:134257
99. Greven FE, Vonk JM, Fischer P, Duijm F, Vink NM, Brunekreef B (2019) Air pollution during new Year's fireworks and daily mortality in the Netherlands. Sci Rep 9:5735
100. Zoran MA, Savastru RS, Savastru DM, Tautan MN (2020) Assessing the relationship between surface levels of PM2.5 and PM10 particulate matter impact on COVID-19 in Milan, Italy. Sci Total Environ 738:139825
101. Chu B, Zhang S, Liu J, Ma Q, He H (2021) Significant concurrent decrease in PM2.5 and NO2 concentrations in China during COVID-19 epidemic. J Environ Sci 99:346–353
102. Hashim BM, Al-Naseri SK, Al-Maliki A, Al-Ansari N (2021) Impact of COVID-19 lockdown on NO_2, O_3, $PM_{2.5}$ and PM_{10} concentrations and assessing air quality changes in Baghdad, Iraq. Sci Total Environ 754:141978
103. Ju MJ, Oh J, Choi YH (2021) Changes in air pollution levels after COVID-19 outbreak in Korea. Sci Total Environ 750:141521
104. Jonson JE, Borken-Kleefeld J, Simpson D, Nyíri A, Posch M, Heyes C (2017) Impact of excess NOx emissions from diesel cars on air quality, public health and eutrophication in Europe. Environ Res Lett 12:094017
105. Olawoyin R, Schweitzer L, Zhang K, Okareh O, Slates K (2018) Index analysis and human health risk model application for evaluating ambient air-heavy metal contamination in Chemical Valley Sarnia. Ecotoxicol Environ Saf 148:72–81

106. Munawer ME (2018) Human health and environmental impacts of coal combustion and post-combustion wastes. J Sustain Min 17:87–96
107. Feng Z, De Marco A, Anav A, Gualtieri M, Sicard P, Tian H, Fornasier F, Tao F, Guo A, Paoletti E (2019) Economic losses due to ozone impacts on human health, forest productivity and crop yield across China. Environ Int 131:104966
108. Zhao Y, Hu J, Tan Z, Liu T, Zeng W, Li X, Huang C, Wang S, Huang Z, Ma W (2019) Ambient carbon monoxide and increased risk of daily hospital outpatient visits for respiratory diseases in Dongguan, China. Sci Total Environ 668:254–260
109. Barn P, Giles L, Héroux ME, Kosatsky T (2018) A review of the experimental evidence on the toxicokinetics of carbon monoxide: the potential role of pathophysiology among susceptible groups. Environ Health 17:13
110. Chossière GP, Malina R, Allroggen F, Eastham SD, Speth RL, Barrett SRH (2018) Country- and manufacturer-level attribution of air quality impacts due to excess NOx emissions from diesel passenger vehicles in Europe. Atmosph Environ 189:89–97
111. Rovira J, Domingo JL, Schuhmacher M (2020) Air quality, health impacts and burden of disease due to air pollution (PM10, PM2.5, NO2 and O3): application of AirQ+ model to the Camp de Tarragona County (Catalonia, Spain). Sci Total Environ 703:135538
112. Thilakaratne RA, Malig MJ, Basu R (2020) Examining the relationship between ambient carbon monoxide, nitrogen dioxide, and mental health-related emergency department visits in California, USA. Sci Total Environ 746:140915
113. Simonsen C, Thorsteinsson K, Nørmark Mortensen R, Torp-Pedersen C, Torp-Pedersen P, Andreasen JJ (2019) Carbon monoxide poisoning in Denmark with focus on mortality and factors contributing to mortality. Plos One 17. https://doi.org/10.1371/journal.pone.0210767
114. Chossière GP, Malina R, Ashok A, Dedoussi IC, Eastham SD, Speth RL, Barrett SRH (2017) Public health impacts of excess NOx emissions from Volkswagen diesel passenger vehicles in Germany. Environ Res Lett 12:034014
115. Lin Y, Jiang F, Zhao J, Zhu G, He X, Ma X, Li S, Sabel CE, Wang H (2018) Impacts of O_3 on premature mortality and crop yield loss across China. Atmosph Environ 194:41–47
116. Wu Y, Cui L, Meng Y, Cheng H, Fu H (2020) The high-resolution estimation of sulfur dioxide (SO2) concentration, health effect and monetary costs in Beijing. Chemosphere 241:125031
117. Liu Q, Harris JT, Chiu LS, Sun D, Houser PR, Yu M, Duffy DQ, Little MM, Yang C (2021) Spatiotemporal impacts of COVID-19 on air pollution in California, USA. Sci Total Environ 750:141592
118. Xiang J, Austin E, Gould T, Larson T, Shirai J, Liu Y, Marshall J, Seto E (2020) Impacts of the COVID-19 responses on traffic-related air pollution in a Northwestern US city. Sci Total Environ 747:141325
119. Gasperi J, Wright SL, Dris R, Collard F, Mandin C, Guerrouache M, Kelly FJ, Tassin B (2018) Microplastics in air: are we breathing it in? Curr Opin Environ Sci Health 1:1–5
120. Prata JC, Costa JP, Lopes I, Duarte AC, Rocha-Santos T (2020) Environmental exposure to microplastics: an overview on possible human health effects. Sci Total Environ 702:134455
121. Gao M, Guttikunda SK, Carmichael GR, Wang Y, Liu Z, Stanier CO, Saide PE, Yu M (2015) Health impacts and economic losses assessment of the 2013 severe haze event in Beijing area. Sci Total Environ 511:553–561
122. Zhao X, Yu X, Wang Y, Fan C (2016) Economic evaluation of health losses from air pollution in Beijing, China. Environ Sci Pollut Res 23:11716–11728
123. Gu Y, Wong TW, Law CK, Dong GH, Ho KF, Yang Y, Yim SHL (2018) Impacts of sectoral emissions in China and the implications: air quality, public health, crop production, and economic costs. Environ Res Lett 13:084008
124. Tai APK, Martin MV (2017) Impacts of ozone air pollution and temperature extremes on crop yields: spatial variability, adaptation and implications for future food security. Atmosph Environ 169:11–21
125. Liu X, Sun H, Feike T, Zhang X, Shao L, Chen S (2016) Assessing the impact of air pollution on grain yield of winter wheat – a case study in the North China plain. Plos One 11:e0162655

126. Brevik E, Slaughter L, Steffan J, Collier D, Barnhardt P, Pereira P (2020) Soil and human health: current status and future needs. Air Soil Water Res 13:1–23
127. Boente C, Matanzas N, Garcia-Gonzalez N, Rodriguez-Valdez E, Gallego JR (2017) Trace elements of concern affecting urban agriculture in industrialized areas: a multivariate approach. Chemosphere 183:546–556
128. Dala-Paula BM, Custodio FB, Knupp EAN, Palmieri HEL, Silva JBB, Gloria MBA (2018) Cadmium, copper and lead levels in different cultivars of lettuce and soil from urban agriculture. Environ Pollut 242:383–389
129. Ercilla-Montserrat M, Munoz P, Montero JI, Gabarrell X, Rieradevall J (2018) A study on air quality and heavy metals content of urban food produced in a Mediterranean city (Barcelona). J Clean Prod 195:385–395
130. Guan Q, Wang F, Xu C, Pan N, Lin J, Zhao R, Yang Y, Luo H (2018) Source apportionment of heavy metals in agricultural soil based on PMF: a case study in Hexi Corridor, northwest China. Chemosphere 193:187–197
131. Huang Y, Chen Q, Deng M, Japenga J, Li T, Yang X, He Z (2018) Heavy metal pollution and health risk assessment of agricultural soils in a typical peri-urban area in Southeast China. J Environ Manag 207:159–168
132. Nam KM, Li M, Wang Y, Wong KKH (2018) Spatio-temporal boundary effects on pollution-health costs estimation: the case of $PM_{2.5}$ pollution in Hong Kong. Int J Urban Sci 23:498–518
133. Zhang M, Song Y, Cai X (2004) A health-based assessment of particulate air pollution in urban areas of Beijing in 2000–2004. Sci Total Environ 376:100–108
134. Du Y, Li T (2016) Assessment of health-based economic costs linked to fine particulate ($PM_{2.5}$) pollution: a case study of haze during January 2013 in Beijing, China. Air Qual Atmos Health 9:439–445
135. Kan H, Chen B (2004) Particulate air pollution in urban areas of Shanghai, China: health-based economic assessment. Sci Total Environ 322:71–79
136. Zhang D, Aunan K, Seip HM, Larssen S, Liu S, Zhang D (2010) The assessment of health damage caused by air pollution and its implication for policy making in Taiyuan, Shanxi, China. Energy Policy 38:491–502
137. Lu X, Yao T, Fung JCH, Lin C (2016) Estimation of health and economic costs of air pollution over the Pearl River Delta region in China. Sci Total Environ 566–567:134–143
138. Renjie C, Bing Heng C, Hai Dong K (2010) A health-based economic assessment of particulate air pollution in 113 Chinese cities. China Environ Sci 30:410–415
139. Lee YJ, Lim YW, Yang JY, Kim CS, Shin YC, Shin DC (2011) Evaluating the PM damage cost due to urban air pollution and vehicle emissions in Seoul, Korea. J Environ Manag 92:603–609
140. Etchie TO, Sivanesan S, Adewuyi GO, Krishnamurthi K, Rao PS, Etchie AT, Pillarisetti A, Arora AT, Smith KR (2017) The health burden and economic costs averted by ambient PM2.5 pollution reductions in Nagpur, India. Environ Int 102:145–156
141. Bayat R, Ashrafi K, Shafiepour Motlagh M, Sadegh Hassanvand M, Daroudi R, Fink F, Künzli N (2019) Health impact and related cost of ambient air pollution in Tehran. Environ Res 176:108547
142. Vlachokostas C, Achillas C, Moussiopoulos N, Kalogeropoulos K, Sigalas G, Kalognomou EA, Banias G (2012) Health effects and social costs of particulate and photochemical urban air pollution: a case study for Thessaloniki, Greece. Air Qual Atmos Health 5:325–334
143. Monzon A, Guerrero MJ (2004) Valuation of social and health effects of transport-related air pollution in Madrid (Spain). Sci Total Environ 334–335:427–434
144. Sanchez Martinez G, Spadaro JV, Chapizanis D, Kendrovski V, Kochubovski M, Mudu P (2018) Health impacts and economic costs of air pollution in the metropolitan area of Skopje. Int J Environ Res Public Health 15:626
145. Curvelo Santana JC, Carvalho Miranda A, Kenji Yamamura CL, Catureba da Silva Filho S, Tambourgi EB, Lee Ho L, Tobal Berssaneti F (2020) Effects of air pollution on human health and costs: current situation in São Paulo, Brazil. Sustainability 12:4875

146. Miraglia SG, Saldiva PHN, Bohm GM (2005) An evaluation of air pollution health impacts and costs in Sao Paulo, Brazil. Environ Manag 35:667–676
147. Levy JI, Buonocore JJ, von Stackelberg K (2010) Evaluation of the public health impacts of traffic congestion: a health risk assessment. Environ Health 9:65
148. FAO (2015) Status of the World's soil resources. FAO, Rome
149. Rodríguez-Eugenio N, McLaughlin M, Pennock D (2018) Soil pollution: a hidden reality. FAO, Rome
150. Pereira P, Ferreira AJD, Pariente S, Cerda A, Walsh R, Keestra S (2016) Urban soils and sediments. J Soils Sediments 16:2493–2499
151. European Union (2013) Evaluation of expenditure and jobs for addressing soil contamination in member states. Final report to the European Commission, Directorate-General Environment
152. Liu L, Liu Q, Ma Q, Wu H, Qu Y, Gong Y, Yang Y, An Y, Zhou Y (2020) Heavy metal(loid)s in the topsoil of urban parks in Beijing, China: concentrations, potential sources, and risk assessment. Environ Pollut 260:114083
153. Qu Y, Gong Y, Ma J, Wei H, Liu Q, Liu L, Wu H, Yang S, Chen Y (2020) Potential sources, influencing factors, and health risks of polycyclic aromatic hydrocarbons (PAHs) in the surface soil of urban parks in Beijing, China. Environ Pollut 260:114016
154. Lee J, Han MH, Kim EH, Lee CW, Jeong HS (2020) Assessment of radionuclide deposition on Korean urban residential area. J Radiat Protect Res 45:101–107
155. Toth G, Hermann T, Szatmári G, Pásztor L (2016) Maps of heavy metals in the soils of the European Union and proposed priority areas for detailed assessment. Sci Total Environ 565:1054–1062
156. Jiang B, Adebayo A, Jia J, Xing Y, Deng S, Guo S, Liang Y, Zhang D (2019) Impacts of heavy metals and soil properties at a Nigerian e-waste site on soil microbial community. J Hazard Mater 362:187–195
157. Xu Y, Seshadri B, Bolan N, Sarkar B, Ok YS, Zhang W, Rumpel C, Sparks D, Farrell M, Hall T, Dong Z (2019) Microbial functional diversity and carbon use feedback in soils as affected by heavy metals. Environ Int 125:478–488
158. Morgado RG, Loureiro S, González-Alcaraz MN (2018) Changes in soil ecosystem structure and functions due to soil contamination. In: Duarte AC, Cachada A, Rocha-Santos T (eds) Soil pollution. From monitoring to remediation. Elsevier, Amsterdam, pp 59–87
159. Ma J, Ullah S, Niu A, Liao Z, Qin Q, Xu S, Lin C (2020) Heavy metal pollution increases CH_4 and decreases CO_2 emissions due to soil microbial changes in a mangrove wetland: microcosm experiment and field examination. Chemosphere:128735
160. Geng S, Cao W, Yuan J, Wang Y, Guo Y, Ding A, Zhu Y, Dou J (2020) Microbial diversity and co-occurrence patterns in deep soils contaminated by polycyclic aromatic hydrocarbons (PAHs). Ecotoxicol Environ Saf 203:110931
161. Hoyos-Hernandez C, Courbert C, Simonucci C, David S, Vogel TM, Larose C (2019) Community structure and functional genes in radionuclide contaminated soils in Chernobyl and Fukushima. FEMS Microbiol Lett 366:fnz180
162. Gałązka A, Grządziel J, Gałązka R, Ukalska-Jaruga A, Strzelecka J, Smreczak B (2018) Genetic and functional diversity of bacterial microbiome in soils with long term impacts of petroleum hydrocarbons. Front Microbiol 9:1923
163. Pino-Otín MR, Muniz S, Val J, Navarro E (2017) Effects of 18 pharmaceuticals on the physiological diversity of edaphic microorganisms. Sci Total Environ 595:441–450
164. Wiedner K, Polifka S (2020) Effects of microplastic and microglass particles on soil microbial community structure in an arable soil (Chernozem). Soil 6:315–324
165. Du Z, Zhu Y, Zhang J, Li B, Wang J, Wang J, Zhang C, Cheng C (2018) Effects of the herbicide mesotrione on soil enzyme activity and microbial communities. Ecotoxicol Environ Saf 164:571–578
166. Tripathi S, Srivastava P, Devi RS, Bhadouria R (2020) Influence of synthetic fertilizers and pesticides on soil health and soil microbiology. In: Vara Prasad MN (ed) Agrochemicals

detection, treatment and remediation pesticides and chemical fertilizers. Elsevier, Amsterdam, pp 25–54
167. Meftaul IM, Venkateswarlu K, Dharmarajan R, Annamalai M, Megharaj M (2020) Pesticides in the urban environment: a potential threat that knocks at the door. Sci Total Environ 711:134612
168. Gholizadeh A, Kopačková V (2019) Detecting vegetation stress as a soil contamination proxy: a review of optical proximal and remote sensing techniques. Int J Environ Sci Technol 16:2511–2524
169. Kumar A, Aery NC (2016) Impact, metabolism, and toxicity of heavy metals in plants. In: Singh A, Prasad S, Singh R (eds) Plant responses to xenobiotics. Springer, Singapore, pp 141–176
170. Rajput VD, Minkina TM, Behal A, Sushkova SN, Mandzhieva SN, Singh R, Gorovtsov A, Tsitsuashvili VS, Purvis WO, Ghazaryan KO, Movsesyan HS (2018) Effects of zinc-oxide nanoparticles on soil, plants, animals and soil organisms: A review. Environ Nanotechnol Monit Manag 9:76–84
171. Qin S, Liu H, Nie Z, Rengel Z, Gao W, Li W, Zhao P (2020) Toxicity of cadmium and its competition with mineral nutrients for uptake by plants: A review. Pedosphere 30:168–180
172. Sachan P, Lal N (2017) An overview of nickel (Ni2+) essentiality, toxicity and tolerance strategies in plants. Asian J Biol 2:1–15
173. Xiong TT, Austruy A, Pierart A, Shahid M, Schreck E, Mombo S, Dumat C (2016) Kinetic study of phytotoxicity induced by foliar lead uptake for vegetables exposed to fine particles and implications for sustainable urban agriculture. J Environ Sci 46:16–17
174. Abbas G, Murtaza B, Bibi I, Sahid M, Niazi NK, Khan MI, Amjad M, Hussain M (2018) Arsenic uptake, toxicity, detoxification, and speciation in plants: physiological, biochemical, and molecular aspects. Int J Environ Res Public Health 15:59
175. Samsøe-Petersen L, Larsen HK, Larsen PB, Bruun P (2002) Uptake of trace elements and PAHs by fruit and vegetables from contaminated soils. Environ Sci Technol 36:3057–3063
176. Zhang S, Yao H, Lu Y, Yu X, Wang J, Sun S, Liu M, Li D, Li YF, Zhang D (2017) Uptake and translocation of polycyclic aromatic hydrocarbons (PAHs) and heavy metals by maize from soil irrigated with wastewater. Sci Rep 7:12165
177. Tian L, Yin S, Ma Y, Kang H, Zhang X, Tan H, Meng H, Liu C (2019) Impact factor assessment of the uptake and accumulation of polycyclic aromatic hydrocarbons by plant leaves: morphological characteristics have the greatest impact. Sci Total Environ 652:1149–1155
178. Kummerová M, Zezukla S, Babula P, Vanova L (2013) Root response in Pisum sativum and Zea mays under fluoranthene stress: morphological and anatomical traits. Chemosphere 90:665–673
179. Desalme D, Binet P, Chiapusio G (2013) Challenges in tracing the fate and effects of atmospheric polycyclic aromatic hydrocarbon deposition in vascular plants. Environ Sci Technol 47:3967–3981
180. Carvalho PN, Basto MCP, Almeida CMR, Brix H (2014) A review of plant–pharmaceutical interactions: from uptake and effects in crop plants to phytoremediation in constructed wetlands. Environ Sci Pollut Res 21:11729–11763
181. Sauvêtre A, Schröder P (2015) Uptake of carbamazepine by rhizomes and endophytic bacteria of *Phragmites australis*. Front Plant Sci 6:83
182. Carter LJ, Harris E, Williams M, Ryan JJ, Kookana RS, Boxall ABA (2014) Fate and uptake of pharmaceuticals in soil–plant systems. J Agric Food Chem 62:816–825
183. Marsoni M, De Mattia F, Labra F, Bruno A, Bracale M, Vannini C (2014) Uptake and effects of a mixture of widely used therapeutic drugs in *Eruca sativa L. and Zea mays L.* plants. Ecotoxicol Environ Saf 108:52–57
184. Carter LJ, Williams M, Böttcher C, Kookana RS (2015) Uptake of pharmaceuticals influences plant development and affects nutrient and hormone homeostases. Environ Sci Technol 49:12509–12518

185. Bartrons M, Peñuelas J (2017) Pharmaceuticals and personal-care products in plants. Trends Plant Sci 22:194–203
186. Sun C, Dudley S, Trumble J, Gan J (2018) Pharmaceutical and personal care products-induced stress symptoms and detoxification mechanisms in cucumber plants. Environ Pollut 234:39–47
187. Valenca DC, Campos de Lelis DC, Pinho CF, Mendes Bezerra AC, Ferreira MA, Gama Junqueira NE, Macrae A, Medici LO, Reinert F, Silva BO (2020) Changes in leaf blade morphology and anatomy caused by clomazone and saflufenacil in Setaria viridis, a model C4 plant. S Afr J Bot 135:365–376
188. Timms KP, Wood LJ (2020) Sub-lethal glyphosate disrupts photosynthetic efficiency and leaf morphology in fruit-producing plants, red raspberry (Rubus idaeus) and highbush cranberry (Viburnum edule). Glob Ecol Conserv 24:e01319
189. Zaller JG, Cantelmo C, Dos Santos G, Muther S, Gruber E, Pallua P, Mandl K, Friedrich B, Hofstetter I, Schmuckenschlager B, Faber F (2018) Herbicides in vineyards reduce grapevine root mycorrhization and alter soil microorganisms and the nutrient composition in grapevine roots, leaves, xylem sap and grape juice. Environ Sci Pollut Res 25:23215–23226
190. Liu N, Zhong G, Zhou J, Liu Y, Pang Y, Cai H, Wu Z (2019) Separate and combined effects of glyphosate and copper on growth and antioxidative enzymes in Salvinia natans (L.) all. Sci Total Environment 655:1448–1456
191. Fei X, Lou Z, Christakos G, Ren Z, Liu Z, Lv X (2018) The association between heavy metal soil pollution and stomach cancer: a case study in Hangzhou City, China. Environ Geochem Health 40:2481–2490
192. Jia Z, Li S, Wang L (2018) Assessment of soil heavy metals for eco-environment and human health in a rapidly urbanization area of the upper Yangtze Basin. Sci Rep 8:3256
193. Ingaramo P, Alarcón R, Muñoz-de-Toro M, Luque EH (2020) Are glyphosate and glyphosate-based herbicides endocrine disruptors that alter female fertility? Mol Cell Endocrinol 518:110934
194. Peillex C, Pelletier M (2020) The impact and toxicity of glyphosate and glyphosate-based herbicides on health and immunity. J Immunotoxicol 17:163–174
195. Alshahri F (2019) Natural and anthropogenic radionuclides in urban soil around non-nuclear industries (Northern Al Jubail), Saudi Arabia: assessment of health risk. Environ Sci Pollut Res 26:36226–36235
196. Gbadamosi MR, Banjoko OO, Abudu KA, Ogunbanjo OO, Ogunneye AL (2017) Radiometric evaluation of excessive lifetime cancer probability due to naturally occurring radionuclides in wastes dumpsites soils in Agbara, southwest, Nigeria. J Assoc Arab Univ Basic Appl Sci 24:315–324
197. Keshavarzi B, Najmeddin A, Moore F, Afshari Moghaddam P (2019) Risk-based assessment of soil pollution by potentially toxic elements in the industrialized urban and peri-urban areas of Ahvaz metropolis, southwest of Iran. Ecotoxicol Environ Saf 167:365–375
198. Xiong TT, Dumat C, Dappe V, Vezin H, Schreck E, Shahid M, Piertart A, Sobanska S (2017) Copper oxide nanoparticle foliar uptake, phytotoxicity, and consequences for sustainable urban agriculture. Environ Sci Technol 51:5242–5251
199. Fazeli G, Karbassi A, Khoramnejadian S, Nasrabadi T (2019) Evaluation of urban soil pollution: A combined approach of toxic metals and polycyclic aromatic hydrocarbons (PAHs). Int J Environ Res 13:801–811
200. Ferreira AJD, Mendes Guilherme RIM, Ferreira CSS, Lorena de Oliveira MFM (2018) Urban agriculture, a tool towards more resilient urban communities? Curr Opin Environ Sci Health 5:93–97
201. Ferreira C, Walsh RPD, Ferreira AJD (2018) Degradation in urban areas. Curr Opin Environ Sci Health 5:19–25
202. UN-Water (2018) Sustainable development goal 6. Synthesis report on water and sanitation. New York. 195 pp

203. EEA (2018) European waters. Assessment of status and pressures 2018. Publications Office of the European Union, Luxembourg
204. Zhang Q, Wu Z, Guo G, Zhang H, Tarolli P (2020) Explicit the urban waterlogging spatial variation and its driving factors: the stepwise cluster analysis model and hierarchical partitioning analysis approach. Sci Total Environ. https://doi.org/10.1016/j.scitotenv.2020.143041
205. Islam Shajib HT, Bruun Hansen HC, Liang T, Holm PE (2019) Metals in surface specific urban runoff in Beijing. Environ Pollut 248:584–598
206. Dietrich M, Wolfe A, Burke M, Krekeler MPS (2019) The first pollution investigation of road sediment in Gary, Indiana: anthropogenic metals and possible health implications for a socioeconomically disadvantaged area. Environ Int 128:175–192
207. García-Rivero AE, Olivera J, Sallinas E, Yuli RA, Bulege W (2017) Use of Hydrogeomorphic indexes in SAGA-GIS for the characterization of flooded areas in Madre de Dios, Peru. Int J Appl Enge Res 12:9078–9086
208. UNEP (2016) A snapshot of the world's water quality: towards a global assessment. United Nations Environment Programme, Nairobi, p 162
209. Liu J, Guo LC, Luo XL, Chen FR, Zeng EY (2014) Impact of anthropogenic activities on urban stream water quality: a case study in Guangzhou, China. Environ Sci Pollut Res 21:13412–13419
210. Wilbers GJ, Becker M, Nga LT, Sebesvari Z, Renaud FG (2014) Spatial and temporal variability of surface water pollution in the Mekong Delta, Vietnam. Sci Total Environ 485–486:653–665
211. Zhu L, Chen B, Wang J, Shen H (2004) Pollution survey of polycyclic aromatic hydrocarbons in surface water of Hangzhou, China. Chemosphere 56:1085–1095
212. Zhou F, Huang G, Guo H, Zhang W, Hao Z (2007) Spatio-temporal patterns and source apportionment of coastal water pollution in eastern Hong Kong. Water Res 41:3429–3439
213. Suthar S, Sharma J, Chabukdhara C, Nema AK (2010) Water quality assessment of river Hindon at Ghaziabad, India: impact of industrial and urban wastewater. Environ Monit Assess 165:103–112
214. Revitt DM, Ellis JB (2016) Urban surface water pollution problems arising from misconnections. Sci Total Environ 551–552:163–174
215. Peng Y, Fang W, Krauss M, Brack W, Wang Z, Li F, Zhang X (2018) Screening hundreds of emerging organic pollutants (EOPs) in surface water from the Yangtze River Delta (YRD): occurrence, distribution, ecological risk. Environ Pollut 241:484–493
216. Montes-Grajales D, Fennix-Agudelo F, Miranda-Castro W (2017) Occurrence of personal care products as emerging chemicals of concern in water resources: A review. Sci Total Environ 595:601–614
217. Mastroianni N, Bleda MJ, Lopez de Alda M, Barcelo D (2016) Occurrence of drugs of abuse in surface water from four Spanish river basins: spatial and temporal variations and environmental risk assessment. J Hazard Mater 316:134–142
218. Rovieri V, Guimaraes LL, Toma V, Correia AT (2020) Occurrence and ecological risk assessment of pharmaceuticals and cocaine in a beach area of Guarujá, São Paulo State, Brazil, under the influence of urban surface runoff. Environ Sci Pollut Res 27:45063–45075
219. Koelmans AA, Mohamed Nor NH, Hermsen E, Kooi M, Mintenig SM, De France J (2019) Microplastics in freshwaters and drinking water: critical review and assessment of data quality. Water Res 155:410–422
220. Sjerps RMA, Kooij PJF, van Loon A, Van Wezel AP (2019) Occurrence of pesticides in Dutch drinking water sources. Chemosphere 235:510–518
221. Kamata M, Matsui Y, Asami M (2020) National trends in pesticides in drinking water and water sources in Japan. Sci Total Environ 744:140930
222. Yang Y, Ok YS, Kim KH, Kwon EE, Tsang YF (2017) Occurrences and removal of pharmaceuticals and personal care products (PPCPs) in drinking water and water/sewage treatment plants: A review. Sci Total Environ 596–597:303–320

223. Putt AE, MacIsaac EA, Herunter HE, Cooper AB, Selbie DT (2019) Eutrophication forcings on a peri-urban lake ecosystem: context for integrated watershed to airshed management. Plos One 14:e0219241
224. Duprey NN, Yasuhara M, Baker DM (2016) Reefs of tomorrow: eutrophication reduces coral biodiversity in an urbanized seascape. Glob Chang Biol 22:3550–3565
225. Wurtsbaugh A, Paerl HW, Dodds WK (2019) Nutrients, eutrophication and harmful algal blooms along the freshwater to marine continuum. WIREs Water 6:e1373
226. Le Moal M, Gascuel-Odoux C, Ménesguen A, Souchon Y, Étrillard E, Levain E, Moatar F, Pannard A, Souchu P, Lefebvre A, Lefebvre G (2019) Eutrophication: a new wine in an old bottle? Sci Total Environ 651:1–11
227. Sinha E, Michalk AM, Balaji V (2017) Eutrophication will increase during the 21st century as a result of precipitation changes. Science 357:405–408
228. Abdelhady AA, Khalil MM, Ismail E, Mohamed RSA, Ali A, Gamal Snousy M, Fan J, Zhang S, Xiao J (2019) Potential biodiversity threats associated with the metal pollution in the Nile–Delta ecosystem (Manzala lagoon, Egypt). Ecol Indic 98:844–853
229. Bonsignore M, Salvagio Manta D, Mirto S, Quinci EM, Ape F, Montalto V, Gristina M, Sprovieri M (2018) Bioaccumulation of heavy metals in fish, crustaceans, molluscs and echinoderms from the Tuscany coast. Ecotoxicol Environ Saf 162:554–562
230. Barchiesi F, Branciari R, Latini M, Roila R, Lediani G, Filippini G, Scortichini G, Piersanti A, Rocchegiani E, Ranucci D (2020) Heavy metals contamination in shellfish: benefit-risk evaluation in Central Italy. Foods 9:1720
231. Huang JS, Koongolla JB, Li HX, Pan LF, Liu S, He WH, Maharana D, Xu XR (2020) Microplastic accumulation in fish from Zhanjiang mangrove wetland, South China. Sci Total Environ 708:134839
232. Clasen B, Loro VL, Murassi CR, Tiecher TL, Moraes B, Zanella R (2018) Bioaccumulation and oxidative stress caused by pesticides in Cyprinus carpio reared in a rice-fish system. Sci Total Environ 626:737–743
233. Ojemaye CY, Onwordi CT, Petrik L (2020) Herbicides in the tissues and organs of different fish species (Kalk Bay harbour, South Africa): occurrence, levels and risk assessment. Int J Environ Sci Technol 17:1637–1648
234. Díaz-Cruz MS, Molins-Delgado D, Serra-Roig MP, Kalogianni E, Skoulikidis NT, Barcelo D (2019) Personal care products reconnaissance in EVROTAS river (Greece): water-sediment partition and bioaccumulation in fish. Sci Total Environ 651:3079–3089
235. Achary MS, Satpathy SS, Panigrahi S, Mohanty AK, Padhi RK, Biswas S, Prabhu RK, Vijayalakshmi S, Panigrahy RC (2017) Concentration of heavy metals in the food chain components of the nearshore coastal waters of Kalpakkam, southeast coast of India. Food Control 72:232–243
236. Waring RH, Harris RM, Mitchell SC (2018) Plastic contamination of the food chain: A threat to human health? Maturitas 155:64–68
237. Jijie R, Solcan G, Niocara M, Micu D, Strungaru SA (2020) Antagonistic effects in zebrafish (Danio rerio) behavior and oxidative stress induced by toxic metals and deltamethrin acute exposure. Sci Total Environ 698:134299
238. Mahboob S, Al-Ghanim KA, Al-Balawi HF, Al-Misned F, Ahmed Z (2020) Toxicological effects of heavy metals on histological alterations in various organs in Nile tilapia (*Oreochromis niloticus*) from freshwater reservoir. J King Saud Univ Eng Sci 32:920–973
239. Green AJ, Planchart A (2018) The neurological toxicity of heavy metals: A fish perspective. Comp Biochem Physiol C Toxicol Pharmacol 208:12–19
240. Qi L, Ma SJ, Li S, Cui X, Peng X, Wang W, Ren Z, Han M, Zhang Y (2017) The physiological characteristics of zebra fish (Danio rerio) based on metabolism and behavior: A new method for the online assessment of cadmium stress. Chemosphere 184:1150–1156
241. Ahrendt C, Perez-Venegas DJ, Urbina M, Gonzalez C, Echeveste P, Aldana M, Pulgar J, Galbán-Malagón C (2020) Microplastic ingestion cause intestinal lesions in the intertidal fish Girella laevifrons. Mar Pollut Bull 151:110795

242. Malinich TD, Chou N, Sepúlveda MS, Höök TO (2018) No evidence of microplastic impacts on consumption or growth of larval *Pimephales promelas*. Integr Environ Assess Manag 37:2912–2918
243. Foley CJ, Feiner ZS, Malinich TD, Höök TO (2018) A meta-analysis of the effects of exposure to microplastics on fish and aquatic invertebrates. Sci Total Environ 631–632:550–559
244. Cherr GN, Fairbairn E, Whitehead A (2017) Impacts of petroleum-derived pollutants on fish development. Annu Rev Anim Biosci 5:185–203
245. Gonçalves C, Teixeira Marins A, Blank do Amaral AM, Medina Nunes ME, Ellwanger Müller T, Severo E, Feijó A, Rodrigues CCR, Zanella R, Damian Prestes D, Clasen B, Loro VL (2020) Ecological impacts of pesticides on Astyanax jacuhiensis (Characiformes: Characidae) from the Uruguay river, Brazil. Ecotoxicol Environ Saf 205:111314
246. Rossi AS, Fanton N, Michlig MP, Repetti MR, Cazenave J (2020) Fish inhabiting rice fields: bioaccumulation, oxidative stress and neurotoxic effects after pesticides application. Ecol Indic 113:106186
247. Sula E, Aliko V, Barcelo D, Faggio C (2020) Combined effects of moderate hypoxia, pesticides and PCBs upon crucian carp fish, Carassius carassius, from a freshwater lake- in situ ecophysiological approach. Aquat Toxicol 228:105644
248. Santana MS, Sandrini-Neto L, Di Domenico N, Mela PM (2020) Pesticide effects on fish cholinesterase variability and mean activity: A meta-analytic review. Sci Total Environ. https://doi.org/10.1016/j.scitotenv.2020.143829
249. Ebele AJ, Abou-Elwafa Abdallah M, Harrad S (2017) Pharmaceuticals and personal care products (PPCPs) in the freshwater aquatic environment. Emerg Contam 3:1–16
250. Martin JM, Bertram MG, Saaristo M, Ecker TE, Hannington SL, Michelangeli M, O'Bryan MK, Wong BBM (2019) Impact of the widespread pharmaceutical pollutant fluoxetine on behaviour and sperm traits in a freshwater fish. Sci Total Environ 650:1771–1778
251. Pico Y, Alvarez-Ruiz R, Alfarhan AH, El-Sheikh M, Alshahrani HO, Barceló D (2020) Pharmaceuticals, pesticides, personal care products and microplastics contamination assessment of Al-Hassa irrigation network (Saudi Arabia) and its shallow lakes. Sci Total Environ 701:135021
252. Jia Z, Bian J, Wang Y (2018) Impacts of urban land use on the spatial distribution of groundwater pollution, Harbin City, Northeast China. J Contam Hydrol 215:29–38
253. Smith M, Cross K, Paden M, Laban P (eds) (2016) Spring – managing groundwater sustainably. IUCN, Gland
254. Singh A, Patel AK, Deka JP, Das A, Kumar A, Kumar M (2019) Prediction of arsenic vulnerable zones in the groundwater environment of a rapidly urbanizing setup, Guwahati, India. Geochemistry. https://doi.org/10.1016/j.chemer.2019.125590
255. Azzellino A, Colombo L, Lombi S, Marchesi V, Piana A, Andrea M, Alberti L (2019) Groundwater diffuse pollution in functional urban areas: the need to define anthropogenic diffuse pollution background levels. Sci Total Environ 656:1207–1222
256. Burri NM, Weatherl R, Moeck C, Schirmer M (2019) A review of threats to groundwater quality in the Anthropocene. Sci Total Environ 684:136–154
257. Howard K, Gerber R (2018) Impacts of urban areas and urban growth on groundwater in the Great Lakes Basin of North America. J Great Lakes Res 44:1–13
258. Costa CW, Lorandi R, Lollo JA, Severino SV (2019) Potential for aquifer contamination of anthropogenic activity in the recharge area of the Guarani aquifer system, southeast of Brazil. Groundw Sustain Dev 8:10–23
259. Li Q, Guo S, Fu K, Liao L, Xu Y, Cheng S (2020) Groundwater pollution source apportionment using principal component analysis in a multiple land-use area in southwestern China. Environ Sci Pollut Res 27:9000–9011
260. Lapworth DJ, Das P, Shaw S, Mukherjee A, Civil W, Petersen JO, Gooddy DC, Wakefield O, Finlayson A, Krishan G, Sengupta P, MacDonald AM (2018) Deep urban groundwater vulnerability in India revealed through the use of emerging organic contaminants and residence time tracers. Environ Pollut 240:938–949

261. Khan A, Michelsen N, Marandi A, Hossain R, Abed Hossain M, Roehl KE, Zahid A, Qumrul Hassan M, Schüth C (2020) Processes controlling the extent of groundwater pollution with chromium from tanneries in the Hazaribagh area, Dhaka, Bangladesh. Sci Total Environ 710:136213
262. Grimmeisen F, Lehmann MF, Liesch T, Goeppert N, Klinger J, Zopfi J, Goldscheider N (2017) Isotopic constraints on water source mixing, network leakage and contamination in an urban groundwater system. Sci Total Environ 583:202–213
263. García-Gil A, Garrido Schneider E, Mejías M, Barcelo D, Vázquez-Suñé E, Díaz-Cruz S (2018) Occurrence of pharmaceuticals and personal care products in the urban aquifer of Zaragoza (Spain) and its relationship with intensive shallow geothermal energy exploitation. J Hydrol 556:629–642
264. Rodriguez F, Le Delliou AL, Andrieu H, Gironas J (2020) Groundwater contribution to sewer network Baseflow in an urban catchment-case study of pin sec catchment, Nantes, France. Water 12:689
265. Busico G, Cuoco E, Sirna M, Mastrocicco M, Tedesco D (2017) Aquifer vulnerability and potential risk assessment: application to an intensely cultivated and densely populated area in Southern Italy. Arab J Geosci 10:222
266. Karges U, Becker J, Püttmann W (2018) 1,4-Dioxane pollution at contaminated groundwater sites in western Germany and its distribution within a TCE plume. Sci Total Environ 619–620:712–720
267. Kringel R, Rechenburg A, Kuitcha D, Fouépé A, Bellenberg S, Kengne IM, Fomo MA (2016) Mass balance of nitrogen and potassium in urban groundwater in Central Africa, Yaounde/Cameroon. Sci Total Environ 547:382–395
268. Verlicchi P, Grillini V (2020) Surface water and groundwater quality in South Africa and Mozambique—analysis of the Most critical pollutants for drinking purposes and challenges in water treatment selection. Water 12:305
269. Gbadebo AM (2020) Assessment of quality and health risk of peri-urban groundwater supply from selected areas of Abeokuta, Ogun State, Southwestern Nigeria. Environ Geochem Health. https://doi.org/10.1007/s10653-020-00746-5
270. Hepburn E, Madden C, Szabo D, Coggan TL, Clarke B, Currell M (2019) Contamination of groundwater with per- and polyfluoroalkyl substances (PFAS) from legacy landfills in an urban re-development precinct. Environ Pollut 248:101–113
271. Moreau M, Hadfield J, Hughey J, Sanders F, Lapworth DJ, White D, Civil W (2019) A baseline assessment of emerging organic contaminants in New Zealand groundwater. Sci Total Environ 686:425–439
272. Akbari M, Najafi Alamdarlo H, Habibollah MS (2020) The effects of climate change and groundwater salinity on farmers' income risk. Ecol Indic 110:105893
273. Mas-Pla J, Mencio A (2019) Groundwater nitrate pollution and climate change: learnings from a water balance-based analysis of several aquifers in a western Mediterranean region (Catalonia). Environ Sci Pollut Res 26:184–222
274. Kabisch N, Frantzeskaki N, Pauleit S, Naumann S, Davis M, Artmann M, Haase D, Knapp S, Korn H, Stadler J, Zaunberger K, Bonn A (2016) Nature-based solutions to climate change mitigation and adaptation in urban areas: perspectives on indicators, knowledge gaps, barriers, and opportunities for action. Ecol Soc 21:39
275. Raymond CM, Frantzeskaki N, Kabisch N, Perry P, Breil M, Razvan Nita M, Geneletti D, Calfapietra C (2017) A framework for assessing and implementing the co-benefits of nature-based solutions in urban areas. Environ Sci Pol 77:15–24
276. Han D, Shen H, Duan W, Chen L (2020) A review on particulate matter removal capacity by urban forests at different scales. Urban For Urban Green 48:126565
277. Sgrigna G, Baldacchini C, Dreveck S, Cheng Z, Calfapietra C (2020) Relationships between air particulate matter capture efficiency and leaf traits in twelve tree species from an Italian urban-industrial environment. Sci Total Environ 718:137310

278. Räsänen JV, Holopainen T, Joutsensaari J, Ndam C, Pasanen P, Rinnan A, Kivimäenpää M (2013) Effects of species-specific leaf characteristics and reduced water availability on fine particle capture efficiency of trees. Environ Pollut 183:64–70
279. Beckett KP, Freer-Smith PH, Taylor G (2000) Particulate pollution capture by urban trees: effect of species and windspeed. Glob Chang Biol 6:995–1003
280. Chen L, Liu C, Zhang L, Zou R, Zhang Z (2017) Variation in tree species ability to capture and retain airborne fine particulate matter ($PM_{2.5}$). Sci Rep 7:3206
281. Nguyen T, Yu X, Zhan Z, Liu M, Liu X (2015) Relationship between types of urban forest and PM2.5 capture at three growth stages of leaves. J Environ Sci 27:33–41
282. Liang D, Ma C, Wang YQ, Wang YJ, Chen-xi Z (2016) Quantifying PM2.5 capture capability of greening trees based on leaf factors analyzing. Environ Sci Pollut Res 23:21176–21186
283. Nowak DJ, Crane DE, Stevens JC (2006) Air pollution removal by urban trees and shrubs in the United States. Urban For Urban Green 4:115–123
284. Nyelele C, Kroll CN, Nowak DJ (2019) Present and future ecosystem services of trees in the Bronx, NY. Urban For Urban Green 42:10–20
285. Tallis M, Taylor G, Sinnet D, Freer-Smith P (2011) Estimating the removal of atmospheric particulate pollution by the urban tree canopy of London, under current and future environments. Landsc Urban Plan 103:129–138
286. Jim CY, Chen WY (2008) Assessing the ecosystem service of air pollutant removal by urban trees in Guangzhou (China). J Environ Manag 88:665–676
287. Wu J, Wang Y, Qiu S, Peng J (2019) Using the modified i-tree eco model to quantify air pollution removal by urban vegetation. Sci Total Environ 688:673–683
288. Yang J, McBride J, Zhou J, Sun Z (2005) The urban forest in Beijing and its role in air pollution reduction. Urban For Urban Green 3:65–78
289. Nowak DJ, Hirabayashi S, Doyle M, McGovern M, Pasher J (2018) Air pollution removal by urban forests in Canada and its effect on air quality and human health. Urban For Urban Green 29:40–48
290. Uni D, Katra I (2017) Airborne dust absorption by semi-arid forests reduces PM pollution in nearby urban environments. Sci Total Environ 598:984–992
291. Rieger I, Kowarik I, Cherubuni P, Cierjacks A (2017) A novel dendrochronological approach reveals drivers of carbon sequestration in tree species of riparian forests across spatiotemporal scales. Sci Total Environ 574:1261–1275
292. Isaifan RJ, Baldauf RW (2020) Estimating economic and environmental benefits of urban trees in desert regions. Front Ecol Evol 8:16
293. Ma Z, Chen HYH, Bork EW, Carlyke CN, Chang SX (2020) Carbon accumulation in agroforestry systems is affected by tree species diversity, age and regional climate: A global meta-analysis. Glob Ecol Biogeogr 29:1817–1828
294. Rytter RM, Rytter L (2020) Carbon sequestration at land use conversion – early changes in total carbon stocks for six tree species grown on former agricultural land. For Ecol Manag 13:484–494
295. Meineke E, Youngsteadt E, Dunn RR, Frank SD (2016) Urban warming reduces aboveground carbon storage. Proc R Soc B 283:20161574
296. Machacova K, Borak L, Agyei T, Schindler T, Soosaar K, Mander U (2020) Trees as net sinks for methane (CH_4) and nitrous oxide (N_2O) in the lowland tropical rain forest on volcanic Réunion Island. New Phytol. https://doi.org/10.1111/nph.17002
297. Safford H, Larry E, McPherson EG, Nowak DJ, Westphal LM (2013) Urban forests and climate change. U.S. Department of Agriculture, Forest Service, Climate Change Resource Center. www.fs.usda.gov/ccrc/topics/urban-forests
298. De la Sota S, Ruffato-Ferreira VJ, Ruiz-Garcia L, Alvarez S (2019) Urban green infrastructure as a strategy of climate change mitigation. A case study in northern Spain. Urban For Urban Green 40:145–151
299. Doukalianou F, Radoglou K, Agnelli AE, Kitikidou K, Milios E, Orfanoudakis M, Lagomarsino A (2019) Annual greenhouse-gas Emissions from Forest soil of a Peri-urban

conifer Forest in Greece under different thinning intensities and their climate-change mitigation potential. For Sci 65:387–400
300. Li Y, Babcock Jr RW (2014) Green roofs against pollution and climate change. A review. Agron Sustain Dev 34:695–705
301. Teotónio I, Matos Silva I, Oliveira CC (2018) Eco-solutions for urban environments regeneration: the economic value of green roofs. J Clean Prod 199:121–135
302. Shafique M, Xue X, Luo X (2020) An overview of carbon sequestration of green roofs in urban areas. Urban For Urbasn Green 47:126515
303. Berardi U, GhaffarianHoseini A, GhaffarianHoseini A (2014) State-of-the-art analysis of the environmental benefits of green roofs. Appl Energy 115:411–428
304. Møller Francis LJ, Bergen JM (2017) Benefits of green roofs: a systematic review of the evidence for three ecosystem services. Urban For Urban Green 28:167–176
305. Yang J, Yu Q, Gong P (2008) Quantifying air pollution removal by green roofs in Chicago. Atmos Environ 42:7266–7273
306. Gourdji S (2018) Review of plants to mitigate particulate matter, ozone as well as nitrogen dioxide air pollutants and applicable recommendations for green roofs in Montreal, Quebec. Environ Pollut 241:378–387
307. Kuronuma T, Watanabe H (2017) Relevance of carbon sequestration to the physiological and morphological traits of several Green roof plants during the first year after construction. Am J Plant Sci 8:72981
308. Charoenkit S, Yiemwattana S (2017) Role of specific plant characteristics on thermal and carbon sequestration properties of living walls in tropical climate. Build Environ 115:67–79
309. Agra H, Klein T, Vasl A, Shalom H, Kadas G, Blaustein L (2017) Sedum-dominated green-roofs in a semi-arid region increase CO2 concentrations during the dry season. Sci Total Environ 584–585:1147–1151
310. Teemusk A, Kull A, Kanal A, Mander U (2019) Environmental factors affecting greenhouse gas fluxes of green roofs in temperate zone. Sci Total Environ 649:133699
311. Karteris M, Theodoridou I, Mallinis G, Tsiros E, Karteris A (2016) Towards a green sustainable strategy for Mediterranean cities: assessing the benefits of large-scale green roofs implementation in Thessaloniki, northern Greece, using environmental modelling, GIS and very high spatial resolution remote sensing data. Renew Sust Energ Rev 58:510–525
312. Azeñas V, Janner I, Medrano H, Gulías J (2018) Performance evaluation of five Mediterranean species to optimize ecosystem services of green roofs under water-limited conditions. J Environ Manag 212:236–247
313. Collazo-Ortega M, Rosas U, Reyes-Santiago J (2017) Towards providing solutions to the air quality crisis in the Mexico City metropolitan area: carbon sequestration by succulent species in Green roofs. PlosOne 31:9
314. Zaid SM, Perisamy E, Hussein H, Myeda NE, Zainon N (2018) Vertical greenery system in urban tropical climate and its carbon sequestration potential: A review. Ecol Indic 91:57–70
315. Hu Y, Zheng J, Kong X, Sun J, Li Y (2019) Carbon footprint and economic efficiency of urban agriculture in Beijing – a comparative case study of conventional and home-delivery agriculture. J Clean Prod 234:615–625
316. Clayden A, Green T, Hockey J, Powell M (2018) Cutting the lawn – natural burial and its contribution to the delivery of ecosystem services in urban cemeteries. Urban For Urban Green 33:99–106
317. Tang Y, Chen A, Zhao S (2016) Carbon storage and sequestration of urban street trees in Beijing, China. Front Ecol Evol 4:53
318. Okunlola I, Ibironke S, Akinbobola T (2019) Net carbon sequestration and emission potential on lawns in the Federal University of Technology, Akure (Futa) Ondo State, Nigeria. Int J Res Agric For 6:1–16
319. Hewitt CN, Ashworth K, MacKenzie AR (2020) Using green infrastructure to improve urban air quality (GI4AQ). Ambio 49:62–73

320. Liu C, Li X (2012) Carbon storage and sequestration by urban forests in Shenyang, China. Urban For Urban Green 11:121–128
321. Zhao M, Kong ZH, Escobedo FJ, Gao J (2010) Impacts of urban forests on offsetting carbon emissions from industrial energy use in Hangzhou, China. J Environ Manag 91:807–813
322. Chen WY (2015) The role of urban green infrastructure in offsetting carbon emissions in 35 major Chinese cities: a nationwide estimate. Cities 44:112–120
323. Kiran GS, Kinnary S (2011) Carbon sequestration by urban trees on roadsides of Vadodara city. Int J Eng Sci Technol 3:3066–3070
324. Ragula A, Chandra KK (2020) Tree species suitable for roadside afforestation and carbon sequestration in Bilaspur, India. Carbon Manage 11:369–380
325. Intasen M, Hauer RJ, Werner LP, Larsen E (2017) Urban forest assessment in Bangkok, Thailand. J Sustain For 36:148–163
326. Tuğluer M, Gül A, Keleş E, Faruk Uzun O (2017) Ecological importance and role in carbon sequestration of urban trees (In case of Isparta Anadolu Neighborhood). In: International symposium on new horizons in forestry, pp 156–164
327. Escobedo F, Varela S, Zhao M, Wagner JE, Zipperer W (2010) Analyzing the efficacy of subtropical urban forests in offsetting carbon emissions from cities. Environ Sci Pol 13:362–372
328. Nowak DJ, Greenfield EJ, Hoehn RE, Lapoint E (2013) Carbon storage and sequestration by trees in urban and community areas of the United States. Environ Pollut 178:229–236
329. Martin NA, Chappekla AH, Loewenstein EF, Keever GJ (2012) Comparison of carbon storage, carbon sequestration, and air pollution removal by protected and maintained urban forests in Alabama, USA. Int J Biodivers Sci Ecosyst Serv Manag 8:265–272
330. Ning ZH, Chambers R, Abdollahi K (2016) Modeling air pollutant removal, carbon storage, and CO2 sequestration potential of urban forests in Scotlandville, Louisiana. USA iForest 9:860–867
331. Pasher J, McGovern M, Khoury M, Duffe J (2014) Assessing carbon storage and sequestration by Canada's urban forests using high resolution earth observation data. Urban For Urban Green 13:484–494
332. Pasher J, McGovern M (2016) Canadian urban tree canopy cover and carbon sequestration status and change 1990–2012. Urban For Urban Green 20:227–232
333. Martinez-Carretero E, Moreno G, Duplancic A, Abud A, Bento B, Alcalá JJ (2017) Urban forest of Mendoza (Argentina): the role of Morus alba (Moraceae) in carbon storage. Carbon Manag 8:237–244
334. Chaparro L, Terradas J (2009) Ecological services of urban forest in Barcelona. Centre de Recerca Ecologica i, Aplicacions Forestals, Universitat Autonoma de Barcelona, Bellaterra
335. Gratani L, Varone L, Bonito A (2016) Carbon sequestration of four urban parks in Rome. Urban For Urban Green 19:184–193
336. Jo HK (2002) Impacts of urban greenspace on offsetting carbon emissions for middle Korea. J Environ Manag 64:115–126
337. Park JO, Baek SG, Kwon MY, Je SM, Woo SY (2018) Volumetric equation development and carbon storage estimation of urban forest in Daejeon, Korea. For Sci Technol 14:97–104
338. Stoffberg GH, van Rooyen MW, van der Linde MJ, Groeneveld HT (2010) Carbon sequestration estimates of indigenous street trees in the City of Tshwane, South Africa. Urban For Urban Green 9:9–14
339. Agbelade DA, Onyekwelu JC (2020) Tree species diversity, volume yield, biomass and carbon sequestration in urban forests in two Nigerian cities. Urban Ecosyst 23:957–970
340. Ferreira CSS, Pereira P, Kalantari Z (2018) Human impact on soil. Sci Total Environ 644:830–834
341. Calzolari C, Tarocco T, Lombardo N, Marchi N, Ungaro F (2020) Assessing soil ecosystem services in urban and peri-urban areas: from urban soils survey to providing support tool for urban planning. Land Use Policy 99:105037

342. Liu R, Wang M, Chen W (2018) The influence of urbanization on organic carbon sequestration and cycling in soils of Beijing. Landsc Urban Plan 169:241–249
343. Sapkota M, Young J, Coldren C, Slaughter L, Longing S (2020) Soil physiochemical properties and carbon sequestration of urban landscapes in Lubbock, TX, USA. Urban For Urban Green 56:126847
344. Lebed-Sharlevich Y, Kulachkova S, Mozharova N (2019) Generation, sink, and emission of greenhouse gases by urban soils at different stages of the floodplain development in Moscow. J Soils Sediments 19:3204–3216
345. Sarzhanov DA, Vasenev VI, Vasenev II, Sotnikova S, Ryzhkov OV, Morin T (2017) Carbon stocks and CO2 emissions of urban and natural soils in central Chernozemic region of Russia. Catena:131–140
346. Lu C, Kotze DJ, Setälä HM (2020) Soil sealing causes substantial losses in C and N storage in urban soils under cool climate. Sci Total Environ 725:138369
347. Milnar M, Ramaswami A (2020) Impact of urban expansion and in situ greenery on community-wide carbon emissions: method development and insights from 11 US cities. Environ Sci Technol. https://doi.org/10.1021/acs.est.0c02723
348. Riches D, Porter I, Dingle G, Gendall A, Grover S (2020) Soil greenhouse gas emissions from Australian sports fields. Sci Total Environ 707:134420
349. Townsend-Small A, Czimczik CI (2010) Carbon sequestration and greenhouse gas emissions in urban turf. Geophys Res Lett 37:L02707
350. Vasanec VI, Kusyakov Y (2018) Urban soils as hot spots of anthropogenic carbon accumulation: review of stocks, mechanisms and driving factors. Land Degrad Dev 29:1607–1622
351. Lv H, Wang W, He X, Wei C, Xiao L, Zang B, Zhou W (2018) Association of urban forest landscape characteristics with biomass and soil carbon stocks in Harbin City, Northeastern China. PeerJ 6:e5825
352. Setälä HM, Francini G, Allen JA, Hui N, Jumpponen A, Kotze DJ (2016) Vegetation type and age drive changes in soil properties, nitrogen, and carbon sequestration in urban parks under cold climate. Front Ecol Evol 4:93
353. Llorach-Massana P, Muñoz P, Riera MR, Gabarrell X, Montero JI, Villalba G (2017) N_2O emissions from protected soilless crops for more precise food and urban agriculture life cycle assessments. J Clean Prod 149:1118–1126
354. Kulak M, Graves A, Chatterton J (2013) Reducing greenhouse gas emissions with urban agriculture: A life cycle assessment perspective. Landsc Urban Plan 111:68–78
355. Häring V, Manka'abusi D, Akoto-Danso EK, Werner S, Atiah K, Steiner C, Lompo DJP, Adiku S, Buerkert A, Marschner B (2017) Effects of biochar, waste water irrigation and fertilization on soil properties in west African urban agriculture. Sci Rep 7:10738
356. Chen H, Ma J, Wang X, Xu P, Zheng S, Zhao Y (2018) Effects of biochar and sludge on carbon storage of urban Green roofs. Forests 9:413
357. Whiting GJ, Chanton JP (2003) Greenhouse carbon balance of wetlands: methane emission versus carbon sequestration. Tellus B 53:521–528
358. Sovik AK, Augustin J, Heikkinen K, Huttunen JT, Necki JM, Karjalainen SM, Klove B, Liikanen A, Mander U, Puustinen M, Teiter S, Wachniew P (2006) Emission of the greenhouse gases nitrous oxide and methane from constructed wetlands in Europe. J Environ Qual 35:2360–2373
359. Mander U, Dotro G, Ebie Y, Towprayoon S, Chiemchaisri C, Furlan Nogueira S, Jamsranjav B, Kasak K, Truu J, Tournebize J, Mitsch WJ (2014) Greenhouse gas emission in constructed wetlands for wastewater treatment: A review. Ecol Eng 66:19–35
360. de Klein JJM, van der Werf JJM (2014) Balancing carbon sequestration and GHG emissions in a constructed wetland. Ecol Eng 66:36–42
361. Wu H, Zhang J, Ngo HH, Guo W, Liang S (2017) Evaluating the sustainability of free water surface flow constructed wetlands: methane and nitrous oxide emissions. J Clean Prod 147:152–156

362. Maucieri C, Barbera AC, Vymazal J, Borin M (2017) A review on the main affecting factors of greenhouse gases emission in constructed wetlands. Agric For Meteorol 236:175–193
363. Gorsky AL, Racanelli GA, Belvin AC, Chambers RM (2019) Greenhouse gas flux from stormwater ponds in southeastern Virginia (USA). Anthropocene 28:100218
364. McPhillips L, Walter MT (2015) Hydrologic conditions drive denitrification and greenhouse gas emissions in stormwater detention basins. Ecol Eng 85:67–75
365. Badiou P, Page B, Ross L (2019) A comparison of water quality and greenhouse gas emissions in constructed wetlands and conventional retention basins with and without submerged macrophyte management for storm water regulation. Ecol Eng 127:292–301
366. Audet J, Vodder Carstensen M, Hoffmann CC, Lavaux L, Thiemer K, Davidson TA (2020) Greenhouse gas emissions from urban ponds in Denmark. Inland Waters 10:373–385
367. Peacock M, Audet J, Jordan S, Smeds J, Wallin MB (2019) Greenhouse gas emissions from urban ponds are driven by nutrient status and hydrology. Ecosphere 10:e02643
368. Stumpner EB, Kraus TEC, Liang YL, Bachand CM, Horwath WR, Bachand PAM (2018) Sediment accretion and carbon storage in constructed wetlands receiving water treated with metal-based coagulants. Ecol Eng 111:176–185
369. Merriman LS, Moore TLC, Wang JW, Osmond DL, Al-Rubaei AM, Smolek AP, Blecken GT, Viklander M, Hunt WF (2017) Evaluation of factors affecting soil carbon sequestration services of stormwater wet retention ponds in varying climate zones. Sci Total Environ 583:133–141
370. Mitsch WJ, Mander U (2018) Wetlands and carbon revisited. Ecol Eng 114:1–6
371. Liu L, Li W, Song W, Guo M (2018) Remediation techniques for heavy metal-contaminated soils: principles and applicability. Sci Total Environ 633:206–219
372. Chibueze Azubuike C, Blaise Chikere C, Chijioke OC (2016) Bioremediation techniques–classification based on site of application: principles, advantages, limitations and prospects. World J Microbiol Biotechnol 32:180
373. Ossai IC, Ahmed A, Hassan A, Shahul HF (2020) Remediation of soil and water contaminated with petroleum hydrocarbon: A review. Environ Technol Innov 17:100526
374. Baldan E, Basaglia M, Fontana F, Shapleigh JP, Casella S (2015) Development, assessment and evaluation of a biopile for hydrocarbons soil remediation. Int Biodeterior Biodegrad 98:66–72
375. Kim T, Hong JK, Jho EH, Kang G, Yang DJ, Lee SJ (2019) Sequential biowashing-biopile processes for remediation of crude oil contaminated soil in Kuwait. J Hazard Mater 378:120710
376. Garcia-Carmona M, Romero-Freire A, Sierra Aragon M, Martinez-Garzon FJ, Martin Peinado FJ (2017) Evaluation of remediation techniques in soils affected by residual contamination with heavy metals and arsenic. J Environ Manag 191:228–236
377. Iturbe R, Flores C, Chavez C, Bautista G, Torres L (2004) Remediation of contaminated soil using soil washing and biopile methodologies at a field level. J Soils Sediments 4:115
378. Llorens-Blanch G, Parladé E, Martinez-Alonso M, Gaju N, Caminal G, Blánquez P (2018) A comparison between biostimulation and bioaugmentation in a solid treatment of anaerobic sludge: drug content and microbial evaluation. Waste Manag 72:206–217
379. Morillo E, Villaverde J (2017) Advanced technologies for the remediation of pesticide-contaminated soils. Sci Total Environ 586:576–597
380. Gidudu B, Nkhalambayausi Chirwa EM (2020) The combined application of a high voltage, low electrode spacing, and biosurfactants enhances the bio-electrokinetic remediation of petroleum contaminated soil. J Clean Prod 276:122745
381. Huang D, Hu C, Zeng G, Cheng M, Xu P, Gong X, Wang R, Xue W (2017) Combination of Fenton processes and biotreatment for wastewater treatment and soil remediation. Sci Total Environ 574:1599–1610
382. Song B, Zeng G, Gong J, Liang J, Xu P, Liu Z, Zhang Y, Zhang C, Cheng M, Liu Y, Ye S, Yi H, Ren X (2017) Evaluation methods for assessing effectiveness of in situ remediation of

soil and sediment contaminated with organic pollutants and heavy metals. Environ Int 105:43–55
383. Kong L, Gao Y, Zhou Q, Zhao X, Sun Z (2018) Biochar accelerates PAHs biodegradation in petroleum-polluted soil by biostimulation strategy. J Hazard Mater 343:276–284
384. Rahman T, Seraj MF (2018) Available approaches of remediation and stabilisation of metal contamination in soil: A review. Am J Plant Sci 9:87344
385. Hatzinger PB, Lippincott DR (2019) Field demonstration of N-Nitrosodimethylamine (NDMA) treatment in groundwater using propane biosparging. Water Res 164:114923
386. EPA (2014) How to evaluate alternative cleanup technologies for underground storage tank sites. A guide for corrective action plan reviewers. Land emergency management 540IR
387. Gaur N, Narasimhulu K, PydiSetty Y (2018) Recent advances in the bio-remediation of persistent organic pollutants and its effect on environment. J Clean Prod 198:1602–1631
388. Zouboulis AI, Moussas PA (2011) Groundwater and soil pollution: bioremediation. In: Nriagu JO (ed) Encyclopedia of environmental health. Elsevier, Amsterdam, pp 1037–1044
389. Das S, Dash HR (2014) Microbial bioremediation: a potential tool for restoration of contaminated areas. In: Das S (ed) Microbial biodegradation and bioremediation. Elsevier, Amsterdam, pp 1–21
390. Mosco MJ, Zytner RG (2017) Large-scale bioventing degradation rates of petroleum hydrocarbons and determination of scale-up factors. Biorem J 21:149–162
391. Priya R, Ramesh D, Khosla E (2020) Biodegradation of pesticides using density-based clustering on cotton crop affected by *Xanthomonas malvacearum*. Environ Dev Sustain 22:1353–1369
392. Miller RR (1996) Bioslurping. Ground-Water Remediation Technologies Analysis Center. Ground-Water Remediation Technologies Analysis Center, p 10
393. Kim S, Krajmalnik-Brown R, Kim JO, Chung J (2014) Remediation of petroleum hydrocarbon-contaminated sites by DNA diagnosis-based bioslurping technology. Sci Total Environ 497–498:250–259
394. Goswami M, Chakraborty P, Mukherjee K, Mitra G, Bhattacharyya P, Dey S, Tribeti P (2018) Bioaugmentation and biostimulation: a potential strategy for environmental remediation. J Microbiol Exp 6:223–231
395. Gopinath Kanissery R, Sims GK (2011) Biostimulation for the enhanced degradation of herbicides in soil. Appl Environ Soil Sci:843450
396. Chen P, Li J, Wang HY, Zheng RL, Sun GX (2017) Evaluation of bioaugmentation and biostimulation on arsenic remediation in soil through biovolatilization. Environ Sci Pollut Res 24:21739–21749
397. Simpanen S, Dahl M, Gerlach M, Mikkonen A, Malk V, Mikola J, Romantschuk M (2016) Biostimulation proved to be the most efficient method in the comparison of in situ soil remediation treatments after a simulated oil spill accident. Environ Sci Pollut Res 23:25024–25038
398. Petsas AS, Vagi NC (2019) Trends in the bioremediation of pharmaceuticals and other organic contaminants using native or genetically modified microbial strains: a review. Curr Pharm Biotechnol 20:787–824
399. Cycoń M, Mrozik A, Piotrowska-Seget Z (2017) Bioaugmentation as a strategy for the remediation of pesticide-polluted soil: a review. Chemosphere 172:52–71
400. Varjani S, Upasani VN (2019) Influence of abiotic factors, natural attenuation, bioaugmentation and nutrient supplementation on bioremediation of petroleum crude contaminated agricultural soil. J Environ Manag 245:358–366
401. Kim JM, Le NT, Chung BS, Park JH, Bae JW, Madsen EL, Jeon CO (2008) Influence of soil components on the biodegradation of benzene, toluene, ethylbenzene, and o-, m-, and p-xylenes by the newly isolated bacterium Pseudoxanthomonas spadix BD-a59. Appl Environ Microbiol 74:7313–73120

402. Zhang X, Yang YS, Lu Y, Wen YJ, Li PP, Zhang G (2018) Bioaugmented soil aquifer treatment for P-nitrophenol removal in wastewater unique for cold regions. Water Res 144:616–627
403. Tran HT, Lin C, Gui XT, Ngo HH, Kiprotich Cheruiyot N, Hoang HG, Vu CT (2021) Aerobic composting remediation of petroleum hydrocarbon-contaminated soil. Current and future perspectives. Sci Total Environ 753:142250
404. Ren X, Zeng G, Tang L, Wang J, Wan J, Deng Y, Liu Y, Peng B (2018) The potential impact on the biodegradation of organic pollutants from composting technology for soil remediation. Waste Manag 72:138–149
405. Zhang Y, Zhu YG, Hout S, Qiao M, Nunan N, Garnier P (2011) Remediation of polycyclic aromatic hydrocarbon (PAH) contaminated soil through composting with fresh organic wastes. Environ Sci Pollut Res 18:1574–1584
406. Chen X, Zhao Y, Zhang C, Zhang D, Yao C, Meng Q, Zhao R, Wei Z (2020) Speciation, toxicity mechanism and remediation ways of heavy metals during composting: A novel theoretical microbial remediation method is proposed. J Environ Manag 272:111109
407. Sanchez-Hernandez JC (2020) Vermiremediation of pharmaceutical-contaminated soils and organic amendments. In: The handbook of environmental chemistry. Springer, Berlin. https://doi.org/10.1007/698_2020_625
408. Kokana RS, Sarmah AK, Van Zwieten L, Krull E, Singh B (2011) Biochar application to soil: agronomic and environmental benefits and unintended consequences. In: Sparks DL (ed) Advances in agronomy. Elsevier, Amsterdam, pp 103–143
409. Kim JY, Oh S, Park YK (2020) Overview of biochar production from preservative-treated wood with detailed analysis of biochar characteristics, heavy metals behaviors, and their ecotoxicity. J Hazard Mater 384:121356
410. Izaurralde RC, McGill WB, Williams JR (2012) Development and application of the EPIC model for carbon cycle, greenhouse gas mitigation, and biofuel studies. In: Liebig MA, Franzluebbers AJ, Follett RF (eds) Managing agricultural greenhouse gases. Elsevier, Amsterdam, pp 293–308
411. Xu RK, Qafoku N, Van Ranst E, Li JY, Jiang J (2016) Adsorption properties of subtropical and tropical variable charge soils: implications from climate change and biochar amendment. In: Sparks DL (ed) Advances in agronomy. Elsevier, Amsterdam, pp 1–58
412. Carlile WR, Raviv M, Prasad M (2019) Organic soilless media components. In: Raviv M, Heinrich Lieth J, Bar-Tal A (eds) Soilless culture theory and practice. Elsevier, Amsterdam, pp 303–378
413. Kavitha B, Venkata Laxma Reddy P, Kim B, Lee SS, Pandey SK, Kim KH (2018) Benefits and limitations of biochar amendment in agricultural soils: a review. J Environ Manag 227:146–154
414. Dai Y, Zheng H, Jiang Z, Xing B (2020) Combined effects of biochar properties and soil conditions on plant growth: a meta-analysis. Sci Total Environ 713:136635
415. Elliston T, Oliver IW (2020) Ecotoxicological assessments of biochar additions to soil employing earthworm species *Eisenia fetida* and *Lumbricus terrestris*. Environ Sci Pollut Res 27:33410–33418
416. Llovet A, Mattana S, Chin-Pampillo C, Otero N, Carrey R, Mondini C, Gasco G, Margalef R, Alcañiz JM, Domene X, Ribas A (2021) Fresh biochar application provokes a reduction of nitrate which is unexplained by conventional mechanisms. Sci Total Environ 755:142430
417. Qi F, Dong Z, Lamb D, Naidu R, Bolan NS, Ok YS, Liu C, Khan N, Johir MAH, Semple KT (2017) Effects of acidic and neutral biochars on properties and cadmium retention of soils. Chemmosphere 180:564–573
418. Liu H, Xu F, Wang C, Zhang A, Li L, Xu H (2018) Effect of modified coconut shell biochar on availability of heavy metals and biochemical characteristics of soil in multiple heavy metals contaminated soil. Sci Total Environ 645:702–709
419. Lu HP, Gasco G, Mendez A, Shen Y, Paz-Ferreiro J (2018) Use of magnetic biochars for the immobilization of heavy metals in a multi-contaminated soil. Sci Total Environ 622–623:899

420. Zielińska A, Oleszczuk P (2016) Attenuation of phenanthrene and pyrene adsorption by sewage sludge-derived biochar in biochar-amended soils. Environ Sci Pollut Res 23:21822–21832
421. Khosla K, Rathour R, Maurya R, Maheshwari N, Gnansounou E, Larroche C, Shekhar Thakur I (2017) Biodiesel production from lipid of carbon dioxide sequestrating bacterium and lipase of psychrotolerant Pseudomonas sp. ISTPL3 immobilized on biochar. Bioresour Technol 245:743–750
422. Khalid S, Shahid M, Murtaza B, Bibi I, Natasha ANM, Khan Niazi N (2020) A critical review of different factors governing the fate of pesticides in soil under biochar application. Sci Total Environ 711:134645
423. Yavari S, Malakahmad A, Sapari NB, Yavari S (2016) Sorption-desorption mechanisms of imazapic and imazapyr herbicides on biochars produced from agricultural wastes. J Environ Chem Eng 4:3981–3989
424. Wu L, Bi E (2019) Sorption of ionic and neutral species of pharmaceuticals to loessial soil amended with biochars. Environ Sci Pollut Res 26:35871–35881
425. Puglisi E, Romaniello F, Galletti S, Boccaleri E, Frache A, Cocconcelli PS (2019) Selective bacterial colonization processes on polyethylene waste samples in an abandoned landfill site. Sci Rep 9:14138
426. Ru J, Huo Y, Yang Y (2020) Microbial degradation and valorization of plastic wastes. Front Microbiol 11:142
427. Bailes G, Lind M, Ely A, Powell M, Moore-Kucera J, Miles C, Inglis D, Brodhagen M (2013) Isolation of native soil microorganisms with potential for breaking down biodegradable plastic mulch films used in agriculture. J Vis Exp 75:50373
428. Sun Z, Brittain JE, Sokolova E, Thygesen E, Jakob Saltveit S, Rauch S, Meland S (2018) Aquatic biodiversity in sedimentation ponds receiving road runoff – what are the key drivers? Sci Toal Environ 610–611:1527–1535
429. Muller A, Österlund H, Marsalek J, Viklander M (2020) The pollution conveyed by urban runoff: A review of sources. Sci Total Environ 709:136125
430. Hsu TTD, Mitsch WJ, Martin JF, Lee J (2017) Towards sustainable protection of public health: the role of an urban wetland as a frontline safeguard of pathogen and antibiotic resistance spread. Ecol Eng 108:547–555
431. Liu G, He T, Liu Y, Chen Z, Li L, Huang Q, Xie Z, Xie Y, Wu L, Liu J (2019) Study on the purification effect of aeration-enhanced horizontal subsurface-flow constructed wetland on polluted urban river water. Environ Sci Pollut Res 26:12867–12880
432. Singh NK, Gupta G, Upadhyay AK, Rai UN (2019) Biological wastewater treatment for prevention of river water pollution and reuse: perspectives and challenges. In: Singh R, Kolok A, Bartelt-Hunt S (eds) Water conservation, recycling and reuse: issues and challenges. Springer, Singapore, pp 81–93
433. Sharley DJ, Sharp SM, Marshal S, Jeppe K, Pettigrove VJ (2017) Linking urban land use to pollutants in constructed wetlands: implications for stormwater and urban planning. Landsc Urban Plan 162:80–91
434. Malaviya P, Singh A (2012) Constructed wetlands for management of urban stormwater runoff. Crit Rev Environ Sci Technol 42:2153–2214
435. Haberl R, Grego S, Langergraber G, Kadlec RH, Cicalini AR, Dias SM, Novais JM, Aubert S, Gerth A, Thomas H, Hebner A (2003) Constructed wetlands for the treatment of organic pollutants. J Soil Sediments 3:109–124
436. Scholz M, Lee BW (2005) Constructed wetlands: a review. Int J Environ Stud 62:421–447
437. Gorito AM, Ribeiro AR, Almeida CMR, Silva AMT (2017) A review on the application of constructed wetlands for the removal of priority substances and contaminants of emerging concern listed in recently launched EU legislation. Environ Pollut 227:428–443
438. Wang M, Zhang DQ, Dong JW, Tan SK (2017) Constructed wetlands for wastewater treatment in cold climate – A review. J Environ Sci 57:293–311

439. Sandoval L, Zamora-Castro SA, Vidal-Álvarez M, Marín-Muñiz JL (2019) Role of wetland plants and use of ornamental flowering plants in constructed wetlands for wastewater treatment: a review. Appl Sci 9:685
440. Koskiaho J, Puustinen M (2019) Suspended solids and nutrient retention in two constructed wetlands as determined from continuous data recorded with sensors. Ecol Eng 137:65–75
441. Varma M, Gupta AK, Ghosal PS, Majumder A (2021) A review on performance of constructed wetlands in tropical and cold climate: insights of mechanism, role of influencing factors, and system modification in low temperature. Sci Total Environ 755:142540
442. Babatunde AO, Zhao YQ, O'Neill M, O'Sullivan O (2008) Constructed wetlands for environmental pollution control: a review of developments, research and practice in Ireland. Environ Int 34:116–126
443. Vymazal J, Březinová T (2016) Accumulation of heavy metals in aboveground biomass of Phragmites australis in horizontal flow constructed wetlands for wastewater treatment: a review. Chem Eng J 290:232–242
444. Liu H, Hu Z, Zhang J, Ngo HH, Guo W, Liang S, Fan J, Lu S, Wu H (2016) Optimizations on supply and distribution of dissolved oxygen in constructed wetlands: a review. Bioresour Technol 214:797–805
445. Arden S, Ma X (2018) Constructed wetlands for greywater recycle and reuse: a review. Sci Total Environ 630:587–599
446. Tang S, Liao Y, Xu Y, Dang Z, Zhu X, Ji G (2020) Microbial coupling mechanisms of nitrogen removal in constructed wetlands: a review. Bioresour Technol 314:123759
447. Bakhshoodeh R, Alavi N, Oldham C, Santos RM, Babaei AA, Vymazal J, Paydary P (2020) Constructed wetlands for landfill leachate treatment: A review. Ecol Eng 146:105725
448. Ilyas H, Masih I (2017) The performance of the intensified constructed wetlands for organic matter and nitrogen removal: a review. J Environ Manag 198:372–383
449. Kadlec RH (2016) Large constructed wetlands for phosphorus control: A review. Water 8:243
450. Jain M, Majumder A, Sarathi Ghosal P, Kumar GA (2020) A review on treatment of petroleum refinery and petrochemical plant wastewater: a special emphasis on constructed wetlands. J Environ Manag 272:111057
451. Türker OC, Vymazal J, Türe C (2014) Constructed wetlands for boron removal: a review. Ecol Eng 64:350–359
452. Karimian N, Johnston SG, Burton ED (2018) Iron and sulfur cycling in acid sulfate soil wetlands under dynamic redox conditions: a review. Chemosphere 197:802–816
453. Guan Y, Wang B, Gao Y, Liu W, Zhao X, Huang X, Yu J (2017) Occurrence and fate of antibiotics in the aqueous environment and their removal by constructed wetlands in China: a review. Pedosphere 27:42–51
454. Ilyas H, van Hullebusch ED (2020) Performance comparison of different types of constructed wetlands for the removal of pharmaceuticals and their transformation products: a review. Environ Sci Pollut Res 27:14342–14364
455. Ilyas H, van Hullebusch ED (2020) A review on the occurrence, fate and removal of steroidal hormones during treatment with different types of constructed wetlands. J Environ Chem Eng 8:103793
456. Vymazal J, Březinová T (2015) The use of constructed wetlands for removal of pesticides from agricultural runoff and drainage: a review. Environ Int 75:11–20
457. Vyzmal J (2018) Does clogging affect long-term removal of organics and suspended solids in gravel-based horizontal subsurface flow constructed wetlands? Chem Eng J 331:663–674
458. Benvenuti T, Hamerski F, Giacobbo A, Bernardes AM, Zoppas-Ferreira J, Rodrigues MAS (2018) Constructed floating wetland for the treatment of domestic sewage: a real-scale study. J Environ Chem Eng 6:5706–5711
459. Khalifa ME, Abou El-Reash YG, Ahmed MI, Rizk FW (2020) Effect of media variation on the removal efficiency of pollutants from domestic wastewater in constructed wetland systems. Ecol Eng 143:105668

460. Gizińska-Górna M, Jóźwiakowski K, Marzec M (2020) Reliability and efficiency of pollutant removal in four-stage constructed wetland of SSVF-SSHF-SSHF-SSVF type. Water 12:3153
461. Zheng Y, Wang X, Dzakpasu M, Ge Y, Xiong J, Zhao Y (2016) Feasibility study on using constructed wetlands for remediation of a highly polluted Urban River in A semi-arid region of China. J Water Sustain 6:139–148
462. Bai X, Zhu X, Jiang H, Wang Z, He C, Sheng L, Zhuang J (2020) Purification effect of sequential constructed wetland for the polluted water in Urban River. Water 12:1054
463. Schwammberger PF, Lucke T, Walker C, Trueman SJ (2019) Nutrient uptake by constructed floating wetland plants during the construction phase of an urban residential development. Sci Total Environ 677:390–403
464. Adyel TM, Hipsey MR, Oldham CE (2017) Temporal dynamics of stormwater nutrient attenuation of an urban constructed wetland experiencing summer low flows and macrophyte senescence. Ecol Eng 102:641–661
465. Zheng Y, Wang XC, Dzakpasu M, Ge Y, Zhao Y, Xiong J (2016) Performance of a pilot demonstration-scale hybrid constructed wetland system for on-site treatment of polluted urban river water in Northwestern China. Environ Sci Pollut Res 23:447–454
466. Griffiths LN, Mitsch WJ (2020) Nutrient retention via sedimentation in a created urban stormwater treatment wetland. Sci Total Environ 727:138337
467. Kabenge I, Ouma G, Aboagye D, Banadda N (2018) Performance of a constructed wetland as an upstream intervention for stormwater runoff quality management. Environ Sci Pollut Res 25:36765–36774
468. Qasaimeh A, AlSharie H, Masoud T (2015) A review on constructed wetlands components and heavy metal removal from wastewater. J Environ Prot 6:58324
469. Walaszek M, Bois P, Laurent J, Lenormand E, Wanko A (2018) Urban stormwater treatment by a constructed wetland: Seasonality impacts on hydraulic efficiency, physico-chemical behavior and heavy metal occurrence. Sci Total Environ 637–638:443–454
470. Maniquiz-Redillas MC, Kim LH (2016) Evaluation of the capability of low-impact development practices for the removal of heavy metal from urban stormwater runoff. Environ Technol 37:2265–2272
471. Hamad MTMH (2020) Comparative study on the performance of Typha latifolia and Cyperus Papyrus on the removal of heavy metals and enteric bacteria from wastewater by surface constructed wetlands. Chemosphere 260:127551
472. Tromp K, Lima AT, Barendregt A, Verhoeven JTA (2012) Retention of heavy metals and poly-aromatic hydrocarbons from road water in a constructed wetland and the effect of de-icing. J Hazard Mater 203–204:290–298
473. Schmitt N, Wanko A, Laurent J, Bois P, Molle P, Mose R (2015) Constructed wetlands treating stormwater from separate sewer networks in a residential Strasbourg urban catchment area: micropollutant removal and fate. J Environ Chem Eng 3:2816–2824
474. Terzakis S, Fountoulakis MS, Georgaki I, Albantakis D, Sabathianakis I, Karathanasis AD, Kalogerakis N, Manios T (2008) Constructed wetlands treating highway runoff in the Central Mediterranean region. Chemosphere 72:141–149
475. Qin Z, Zhao Z, X Adam A, Li Y, Chen D, Mela SM, Li H (2019) The dissipation and risk alleviation mechanism of PAHs and nitrogen in constructed wetlands: the role of submerged macrophytes and their biofilms-leaves. Environ Int 131:104940
476. Li Y, Zhu G, Ng WJ, Tan SK (2014) A review on removing pharmaceutical contaminants from wastewater by constructed wetlands: design, performance and mechanism. Sci Total Environ 468–469:908–932
477. Vo HNP, Bui XT, Nguyen TMH, Koottatep T, Bandyopadhyay A (2018) Insights of the removal mechanisms of pharmaceutical and personal care products in constructed wetlands. Curr Pollut Rep 4:93–103
478. Zhang X, Jing R, Feng X, Dai Y, Tao R, Vymazal J, Cai J, Yang Y (2018) Removal of acidic pharmaceuticals by small-scale constructed wetlands using different design configurations. Sci Total Environ 639:640–647

479. Vystavna Y, Frkova Z, Marchand L, Vergeles Y, Stolberg F (2017) Removal efficiency of pharmaceuticals in a full-scale constructed wetland in East Ukraine. Ecol Eng 108:50–58
480. Park J, Cho KH, Lee E, Lee S, Cho J (2018) Sorption of pharmaceuticals to soil organic matter in a constructed wetland by electrostatic interaction. Sci Total Environ 635:1345–1350
481. Hijosa-Valsero M, Reyes-Contreras C, Domínguez C, Bécares E, Bayona JM (2016) Behaviour of pharmaceuticals and personal care products in constructed wetland compartments: influent, effluent, pore water, substrate and plant roots. Chemosphere 145:508–517
482. Berberidou C, Kitsiou V, Lambropoulou DA, Antoniadis A, Ntonou E, Zalidis GC, Poulios I (2017) Evaluation of an alternative method for wastewater treatment containing pesticides using solar photocatalytic oxidation and constructed wetlands. J Environ Manag 195:133–139
483. Gikas GD, Vryzas Z, Tsihrintzis VA (2018) S-metolachlor herbicide removal in pilot-scale horizontal subsurface flow constructed wetlands. Chem Eng J 339:108–116

The Role of Floods on Pathogen Dispersion

Bernard Bett, Dan Tumusiime, Johanna Lindahl, Kristina Roesel, and Grace Delia

Contents

B. Bett (✉) and K. Roesel
International Livestock Research Institute, Nairobi, Kenya
e-mail: B.BETT@cgiar.org; K.Roesel@cgiar.org

D. Tumusiime
International Livestock Research Institute, Nairobi, Kenya

Ministry of Agriculture, Animal Industries and Fisheries, Entebbe, Uganda

J. Lindahl
International Livestock Research Institute, Nairobi, Kenya

Uppsala University, Uppsala, Sweden

Swedish University of Agricultural Sciences, Uppsala, Sweden
e-mail: J.Lindahl@cgiar.org

G. Delia
International Livestock Research Institute, Nairobi, Kenya

Natural Resources Institute, University of Greenwich, London, UK
e-mail: D.Randolph@cgiar.org

Carla S. S. Ferreira, Zahra Kalantari, Thomas Hartmann, and Paulo Pereira (eds.),
Nature-Based Solutions for Flood Mitigation: Environmental and Socio-Economic Aspects,
Hdb Env Chem (2022) 107: 139–158, DOI 10.1007/698_2021_761,

Published online: 9 May 2021

Abstract Floods precipitate many infectious disease epidemics in humans and animals. These incidences are more prevalent in developing countries where about 80% of illnesses and deaths in humans are water related. This chapter identifies three categories of flood-borne infections based on how floods influence their occurrence patterns. The first category includes acute infections such as cholera and leptospirosis, caused by bacteria that are carried mechanically by water and are often ingested with water or food. These infections thrive in areas with high human population densities with poor sanitation. In these settings, floods enhance transmission of infectious agents between hosts. The second category is vector-borne infections such as malaria, Rift Valley fever, and schistosomiasis. They are transmitted by vectors that breed in inundated areas. Their epidemics often follow flood events by weeks or months depending on the duration of their development cycles. The last category is skin and eye infections that occur following direct contact with contaminated water. All these diseases can be controlled more effectively if the standard surveillance and control measures are integrated with nature-based solutions (NBS) for flood management. Examples the NBS that can be used include re-forestation, tree planting especially along streams, and development of green infrastructure in cities to enhance water retention, infiltration, and replenishment of groundwater.

Keywords Deforestation, Environment, Flood management, Flood-borne agents, Global warming, Hotspots, Re-forestation, Urbanization

1 Introduction

Flooding is defined as a temporary overflow or submerging of land, which would otherwise have been dry [1]. It is a common disaster caused by multiple environmental and socioeconomic factors, including above normal and persistent precipitation especially in low and flat terrains with soils comprising poor draining properties [2]. It also derives from an un-anticipated incursion of water such as when a dam breaks or during a tidal surge. Landscapes with wide vegetation cover are more likely to reduce surface runoff and promote infiltration or groundwater recharge especially in sloping terrain [3]. The frequency and severity of floods are expected to rise with global warming, which, among other things, would increase the frequency of above normal precipitation events [4].

Floods have devastating effects on human health and wellbeing. These effects may be short- or long-term depending on flood characteristics and the vulnerability of affected populations. Floods may trigger epidemics of many infectious diseases in humans and animals. In humans, up to 80% of illnesses and deaths in the developing countries are water related, and half of the world's hospital beds are occupied by people with water-related diseases [5]. The World Health Organization identifies several infectious diseases promoted by flooding such as leptospirosis, mosquito-borne infections, and water-borne gastrointestinal parasites [6]. Floods also disrupt infrastructure, therefore limiting the delivery or access to health services especially in remote locations with limited transportation networks [7]. In animals, floods cause physical injury and deaths, limit access to feed, and hasten the transmission of diseases such as botulism, foot rot, erysipelas, pneumonia, and mastitis among others [8].

This chapter illustrates how floods trigger emergence and spread of infectious diseases in humans and animals, using selected infectious diseases such as cholera, leptospirosis, Rift Valley fever (RVF), and malaria as case studies. It starts by describing common multiscale drivers of flooding and infectious disease outbreaks, followed by a description of the processes through which flooding trigger epidemics. Nature-based interventions that can be used to manage infectious diseases caused by floods are presented together with other measures for floods and associated diseases at the end of the chapter.

2 Multiscale Drivers for Floods and Associated Infectious Diseases

Floods are a major risk factor for water-borne diseases such as cholera, typhoid, and leptospirosis [9]. Many water-borne diseases are transmitted between humans and animals, and the five most important zoonotic pathogens in this regard are *Cryptosporidium parvum, Giardia duodenalis, Escherichia coli* O157:H7, *Salmonella* sp., and *Campylobacter* sp. [10]. Floods and infectious diseases are likely to have common antecedent multiscale drivers including climate change, land use change (e.g., urbanization and agricultural intensification), and changes in human population. A clearer understanding on these relationships could provide better knowledge of underlying processes on how flooding is associated with pathogen transmission, as well as potential ways of controlling the transmission.

2.1 Climate Change

The mean global temperature has been on a steady increase since the 1880s, at a rate of 0.07°C per decade, due to the rising levels of greenhouse gases (GHGs) in the atmosphere [11]. The five most important GHGs that are thought to have substantial effect on global warming are carbon dioxide, methane, nitrous oxide, sulfur oxide, and chlorofluorocarbons. These gases are largely generated from anthropogenic activities such as the use of fossil fuel for transport and heating, manufacturing and construction, deforestation, use of fertilizers, and other food production activities [12]. The rising human population is expected to hasten the production of GHGs due to expansion of agricultural and other socioeconomic activities that are needed to boost food and energy requirements [13].

2.1.1 Climate Change and Flooding

Climate change is expected to increase the frequency and intensity of floods. Climate simulation models, for example, demonstrate that if historical trends continue in the future, the most intense precipitation events observed nowadays are likely to almost double in occurrence for each additional degree of global warming [4]. Both frequency and intensity of precipitation events are expected to increase [4], dictating the need to enhance the resilience of societies to these changes. In addition, climate change would increase the frequency of extreme *El Niňo* events [14], which trigger intense precipitation and associated flooding and cyclones, among other natural hazards, with severe socioeconomic consequences. *El Niňo* events recorded in 1982/83 and 1997/98 have been cited as the most intense globally, as they were associated with a rise in sea surface temperatures above 28°C in the equatorial eastern Pacific [15]. Following these events, extensive floods ensued in various parts of the world including south America and eastern Africa.

2.1.2 Climate Change as a Predictor of Infectious Diseases

Climate change has direct and indirect effects on pathogen transmission in humans [16] and animals [17]. A rise in temperature within a specific physiological range of biological vectors such as mosquitoes and ticks hasten the transmission dynamics of vector-borne pathogens. This is because temperatures amplify the development rates of both the pathogens and vectors that are involved in their transmission [18]. Some of the vector-borne diseases, such as malaria (transmitted by mosquitoes) and schistosomiasis (transmitted by freshwater snails) are particularly important in this regard because their occurrence and distribution are not only influenced by flooding but also by global warming [18].

2.2 *Vegetation Cover Changes*

2.2.1 Effects of Land Use Changes on Flooding

The conversion of natural environments to cropland or urban settlements leads to the removal of vegetation that often increases infiltration rates and reduces the speed at which surface runoff flows [19]. Trees and shrubs improve the porosity of the soils which allows rainwater to percolate into the subsurface [20]. A similar mechanism operates when debris from shrubs and trees pile up and get buried in soils to increase soil porosity. Natural landscapes have high water storage capacity provided, for example, by trees and depressions on the surface. All these terrain features provide a natural flood regulation ecosystem service, which is minimal or absent in cultivated areas [21].

2.2.2 Effect of Land Use Changes on Infectious Diseases

Apart from increasing the risk of flooding, land use changes can directly influence the transmission of infectious diseases. Using deforestation as an example, a 10% increase in deforestation has been associated with a 3.3% increase in malaria incidence in the Amazon [22]. This is because deforestation disrupts the ecology of mosquito vectors, prompting their dispersal to human settlements where they can transmit the disease. Fragmentation of habitats and deforestation have been reported to favor Ebola transmission [23].

2.2.3 Urbanization

Urbanization is rapidly expanding, particularly in developing countries, as people move from rural to urban centers to pursue improved livelihood options. It is predicted that the proportion of people living in urban centers will reach 60% by 2030, with most of this growth occurring in developing countries [24]. Unplanned urban settlements, however, present additional challenges, not only in the delivery of services to the population, but also with the maintenance, for example, of the drainage systems. Poor sanitation and drainage systems are associated with a number of public health risks such as water-related vector-borne diseases and fecal–oral bacterial infections [25].

2.2.4 Urbanization and Flooding

Urban settlements experience higher risk of flooding than rural areas, partly because permeable surface soils are often compacted or sealed during the construction of buildings, pavements, and other structures. Large soil surfaces are covered by concrete, asphalt, or other impermeable materials that seal the surface, therefore reducing water absorption and increasing runoff and flood hazard [26]. When drainage systems are appropriately designed, surface flow is efficiently managed, but in poor urban settlements, drainage systems are often not well designed and frequently get clogged with sediments and solid waste, therefore increasing chances of flooding [27]. In addition, buildings are often constructed in floodplains and in every open space to provide room for increasing population who are seeking for accommodation. In addition, human settlements also develop in wetlands that are prone to inundation especially during the wet seasons.

2.2.5 Urbanization and Infectious Diseases

Informal settlements have been recognized as hotspots for some infectious diseases such as cholera [28] and antimicrobial resistant bacteria [29]. The dense population in these settings enhance effective host contact which enables faster pathogen transmission than in remote or sparsely populated areas. Sewage and other solid waste disposal services are not always efficient, and flooding provides opportunities for infectious pathogens from uncollected wastewater and solid waste to multiply and contaminate the environment, potentially leading to human infections [30]. Uncollected wastewater also can serve as breeding points for a wide range of vectors such as mosquitoes.

3 Processes Through Which Flooding Enhances Infectious Diseases Occurrence

Floods trigger epidemics of several infectious diseases in humans and animals by amplifying pathogen transmission processes. In human settlements and urban settings, flood waters are usually contaminated by bacteria, viruses, protozoa, and helminths carried from sewers, dumpsites, and other contaminated surfaces [31]. Exposure to these pathogens through water or food can cause various forms of infections, ranging from a mild disease to severe gastroenteritis [32]. Floods not only carry these pathogens from one area to another, but may also promote their multiplication (especially in the tropics and subtropics with favorable ambient temperatures) amplifying vector populations, causing catastrophic [31] outbreaks [33], and/or displacement and concentration of people and animals, therefore promoting infectious disease spread and mental health problems [34].

Flood-borne infectious diseases can be classified into three categories according to the role of floods in their transmission processes. The first group includes acute infections caused by pathogens that are carried mechanically by water and are often ingested with contaminated water or food. They usually cause widespread epidemics with far-reaching human health burden. Examples of these diseases include cholera, leptospirosis, typhoid, amoebiasis, polio, cryptosporidiosis, and hepatitis A. The second category is the sub-acute, vector-borne diseases, whose epidemics occur when their vectors multiply to large population densities in standing water masses. Examples of these diseases include malaria, dengue fever, Chikungunya, Rift Valley fever, and schistosomiasis. The third category is skin and eye infections that occur following direct contact with contaminated water. These include fungal and bacterial skin infections caused by a variety of pathogens such as *Streptococcus pyogenes*, *Staphylococcus aureus*, *Vibrio* spp., insect bite reactions [35], and eye infections [36]. However, this classification of diseases based on severity (i.e., acute or sub-acute) is simplistic and used here only for illustration purposes.

Floods increase disease burden through other indirect ways such as physical injuries, crowding, and reduced access to public and animal health services due to destruction of infrastructures [37]. Not much data, though, have been published to show the relative contribution of the indirect effects of flooding on the overall health burden. A morbidity study performed after a flood in Missouri in 1993, which submerged over two million acres of land, reported that among people presenting at hospital emergency departments, most had injuries (48%) and illnesses (46%) [38].

3.1 Water- and Food-Borne Diseases

Cholera, leptospirosis, and cryptosporidiosis are caused by water-borne infectious agents. After flooding, affected areas remain infectious since pathogens can persist in the soils for variable periods of time, depending on temperature, sunlight, soil pH, and other conditions such as soil organic matter [32].

3.1.1 Cholera

Cholera is a water-borne infection triggered by floods, especially in densely populated areas. It is caused by a gram-negative bacterium called *Vibrio cholerae*. Two strains – 01 and 0139 – are recognized as being the main causes of cholera epidemics [39]. Bacteria in the feces of infected people, which contaminate water and food are the main sources of infection. The disease causes acute diarrhea and vomiting due to the production of a cholera toxin by the agent [40]. The toxin prevents digestion and absorption of water in the intestines, leading to acute dehydration. The disease affects people of all ages although most of the cases are reported in children with less than 5 years, which are also the most vulnerable to dehydration [41]. The World Health Organization estimates that about 1.4–4 million cholera cases occur globally every year, and about 21,000 to 143,000 cases result in death [42]. The number of cases, however, is expected to be higher given the low levels of reporting.

The agent (*Vibrio cholerae*) is a free-living organism which inhabits in surface waters, ponds, lakes, and rivers. Floods displace such contaminated waters and in areas with high human population densities, increase rates of contact and transmission leading to devastating epidemics. Up to 7 cholera pandemics have been observed since the disease was first identified in Bengal, India, in the early 1800s [43]. Each pandemic runs a long course that may exceed 3–4 years, with millions of cases. In addition, many smaller epidemics affected thousands of people.

Table 1 Examples of leptospirosis outbreaks associated with heavy rainfall and flooding (modified from Lau, Smythe and Weinstein, 2010)

Region/ country	Year	Flooding event
New Caledonia	2008	High rainfall and flooding associated with La Niña in early 2008. Epidemic of leptospirosis in 135 people. Incidence of 500/100000 population in Bourail region
Guyana	2008	Epidemic followed severe flooding with 30% of Guyana inhabitants displaced from their homes
Laos	2006	Flooding in home property associated with seropositivity for leptospirosis (odds ratio 2.12)
Mumbai, India	2005	944 mm of rain in 24 h resulted in an eight-fold rise in the number of cases of leptospirosis compared with the previous 4 years
Kerala, India	2002	Peaks in leptospirosis incidence recorded 7–10 days after heavy rainfall peaks
Indonesia	2002	Outbreak followed massive flooding in Jakarta, in January
Italy	2002	Devastating flooding in suburban area resulted in 6.8% seroconversion rate for leptospirosis
Orissa, India	1999	19.2% of studied subjects in flooded villages after a cyclone were found to have serological evidence of symptomatic leptospiral infection
Puerto Rico	1996	Leptospirosis diagnosed in 6% of non-dengue febrile illnesses pre-hurricane versus 24% of non-dengue febrile illnesses post-hurricane
Nicaragua	1995	Epidemic of leptospirosis followed severe rainfall and flooding in 1995. Over 5,000 mm of annual rainfall compared with annual average of 1,300 mm

3.1.2 Leptospirosis

Leptospirosis is another important zoonotic disease associated with flooding. It affects animals and man and is caused by pathogenic spirochetes of the genus *Leptospira* spp. [44]. Leptospires have been found in more than 180 species of animals, including domestic and wild animals such as cattle, pigs, horses, dogs, rats, and other rodents [45]. Animals that survive acute infection may continue shedding the bacterium for many months to years. Once excreted via urine, pathogenic leptospires can survive in moist environments for about a month [46]. Humans get exposed through (i) direct contact with urine, blood, or tissues of infected animals, or urine-contaminated environments such as surface water, soil, and plants; or (ii) ingestion of the organism through contaminated food or water [47]. The World Health Organization's Leptospirosis Burden Reference Group (LERG) indicates that leptospirosis is common in urban slum areas with poor sewage disposal and poor water supply. They also indicate that outbreaks are often triggered by floods. Table 1 provides an outline of leptospirosis outbreaks that have been associated with flooding [48].

Environmental conditions that favor the survival of the pathogen are pH, ranging between 6.2–8.0, and temperature between 28–38°C [49]. A survey conducted in Argentina from 1999 to 2005 revealed that 76% of leptospirosis cases occurred in warmer and wetter months [50]. Survival in water is inhibited by high acidity, high salinity and sewage [51]. Risk factors that are associated with *Leptospira* spp. infection include wet surfaces, streams near contaminated environments such as industrial areas, contact with animals, and rat infestations [52].

Floods also bring to surface any leptospires that could have been lodged in drainage systems. If the floods join rivers or fill other reservoirs used by humans as sources of water for domestic use, cross contamination and subsequent infection ensues. In addition, floods, and also flood irrigated fields, increase the time people are in contact with water (mainly wading) and since the pathogen can penetrate skin, this increases the infection rate [53].

There is insufficient information on the burden of leptospirosis given its poor reporting levels. Globally, it is estimated that seven to ten million people are infected by the disease, leading to about 1.03 million clinical cases and 58,900 deaths annually. This translates to approximately 2.90 million Disability Adjusted Life Years, a metric used to quantify disease burden in humans [54]. A large proportion of cases (48%) and deaths (42%) occur in adult males within age of 20–49 years old [55]. In Africa, *Leptospira* incidence has been estimated to be 95.5 cases per 100,000 and prevalence ranges from 2.3% to 19.8% [56].

3.2 *Vector-Borne Diseases*

Floods enhance population densities of many vectors such as mosquitoes, snails, and biting flies which transmit many pathogens. There are over 3,500 species of mosquitoes, but a few are recognized as critical vectors of important pathogens in humans and animals [53]. They all have a common life cycle where eggs laid on water or moist surfaces hatch into larvae, which develop into pupae before emerging as adults. There is a lot of variation on breeding requirements across species, but in general, the immature stages (i.e., egg, larvae, and pupae) require water [57].

Schistosoma japonicum is the causative agent for schistosomiasis that occurs in at least 76 countries throughout the world, with a disability adjusted life years of about 1.53 million [58]. The distribution of the disease is affected by environmental factors such as vegetation coverage, temperature, soil type, and water level [59]. A study conducted in China showed that flooding or inundation lasting from 2 to 7 months was needed for the development of the snails [60]. This would be more relevant for floods that take longer time to propagate, such as in marshy areas.

Surface runoff can destroy established breeding grounds for these vectors, and thus in some cases, a reduced number of mosquitoes may be observed [61].

3.2.1 Malaria

Malaria is one of the mosquito-borne parasitic diseases that are prevalent in the tropics. The World Malaria Report (2019) by the World Health Organization showed that a total of 228 million cases of malaria and 405,000 deaths were reported worldwide in 2018 [62]. Africa contributes to 90% of malaria cases [63]. The disease is transmitted by *Anopheles arabiensis* and *Anopheles gambiae*. Typically, these vectors breed in shallow waters, puddles, ponds, burrow-pits, tire tracks, or hoof prints [64]. Apart from the presence of water, other critical meteorological parameters such as temperature, clear (non-turbid) water, pH, and oxygen concentration levels are required for optimal development of malaria vectors [65].

Malaria epidemics associated with flooding have been reported in many countries such as Sudan [66], Uganda [67], and Pakistan. Studies conducted in Uganda by Boyce et al. [67] revealed that malaria incidence was observed during the post-flooding periods, mainly along flood-affected rivers. In that study, a 3-month lag period between flooding and malaria incidence was reported. Shorter lag periods have been reported in other endemic areas. These discrepancies could be attributed to environmental differences that permit flushing or re-colonization of mosquito breeding sites.

3.2.2 Rift Valley Fever

RVF is another example of a mosquito-borne zoonoses which occurs periodically in Africa [68] and the Middle East [69], following periods of above-normal precipitation and flooding. The disease affects sheep, goats, cattle, and camels causing abortion in pregnant animals and heavy mortalities in the young animals [70]. Humans may get infected while handling body fluids or tissues from infected animals, for example, when relieving dystocia, disposing aborted fetuses, or slaughtering an infected animal [71]. Humans may also get infected if they get bitten by an infectious mosquito. RVF cases in humans have variable manifestations; in the majority of cases (>80%) it presents as mild influenza-like symptoms, but a few presents hemorrhagic febrile illnesses with high mortality rates [72]. Estimates of the health burden of the disease were recently generated in Kenya following the 2006/2007 outbreaks. A total of 3,974 unweighted disability adjusted life years were determined [73]. This could be underestimated given that many other cases occur during inter-epidemic periods which are unreported.

The processes involved in the transmission of RVF are poorly understood, but there are two recognized hypotheses about it. The first suggests that flood water mosquitoes in the subgenera *Neomelaniconion* – e.g., *Aedes mcintoshi* and *Aedimorphus vexans* – maintain the virus over the inter-epidemic periods via transovarial transmission [74]. Infectious female mosquitoes lay infected eggs which remain viable in the soils for many years. Eggs are usually laid at the edges of flooded depressions just above the water surface. These eggs hatch when they get inundated for at least 10–14 days, and adult mosquitoes that emerge are infectious and can transmit the virus to susceptible animals. However, for an RVF outbreak to occur, floods must persist for 4–6 weeks to allow the development of other mosquitoes such as *Culex* sp., *Mansonia* sp., and *Anopheles* sp. to amplify the transmission processes. Domestic animals including sheep, goats, and cattle must be present in the

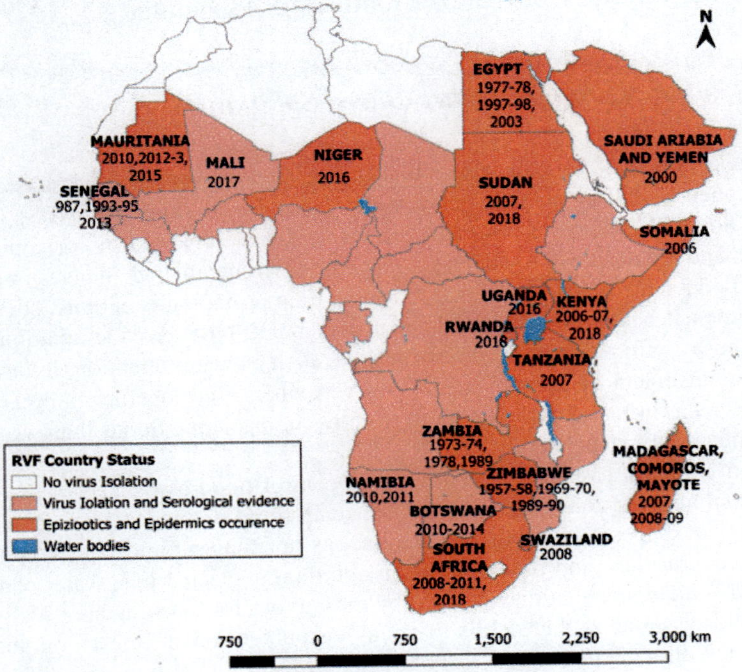

Fig. 1 A map of Africa and the Middle East showing countries that have reported epidemics, the year when outbreaks were reported as well as countries where evidence of exposure has been determined based on serological findings

area since when these animals are infected, their viremia levels rise to such high levels that would enable many mosquitoes to be infected while feeding [75].

The second hypothesis assumes that low grade RVF virus transmissions occur usually in riverine and forested habitats, and heavy rainfall and flooding cause an exponential rise in the mosquito populations which leads to the amplification of the transmission processes and epidemics. This has been supported by findings of seroconversions during inter-epidemic periods [76]. RVF outbreaks are common in southern, eastern, and western Africa. Figure 1 shows countries that have reported epidemics, those where the virus has been isolated from, and with serological evidence of exposure but no epidemics. A common feature in all the epidemics identified is the above normal and persistent rainfall, which creates floods where mosquito vectors develop. A few of the incidences, e.g. the 1987 RVF outbreak in the Senegal River basin, have occurred following the construction of a dam and subsequent flooding along the banks of the Senegal river in Mauritania-Senegal border [77].

4 Solutions for the Control of Floods and Associated Diseases

4.1 Prevention Through Nature-Based Solutions

The traditional methods for controlling flood-associated diseases can be substantially strengthened if they are integrated with Nature-based solutions (NBS) for controlling floods. NBS are "actions that are inspired by, supported by, or copied from nature, and are designed to address a range of environmental challenges in an efficient and adaptable manner, while at the same time providing economic, social and environmental benefits" [78]. On flood management, NBS provide additional benefits to those conferred by grey infrastructures, such as water retention, infiltration, and replenishment of groundwater. However, when used together, grey and NBS-based (green) infrastructures are more effective in managing floods than when each of them is used independently [79].

The tradition measures that are often used to control flood-associated diseases include maintaining environmental cleanliness, reinforcing drainage systems, and ensuring efficient disposal of debris that may act as breeding grounds for insects, mosquitoes, cockroaches, and rodents [80]. In addition, neglected tins, water containers, or old vehicle tires that may be used by mosquitoes for breeding are usually removed. Solid waste and other debris that may encourage rodent infestation and proliferation, therefore increasing the risk of *Leptospira* spp., are also disposed [81].

However, sanitation measures highlighted above are not usually enough if implemented alone, without longer term solutions for managing floods. These integrated solutions ought to be implemented through multisectoral and multidisciplinary partnerships to bring all the required expertise under the One Health framework. Examples of NBS measures that could be used to prevent floods include re-forestation and development of green cities, while the standard disease prevention measures include pest management, surveillance, vaccination, and other medical interventions. Community sensitization and training to recognize linkages between NBS, flood control, and infectious diseases are also needed to encourage stakeholder participation.

Other NBS measures that can promote better landscape management include tree planting alongside streams. They prevent contamination of water by infectious pathogens by filtering *Cryptosporidium* oocysts shed from livestock, and reducing transmission to water sources [82]. Likewise, re-forestation prevents floods and serves as water catchment "sponge."

4.2 Surveillance and Emergency Preparedness

Surveillance can act as an early warning system and should encompass hydrology (flood forecasting) and epidemiology (incidence of flood-associated diseases in animals and people). There should be a sharing of information across the relevant authorities. Technology can improve surveillance and reduce costs, for example, through gauges, drones, GIS data, and modeling used for flood forecasting [83]. Epidemiological innovations in surveillance include drone mapping for improving response to cholera, development of risk maps to identify areas prone to flooding and relevant diseases, and greater involvement of the community in reporting diseases [84]. Predictive models have also been developed for some diseases such as cholera [85] and RVF [86], using temperature, and precipitation and some geographical factors such as soil types and land cover to predict disease dynamics.

Contingency plans and emergency response planning are important to support early and effective actions. These should be written, tested, and refined through simulation exercises. They should ensure that required equipment, material, and supplies are available; that emergency centers are identified and equipped; that a system is in place to communicate with those at risk, and that personnel are trained and capable. Emergency response is often graded, with different levels of risk identified and communicated to the public.

4.3 *Response to Flood-Borne Diseases*

The World Health Organization recommends that a rapid disease assessment should be carried out at the beginning of a flood disaster [87]. It should focus on water, sanitation, nutrition, shelter, exposure to disease, and access to health services. To mitigate death and injuries, and to minimize disease outbreaks, affected communities must be translocated to safer and higher areas, and be supplied with food, clean, and treated water, clothing and medicine. In developed countries, increasing emergency centers allow pets, while livestock may be allowed to stay outdoors, with potential secondary effects on disease spread. Livestock may also need to be moved, rescued, or carcasses safely disposed.

Unfortunately, relocating people into relief camps or emergency shelters also involves health risks. Crowding, insufficient resources, and other stressors have led to outbreaks of disease and violence. The World Health Organization [88] recognizes that the standard control measures such as rehydration, antibiotic treatments, and improved sanitation measures work well in selected settings, but sustaining these measures throughout the vulnerable regions of the world would be costly [89]. Moreover, drug resistance is threatening to reduce the effectiveness of antibiotic treatments that are available for use during epidemics. Vector control measures such as mosquito nets and insecticide sprays are also used, especially in the tropics, to prevent upsurge in vector-borne diseases. In addition, some zoonotic diseases such as RVF would require livestock vaccination to minimize further spread of the disease in both livestock and human populations.

5 Conclusions

Floods are associated with many infectious diseases. Some of these are acute infections that are caused by flood-borne agents that contaminated food and water, while others are sub-acute infections transmitted by vectors that develop in inundated areas. As the frequency of flooding is expected to increase with climate, land use changes, and demographic changes, the incidence and impacts of flood-borne infectious diseases are also expected to increase. In this case, nature-based solutions that are known to be effective in controlling floods (e.g., re-forestation, tree planting along streams, and the development of green infrastructure) should be integrated with standard measures for managing infectious disease epidemics (i.e., disease surveillance, sanitation measures, and case management), under the One Health framework, to prevent these diseases. These actions can only be implemented if effective partnerships between professionals, development actors, and local communities are established.

References

1. OECD (2016) Financial management of flood risk. Financial management of flood risk. OECD Publishing, Paris. https://doi.org/10.1787/9789264257689-en
2. O'Connor J, Costa J (2004) The world's largest floods, past and present – their causes and magnitudes. US Geological Survey
3. Suprayogo D, van Noordwijk M, Hairiah K, Meilasari N, Rabbani AL, Ishaq RM et al (2020) Infiltration-friendly agroforestry land uses on. Land 9
4. Myhre G, Alterskjær K, Stjern CW, Hodnebrog, Marelle L, Samset BH et al (2019) Frequency of extreme precipitation increases extensively with event rareness under global warming. Sci Rep 9:1–10. https://doi.org/10.1038/s41598-019-52277-4
5. Toepfer K (2004) Water and sustainable development. In: Schiffries C, Brewster A (eds). National Council for Science and the Environment, Washington
6. WHO (2015) Floods and health – fact sheets for health professionals. 12. https://www.euro.who.int/__data/assets/pdf_file/0016/252601/Floods-and-health-Fact-sheets-for-health-professionals.pdf
7. Du W, FitzGerald GJ, Clark M, Hou X-Y (2010) Health impacts of floods. Prehosp Disaster Med 25:265–272. https://doi.org/10.1017/S1049023X00008141
8. Powell J, Pennington J, Jonnes F (2006) Flood-related diseases in poultry and livestock. https://www.jumpjet.info/Emergency-Preparedness/Disaster-Mitigation/Water/Flood-Related_Diseases_in_Poultry_and_Livestock.pdf. Accessed 30 Dec 2020
9. World Health Organization(WHO) (2021) Water-related diseases: information sheets. https://www.who.int/water_sanitation_health/diseases-risks/diseases/diseasefact/en/. Accessed25 Dec 2020
10. Atwill E, Xunde L, Grace D, Gannon V, Ángel JC (2002) Zoonotic waterborne pathogen loads in livestock. In: Dufour A, Bartram J (eds.) Animal waste, water quality and human health. World Health Organisation, Geneva, Unites States Environmental Protection Agency, and IWA Publishing
11. Smith P, Davis SJ, Creutzig F, Fuss S, Minx J, Gabrielle B et al (2016) Biophysical and economic limits to negative CO2 emissions. Nat Clim Chang 6:42–50. https://doi.org/10.1038/nclimate2870

12. Agency EP (2019) Sources of greenhouse gas emissions. In: Climate change. pp 1–2. https://www.epa.gov/ghgemissions/sources-greenhouse-gas-emissions%0A; http://www.epa.gov/climatechange/ghgemissions/sources/transportation.html
13. Vergé XPC, De Kimpe C, Desjardins RL (2007) Agricultural production, greenhouse gas emissions and mitigation potential. Agric For Meteorol 142:255–269. https://doi.org/10.1016/j.agrformet.2006.06.011
14. Cai W, Borlace S, Lengaigne M, van Rensch P, Collins M, Vecchi G et al (2014) Increasing frequency of extreme El Niño events due to greenhouse warming. Nat Clim Chang 5:1–6. https://doi.org/10.1038/nclimate2100
15. Rein B (2007) How do the 1982/83 and 1997/98 El Niños rank in a geological record from Peru? Quat Int 161:56–66. https://doi.org/10.1016/j.quaint.2006.10.023
16. Booth M (2018) Climate change and the neglected tropical diseases. In: Advances in parasitology1st edn. Elsevier. https://doi.org/10.1016/bs.apar.2018.02.001
17. Bett B, Kiunga P, Gachohi J, Sindato C, Mbotha D, Robinson T et al (2016) Effects of climate change on the occurrence and distribution of livestock diseases. Prev Vet Med. https://doi.org/10.1016/j.prevetmed.2016.11.019
18. Yi L, Xu X, Ge W, Xue H, Li J, Li D et al (2019) The impact of climate variability on infectious disease transmission in China: current knowledge and further directions. Environ Res 173:255–261. https://doi.org/10.1016/j.envres.2019.03.043
19. Yang C, Yu Z, Hao Z, Lin Z, Wang H (2013) Effects of vegetation cover on hydrological processes in a large region: Huaihe River Basin, China. J Hydrol Eng 18:1477–1483. https://doi.org/10.1061/(ASCE)HE.1943-5584.0000440
20. Bartens J, Day SD, Harris JR, Dove JE, Wynn TM (2008) Can urban tree roots improve infiltration through compacted subsoils for Stormwater management? J Environ Qual 37:2048–2057. https://doi.org/10.2134/jeq2008.0117
21. Qiu J (2019) Effects of landscape pattern on pollination, Pest control, water quality, flood regulation, and cultural ecosystem services: a literature review and future research prospects. Curr Landsc Ecol Rep 4:113–124. https://doi.org/10.1007/s40823-019-00045-5
22. Macdonald AJ, Mordecai EA ((2019)) Erratum: Amazon deforestation drives malaria transmission, and malaria burden reduces forest clearing. Proceedings of the National Academy of Sciences of the United States of America 116, 22212–22218. https://doi.org/10.1073/pnas.1905315116. Proc Natl Acad Sci U S A. 2020;117: 20335. https://doi.org/10.1073/PNAS.2014828117
23. Rulli MC, Santini M, Hayman DTS, D'Odorico P (2017) The nexus between forest fragmentation in Africa and Ebola virus disease outbreaks. Sci Rep 7:41613. https://doi.org/10.1038/srep41613
24. Bloom D, Khanna T (2008) The urban revolution. https://perencanamuda.wordpress.com/. Accessed 5 Dec 2020
25. Fritsch M (1997) Health issues related to drainage water management. In: Management of agricultural drainage water quality, Water reports 13. http://www.fao.org/3/w7224e/w7224e0b.htm. Accessed 11 Nov 2020
26. Konrad C (2016) Effects of urban development on floods. https://pubs.usgs.gov/fs/fs07603/. Accessed 5 Nov 2020
27. Zambrano L, Pacheco-Muñoz R, Fernández T (2018) Influence of solid waste and topography on urban floods: the case of Mexico City. Ambio 47:771–780. https://doi.org/10.1007/s13280-018-1023-1
28. Azman AS, Luquero FJ, Salje H, Mbaïbardoum NN, Adalbert N, Ali M, et al (2018) Micro-hotspots of risk in urban cholera epidemics. J Infect Dis 218: 1164–1168. https://doi.org/10.1093/infdis/jiy283. Environmental conditions that favor the survival of the pathogen are pH, ranging between 6.2–8.0, and temperature between 28–38°C
29. Nadimpalli ML, Marks SJ, Montealegre MC, Gilman RH, Pajuelo MJ, Saito M et al (2020) Urban informal settlements as hotspots of antimicrobial resistance and the need to curb

environmental transmission. Nat Microbiol 5:787–795. https://doi.org/10.1038/s41564-020-0722-0
30. Berendes DM, Leon JS, Kirby AE, Clennon JA, Raj SJ, Yakubu H et al (2019) Associations between open drain flooding and pediatric enteric infections in the MAL-ED cohort in a low-income, urban neighborhood in Vellore, India. BMC Public Health 19:1–11. https://doi.org/10.1186/s12889-019-7268-1
31. ten Veldhuis JAE, Clemens FHLR, Sterk G, Berends BR (2010) Microbial risks associated with exposure to pathogens in contaminated urban flood water. Water Res 44:2910–2918. https://doi.org/10.1016/j.watres.2010.02.009
32. CDC (2011) Guidance on microbial contamination in previously flooded outdoor areas. Atlanta. https://www.cdc.gov/nceh/ehs/docs/guidance_contamination_of_flooded_areas.pdf
33. Yomwan P, Cao C, Rakwatin P, Suphamitmongkol W, Tian R, Saokarn A (2015) A study of waterborne diseases during flooding using Radarsat-2 imagery and a back propagation neural network algorithm. Geomat Nat Hazards Risk 6:289–307. https://doi.org/10.1080/19475705.2013.853325
34. Tong S (2017) Flooding-related displacement and mental health. Lancet Planet Health 1:e124–e125. https://doi.org/10.1016/S2542-5196(17)30062-1
35. Tempark T, Lueangarun S, Chatproedprai S, Wananukul S (2013) Flood-related skin diseases: a literature review. Int J Dermatol 52:1168–1176. https://doi.org/10.1111/ijd.12064
36. Huang LY, Wang YC, Wu CC, Chen YC, Huang YL (2016) Risk of flood-related diseases of eyes, skin and gastrointestinal tract in Taiwan: a retrospective cohort study. PLoS One 11:1–11. https://doi.org/10.1371/journal.pone.0155166
37. Okaka FO, Odhiambo BDO (2018) Relationship between flooding and out break of infectious diseases in Kenya: a review of the literature. J Environ Public Health 2018. https://doi.org/10.1155/2018/5452938
38. CDC (1993) Public health consequences of a flood disaster, Iowa. MMWR Morb Mortal Wkly Rep. 1993;42: 653–656. https://www.cdc.gov/mmwr/preview/mmwrhtml/00021451.htm
39. Faruque SM, Nair GB (2002) Molecular ecology of toxigenic Vibrio cholerae. Microbiol Immunol 46:59–66. https://doi.org/10.1111/j.1348-0421.2002.tb02659.x
40. LaRocque R, Harris J (2020) Cholera: clinical features, diagnosis, treatment, and prevention. https://www.uptodate.com/contents/cholera-clinical-features-diagnosis-treatment-and-prevention/print. Accessed 3 Dec 2020
41. Colombara DV, Cowgill KD, Faruque ASG (2013) Risk factors for severe cholera among children under five in rural and urban Bangladesh, 2000-2008: a hospital-based surveillance study. PLoS One 8:2000–2008. https://doi.org/10.1371/journal.pone.0054395
42. World Health Organization(WHO) (2021) Number of reported cholera cases. https://www.who.int/gho/epidemic_diseases/cholera/cases_text/en/
43. Harris J, LaRocque R, Qadri F, Ryan E, Calderwood S (2012) NIH public access - cholera. Lancet 379:2466–2476. https://doi.org/10.1016/S0140-6736(12)60436-X.Cholera
44. Levett PN (2001) Leptospirosis. Clin Microbiol Rev 14:296–326. https://doi.org/10.1128/CMR.14.2.296-326.2001
45. Guernier V, Goarant C, Benschop J, Lau CL (2018) A systematic review of human and animal leptospirosis in the Pacific Islands reveals pathogen and reservoir diversity. PLoS Negl Trop Dis. https://doi.org/10.1371/journal.pntd.0006503
46. Casanovas-Massana A, Pedra G, Wunder E, Diggle P, Begon M, Ko A (2018) Quantification of Leptospira interrogans survival in soil and water microcosms. Appl Environ Microbiol 84:1–11
47. Vijayachari P, Sugunan AP, Shriram AN (2008) Leptospirosis: an emerging global public health problem. J Biosci 33:557–569. https://doi.org/10.1007/s12038-008-0074-z
48. Lau C, Smythe L, Weinstein P (2010) Leptospirosis: an emerging disease in travellers. Travel Med Infect Dis 8:33–39. https://doi.org/10.1016/j.tmaid.2009.12.002
49. Bierque E, Thibeaux R, Girault D, Soupé-Gilbert ME, Goarant C (2020) A systematic review of Leptospira in water and soil environments. PLoS One 15:1–22. https://doi.org/10.1371/journal.pone.0227055

50. Vanasco NB, Schmeling MF, Lottersberger J, Costa F, Ko AI, Tarabla HD (2008) Clinical characteristics and risk factors of human leptospirosis in Argentina (1999–2005). Acta Trop 107:255–258. https://doi.org/10.1016/j.actatropica.2008.06.007
51. Batterman S, Elsenberg J, Hardin R, Kruk ME, Lemos MC, Michalak AM et al (2009) Sustainable control of water-related infectious diseases: a review and proposal for interdisciplinary health-based systems research. Environ Health Perspect 117:1023–1032. https://doi.org/10.1289/ehp.0800423
52. Zala DB, Khan V, Sanghai AA, Dalai SK, Das VK (2018) Leptospira in the different ecological niches of the tribal union territory of India. J Infect Dev Ctries 12:849–854. https://doi.org/10.3855/jidc.10541
53. Wynwood SJ, Graham GC, Weier SL, Collet TA, McKay DB, Craig SB (2014) Leptospirosis from water sources. Pathog Glob Health 108:334–338. https://doi.org/10.1179/2047773214Y.0000000156
54. Torgerson PR, Hagan JE, Costa F, Calcagno J, Kane M, Martinez-Silveira MS et al (2015) Global burden of leptospirosis: estimated in terms of disability adjusted life years. PLoS Negl Trop Dis 9:e0004122. https://doi.org/10.1371/journal.pntd.0004122
55. Costa F, Hagan JE, Calcagno J, Kane M, Torgerson P, Martinez-Silveira MS et al (2015) Global morbidity and mortality of leptospirosis: a systematic review. PLoS Negl Trop Dis:1–19. https://doi.org/10.1371/journal.pntd.0003898
56. Allan KJ, Biggs HM, Halliday JEB, Kazwala RR, Maro VP, Cleaveland S et al (2015) Epidemiology of Leptospirosis in Africa: A Systematic Review of a Neglected Zoonosis and a Paradigm for 'One Health' in Africa. Zinsstag J, editor. PLoS Negl Trop Dis 9:e0003899. https://doi.org/10.1371/journal.pntd.0003899
57. Ermert V, Fink AH, Jones AE, Morse AP (2011) Development of a new version of the Liverpool Malaria Model. I. Refining the parameter settings and mathematical formulation of basic processes based on a literature review. Malar J 10:35. https://doi.org/10.1186/1475-2875-10-35
58. Gryseels B, Polman K, Clerinx J, Kestens L (2006) Human schistosomiasis. Lancet 368:1106–1118. https://doi.org/10.1016/S0140-6736(06)69440-3
59. Mas-Coma S, Valero MA, Bargues MD (2009) Climate change effects on trematodiases, with emphasis on zoonotic fascioliasis and schistosomiasis. Vet Parasitol 163:264–280. https://doi.org/10.1016/j.vetpar.2009.03.024
60. Yang Y, Zheng SB, Yang Y, Cheng WT, Pan X, Dai QQ et al (2018) The three gorges dam: does the flooding time determine the distribution of schistosome-transmitting snails in the middle and lower reaches of the Yangtze river, China? Int J Environ Res Public Health 15. https://doi.org/10.3390/ijerph15071304
61. Tompkins AM, Ermert V (2013) A regional-scale, high resolution dynamical malaria model that accounts for population density, climate and surface hydrology. Malar J 12:1–24. https://doi.org/10.1186/1475-2875-12-65
62. World Health Organization (WHO) (2021) The "World malaria report 2019" at a glance. https://www.who.int/news-room/feature-stories/detail/world-malaria-report-2019. Accessed 5 Oct 2020
63. Garcia LS (2010) Malaria. Clin Lab Med 30:93–129. https://doi.org/10.1016/j.cll.2009.10.001
64. Williams J, Pinto J (2012) Training manual on malaria entomology for entomology and vector control technicians (basic level). https://www.paho.org/hq/dmdocuments/2012/2012-Training-manual-malaria-entomology.pdf
65. Gopalakrishnan R, Das M, Baruah I, Veer V, Dutta P (2013) Physicochemical characteristics of habitats in relation to the density of container-breeding mosquitoes in Asom, India. J Vector Borne Dis 50:215–219. http://www.ncbi.nlm.nih.gov/pubmed/24220081
66. Elsanousi YEA, Elmahi AS, Pereira I, Debacker M (2018) Impact of the 2013 Floods on the incidence of Malaria in Almanagil Locality, Gezira State, Sudan. PLoS Curr:10. https://doi.org/10.1371/currents.dis.8267b8917b47bc12ff3a712fe4589fe1

67. Boyce R, Reyes R, Matte M, Ntaro M, Mulogo E, Metlay JP et al (2016) Severe flooding and malaria transmission in the Western Ugandan highlands: implications for disease control in an era of global climate change. J Infect Dis 214:1403–1410. https://doi.org/10.1093/infdis/jiw363
68. Clements AC, Pfeiffer DU, Martin V, Otte MJ (2007) A Rift Valley fever atlas for Africa. Prev Vet Med 82:72–82. https://doi.org/10.1016/j.prevetmed.2007.05.006
69. Jupp PG, Kemp A, Grobbelaar A, Leman P, Burt FJ, Alahmed AM et al (2002) The 2000 epidemic of Rift Valley fever in Saudi Arabia: mosquito vector studies. Med Vet Entomol 16:245–252. https://doi.org/10.1046/j.1365-2915.2002.00371.x
70. Ikegami T, Makino S (2011) The pathogenesis of rift valley fever. Viruses 3:493–519. https://doi.org/10.3390/v3050493
71. Nguku PM, Sharif SK, Mutonga D, Amwayi S, Omolo J, Mohammed O et al (2010) An investigation of a major outbreak of Rift Valley fever in Kenya: 2006–2007. Am J Trop Med Hyg 83:5–13. https://doi.org/10.4269/ajtmh.2010.09-0288
72. Soumare B, Tempia S, Cagnolati V, Mohamoud A, Van Huylenbroeck G, Berkvens D (2007) Screening for Rift Valley fever infection in northern Somalia: a GIS based survey method to overcome the lack of sampling frame. Vet Microbiol 121:249–256. https://doi.org/10.1016/j.vetmic.2006.12.017
73. Kimani T, Schelling E, Bett B, Ngigi M, Randolph T (2016) Public health benefits from livestock Rift Valley fever control: a simulation of two epidemics in Kenya. Ecohealth. https://doi.org/10.1007/s10393-016-1178-9
74. Linthicum KJ, Davies FG, Kairo A, Bailey CL (1985) Rift Valley fever virus (family Bunyaviridae, genus Phlebovirus). Isolations from Diptera collected during an inter-epizootic period in Kenya. J Hyg (Lond) 95:197–209. http://www.ncbi.nlm.nih.gov/pubmed/2862206
75. Anon (2005) The risk of a Rift Valley fever incursion and its persistence within the community. EFSA J 238:1–128
76. Mbotha D, Bett B, Kairu-Wanyoike S, Grace D, Kihara A, Wainaina M et al (2017) Inter-epidemic Rift Valley fever virus seroconversions in an irrigation scheme in Bura, South-East Kenya. Transbound Emerg Dis. https://doi.org/10.1111/tbed.12674
77. Thonnon J, Picquet M, Thiongane Y, Lo M, Sylla R, Vercruysse J (1999) Rift valley fever surveillance in the lower Senegal river basin: update 10 years after the epidemic. Trop Med Int Health 4:580–585. http://www.ncbi.nlm.nih.gov/pubmed/10499082
78. ECDG (2015) Towards an EU research and innovation policy agenda for nature-based solutions and re-Naturing cities. Luxemborg
79. Pamungkas A, Purwitaningsih S (2019) Green and grey infrastructures approaches in flood reduction. Int J Disaster Resil Built Environ 10:343–362. https://doi.org/10.1108/IJDRBE-03-2019-0010
80. World Health Organization (WHO) (1982) Manual on environmental management for mosquito control. Geneva. https://apps.who.int/iris/bitstream/handle/10665/37329/9241700661_eng.pdf;sequence=1
81. Goarant C (2016) Leptospirosis: risk factors and management challenges in developing countries. Res Rep Trop Med 7:49–62. https://doi.org/10.2147/RRTM.S102543
82. Kay D (2012) Effectiveness of best management practices for attenuating the transport of livestock derived pathogens within catchments. In: Dufour A, Bartram J (eds) Animal waste water quality and human health. IWA Publishing, London, pp 195–255
83. Merkuryeva G, Merkuryev Y, Sokolov BV, Potryasaev S, Zelentsov VA, Lektauers A (2015) Advanced river flood monitoring, modelling and forecasting. J Comput Sci 10:77–85. https://doi.org/10.1016/j.jocs.2014.10.004
84. Phwitiko R (2018) Drones for cholera response: innovating for children in Malawi. https://medium.com/@unicef_malawi/drones-for-cholera-response-innovating-for-children-in-malawi-6dcab2c4de53#:~:text=UNICEF has introduced drones as,to be affected by cholera. Accessed 8 Feb 2021

85. Pasetto D, Finger F, Rinaldo A, Bertuzzo E (2017) Real-time projections of cholera outbreaks through data assimilation and rainfall forecasting. Adv Water Resour 108:345–356. https://doi.org/10.1016/j.advwatres.2016.10.004
86. Anyamba A, Linthicum KJ, Tucker CJ (2001) Climate-disease connections: Rift Valley fever in Kenya. Cad Saude Publica 17:S133–S140. https://doi.org/10.1590/S0102-311X2001000700022
87. WHO (2008) Communicable disease risk assessment and interventions: cyclone Nagri, Myanma. World Health Organization, Geneva
88. WHO (2018) Cholera control interim guidance. pp 1–12. https://www.who.int/immunization/monitoring_surveillance/burden/vpd/WHO_SurveillanceVaccinePreventable_02_Cholera_R2.pdf?ua=1
89. World Health Organization (2018) Managing epidemics. https://www.who.int/emergencies/diseases/managing-epidemics/en/

The Role of Plants in Water Regulation and Pollution Control

Zorana Hrkić Ilić, Marijana Kapović Solomun, Nada Šumatić, Ratko Ristić, and Željka Marjanović-Balaban

Contents

Abstract Development of the human society has a conspicuous negative influence on water resources and causes serious environmental contamination that is nowadays reaching a critical level. The quality of water is one of the vital components of the overall environment. Thus, water pollution can lead to human health issues, poisoned wildlife, and to long-term ecosystem damages. Plants are the first organisms that react to negative environmental changes and they are often used as bioindicators of water and air pollution. In addition, a significant number of plant species have the ability to accumulate harmful pollutants from soils and water. Recently, special attention has been paid to investigating the potential of plants to absorb toxic substances and reduce their negative impact on water resources. Besides, proper management of water resources depends upon understanding how plants regulate the use and retention of water. Environmental pollutants such as heavy metals can cause disturbance in root structure and function, thus having a negative effect on the water uptake. This chapter will review and discuss the role of the plants in water regulation and the control of water pollution in urban and mining

Z. Hrkić Ilić (✉), M. Kapović Solomun, N. Šumatić, and Ž. Marjanović-Balaban
Faculty of Forestry, University of Banja Luka, Banja Luka, Bosnia and Herzegovina
e-mail: zorana.hrkic-ilic@sf.unibl.org

R. Ristić
Faculty of Forestry, University of Belgrade, Beograd, Republic of Serbia

Carla S. S. Ferreira, Zahra Kalantari, Thomas Hartmann, and Paulo Pereira (eds.),
Nature-Based Solutions for Flood Mitigation: Environmental and Socio-Economic Aspects,
Hdb Env Chem (2022) 107: 159–186, DOI 10.1007/698_2021_774,

Published online: 30 May 2021

areas. Information presented in this chapter will provide better insights into the plant-based technologies aimed at contributing to the purification and remediation of polluted water resources.

Keywords Macrophytes, Mining industry, Phytoremediation, Pollution, Urban areas, Water resources, Wetlands, Woody plants

1 Introduction

Water represents a renewable source and is an essential element for life [1]. About 35% of the Earth's available and renewable freshwater is used for industrial, domestic, and agricultural purposes [2]. Water pollution is a serious global problem that negatively affects the environment settings, human and animal health, agriculture and food production, and the overall quality of life [1, 3, 4]. Water pollution can originate from natural sources, since toxic chemicals can occur naturally (e.g. arsenic, fluorine) and contaminate groundwater and surface waters [3, 5]. However, water pollution is mainly caused by human activities [6]. The main sources of water pollution include urbanization, industrialization, mining and smelting, agriculture, paper mills, petrochemical manufacturing, and textile, leather, and tanning industries [1–7]. Among them, urban activities and mining industry are two sources of particular interest for water pollution and assessment of the role that plants play in water purification. The main pollutants released from urban sources are pathogenic microorganisms, heavy metals, dioxins, suspended solids, oils, grease, and phenols. In comparison, pollutants originating from industrial sources include benzene, ammonium, acids, sulfides, chlorides, fertilizers, herbicides, and insecticides [1, 3–5, 7]. Over the last years, particular concerns have been raised about emerging water contaminants that are globally persistent and released from urban areas and mining industry, such as drugs, biocides, fragrances, pesticides, hormones, plasticizers, detergents, metabolites, antioxidants, combustion indicators, and flame retardants [1, 7].

The Stockholm Convention listed 12 substances as persistent organic pollutants, characterized by long-term toxicity and accumulation in fatty tissue of living organisms, which can cause nervous system damage, immune system diseases, reproductive or developmental disorders, and cancer, even if present in the organisms in small quantities [7]. The United Nations 2030 Agenda for Sustainable Development calls for availability and access to safe and clean drinking water and sanitation through the Sustainable Development Goal (SDG) 6: Clean Water and Sanitation. Wastewater treatment systems are fundamental in terms of mitigating pollution and preventing the spread of diseases and thus preserving the health and well-being of millions. In addition, the Sustainable Development Goal 9 (SDG 9) targets resilient infrastructure, sustainable industrialization, and introduction and promotion of new technologies enabling the efficient use of resources. Striving toward these aims, a new concept known as nature-based solution (NBS) has drawn a lot of attention in recent decades. Nature-base solutions refer to actions that are "inspired by, supported by, or copied from nature" [8, 9]. A few examples include phytoremediation and natural and constructed wetlands, which offer a notable potential for application in the control of water quality and mitigation of water pollution [10, 11], with multiple environmental, social, and economic benefits [12]. Dendroremediation, which is the use of trees to clean up polluted soil and water, represents one of the phytoremediation techniques with great potential [13]. Fast-growing tree species can be considered as ideal low-cost candidates for phytoremediation applications [14]. Trees act as water filters and improve water quality by conducting the uptake of heavy metals through their extensive root system [15]. Besides, the phytoremediation technique has been extensively used in constructed wetlands, given the cost-effective and environmental-friendly approach and economic benefits associated with the maintenance of natural wastewater treatment systems [10, 16]. Constructed wetlands are suitable for water retention and degradation of contaminants both in urban and mining areas [17], while phytoremediation can support urban water management [12]. Aquatic plant species (also designated as macrophytes or hydrophytes) are highly recommended options for the phytoremediation of heavy metals and other water contaminants [10, 11]. The use of aquatic macrophytes to mitigate water pollution is one of the most researched issues all over the world. Aquatic macrophytes occur naturally and they are well adapted to their surroundings. The management of water quality with aquatic macrophytes is based on their capability to remove excessive nutrient loads, susceptible of causing eutrophication of surface waters. Aquatic macrophytes are able to retain metals by acting as traps of suspended particles [18]. This chapter aims to review the importance of phytoremediation and wetland systems as NBS for water regulation and pollution control, especially in the urban areas and mining industry.

2 Phytoremediation as a Nature-Based Solution to Control Pollution

The NBS has relevant potential with regard to effective environmental management, namely of water resources, including to reduce and mitigate surface water and groundwater pollution, and also contamination of the soil, sediments, sludges, and even air [19, 20]. Plant organisms have been proposed to mitigate the dangerous effects of pollutants in the environment, having a major role in nature-based phytotechnologies usually named phytoremediation [20–22]. Phytotechnologies have been used worldwide to remediate and restore damaged ecosystems, especially those caused by industrial byproducts affecting rivers and other waterways [23]. The term phytoremediation is derived from the Greek prefix "*phyto*" meaning plant and the Latin suffix "*remedium*" meaning to clean or restore [19, 24, 25]. The use of the term phytoremediation was introduced by the United States Environmental

Protection Agency (EPA) in 1991, and it was first used in literature in 1993 by Cunningham and Berti [26].

Phytoremediation is a plant-based technology that uses naturally occurring or genetically engineered plants (trees, shrubs, and grasses) and/or their associated rhizosphere microorganisms, performing as soil amendments and being applied in agronomic techniques for degradation and sequestration of inorganic and organic pollutants, thus cleaning contaminated environments [27–33]. Phytoremediation represents an environmentally friendly and sustainable solution, since it is a solar-driven and a cost-effective alternative approach to the rehabilitation of hazardous waste sites in the industrial, agricultural, and urban territories [20, 22, 34–39].

Phytoremediation as the green phytotechnology consists of several different plant-based technologies, each one having a different mechanism for remediation of polluted soil, sediment, and water [19, 24, 37, 39, 40]:

- Phytoextraction is the use of the plants in waste sites to absorb metals from the soil and translocate and accumulate them into the aboveground plant organs;
- Phytostabilization uses plants tolerant to the target element to reduce its environmental mobility, thus stabilizing rather than cleaning contaminated soils, sediments, and groundwater, through the absorption and accumulation in the roots, adsorption onto the roots, or precipitation or immobilization within the root zone or in the substrate;
- Phytovolatilization (also called phytodegradation or phytotransformation) is the process of uptaking contaminants from soils, sediments, and water through the plant roots, subsequent transformation by the plant, and releasing of the contaminant or a modified form of the contaminant to the atmosphere, through the plant shoots or leaves by transpiration. It occurs as growing trees and other plants take up water and organic contaminants;
- Rhizofiltration (also called phytofiltration) involves the use of aquatic plants to clean various types of water environment; it is the process in which roots or the whole plants absorb, concentrate, and/or precipitate hazardous compounds from aqueous solutions and are later harvested to remove the pollutants;
- Rhizodegradation, also called enhanced rhizosphere biodegradation, phytostimulation, or plant-assisted bioremediation or degradation, is the breakdown of soil contaminants through the activity of rhizosphere microorganisms.

The selection of the phytoremediation approach depends on several factors, including the natural capability of plants to tolerate, accumulate, and translocate a high content of the pollutant; the high production of biomass; the climatic conditions; the additional technologies available for the recovery of pollutants from the harvested plant biomass; and the amendments applied. Phytoextraction and phytostabilization are the most widespread methods used for soil phytoremediation, while rhizofiltration is a phytoremediation technology more suitable for the removal of pollutants from contaminated waters [11, 37, 41].

In comparison with the conventional technologies for water and wastewater treatment (e.g. flotation, adsorption, and electrochemical oxidation) [42, 43], phytoremediation has its own advantages and limitations. Among the evident advantages of phytoremediation are the economic efficiency, ecological safety, esthetic attractiveness, and public recognitions [12, 37]. It can be applied both in situ and ex situ, in remote locations, for one or multiple contaminants, and complementary to other remediation methods. The use of plant species for the cleanup of contaminated water and soil, with both inorganic and organic substances, also contributes to the control of soil erosion and surface runoff and climate change mitigation through greenhouse gases sequestration (e.g. carbon dioxide) and increases the esthetic function of restored habitats [20, 26]. An important advantage of the phytoremediation is the application of different amendments (for example, phyllosilicates, zeolites, peat, ash, lignin, compost). It promotes the stabilization of heavy metals in soils and the availability of metals for plants [37]. Conversely, there are some disadvantages of phytoremediation, such as the concentration of contaminants, toxicity, and bioavailability. The plant choice and stress tolerance are consequential disadvantages in terms of the accumulation of pollutant in fruit and other edible plant parts, low biomass production that requires several planting and harvesting decontamination, and the handling and disposal of contaminated plants [44]. The efficiency of phytoremediation is limited by the depth of the root systems and the solubility and availability of the pollutants; it requires a long time to clean the water and is highly affected by climatic and seasonal conditions. Furthermore, the introduction of nonlocal plant species in the environment can affect the biodiversity. Operation and maintenance costs, involving moving, replanting, pruning, harvesting, fertilization, and monitoring the status of vegetation and performance of the system, should be considered in the design of phytoremediation solutions [45].

The plant species widely used in phytoremediation can be grouped into three categories. The first category consists of excluders, which include plant species that prevent the absorption of pollutants by roots and their translocation to the aboveground organs. Excluders are used for the stabilization of soils and the prevention of spreading toxic metals due to erosion. The second category consists of accumulators, i.e. plants that concentrate pollutants in the aboveground parts without signs of toxicity, and can be used for the cleanup of contaminated environments. Among them, the hyperaccumulators are the plants capable of absorbing 50–500 times greater amounts of pollutants than usual. They are mainly included into the Asteraceae, Brassicaceae, Caryophyllaceae, Euphorbiaceae, Fabaceae, Flacourtiaceae, Lamiaceae, Poaceae, and Violaceae families. Brassicaceae contains the highest number of hyperaccumulator plant species. The third category consists of indicators, plants used in phytomonitoring and phytoindication. Plants naturally growing in polluted environments have the best potential for phytoremediation in aquatic or terrestrial environments [37, 44].

3 The Role of Trees and Aquatic Plants in Water Regulation and Pollution Control

Dendroremediation represents the use of trees to clean up polluted environments [13]. Trees have been suggested as the appropriate plants for the phytoremediation of contaminated land and water, because they provide a number of beneficial attributes. The most important phytoremediation traits of trees include their large biomass and genetic variability and comprise established agronomic and management practices with economic value, with a high degree of public acceptability [46, 47]. Their natural growth in conditions of highly variable biotic and abiotic stresses, their deep root system, intense transpiration, and high productivity of biomass have been proposed as the traits for the effective phytoextraction. A combination of trees and grasses promotes effective phytostabilization [13, 37].

The most phytoremediation-promising woody plants are the fast-growing species of the Salicaceae family (*Salix* spp., *Populus* spp.) [48–63]. The extensive root system allows willows (Fig. 1) and poplars to tolerate and uptake high concentrations of pollutants and to play an important role in the phytostabilization technique [52–58]. Those plant species have the ability to uptake different heavy metals and organic pollutants from contaminated substrates (Table 1), accumulate pollutants in the root tissues, and translocate them to the aboveground organs [52, 56–62].

Fig. 1 Traits of woody *Salix* sp. important for phytoremediation purpose (Zorana Hrkić Ilić 2020)

Table 1 Review of heavy metal (HM) concentrations accumulated in the vegetative organs of *Salix* L. and *Populus* L. genotypes, used in phytoremediation of contaminated urban and industrial sites

Genotype	Plant organ	HM	Concentration of HM (mg/kg dry weight)	Ref.
10 *Populus* clones	Roots	Cd	1.8–7.4	[63]
10 *Populus* clones	Leaves	Cd	0.03–0.71	[63]
Populus nigra L. clone "Poli"	Root	Cd	9.7	[64]
Populus nigra clone "Poli"	Leaves	Cd	9.7	[64]
Populus × *canadensis* Moench clone "I-214"	Root	Zn	3.6	[65]
Salix alba L. clone "SS5"	Root	Cd	4.1	[64]
Salix alba L. clone "SS5"	Leaves	Cd	376.0	[64]
Salix × *rubra* Huds.	Roots	Ni	14.6	[66]
Salix × *fragilis* L.	Leaves	Cd	168	[67]
Salix × *smithiana* Willd.	Roots	Cd	3.05	[68]
12 *Salix* clones	Roots	Cu	2,065.3–2,876.2	[69]
12 *Salix* clones	Leaves	Cu	391.1–660.8	[69]
Salix alba clone "68/53/1"	Roots	Ni	2970.6	[52]
Salix nigra Marshall clone "0408"	Roots	Ni	4,161.9	[52]

Fig. 2 (**a**) Short rotation plantations of *Salix viminalis* L. clone "Inger" (in Romania, near Ghilad), willow genotype significant for use in the phytoremediation; (**b**) width of clone "Inger" plantation belts (Photo by Ratko Ristić 2020)

Poplars and willows, as flood tolerant species [70, 71], might be a good solution for the remediation of contaminated waters [37, 48–51]. The root systems of willows and poplars uptake large amounts of water and thus may contribute to flood mitigation [65] while decreasing the metal contents available for leaching from the soil to water [46, 47].

Willows and poplars with a rapid and high production of biomass might be a proper solution for different sources of pollution [65, 72–74]. For example, energy willow (*Salix viminalis* L. clone "Inger," Fig. 2a, b) has valuable phytoremediation properties. Due to high transpiration rates (15–20 l/m^2/day), it is suitable to drain soils where surface water is retained or groundwater levels are high. The "Inger" clone grows successfully on polluted and eroded surfaces, enriching them with organic contents and binding pollutants. It can be planted in the belts for the protection of soils against wind erosion and pollution in canals and settlements. Sludge and wastewater from treatment facilities can be deposited and very quickly decomposed under "Inger" plantations, which contribute to the enhancement of soil structure and improve its quality (https://poljoprivreda.info/tekst/energetska-vrba-revolucija-u-proizvodnji-zelene-energije; https://balkangreenenergynews.com/energy-willow-salix-viminalis-biomass-where-you-want-it/).

Aquatic plants are also important in phytoremediation since they play a crucial role in the active and passive biogeogenic cycling of trace elements [23]. The properties of intensive and nonselective uptake and hyperaccumulation of various chemical elements and substances determine the importance of aquatic plants as bioindicators [75, 76]. The chemical composition of aquatic plants can significantly reflect the quality of water and sediments within river basins. Therefore, it is possible to monitor the pollution of aquatic ecosystems based on the degree of accumulation of nutrients and heavy metals in their tissues. For the purpose of bioindication, data on the concentrations of certain chemical elements in different parts of plant tissue can be used as indicators for the chemical load of natural resources [77–79].

Changes in the distribution and structure of macrophytic communities are considered very reliable biological indicators of water quality, and their role in aquatic ecosystems is manifested through the processes of chemical bioconcentration, i.e. phytoextraction [80]. Aquatic plants act as the bioaccumulators given their properties of nonselective absorption, degradation, and mineralization of nutrients, heavy metals, and organic pollutants within the aquatic environments, including sediments [81]. These species are known as hyperaccumulators, since they maintain normal metabolism even in the presence of larger amounts of some pollutants [82, 83]. The ability of hyperaccumulation is particularly important for removing heavy metals from the environment and successful purification of natural waters [76]. Especially, rooted aquatic plants accumulate heavy metals in their roots, thus playing a significant role in the immobilization and hyperaccumulation of pollutants and acting as bioremediation agents [73, 77]. The increased concentration of nutrients and heavy metals in tissues of the aquatic plants may be due to their high concentration in the aquatic environment [80]. However, metal concentration in aquatic plants is usually significantly larger than in the surrounding water [82] and varies depending on the plant species, age, organs, and other environmental factors such as temperature, salinity, and pH [83].

For example, a case study was conducted in the swamp-marsh ecosystem complex Bardača located in the far North-East of the Lijevče field, Republic of Srpska, Bosnia and Herzegovina (45 06′ 06″ North latitude and 17 26′ 26″ East longitude), covering 2.810 ha. On 2 March 2007, Bardača was declared a Ramsar area (number 1858) and an "Important Bird Area" (Fig. 3a, b). As it is, this sensitive ecosystem is under strong anthropogenic influence today, and it needs constant monitoring of water quality, in particular heavy metal contents in aquatic macrophytes. The measurements of heavy metals (Fe, Mn, Cu, Zn, and Pb) accumulated in the aquatic *Utricularia vulgaris* L. and *Salvinia natans* (L.). All plants, sampled in Bardača area, revealed that those species accumulated high concentrations of Fe, Mn, and Zn and

Fig. 3 (**a**) Nature wetland in Bardača Ramsar site, (**b**) Bardača aquatic plant community (Photo by Zorana Hrkić Ilić 2020)

could be used in the phytoremediation of water-polluted ecosystems [77]. Several studies of heavy metal and macronutrients accumulation in macrophytes that grow in urban rivers reveal their potential use as indicators of aquatic ecosystems and their remediation potential. These species include *Ceratophyllum demersum* L., *Phragmites australis*, *Hydrocharis morsus-ranae* L., *Potamogeton natans* L., *Stuckenia pectinata* (L.) Börner, *Potamogeton crispus* L., *Myriophyllum spicatum* L., *Trapa longicarpa* Janković, *Typha latifolia* L., *Elodea canadensis* Michx., *Callitriche palustris* L., *Persicaria amphibia* (L.) Delarbre, *Vallisneria natans* (Lour.) H. Hara, *Hydrilla verticillata* (L. f.) Royle, *Myriophyllum aquaticum* (Vell.) Verdc., and *Mentha aquatica* L. [80, 84, 85].

4 NBS Strategies for Water Regulation and Pollution Control in Urban Areas

The contamination of water in urban areas is receiving an increasing attention worldwide. Urban water is particularly exposed to contamination due to overpopulation and urban expansion in recent decades [86, 87]. The rapid growth of urban areas results in an increasing demand for freshwater resources and asserts the importance of reliable, resilient, and sustainable water management [88–90]. Water resources like rivers and streams are a natural connection between watersheds and seas and also a medium for the transport of pollutants from anthropogenic sources. Unfortunately, the research of the ecology of urban streams is mainly related to large cities and developed countries [91], even though serious threat has also been identified in urban areas of developing countries as well [92]. The effects of urbanization on water resources are defined as the urban stream syndrome [93], characterized by the altered channel morphology and stability, higher concentrations of pollutants, nutrients, and toxicants, and reduced species richness [94].

The status of water quality is difficult to assess in urban areas because of the complexity of pollution sources and type of pollutants [95–97], but most of them are a result of human activity. The identification of pollution sources and pollutants is a continuous process. Vehicular transportation sector and the atmospheric deposition have been previously identified as the major sources of water pollution within urban areas [95, 96]. The early studies addressed conventional pollutants such as total suspended solids, chemical or biochemical oxygen demand, trace metals, and various nitrogen and phosphorus elements. Nowadays, polluted urban stormwater and snowmelt contributed significantly to the deterioration of water quality in many urban areas. Recently, the sources of stormwater pollution in the urban areas are grouped into four main categories: atmospheric deposition, drainage surfaces, anthropogenic activities, and urban drainage systems. Industry is among the most important anthropogenic sources of pollution that endanger long term sustainability of nature and ecosystem services provided by streams and water [96].

Water pollutants can be divided into three main groups: (1) physical contaminants, which primarily impact the physical appearance or other physical properties of water, including sediments or organic material suspended in the water and originating from soil erosion; (2) chemical contaminants are elements or compounds, including organic compounds, nutrients, pesticides, metals, toxins, and human or animal drugs; and (3) biological contaminants are microbial organisms in water, including bacteria, viruses, protozoan, and parasites [96–99].

The growing population in urban areas has serious negative impacts on urban water resources due to inappropriate infrastructure to convey sewage or drinking water [100]. Contaminated drinking water and inadequate sanitation infrastructure exacerbate health risks in urban centers, as the major source of the diseases caused by bacterial and viral pathogens [5, 101]. Organic pollutants can also directly contaminate urban water and soils, through deposition after initial emission into the atmosphere followed by transport [102]. Inputs of metals and organic contaminants to the urban wastewater system occur from three sources: domestic, commercial, and urban runoff. The main persistent organic pollutants of global concern are polycyclic aromatic hydrocarbons (PAHs), polychlorinated biphenyls (PCBs), and polychlorinated dibenzo-*p*-dioxins and polychlorinated dibenzo-*p*-furan (PCDD/Fs), originating from atmospheric deposition onto paved surfaces and runoff [103, 104].

Increased concentration and loads of several chemical pollutants in stream water appear universal in urban streams, often occurring even at low levels of catchment urbanization [105]. Plants are useful sensors to identify environmental contamination and potential exposures to pollutants [106]. Green infrastructure in urban areas is exposed to the negative influences of pollution and deterioration of physical properties of urban soils, sealing, and traffic loads [107]. Sustainable management and stewardship of green urban infrastructure should be a part of urban planning systems and urban development [108], where a local administration should consider needs and prioritize green infrastructure in environmental maintenance, in alignment with urban development. In that sense, it is quite important to identify those plants that can be used to assess the status and trends related to human health [109], suitable for the improvement of environment in urban areas and featuring a high phytoremediation potential. Water management in urban areas can be achieved through nature-based and sustainable technologies for water remediation, such as phytoremediation and constructed wetlands [12, 110, 111]. By improving water quality, both types of technologies reveal potentials in water sanitation, treatment and reuse of the wastewater and rainwater [18, 36, 112, 113]. They comprise different types of NBS to address water pollution, but provide similar social, economic, and environmental benefits when applied. Phytoremediation can help address urban challenges in water management, while constructed wetlands in urban areas allow degradation of the contaminants and retention of the water [12]. The difference between the terrestrial and wetlands phytoremediation is reflected in the fact that wetland phytoremediation is still new and not a fully developed approach and is generally restricted by the shallow distribution of aquatic plant roots [114]. In addition, wetlands differ from traditional phytoremediation in that manipulation of the partially decomposed litter layer and sediments in wetlands has primary significance in the remediation processes, whereas uptake of pollutants by plants has a secondary role [115].

Urban vegetation, particularly woody plants (trees and shrubs), is a base source for the nature-inspired approach that provides various ecosystem services crucial for the people living in urban zones. Urban greening (Fig. 4a–e) has become a particular interest to many cities and municipalities, in terms of sustainable water management through the improvement of water quality, reduced stormwater, and the enhancement of water storage and interception of rainfall at the source. It is also involved in the reduction of pollution, by enhancing sediment retention and cleaning soil and water from pollutants accumulated in urban areas [116–118]. Numerous studies revealed that tree species commonly planted in urban green areas have a strong phytoremediation potential, such as *Betula pendula* Roth, *Robinia pseudoacacia* L., *Tilia cordata* Mill [14], *Aesculus hippocastanum* L. [119], *Celtis occidentalis* L., *Quercus robur* L., *Tilia argentea* L., and *Platanus* x *acerifolia* (Aiton) Willd [120].

Aquatic plant species in constructed wetlands have proved to be successful in retaining and transforming various pollutants in urban wastewater, stormwater, and sediments [121, 122], including nutrients, heavy metals, volatile organic compounds, pesticides, explosives, and petroleum hydrocarbons and additives [123–130]. However, assessing the phytoremediation potential of wetland plants is complex due to variable conditions of hydrology, plant species diversity, and water chemistry. Biological and physico-chemical properties of wetlands are important for the remediation of contaminants in water bodies, particularly the expansive rhizosphere of wetland plant species, the seasonality of wetland ecosystem, and the productivity and succession of the wetlands plant communities [121].

Compared to centralized sewer systems in cities, constructed wetlands are decentralized systems with an important role in the mitigation of flooding in urban areas, caused by excessive rain and stormwater [131]. Physico-chemical and biological features of constructed wetlands enable the removal of numerous pollutants from urban waters, including viruses and bacteria [132, 133], nutrients, heavy

Fig. 4 Green infrastructure in different cities of Republic of Srpska, in Bosnia and Herzegovina: (**a**) Public park in the city of Trebinje; (**b**) Urban recreation zone in the city of Trebinje; (**c, d**) Different woody species important for phytoremediation in the cities of Doboj and Prijedor; (**e**) Bushes as important elements of the green infrastructure in the town of Prnjavor (Photo by Marijana Kapović Solomun and Zorana Hrkić Ilić 2019)

metals, organic matter, herbicides, and polycyclic aromatic hydrocarbons [134], pharmaceuticals and personal care products [135].

The selection of suitable macrophyte species to be used in constructed wetland systems depends on the species tolerance to pollution, type of wetland design and mode of operation, and the composition of the urban waters [126, 128]. In most cases, constructed wetlands are planted with competitive, resistant, and proliferative species, such as *Phragmites australis* (Cav.) Trin. ex Steud. or *Typha spp.* [125, 127]. Recent studies, however, have shown that using native, common, and abundant plant species growing near water pollution sources can maximize the removal rate of pollutants in constructed wetlands [128, 130]. Species with high biomass production and ability to add oxygen in the root zone provide positive conditions for microbe development and bioremediation [124, 129, 136]. Plants that can adapt to wetland conditions are useful for phytofiltration [136].

Constructed wetlands, another type of NBS, can provide a number of environmental, social, and economic benefits [12]. Several case studies pointed out the use of constructed wetlands for both contaminant degradation and water retention in cities worldwide. A few examples include:

- *China*, where an experiment with ornamental hydrophytes for potential phytoremediation of urban wastewater was conducted in the arboretum of Zhejiang Normal University in the city of Jinhua, the Zhejiang Province. The results showed that *Iris pseudacorus* and *Acorus gramineus* species are outstanding either in adapting or cleaning the urban sewage, and that wetlands with ornamental hydrophytes were an appropriate choice for treatment of polluted urban wastewater [10].
- *Collateral Channel*, a 485 meter long slip channel constructed in Chicago, Illinois, USA, in the late nineteenth century. Over the years, it has received large quantities of organic and inorganic contaminants from municipal and industrial discharges. The main pollutants identified in sediment samples collected within the channel include lead (Pb), in concentrations up to 483 mg/kg, and PAHs with the total of 16 compounds attaining 1,500 mg/kg. A method based on active capping with an emergent wetland was applied in this channel to provide nutrient removal and habitat conservation. The wetland decreased the amount of contaminants and removed the nutrient content of the water flowing through it, based on plant uptake and microbial processes developed within the sediment layer. The wetland cap is also intended to provide a public recreational space [137].
- *Brisbane, Australia*, where two constructed wetlands provide harvesting of rainwater, used for irrigation purposes, a habitat for various macroinvertebrate species, and recreational and educational opportunities [138].

However, it is important to perform adequate maintenance of the wetland system to avoid any potential bioavailability, bioconcentration, and biomagnification of contaminants through aquatic animals [121, 139].

5 NBS Strategies for Water Regulation and Pollution Control in Mining Industry

Metal mining is another very important source of environmental pollution and poses a risk to human health [140, 141]. Abandoned mine sites and unsecured mine tailings are an important source of many toxic elements [140, 141], which attain concentrations usually significantly higher than in the Earth's crust [141, 142]. The significant increases of toxic metals in the air, soil, water, and sediments in areas near mine sites are typically driven by factors such as high precipitation, floods, mine raw sewage spills, mismanagement in the wastewater treatment, and outdated and obsolete equipment [141, 143].

The lack of wastewater control in abandoned hard rock mines that become flooded over time is a major environmental pollution issue, directly associated with an acid mine drainage (AMD) [144, 145]. The AMD represents water draining from mine sites, often highly acidic and contaminated with toxic metals (usually Cd, Pb, Zn, Fe, Mn) and sulfates. The AMD is produced when sulfide-bearing material is exposed to oxygen and water [146, 147]. Metal concentrations in AMD are often several hundred times greater than water quality standards. Thus, the AMD has prominent negative and toxic effects on soil, water, plants, animals, and humans in the affected region [148–150].

The soluble fraction of metals can spread to adjacent water resources such as streams, rivers, or groundwater [151, 152]. Soil surrounding mines can be highly contaminated with heavy metals due to irrigation with water from rivers receiving mine wastewater discharge. If mine sites are affected by flooding, the surrounding agricultural lands, crops, fields, and gardens can record high heavy metal concentrations. Heavy metals can enter the food chain, posing a serious threat to human health and the environment [141, 153]. Effective and affordable measures of flood management and remediation of land contaminated with toxic heavy metals are necessary for the reduction of environmental pollution [141, 142]. The remediation of mine soil and wastewater with appropriate plant species and communities, that have hyperaccumulating and tolerant traits to toxic levels of heavy metals and trace elements, can contribute to the control of the soil and water pollution caused by toxic elements [142, 149, 154–158]. Considering the characteristics of mine areas and their level of contamination, phytostabilization is the most preferable phytoremediation technique, while phytoextraction could be applied when contamination of mine soil is limited. In addition, the introduction of native plant species, which are tolerant to local conditions, is preferable in comparison to the introduction of invasive species in order to reduce possible impacts on the ecosystem [156–159]. Many case studies conducted worldwide emphasized the significance of phytoremediation in mine restoration and stabilization of contaminated soil and water. Phytomining, a phytoremediation technique that uses heavy metal-tolerant plants for extraction of inorganic substances from mine ore has been successfully used in mine areas [155, 160, 161]. Woody species like *Acacia auriculiformis* Benth., *Acacia confuse* Merr., *Jatropha curcas* L., and *Melaleuca armillaris* (Sol. ex Gaertn.) Sm. are a viable option for remediation of acid mine waters [1]. There are many examples of successful herbaceous and woody plants application in the cleaning of soils and water in contaminated mine sites. Some of these examples are summarized below.

- *São Domingos mine (south-east Portugal)*, an abandoned copper mine located in the Iberian Pyrite Belt, with massive sulfides exploitation from 1868 until 1966, is a focus of great environmental concern. Deposits of waste materials have high acid generating potential and can release enormous quantities of Al, Fe, SO_4^{2-}, Pb, and Cu, which contaminate the river bank soils and stream system. Also, AMD water occurs several kilometers downstream the mine and causes acidification and contamination of the sediments and soils along several kilometers of the water course. Several studies have shown that phytostabilization should be considered as one of the best solutions for recovering the soil of the São

Domingos mining area to stabilize metal contaminants in the soil and reduce the risks to human health and the environment. This goal is currently being achieved with two spontaneously grown and tolerant plant species, *Erica australis* L. and *Erica mackaiana subsp. andevalensis* (Cabezudo & Rivera) D.C. McClint. & E.C. Nelson. Notably, *Erica mackaiana subsp. andevalensis* had the ability to grow in very hostile conditions, such as soils with pH values between 3 and 4 and high contents of Al, Pb, Fe, As, and Sb. Although in the São Domingos mine site the concentration of most trace elements in *Erica mackaiana subsp. andevalensis* and *E. australis* plant samples were within the normal levels, some plant samples displayed excessive or toxic levels of Mn, Pb, and As. The concentrations of these heavy metals in the analyzed plants tissues were 1129.7 mg/kg Mn (toxic limits 400–1,000 mg/kg Mn), 262.81 mg/kg Pb (toxic limits 30–300 mg/kg Pb), and 42.99 mg/kg As (toxic limits 5–20 mg/kg As), respectively. Both *Erica* species are known as Al-tolerant and Mn-accumulators and may be of great importance for the recovery of sulfide mining areas [162].

- *Papua (New Guinea)* is well-known for its rainforest with rich biodiversity but also for the large number of operating mine sites. The risk of environmental pollution is very high and phytoremediation using selected plant species is one of the proven NBS solutions proven suitable for the rehabilitation of the closed mine sites. In the closed Namie mine, in Wau district of Morobe province, Papua, a study investigated the efficiency of local plant species in the remediation of soil and surface water polluted by several heavy metals (Cd, Cu, Fe, Hg, Pb, Zn), after 70 years of mining operations, was investigated. Local contamination is linked with the poisoning of water by riverine tailings disposal. Samples of water, soil, and garden food were collected randomly from areas close to the Namie mine after 19 years of closure and revealed that the soil has high concentrations of heavy metals (9.5143 μg/g Cd, 39.2857 μg/g Cu, 38,742.86 μg/g Fe, 33.14290 μg/g Hg, 56 μg/g Pb, and 249.5714 μg/g Zn, respectively). Other samples from creeks, ponds, and garden foods have shown lower heavy metal contents. The local plant species *Piper aduncum* L., *Brachiaria reptans* (L.) C.A. Gardner & C.E.Hubb. and *Phragmites karka* (Retz.) Trin. ex Steud. can easily adapt into the harsh conditions of the infertile mine soils, and they turned out to be suitable for phytoextraction [163].
- *The region of Bor (Serbia)* represents one of the largest active copper mine basins in Europe that caused severe environmental degradation. It is characterized by large open pits, flotation tailings, waste dumps, coarse texture soil, and high concentrations of As (44.5–271 mg/kg) and Cu (311–2,820 mg/kg) that 10 times exceed the legal Serbian limit values. Endangered and protected plant species, such as *Epilobium dodonaei* Vill., spontaneously colonized the mine slopes and mine waste sites. A study conducted on the copper mine waste sites, located within the industrial area of the copper mining and smelting facilities, revealed that the content of As, Cu, Pb, and Zn in roots (3.98 mg/kg As, 140 mg/kg Cu, 3.19 mg/kg Pb, and 72.8 mg/kg Zn, respectively) and shoots (4.69 mg/kg As, 57.7 mg/kg Cu, 1.17 mg/g Pb, and 59.3 mg/kg Zn, respectively) of *E. dodonaei* reflected the multi-metal pollution of the investigated site. *E. dodonaei* largely

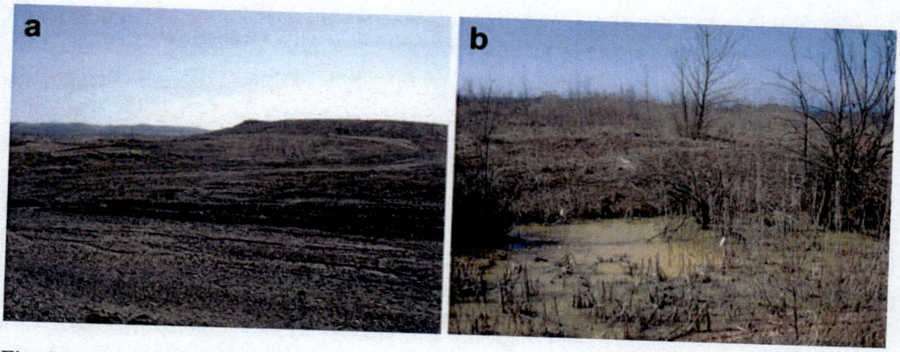

Fig. 5 (**a**) A dump site for residues from the Mital iron mine near Prijedor in the Republic of Srpska, Bosnia and Herzegovina; (**b**) The swampy area of the Mital mine landfill encroached by pioneer plants (photo by Marijana Kapović Solomun & Zoran Stanivuković, 2005)

retains Cu, Pb, and Zn in roots than in shoots and has the potential for the phytoremediation of mine wastes. Considering that *E. dodonaei* is an endangered species within the Carpatho-Balkan metallogenic province, such approach can link ecological restoration with biodiversity preservation [158].

- *The region of the northern Chile*, with numerous copper mine tailing deposits, also represents a risk to the human health and the environment, given the high content of metals. Similar to the abovementioned mining areas, plants also play a significant role in reducing the metal contamination of the region. The evaluation of total metal concentrations of Cu, Mn, Fe, Pb, Zn, and Cd in roots and shoots of three plant species grown in the selected mine tailing, *Prosopis tamarugo* Phil., *Schinus molle* L., and *Atriplex nummularia* Lindl., has shown that native plants can have a phytoremediation potential. Those species were considered as excluders of Cu, Mn, Fe, Pb, and Zn. *A. nummularia* was found to be an accumulator of Mn, Pb, and Zn and the most promising species for the phytostabilization of Cd in tailings. The *S. molle* plant species is included as an accumulator of Cu, Mn, Pb, and Zn [159].

The remediation of mine sites through the use of natural or constructed wetlands (Fig. 5a, b) can be difficult, but comprises one of the most cost-effective methods of phytoremediation of water and soils impacted by AMD [136, 164]. Phytoextraction and phytostabilization are two common phytoremediation techniques for remediation of AMD. A wide range of wetland plant species have an important role in the remediation of heavy metals from the soil and water affected by acid mine drainage [124, 130, 136, 165–167].

Constructed wetlands have been common for the treatment of various types of contaminated waters [122, 136, 164, 168]. However, little is known about the remediation abilities of wetlands [164] that are naturally formed on mine tailings [136]. Natural wetlands are transitional ecosystems between aquatic and terrestrial systems and frequent recipients of stormwater runoff [169]. Native wetland plants with high hyperaccumulating and phytoremediation ability could have a significant role in the effective treatment of contaminated waters generated from mines. For example, in an old natural wetland in Wales, UK, connected with a river contaminated with wastewater from an abandoned copper mine, the levels of toxic dissolved metals in the water were significantly reduced (92% reduction of Fe concentration, 83% of Zn, and 94% of Cu). The wetland also reduced the acidity of the river water. The three most common plants found in this natural wetland were soft rush (*Juncus effusus*), common reed (*Phragmites australis*), and common cottongrass (*Eriophorum angustifolium* Honck.). The highest levels of the three mentioned metals are found in soft rush (with the concentration of Fe reaching 88.7 mg/g in the plant tissue), as well as in the organic sediments around the roots of the wetland plants, confirming their vital role in capturing the metals [164].

6 Final Considerations

Plants are very important for providing many ecosystem service, but their existence and functioning are strongly dependent on human interventions and water availability. There are many woody and aquatic plants with phytoremediation potential, capable to remove different contaminants from soils and water, and therefore of great importance for the reduction of water pollution, particularly in urban and mining areas. Water, as a resource of vital importance, is exposed to constant degradation, particularly in urban areas, where water is managed only by blue and gray infrastructures. The use of water resources is continuous and globally increasing, while the availability of water resources is becoming limited due to environmental degradation through pollution. Therefore, sustainable water management urges the implementation of innovative and nature-based solutions that provide clean and safe water as priority issue. The risk of environmental pollution is high and phytoremediation is one of the promising green technologies that has proven suitable for the rehabilitation of many contaminated sites. The introduction of woody or aquatic plants with phytoremediation potential could be of great importance for sustainable water management, as it provides several additional benefits such as clean soils, biomass production, and genetic variability while also receiving a high degree of public acceptability.

Acknowledgment This chapter was supported by the Ministry of Scientific-Technological Development, Higher Education and Information Society of the Republic of Srpska, Grant No. 19/6-020/961-43/18, and by the EU Cost Action LAND4FLOOD: Natural Flood Retention on Private Land (CA16209), through the Short Term Scientific Mission (STSM) grant entitled "Forests and floods – Public Perception and Institutional Approaches."

References

1. Inyinbor Adejumoke A, Adebesin Babatunde O, Oluyori Abimbola P, Adelani-Akande Tabitha A, Dada Adewumi O, Oreofe Toyin A (2018) Water pollution: effects, prevention, and climatic impact. In: Water challenges of an urbanizing world. https://doi.org/10.5772/intechopen.72018
2. Schwarzenbach RP (2006) The challenge of micropollutants in aquatic systems. Science 313 (5790):1072–1077. https://doi.org/10.1126/science.1127291
3. Kjellstrom T, Lodh M, McMichael T, Ranmuthugala G, Shrestha R, Kingsland S (2006) Air and water pollution: burden and strategies for control. In: Jamison DT, Breman JG, Measham AR, Alleyne G, Claeson M, Evans DB, Jha P, Mills A, Musgrove P (eds) Disease control priorities in developing countries, 2nd edn. The International Bank for Reconstruction and Development/The World Bank, Washington, pp 817–832. Co-published by Oxford University Press, New York, https://www.ncbi.nlm.nih.gov/books/NBK11769/
4. Azizullah A, Khattak MNK, Richter P, Hader DP (2011) Water pollution in Pakistan and its impact on public health – a review. Environ Int 37(2):479–497. https://doi.org/10.1016/j.envint.2010.10.007
5. Schwarzenbach RP, Egli T, Hofstetter TB, von Gunten U, Wehrli B (2010) Global water pollution and human health. Annu Rev Env Resour 35(1):109–136. https://doi.org/10.1146/annurev-environ-100809-125342
6. Owa FD (2013) Water pollution: sources, effects, control and management. Mediterr J Soc Sci 4(8):65. https://www.mcser.org/journal/index.php/mjss/article/view/1760/1764. Accessed 27 July 2020
7. Fuerhacker M (2009) EU water framework directive and Stockholm convention. Environ Sci Pollut Res 16(S1):92–97. https://doi.org/10.1007/s11356-009-0126-4
8. Van den Bosch M, Ode Sang Å (2017) Urban natural environments as nature-based solutions for improved public health – a systematic review of reviews. Environ Res 158:373–384. https://doi.org/10.1016/j.envres.2017.05.040
9. European Commission (2017) Environment – research & innovation policy topics – nature based solutions
10. Zhang H, Liu P, Yang Y, Chen W (2007) Phytoremediation of urban wastewater by model wetlands with ornamental hydrophytes. J Environ Sci 19:902–909
11. Lone MI, He Z, Stoffella PJ, Yang X (2008) Phytoremediation of heavy metal polluted soils and water: progresses and perspectives. J Zhejiang Univ Sci B 9(3):210–220. https://doi.org/10.1631/jzus.b0710633
12. Song Y, Kirkwood N, Maksimović Č, Zhen X, O'Connor D, Jin Y, Hou D (2019) Nature based solutions for contaminated land remediation and brownfield redevelopment in cities: a review. Sci Total Environ. https://doi.org/10.1016/j.scitotenv.2019.01.347
13. González-Oreja JA, Rozas M, Alkorta I, Garbisu C (2008) Dendroremediation of heavy metal polluted soils. Rev Environ Health 23(3). https://doi.org/10.1515/reveh.2008.23.3.223
14. Dadea C, Russo A, Tagliavini M, Mimmo T, Zerbe S (2017) Tree species as tools for biomonitoring and phytoremediation in urban environments. Arboricult Urban For 43 (4):155–167
15. Bose S, Vedamati J, Rai V, Ramanathan AL (2008) Metal uptake and transport by Tyaha angustata L. grown on metal contaminated waste amended soil: an implication of phytoremediation. Geoderma 145:136–142
16. Smith GR (2015) Phytoremediation-by-design: community-scale landscape systems design for healthy communities. Int J Sustain Dev World Ecol:1–7. https://doi.org/10.1080/13504509.2015.1079276
17. Kadlec RH, Wallace S (2008) Treatment wetlands. CRC Press, Boca Raton
18. Srivastava J, Gupta A, Chandra H (2008) Managing water quality with aquatic macrophytes. Rev Environ Sci Biotechnol 7(3):255–266. https://doi.org/10.1007/s11157-008-9135-x

19. Tangahu BV, Sheikh Abdullah SR, Basri H, Idris M, Anuar N, Mukhlisin M (2011) A review on heavy metals (As, Pb, and Hg) uptake by plants through phytoremediation. Int J Chem Eng 2011:1–31. https://doi.org/10.1155/2011/939161
20. Aisien FA, Oboh IO, Aisien ET (2013) Phytotechnology-remediation of inorganic contaminants. In: Anjun NA, Pereira ME, Ahmad I, Duarte AC, Umar S, Khan NA (eds) Phytotechnologies: remediation of environmental contaminants. CRC Press, Boca Raton, pp 75–82
21. Salt D, Blaylock M, Kumar N, Dushenkov V, Ensley B, Chet I, Raskin I (1995) Phytoremediation: a novel strategy for the removal of toxic metals from the environment using plants. Nat Biotechnol 13:468–474. https://doi.org/10.1038/nbt0595-468
22. Jadia CD, Fulekar MH (2009) Phytoremediation of heavy metals: recent techniques. Afr J Biotechnol 8(6):921–928. https://www.ajol.info/index.php/ajb/article/view/59987. Accessed 29 July 2020
23. Pilipović A, Zalesny Jr RS, Rončević S, Nikolić N, Orlović S, Beljin J, Katanić M (2019) Growth, physiology, and phytoextraction potential of poplar and willow established in soils amended with heavy-metal contaminated, dredged river sediments. J Environ Manage 239:352–365. https://doi.org/10.1016/j.jenvman.2019.03.072
24. Prasad MNV, Freitas H (2003) Metal-tolerant plants: biodiversity prospecting for phytoremediation technology. Electron J Biomed 6:110–146
25. Jagdale S, Chabukswar A (2016) Phyto-remediation: using plants to clean up soils: Phyto-remediation. In: Rathoure AK, Dhatwalia VK (eds) Toxicity and waste management using bioremediation. IGI Global, pp 215–235. https://doi.org/10.4018/978-1-4666-9734-8.ch011
26. Russell K (2005) The use and effectiveness of phytoremediation to treat persistent organic pollutants. US Environmental Protection Agency, Washington. https://clu-in.org/download/studentpapers/phyto_to_treat_pops_russell.pdf. Accessed 29 July 2020
27. Cunningham SD, Shann JR, Crowley DE, Anderson TA (1997) Phytoremediation of contaminated water and soil. In: Kruger EL, Anderson TA, Coats JR (eds) Phytoremediation of soil and water contaminants. ACS symposium series. American Chemical Society, Washington, pp 2–17. https://doi.org/10.1021/bk-1997-0664.ch001
28. Flathman PE, Lanza GR (1998) Phytoremediation: current views on an emerging green technology. J Soil Contamin 7(4):415–432. https://doi.org/10.1080/10588339891334438
29. Vysloužilová M, Tlustoš P, Száková J (2003) Cadmium and zinc phytoextraction potential of seven clones of Salix spp. planted on heavy metal contaminated soils. Plant Soil Environ 49 (12):542–547. https://doi.org/10.17221/4191-PSE
30. Vysloužilová M, Tlustoš P, Száková J, Pavliková D (2003) As, Cd, Pb and Zn uptake by Salix spp clones grown in soils enriched by high loads of these elements. Plant Soil Environ 49 (5):191–196. https://doi.org/10.17221/4112-PSE
31. Pilon-Smits E (2005) Phytoremediation. Annu Rev Plant Biol 56(1):15–39. https://doi.org/10.1146/annurev.arplant.56.032604.144214
32. Helmisaari H-S, Salemaa M, Derome J, Kiikkilä O, Uhlig C, Nieminen TM (2007) Remediation of heavy metal–contaminated forest soil using recycled organic matter and native woody plants. J Environ Qual 36(4):1145. https://doi.org/10.2134/jeq2006.0319
33. Cioica N, Tudora C, Iuga D, Deak G, Matei M, Nagy EM, Gyorgy Z (2019) A review on phytoremediation as an ecological method for in situ clean up of heavy metals contaminated soils. E3S Web Conf 112:03024
34. Pulford ID, Watson C (2003) Phytoremediation of heavy metal contaminated land by trees – a review. Environ Int 29(4):529–540. https://doi.org/10.1016/S0160-4120(02)00152-6
35. Vangronsveld J, Herzig R, Weyens N, Boulet J, Adriaensen K, Ruttens A, Thewys T, Vassilev A, Meers E, Nehnevajova E (2009) Phytoremediation of contaminated soils and groundwater: lessons from the field. Environ Sci Pollut Res 16(7):765–794. https://doi.org/10.1007/s11356-009-0213-6

36. Zhang BY, Zheng JS, Sharp RG (2010) Phytoremediation in engineered wetlands: mechanisms and applications. Procedia Environ Sci 2:1315–1325. https://doi.org/10.1016/j.proenv.2010.10.142
37. Koptsik GN (2014) Problems and prospects concerning the phytoremediation of heavy metal polluted soils: a review. Eurasian Soil Sci 47(9):923–939
38. Muehe EM, Weigold P, Adaktylou IJ, Planer-Friedrich B, Kraemer U, Kappler A, Behrens S (2015) Rhizosphere microbial community composition affects cadmium and zinc uptake by the metal-hyperaccumulating plant *Arabidopsis halleri*. Appl Environ Microbiol 81 (6):2173–2181. https://doi.org/10.1128/aem.03359-14
39. Pajević S, Borišev M, Nikoli, N, Arsenov DD, Orlović S, Župunski, M. (2016) Phytoextraction of heavy metals by fast-growing trees: a review. In: Ansari AA, Gill SS, Gill R, Lanza GR, Newman L (eds) Phytoremediation: management of environmental contaminants 3. Springer, Cham, pp. 29–64. https://doi.org/10.1007/978-3-319-40148-5_2
40. Adams PW, Lamoureux S (2005) A literature review of the use of native northern plants for the re-vegetation of arctic mine tailings and mine waste. https://www.enr.gov.nt.ca/sites/enr/files/wkss_northern_plants_re-vegetation-2005.pdf. Accessed 29 July 2020
41. Javed MT, Tanwir K, Akram MS, Shahid M, Niazi NK, Lindberg S (2019) Chapter 20-phytoremediation of cadmium-polluted water/sediment by aquatic macrophytes: role of plant-induced pH changes. In: Hasanuzzaman M, Prasad MNV, Fujita M (eds) Cadmium toxicity and tolerance in plants. Academic Press, London, pp 495–529. https://doi.org/10.1016/B978-0-12-814864-8.00020-6
42. Rubio J, Carissimi E, Rosa JJ (2007) Flotation in water and wastewater treatment and reuse: recent trends in Brazil. Int J Environ Pollut 30(2)
43. Elena A, Orbeci C, Lazau C, Sfirloaga P, Vlazan P, Bandas C, Grozescu I (2013) Waste water treatment methods. Water Treat. https://doi.org/10.5772/53755
44. Hossein F, Nastaein Z, Ramlah T, Hamed F (2016) Advantages and disadvantages of phytoremediation A concise review. Int J Environ Technol Sci 2:69–75
45. Schnoor JL (1997) Phytoremediation. https://clu-in.org/download/toolkit/phyto_e.pdf
46. Chibiuke GU, Obiora SC (2014) Heavy metal polluted soils: effect on plants and bioremediation methods. Appl Environ Soil Sci 2014:1–12. https://doi.org/10.1155/2014/752708
47. Pulford ID, Dickinson NM (2005) Phytoremediation technologies using trees. In: Prasad MNV, Sajwan KS, Naidu R (eds) Trace elements in the environment: biogeochemistry, biotechnology and bioremediation. CRC Press, Boca Raton, Now Taylor and Francis, pp 383–403. https://doi.org/10.1201/9781420032048.sec4
48. Kuzovkina YA, Knee M, Quigley MF (2004) Effects of soil compaction and flooding on the growth of 12 willow (*Salix* L.) species. J Environ Hortic 22(3):155–160
49. Kuzovkina YA, Quigley MF (2005) Willows beyond wetlands: uses of Salix L. species for environmental projects. Water Air Soil Pollut 162(1–4):183–204. https://doi.org/10.1007/s11270-005-6272-5
50. Pilipović A, Orlović S, Nikolić N, Galić Z (2006) Investigating potential of some poplar (*Populus* sp.) clones for phytoremediation of nitrates through biomass production. Environmental applications of poplar and willow working party, Northern Ireland, May 2006, pp 18–20
51. Kuzovkina YA, Volk TA (2009) The characterization of willow (*Salix* L.) varieties for use in ecological engineering applications: co-ordination of structure, function and autecology. Ecol Eng 35(8):1178–1189. https://doi.org/10.1016/j.ecoleng.2009.03.010
52. Hrkić Ilić Z, Pajević S, Borišev M, Luković J (2020) Assessment of phytostabilization potential of two *Salix* L. clones based on the effects of heavy metals on the root anatomical traits. Environ Sci Pollut Res Int 27(23):29361–29383. https://doi.org/10.1007/s11356-020-09228-8
53. Marmiroli M, Pietrini F, Maestri E, Zacchini M, Marmiroli N, Massacci A (2011) Growth, physiological and molecular traits in Salicaceae trees investigated for phytoremediation of

heavy metals and organics. Tree Physiol 31(12):1319–1334. https://doi.org/10.1093/treephys/tpr090
54. Sylvain B, Mikael M-H, Florie M, Emmanuel J, Marilyne S, Sylvain B, Domenico M (2016) Phytostabilization of As, Sb and Pb by two willow species (*S. viminalis* and *S. purpurea*) on former mine technosols. Catena 136:44–52. https://doi.org/10.1016/j.catena.2015.07.008
55. Lebrun M, Miard F, Nandillon R, Léger J-C, Hattab-Hambli N, Scippa GS, Bourgerie S, Morabito D (2018) Assisted phytostabilization of a multicontaminated mine technosol using biochar amendment: early stage evaluation of biochar feedstock and particle size effects on as and Pb accumulation of two Salicaceae species (*Salix viminalis* and *Populus euramericana*). Chemosphere 194:316–326. https://doi.org/10.1016/j.chemosphere.2017.11.113
56. Lux A, Luxová M, Abe J, Morita S (2004) Root cortex: structural and functional variability and responses to environmental stress. Root Res 13(3):117–131. https://doi.org/10.3117/rootres.13.117
57. Lux A, Šottníková A, Opatrná J, Greger M (2004) Differences in structure of adventitious roots in Salix clones with contrasting characteristics of cadmium accumulation and sensitivity. Physiol Plant 120(4):537–545. https://doi.org/10.1111/j.0031-9317.2004.0275.x
58. Lux A, Martinká M, Vaculík M, White PJ (2011) Root responses to cadmium in the rhizosphere: a review. J Exp Bot 62(1):21–37. https://doi.org/10.1093/jxb/erq281
59. Tlustoš P, Száková J, Vysloužilová M, Pavlíková D, Weger J, Javorská H (2007) Variation in the uptake of arsenic, cadmium, lead, and zinc by different species of willows *Salix* spp. grown in contaminated soils. Cent Eur J Biol 2:254–275. https://doi.org/10.2478/s11535-007-0012-3
60. Borišev M, Pajević S, Nikolić N, Krstić B, Župunski M, Kebert M, Pilipović A, Orlović S (2012) Response of *Salix alba* L. to heavy metals and diesel fuel contamination. Afr J Biotechnol 11:14313–14319. https://www.ajol.info/index.php/ajb/article/view/129437. Accessed 9 July 2020
61. Vaculík M, Konlechner C, Langer I, Adlassnig W, Puschenreiter M, Alexander Lux A, Hauser M-T (2012) Root anatomy and element distribution vary between two *Salix caprea* isolates with different Cd accumulation capacities. Environ Pollut 163:117–126. https://doi.org/10.1016/j.envpol.2011.12.031
62. Tőszér D, Magura T, Simon E (2017) Heavy metal uptake by plant parts of willow species: a meta-analysis. J Hazard Mater 336:101–109. https://doi.org/10.1016/j.jhazmat.2017.03.068
63. Pietrini F, Zacchini M, Iori V, Pietrosanti L, Bianconi D, Massacci A (2010) Screening of poplar clones for cadmium phytoremediation using photosynthesis, biomass and cadmium content analyses. Int J Phytoremediation 12:105–120. https://doi.org/10.1080/15226510902767163
64. Zacchini M, Iori V, Mugnozza G, Pietrini F, Massacci A (2011) Cadmium accumulation and tolerance in *Populus nigra* and *Salix alba*. Biol Plant 55:383–386. https://doi.org/10.1007/s10535-011-0060-4
65. Stoláriková M, Vaculík M, Lux A, Baccio D, Minnocci A, Andreucci A, Sebastiani L (2012) Anatomical differences of poplar (*Populus* × *euramericana* clone I-214) roots exposed to zinc excess. Biologia 67(3):483–489. https://doi.org/10.2478/s11756-012-0039-4
66. Drzewiecka K, Mleczek M, Gąsecka M, Magdziak Z, Goliński P (2012) Changes in *Salix viminalis* L. cv. 'Cannabina' morphology and physiology in response to nickel ions – hydroponic investigations. J Hazard Mater 217–218:429–438. https://doi.org/10.1016/j.jhazmat.2012.03.056
67. Luković J, Merkulov L, Pajević S, Zorić L, Nikolić N, Borišev M, Karanović D (2012) Quantitative assessment of effects of cadmium on the histological structure of poplar and willow leaves. Water Air Soil Pollut 223(6):2979–2993. https://doi.org/10.1007/s11270-012-1081-0
68. Kacálková L, Tlustoš P, Száková J (2014) Chromium, nickel, cadmium, and lead accumulation in maize, sunflower, willow, and poplar. Pol J Environ Stud 23(3):753–761

69. Yang W, Wang Y, Zhao F, Ding Z, Zhang X, Zhu Z, Yang X (2014) Variation in copper and zinc tolerance and accumulation in 12 willow clones: implications for phytoextraction. J Zhejiang Univ Sci B 15:788–800. https://doi.org/10.1631/jzus.B1400029
70. Nakai A, Yurugi Y, Kisanuki H (2010) Stress responses in *Salix gracilistyla* cuttings subjected to repetitive alternate flooding and drought. Trees 24:1087–1095. https://doi.org/10.1007/s00468-010-0481-2
71. Rood SB, Nielsen JL, Shenton L, Gill KM, Letts MG (2009) Effects of flooding on leaf development, transpiration, and photosynthesis in narrowleaf cottonwood, a willow-like poplar. Photosynth Res 104(1):31–39. https://doi.org/10.1007/s11120-009-9511-6
72. Sebastiani L, Scebba F, Tognetti R (2004) Heavy metal accumulation and growth responses in poplar clones Eridano (*Populus deltoides* × *maximowiczii*) and I-214 (*P.* × *euramericana*) exposed to industrial waste. Environ Exp Bot 52(1):79–88. https://doi.org/10.1016/j.envexpbot.2004.01.003
73. Tognetti R, Sebastiani L, Minnocci A (2004) Gas exchange and foliage characteristics of two poplar clones grown in soil amended with industrial waste. Tree Physiol 24(1):75–82. https://doi.org/10.1093/treephys/24.1.75
74. Borišev M, Pajević S, Nikolić N, Pilipović A, Krstić B, Orlović S (2009) Phytoextraction of Cd, Ni, and Pb using four willow clones (*Salix* spp.). Pol J Environ Stud 18(4):553–561
75. Ginn BK (2011) Distribution and limnological drivers of submerged aquatic plant communities in Lake Simcoe (Ontario, Canada): utility of macrophytes as bioindicators of lake trophic status. J Great Lakes Res 37:83–89. https://doi.org/10.1016/j.jglr.2011.03.015
76. Favas PJC, Pratas J, Prasad MNV (2012) Accumulation of arsenic by aquatic plants in large-scale field conditions: opportunities for phytoremediation and bioindication. Sci Total Environ 433:390–397. https://doi.org/10.1016/j.scitotenv.2012.06.091
77. Maksimović T, Rončević S, Kukavica B (2019) *Utricularia vulgaris* L. and *Salvinia natans* (L.) All. Heavy metal (Fe, Mn, Cu, Zn and Pb) bioaccumulation specificity in the area of Bardača fishpond. Ekológia (Bratislava) 38(3):201–213. https://doi.org/10.2478/eko-2019-0016
78. Borišev M, Pajević S, Stanković Ž, Krstić B (2008) Macrophytes as indicators and potential remediators in aquatic ecosystems: a case study. Large Rivers 18(1–2):107–115. https://doi.org/10.1127/lr/18/2008/107
79. Bolpagni R, Fanelli G, Oggioni A, Testi A (2012) Macrophyte indicators of environmental quality of rivers in Italy at local, regional and geographical scales. In: Sridhar KR (ed) Aquatic plants and plant diseases. Nova Science Publishers, Inc, Hauppauge, pp 147–171
80. Pajević S, Borišev M, Rončević S, Vukov D, Igić R (2008) Heavy metal accumulation of Danube river aquatic plants – indication of chemical contamination. Cent Eur J Biol 3:285–294. https://doi.org/10.2478/s11535-008-0017-6
81. Samecka-Cymerman A, Kempers AJ (1996) Bioaccumulation of heavy metals by aquatic macrophytes around Wrocław, Poland. Ecotoxicol Environ Saf 35(3):242–247. https://doi.org/10.1006/eesa.1996.0106
82. Lewander M, Greger M, Kautsky L, Szarek E (1996) Macrophytes as indicators of bioavailable Cd, Pb and Zn flow in the river Przemsza, Katowice Region. Appl Geochem 11 (1–2):169–173. https://doi.org/10.1016/0883-2927(95)00074-7
83. Pajević S, Kevrešan, Ž, Vučković M, Radulović S, Frontasyeva M, Pavlov S, Galinskaya T (2004) Aquatic macrophytes as biological resources for monitoring the impacts of heavy metals on the aquatic environment. Internat Assoc Danube Res (IAD), limnological reports 35 (Proceedings of the 35th conference, Novi Sad, Serbia and Montenegro), pp 323–330
84. Lu G, Wang B, Zhang C, Li S, Wen J, Lu G, Zhu C, Zhou Y (2018) Heavy metals contamination and accumulation in submerged macrophytes in an urban river in China. Int J Phytoremediation 20(8):839–846. https://doi.org/10.1080/15226514.2018.1438354
85. Samecka-Cymerman A, Kempers AJ (2007) Heavy metals in aquatic macrophytes from two small rivers polluted by urban, agricultural and textile industry sewages SW Poland. Arch Environ Contam Toxicol 53(2):198–206. https://doi.org/10.1007/s00244-006-0059-6

86. Fletcher TD, Andrieu H, Hamel P (2013) Understanding, management and modelling of urban hydrology and its consequences for receiving waters: a state of the art. Adv Water Resour 51:261–279. https://doi.org/10.1016/j.advwatres.2012.09.001
87. Kuriata-Potasznik A, Szymczyk S (2015) Magnesium and calcium concentrations in the surface water and bottom deposits of a river-lake system. J Elem 20:677–692. https://doi.org/10.5601/jelem.2015.20.1.788
88. Oral HV, Carvalho P, Gajewska M, Ursino N, Masi F, van Hullebusch ED, Kazak JK, Exposito A, Cipolletta G, Andersen TR, Finger DC, Simperler L, Regelsberger M, Rous V, Radinja M, Buttiglieri G, Krzeminski P, Rizzo A, Dehghanian K, Nikolova M, Zimmermann M (2020) A review of nature-based solutions for urban water management in European circular cities: a critical assessment based on case studies and literature. Blue-Green Syst 2 (1):112–136. https://doi.org/10.2166/bgs.2020.932
89. McDonald RI, Weber K, Padowski J, Flörke M, Schneider C, Green PA, Gleeson T, Eckman S, Lehner B, Balk D, Boucher T, Grill G, Montgomery M (2014) Water on an urban planet: urbanization and the reach of urban water infrastructure. Glob Environ Chang 27:96–105. https://doi.org/10.1016/j.gloenvcha.2014.04.022
90. Butler D, Ward S, Sweetapple C, Astaraie-Imani M, Diao K, Farmani R, Fu G (2016) Reliable, resilient and sustainable water management: the safe & SuRe approach. Global Chall 1 (1):63–77. https://doi.org/10.1002/gch2.1010
91. Ramírez A, Rosas KG, Lugo AE, Ramos-González OM (2014) Spatio-temporal variation in stream water chemistry in a tropical urban watershed. Ecol Soc 19(2):45. https://doi.org/10.5751/ES-06481-190245
92. Gunawardena J, Ziyath AM, Bostrom TE, Bekessy LK, Ayoko GA, Egodawatta P, Goonetilleke A (2013) Characterisation of atmospheric deposited particles during a dust storm in urban areas of eastern Australia. Sci Total Environ 2013(461–462):72–80. https://doi.org/10.1016/j.scitotenv.2013.04.080
93. Meyer JL, Paul MJ, Taulbee WK (2005) Stream ecosystem function in urbanizing landscapes. J N Am Benthol Soc 24(3):602–612. https://doi.org/10.1899/04-021.1
94. Biasioli MR, Barberis R, Ajmone-Marsan F (2006) The influence of a large city on some soil properties and metals content. Sci Total Environ 356(1–3):154–164. https://doi.org/10.1016/j.scitotenv.2005.04.033
95. Duda AM, Lenat DR, Penrose DL (1982) Water quality in urban streams: what we can expect. Water Pollut Control Fed 54(7):1139–1147. http://www.jstor.org/stable/25041633?origin=JSTOR-pdf
96. Müller A, Österlund H, Marsalek J, Viklander M (2020) The pollution conveyed by urban runoff: a review of sources. Sci Total Environ 709:1–18. https://doi.org/10.1016/j.scitotenv.2019.136125
97. Salmore AK, Hollis EJ, McLellan SL (2006) Delineation of a chemical and biological signature for stormwater pollution in an urban river. J Water Health 4:247–262. https://doi.org/10.2166/wh.2006.006
98. Luqman M, Butt TM, Tanvir A, Atiq M, Hussan MZY, Yaseen M (2013) Phytoremediation of polluted water by trees: a review. Afr J Agric Res 8(17):1591–1595. https://doi.org/10.5897/AJAR11.1111
99. Martins RT, Melo AS, Gonçalves Jr JF, Hamada N (2015) Leaf-litter breakdown in urban streams of Central Amazonia: direct and indirect effects of physical, chemical, and biological factors. Freshwater Sci 34(2). https://doi.org/10.1086/681086
100. Obolewski K (2013) Use of macrozoobenthos for biological assessment of water quality in oxbow lakes of varying hydrological connectivity to the main river channel in the example of Łyna river valley. Ochrona Środowiska 35:19–26
101. Loucks P, van Beek E (2017) Water resource systems planning and management – an introduction to methods, models, and applications. Springer, Cham. https://doi.org/10.1007/978-3-319-44234-1

102. Cachada A, Pato P, Rocha-Santos T, da Silva EF, Duarte AC (2012) Levels, sources and potential human health risks of organic pollutants in urban soils. Sci Total Environ 430:184–192. https://doi.org/10.1016/j.scitotenv.2012.04.075
103. European Communities (2001) Pollutants in urban waste water and sewage sludge. Final report Luxembourg: Office for Official Publications of the European Communities http://europa.eu.int
104. Campanella BE, Bock C, Schröder P (2002) Phytoremediation to increase the degradation of PCBs and PCDD/Fs – potential and limitations. Environ Sci Pollut Res 9(1):73–85. https://doi.org/10.1065/esDr2001.09.084.6
105. Hatt BE, Fletcher TD, Walsh CJ, Taylor SL (2004) The influence of urban density and drainage infrastructure on concentrations and loads of pollutants in small streams. J Environ Manage 34(1):112–124. https://doi.org/10.1007/s00267-004-0221-8
106. Henry HF, Burken JG, Maier RM, Newman LA, Rock S, Schnoor JL, Suk WA (2013) Phytotechnologies–preventing exposures, improving public health. Int J Phytoremediation 15(9):889–899. https://doi.org/10.1080/15226514.2012.760521
107. Widney S, Fischer B, Vogt J (2016) Tree mortality undercuts ability of tree-planting programs to provide benefits: results of a three-city study. Forests 7(12):65. https://doi.org/10.3390/f7030065
108. Young RF (2011) Planting the living city. J Am Plann Assoc 77(4):368–381. https://doi.org/10.1080/01944363.2011.616996
109. Burger J (2006) Bioindicators: types, development, and use in ecological assessment and research. Environ Bioindic 1(1):22–39. https://doi.org/10.1080/15555270590966483
110. Funk A, Reckendorfer W, Kucera-Hirzinger V, Raab R, Schiemer F (2009) Aquatic diversity in a former floodplain: remediation in an urban context. Ecol Eng 35:1476–1484. https://doi.org/10.1016/j.ecoleng.2009.06.013
111. Bhatia M, Goyal D (2013) Analyzing remediation potential of wastewater through wetland plants: a review Environmental Progress & Sustainable Energy, 1-19. https://doi.org/10.1002/ep.11822
112. Yadav BK, Siebel MA, van Bruggen JJA (2011) Rhizofiltration of a heavy metal (lead) containing wastewater using the wetland plant *Carex pendula*. Clean (Weinh) 39 (5):467–474. https://doi.org/10.1002/clen.201000385
113. Ali S, Abbas Z, Muhammad Rizwan M, Zaheer IE, Yava I, Ünay A, Abdel-Daim MM, Bin-Jumah M, Hasanuzzaman M, Kalderis D (2020) Application of floating aquatic plants in phytoremediation of heavy metals polluted water: a review. Sustainability 12:1–33. https://doi.org/10.3390/su12051927
114. Herath I, Vithanage M (2015) Phytoremediation in constructed wetlands. In: Ansari AA et al (eds) Phytoremediation: management of environmental contaminants, vol 2. Springer, Cham, pp 243–263. https://doi.org/10.1007/978-3-319-10969-5_21
115. Horne AJ (2000) Phytoremediation by constructed wetlands. In: Terry N, Bañuelos G (eds) Phytoremediation of contaminated soil and water. CRC Press, Boca Raton
116. Eisenman TS (2016) Greening cities in an urbanizing age: the human health bases in the nineteenth and early twenty-first centuries. Changing Times 6(2):216–246. http://scholarworks.umass.edu/larp_faculty_pubs/69. Accessed 29 July 2020
117. Beatley T (2017) Handbook of biophilic city planning & design. Island Press, Washington
118. Kapović Solomun M (2019) Commentary: small retention in polish forests from a forest management perspective – copying of existing could be right path. In: Hartmann T, Slavíková L, McCarthy S (eds) Nature-based flood risk management on private land-disciplinary perspectives on a multidisciplinary challenge. Springer, Cham, pp 45–51. https://doi.org/10.1007/978-3-030-23842-1_5
119. Pavlović M, Rakić T, Pavlović D, Kostić O, Jarić S, Mataruga Z, Pavlović P, Mitrović M (2017) Seasonal variations of trace element contents in leaves and bark of horse chestnut (*Aesculus hippocastanum* L.) in urban and industrial regions in Serbia. Arch Biol Sci 69 (2):201–214. https://doi.org/10.2298/ABS161202005P

120. Greksa A, Ljevnaić-Mašić B, Grabić J, Benka P, Radonić V, Blagojević B, Sekulić M (2019) Potential of urban trees for mitigating heavy metal pollution in the city of Novi Sad, Serbia. Environ Monit Assess 191(10):636. https://doi.org/10.1007/s10661-019-7791-7
121. Helfield J, Diamond M (1997) Use of constructed wetlands for urban stream restoration: a critical analysis. Environ Manag 21:329–341. https://doi.org/10.1007/s002679900033
122. Vymazal J (2001) Constructed wetlands for wastewater treatment in the Czech Republic. Water Sci Technol 44(11–12):369–374
123. Williams JB (2002) Phytoremediation in wetland ecosystems: progress, problems, and potential. Crit Rev Plant Sci 21(6):607–635. https://doi.org/10.1080/0735-260291044386
124. Stottmeister U, Wießner A, Kuschk P, Kappelmeyer U, Kästner M, Bederski O, Müller RA, Moormann H (2003) Effects of plants and microorganisms in constructed wetlands for wastewater treatment. Biotechnol Adv 22(1–2):93–117. https://doi.org/10.1016/j.biotechadv.2003.08.010
125. Calheiros CSC, Rangel OSS, Castro PML (2009) Treatment of industrial wastewater with two-stage constructed wetlands planted with *Typha latifolia* and *Phragmites australis*. Bioresour Technol 100(13):3205–3213. https://doi.org/10.1016/j.biortech.2009.02.017
126. Ranieri E, Young TM (2012) Clogging influence on metals migration and removal in sub-surface flow constructed wetlands. J Contam Hydrol 129–130:38–34. https://doi.org/10.1016/j.jconhyd.2012.01.002
127. Vymazal J (2013) The use of hybrid constructed wetlands for wastewater treatment with special attention to nitrogen removal: a review of a recent development. Water Res:47. https://doi.org/10.1016/j.watres.2013.05.029
128. Guittonny-Philippe A, Petit M-E, Masotti V, Monnier Y, Malleret L, Coulomb B, Combroux I, Baumberger T, Viglione J, Laffont-Schwob I (2015) Selection of wild macrophytes for use in constructed wetlands for phytoremediation of contaminant mixtures. J Environ Manage 147:108–123. https://doi.org/10.1016/j.jenvman.2014.09.009
129. Vymazal J, Březinová T (2015) The use of constructed wetlands for removal of pesticides from agricultural runoff and drainage: a review. Environ Int 75:11–20. https://doi.org/10.1016/j.envint.2014.10.026
130. Türker OC, Türe C, Böcük H, Yakar A (2016) Phyto-management of boron mine effluent using native macrophytes in mono-culture and poly-culture constructed wetlands. Ecol Eng 94:65–74. https://doi.org/10.1016/j.ecoleng.2016.05.043
131. Dou T, Troesch S, Petitjean A, Gábor PT, Esser D (2017) Wastewater and rainwater management in urban areas: a role for constructed wetlands. Procedia Environ Sci 37:535–541. https://doi.org/10.1016/j.proenv.2017.03.036
132. Stenström T, Carlander A (2001) Occurrence and die-off of indicator organisms in the sediment in two constructed wetlands. Water Sci Technol J Int Assoc Water Pollut Res 44:223–230. https://doi.org/10.2166/wst.2001.0833
133. Morató J, Codony F, Sánchez O, Pérez LM, García J, Mas J (2014) Key design factors affecting microbial community composition and pathogenic organism removal in horizontal subsurface flow constructed wetlands. Sci Total Environ 481:81–89. https://doi.org/10.1016/j.scitotenv.2014.01.068
134. Malaviya P, Singh A (2012) Constructed wetlands for management of urban stormwater runoff. Crit Rev Environ Sci Technol 42(20):2153–2214. https://doi.org/10.1080/10643389.2011.574107
135. Reyes-Contreras C, Hijosa-Valsero M, Sidrach-Cardona R, Bayona JM, Bécares E (2012) Temporal evolution in PPCP removal from urban wastewater by constructed wetlands of different configuration: a medium-term study. Chemosphere 88(2):161–167. https://doi.org/10.1016/j.chemosphere.2012.02.064
136. Prasad MNV, Greger M, Aravind P (2005) Biogeochemical cycling of trace elements by aquatic and wetland plants: relevance to phytoremediation. In: Prasad MNV, Sajwan KS, Naidu R (eds) Trace elements in the environment: biogeochemistry, biotechnology and bioremediation. CRC Press, Boca Raton, Now Taylor and Francis, Chap. 24, pp 443–474

137. Yin K, Viana P, Zhao X, Rockne K (2010) Characterization, performance modeling, and design of an active capping remediation project in a heavily polluted urban channel. Sci Total Environ 408:3454–3463. https://doi.org/10.1016/j.scitotenv.2010.03.053
138. Greenway M (2017) Stormwater wetlands for the enhancement of environmental ecosystem services: case studies for two retrofit wetlands in Brisbane, Australia. J Clean Prod 163:S91–S100
139. Gimbert F, Petitjean Q, Al-Ashoor A, Cretenet C, Aleya L (2018) Encaged *Chironomus riparius* larvae in assessment of trace metal bioavailability and transfer in a landfill leachate collection pond. Environ Sci Pollut Res 25:11303–11312
140. Wahsha M, Bini C, Argese E, Minello F, Fontana S, Wahsheh H (2012) Heavy metals accumulation in willows growing on spolic technosols from the abandoned Imperina Valley mine in Italy. J Geochem Explor 123:19–24. https://doi.org/10.1016/j.gexplo.2012.07.004
141. Jovanović VS, Mitić V, Mandić SN, Ilić M, Simonović S (2015) Heavy metals in the post-catastrophic soils. In: Sherameti I, Varma A (eds) Heavy metal contamination of soils. Soil biology, vol 44. Springer, Cham. https://doi.org/10.1007/978-3-319-14526-6_1
142. Prasad MNV (2005) Stabilization, remediation, and integrated management of metal – contaminated ecosystems by grasses (Poaceae). In: Prasad MNV, Sajwan KS, Naidu R (eds) Trace elements in the environment (biogeochemistry, biotechnology and bioremediation). CRC Press, Boca Raton. Now Taylor and Francis, pp 405–424. https://doi.org/10.1201/9781420032048.ch21
143. Krzaklewski W, Barszcz J, Małek S, Kozioł K, Pietrzykowski M (2004) Contamination of forest soils in the vicinity of the sedimentation pond after zinc and lead ore flotation (in the region of Olkusz, Southern Poland). Water Air Soil Pollut 159(1):151–164. https://doi.org/10.1023/B:WATE.0000049173.18935.71
144. White S (2003) Wetland use in acid mine drainage remediation. http://home.eng.iastate.edu/~tge/ce421-521/Steven%20White.pdf. Accessed 29 July 2020
145. Obreque-Contreras J, Pérez-Flores D, Gutiérrez P, Chávez-Crooker P (2015) Acid mine drainage in Chile: an opportunity to apply bioremediation technology. Hydrol Curr Res 6 (3):215. https://doi.org/10.4172/2157-7587.1000215
146. Holmström H, Ljungberg J, Öhlander B (2000) The character of the suspended and dissolved phases in the water cover of the flooded mine tailings at Stekenjokk, northern Sweden. Sci Total Environ 247(1):15–31. https://doi.org/10.1016/s0048-9697(99)00454-4
147. Sheoran AS, Sheoran V (2006) Heavy metal removal mechanism of acid mine drainage in wetlands: a critical review. Miner Eng 19:105–116
148. Nyquist J, Greger M (2009) A field study of constructed wetlands for preventing and treating acid mine drainage. Ecol Eng 35:630–642. https://doi.org/10.1016/j.ecoleng.2008.10.018
149. Kersten G, Majestic B, Quigley M (2017) Phytoremediation of cadmium and lead-polluted watersheds. Ecotoxicol Environ Saf 137:225–232. https://doi.org/10.1016/j.ecoenv.2016.12.001
150. Kiiskila JD, Sarkar D, Panja S, Sahi SV, Datta R (2019) Remediation of acid mine drainage-impacted water by vetiver grass (*Chrysopogon zizanioides*): a multiscale long-term study. Ecol Eng 129:97–108. https://doi.org/10.1016/j.ecoleng.2019.01.018
151. Das B, Nordin R, Mazumder A (2009) Watershed land use as a determinant of metal concentrations in freshwater systems. Environ Geochem Health 31:595–607. https://doi.org/10.1007/s10653-008-9244-z
152. Pourang N, Noori AS (2014) Heavy metals contamination in soil, surface water and ground-water of an agricultural area adjacent to Tehran oil refinery. Iran Int J Environ Res 8 (4):871–886
153. Liao J, Wen Z, Ru X, Chen J, Wu H, Wei C (2016) Distribution and migration of heavy metals in soil and crops affected by acid mine drainage: public health implications in Guangdong Province, China. Ecotoxicol Environ Saf 124:460–469. https://doi.org/10.1016/j.ecoenv.2015.11.023

154. Festin ES, Tigabu M, Chileshe MN, Syampungani S, Odén PC (2019) Progresses in restoration of post-mining landscape in Africa. J For Res 30:381–396. https://doi.org/10.1007/s11676-018-0621-x
155. Prasad M, Pratas J, Freitas H (2005) Trace elements in plants and soils of abandoned mines in Portugal. In: Prasad MNV, Sajwan KS, Naidu R (eds) Trace elements in the environment (biogeochemistry, biotechnology and bioremediation). CRC Press, Boca Raton. Now Taylor and Francis, pp 507–519. https://doi.org/10.1201/9781420032048.ch26
156. RoyChowdhury A, Sarkar D, Datta R (2015) Remediation of acid mine drainage-impacted water. Curr Pollut Rep 1:131–141. https://doi.org/10.1007/s40726-015-0011-3
157. Lai T, Cappai G, Carucci A (2016) Phytoremediation of mining areas: an overview of application in lead- and zinc-contaminated soils. In: Ansari A, Gill S, Gill R, Lanza G, Newman L (eds) Phytoremediation. Springer, Cham, pp 3–27
158. Ranđelović D, Gajić G, Mutić J, Pavlović P, Mihailović N, Jovanović S (2016) Ecological potential of *Epilobium dodonaei* Vill. for restoration of metalliferous mine wastes. Ecol Eng 95:800–810. https://doi.org/10.1016/j.ecoleng.2016.07.015
159. Lam EJ, Cánovas M, Gálvez ME, Montofré ÍL, Keith BF, Faz Á (2017) Evaluation of the phytoremediation potential of native plants growing on a copper mine tailing in northern Chile. J Geochem Explor 182:210–217. https://doi.org/10.1016/j.gexplo.2017.06.015
160. Sheoran V, Sheoran AS, Poonia P (2009) Phytomining: a review. Miner Eng 22 (12):1007–1019. https://doi.org/10.1016/j.mineng.2009.04.001
161. Chaney RL, Reeves RD, Baklanov IA, Centofanti T, Broadhurst CL, Baker AJM, Van der Ent A, Roseberg RJ (2014) Phytoremediation and phytomining: using plants to remediate contaminated or mineralized environments. In: Rajakaruna N, Boyd RS, Harris TB (eds) Plant ecology and evolution in harsh environment. Nova Science Publishers, Hauppauge, pp 365–392. https://doi.org/10.13140/2.1.3750.2721
162. Abreu MM, Tavares MT, Batista MJ (2008) Potential use of *Erica andevalensis* and *Erica australis* in phytoremediation of sulphide mine environments: São Domingos, Portugal. J Geochem Explor 96:210–222. https://doi.org/10.1016/j.gexplo.2007.04.007
163. Rungwa S, Arpa G, Sakulas H, Harakuwe A, Timi D (2013) Phytoremediation – an eco-friendly and sustainable method of heavy metal removal from closed mine environments in Papua New Guinea. Procedia Earth Planet Sci 6:269–277. https://doi.org/10.1016/j.proeps.2013.01.036
164. Dean AP, Lynch S, Rowland P, Toft BD, Pittman JK, White KN (2013) Natural wetlands are efficient at providing long-term metal remediation of fresh water systems polluted by acid mine drainage. Environ Sci Technol 47(21):12029–12036. https://doi.org/10.1021/es4025904
165. Das PK (2018) Phytoremediation and nanoremediation : emerging techniques for treatment of acid mine drainage water. Def Life Sci J 3(2):190–196. https://doi.org/10.14429/dlsj.3.11346
166. Karathanasis A, Johnson C (2003) Metal removal potential by three aquatic plants in an acid mine drainage wetland. Mine Water Environ 22:22–30. https://doi.org/10.1007/s102300300004
167. Pat-Espadas AM, Loredo Portales R, Amabilis-Sosa LE, Gomez G, Vidal G (2018) Review of constructed wetlands for acid mine drainage treatment. Water 10(11):1685. https://doi.org/10.3390/w10111685
168. Sheoran AS (2006) A laboratory treatment study of acid mine water of wetlands with emergent macrophyte (*Typha angustata*). Int J Min Reclam Environ 20(3):209–222. https://doi.org/10.1080/13895260600564695
169. Carleton JN, Grizzard TJ, Godrej AN, Post HE, Lampe L, Kenel PP (2000) Performance of a constructed wetlands in treating urban stormwater runoff. Water Environ Res 72(3):295–304. https://doi.org/10.2175/106143000x137518

Part II
NBS Effectiveness in Flood Mitigation: Assessment Methods and Case Studies

Impacts of Land Abandonment on Flood Mitigation in Mediterranean Mountain Areas

E. Nadal-Romero, C. Juez, M. Khorchani, D. Peña-Angulo, N. Lana-Renault, D. Regüés, T. Lasanta, and J. M. García-Ruiz

Contents

Abstract From the mid-twentieth century, Mediterranean mountains were affected by a rapid and generalized land abandonment process. This chapter (1) summarizes the impacts of land abandonment on the hydrological dynamics in Mediterranean mountain areas; (2) evaluates post-land abandonment management practices (LMPs) for flood mitigation based on nature-based solutions (NBS); and (3) briefly discusses some examples in the Central Pyrenees. In general, land abandonment resulted in a

E. Nadal-Romero (✉), C. Juez, M. Khorchani, D. Peña-Angulo, D. Regüés, T. Lasanta, and J. M. García-Ruiz
Instituto Pirenaico de Ecología, Procesos Geoambientales y Cambio Global, IPE-CSIC, Zaragoza, Spain
e-mail: estelanr@ipe.csic.es; carmelo@ipe.csic.es; makki.khorchani@ipe.csic.es; dpa@ipe.csic.es; dregues@ipe.csic.es; fm@ipe.csic.es; humberto@ipe.csic.es

N. Lana-Renault
Departamento de Ciencias Humanas, Universidad de La Rioja, Logroño, Spain
e-mail: noemi-solange.lana-renault@unirioja.es

Carla S. S. Ferreira, Zahra Kalantari, Thomas Hartmann, and Paulo Pereira (eds.), *Nature-Based Solutions for Flood Mitigation: Environmental and Socio-Economic Aspects*, Hdb Env Chem (2022) 107: 189–214, DOI 10.1007/698_2021_772,

Published online: 30 May 2021

natural colonization by shrubs and forests, which, in turn, led to a decrease in river discharges and sediment yields. NBS, as mitigation measures to control flood occurrence, have been carried out including afforestation, shrub clearing, landscape changes, and recovery of terraces and stonewalls. In addition, in the most hydrogeomorphologically active areas, a combination of NBS and grey infrastructures was used to control floods. Grey infrastructures produce immediate effects but they are short-lived and expensive solutions. LMPs based on NBS present advantages and disadvantages: (1) afforestation was the most common practice, reducing floods, hydrological connectivity, peak flows, and sediment yields, in spite of their lower impact in extreme events; (2) shrub clearing decreases the number of forest fires and maintains the occurrence of the most frequent floods; (3) the recovery of mosaic landscapes produces environmental consequences, being important sources of ecosystem services, such as the regulation of floods; and (4) the recovery of agricultural terraces and stonewalls is highly expensive, but presents social and cultural benefits, reducing hydrological connectivity, peak flows, and runoff. In the near future, NBS in abandoned lands should be based on cost-effective and long-term strategies to mitigate flood risk. NBS should be cost-effective, ensure longer lifetime than grey infrastructures, and be adapted to different local objectives and global scenarios.

Keywords Afforestation, Agricultural terraces, Floods, Land abandonment, Nature-based solutions, Revegetation

1 Introduction: Land Abandonment in Mediterranean Mountain Areas

Large areas worldwide have been affected by farmland abandonment particularly in temperate and developed regions [1–4]. Campbell et al. [5] calculated that around 385–472 million hectares of agricultural land were abandoned worldwide between 1700 and 2000 m a.s.l., a process that has not finished yet, and that is forecasted to continue for the next few decades [6]. Particularly, in Mediterranean mountain areas a generalized farmland abandonment occurred during the second half of the twentieth century. For instance, more than 80% of cultivated land was abandoned in the Spanish Pyrenees [7], and around 70% in the eastern Alps [8]. Land abandonment is a major land use change and has resulted in dramatic landscape changes, leading to a massive invasion of shrublands and secondary succession forests [9, 10]. A high percentage of the agricultural land was located in steeply sloping hillsides, frequently exploited with no conservation measures [11]. When they were cultivated, these areas generated high runoff coefficients and high sediment concentrations [12], ultimately leading to thin stony soils and the activation of sheet wash erosion, gullying and rilling processes, as confirmed by experimental studies [3]. However, the management of these areas after land abandonment is still an unresolved "hot-topic" [13], with different practices proposed by public administrations, land managers, and scientists. The choice of a particular practice is of crucial importance since, most of the abandoned areas in the Mediterranean region are located in mountain areas that are the main source of water for the demanding lowlands [14], and also represent opportunities for the development of biodiversity and land management programmes.

Post-land abandonment management practices (LMPs) can produce important changes in landscape, biodiversity, soil quality, and water resources. Each management practice is based on a particular treatment of the plant cover and, consequently, will likely have important environmental implications, particularly in water resources generation and overland flow. This makes water availability in the Mediterranean region challenging for the next decades [15]. However, most of the abandoned lands have been considered marginal, i.e. without direct interest to incorporate them into the global productive system, because of steep slopes, inaccessibility, and soil degradation. Hunting and light tourism should be the only perspective for these marginal lands, together with extensive stockbreeding, although this latter has been progressively displaced to a secondary interest, given the population decline [16–18]. For this reason, there were no specific plans for management, leading to a natural "rewilding" process or landscape naturalization (the process of passively woody encroachment). Until the beginning of the 1970s the most degraded landscapes, affected by gullying and sheet wash erosion, open plant cover, and high rates of overland flow were subjected to land rehabilitation plans, summarized in the afforestation of large areas, accompanied in the most extreme cases with the construction of certain infrastructures. With these plans, forest engineers tried to reduce sediment yield and floods and, occasionally, to increase the extent of forest for wood production. Consequently, since the end of the nineteenth century, the land rehabilitation plans consisted of two strategies intimately related: (1) civil (grey) infrastructures (the traditional ones made with cement) and (2) afforestation programmes mainly using conifers to control hydrological dynamics and reduce flood risk, soil erosion and land degradation, and decrease the rate of silting in reservoirs [19]. Thus, until the 1970s, most of the LMPs and restoration and rehabilitation projects were focused on artificial, man-made strategies (grey infrastructures) that are costly and that frequently are not efficient over a long period of time [20] and on afforestation programmes. More recently, different green infrastructures and nature-based solutions (NBS) have been also considered as sustainable and successful strategies. This is the case, for instance, of the active management of abandoned lands in order to promote and improve ecosystem services [13, 21], i.e. grazing in forest areas and shrubland clearing in order to favour extensive livestock systems, human presence in mountain areas, enhancing biodiversity, heterogeneity, and the control of overland flow and soil erosion. Therefore, for stakeholders, regional and national administrations, policymakers, and scientists there are three critical questions: (1) What can we do with abandoned lands? (2) How should be managed land abandoned areas in Mediterranean mountains to optimize ecosystem services and reduce environmental risks? and (3) Which are the most adequate long-term strategies in order to integrate the marginal mountains in the spatial organization of landscapes to emphasize holistic land conservation policies?

Abandoned agricultural lands and post-land abandonment management practices deserve special attention due to their influence on the water cycle, hydrological dynamics and their role in flood mitigation risks [22]. This chapter examines NBS compared to grey initiatives in abandoned farmland areas of Mediterranean mountains, and evaluates their effects in flood dynamics. The way in which LMPs build on NBS can manage flood mitigation in Mediterranean mountains is briefly discussed in a context of the Central Spanish Pyrenees.

2 Nature-Based Solutions for Flood Mitigation from an Environmental and Socioeconomic Point of View

The NBS concept is a new approach that was introduced in the late 2000s. In the last years, several definitions have been proposed for the NBS concept from different perspectives. Lilli et al. [23] indicate that NBS are actions that "use natural processes in a resource efficient manner to solve societal challenges". Other broader definitions proposed that NBS is a holistic approach integrating both the "engineering and ecosystem components in its implementation, being encouraged in both research and practice and policy-decision making processes" [24]. Similarly, Turconi et al. [25] indicate that NBS are usually defined as complementary or alternative solutions to grey infrastructures with the main aim of conserving and regenerating the functionality of a natural or semi-natural ecosystem. Cohen-Shacham et al. [26] provide a complete definition indicating that NBS are "actions to protect, sustainable management, and restore natural or modified ecosystems, that address societal challenges effectively and adaptively, simultaneously providing human well-being and biodiversity benefits" based on principles such as the maintenance of biological and cultural diversity, the ability of ecosystems to evolve over time, and the application at a landscape scale. The European Commission [27] indicated that NBS is a concept that defined "the actions inspired by, supported by, or copied from nature, and uses complex system processes of nature" or "solutions that aim to help societies address a variety of environmental, social and economic challenges in sustainable way" [27]. Thus, the European Union has adapted NBS as a strategy for achieving the restoration of degraded ecosystems, climate change adaptation and mitigation, the improvement of the risk management, and resilience to extreme events, among other topics.

What can be considered as a nature-based solution? There are different responses depending on the authors. For example, Debele et al. [28] include (1) conservation and land management practices, (2) green and blue approaches mixed with hard engineering structures, and (3) large-scale climate adaptation and mitigation approaches as afforestation and bioengineering practices. Concerning flood risk management, NBS should be natural or semi-natural structures that reduce the prevalence of flooding events. These flood events can have serious impacts from an economic, social, and environmental point of view. Most of the literature and examples of NBS and green infrastructure related to flood risk are focused in urban and peri-urban contexts [29]. Conversely, few studies have been focused on land abandoned environments [30]. Other authors prefer the use of natural flood management (NFM) schemes (similar to NBS) to work with hydromorphological processes through soil and water management to reduce flood risk, including restoration [31].

During decades, grey infrastructures have been used to control flood risk and soil erosion in abandoned Mediterranean mountains. A variety of infrastructures have served, and still serve, to reduce peakflows, particularly reservoirs, which usually have been constructed with other purposes than flood control (mainly hydropower production and irrigation, in some cases water supply for large cities), although they are also used to reduce the peakflow [32]. The efficiency of reservoirs is controlled by their volume in relation to the total river discharge, the season of the year, and the volume of water accumulated in the reservoir prior to the peakflow. For instance, in Pyrenean rivers, floods that occur in autumn are almost absent, because the increase in discharge is used to recover the reservoir after the summer exhaustion, and those occurred in spring are also reduced to finally fulfil the reservoir immediately before the irrigation campaign [32]. In any case, reservoirs act as large sediment traps during the occurrence of floods, and therefore they contribute to the shortening of the lifetime of reservoirs, particularly in areas affected by strong plant cover disturbance [33].

In the case of small torrential ravines, the construction of a series of check dams is a strategy to reduce the velocity of the peakflow wave, although their main finality is the interruption of sediment transfer (particularly bedload) downstream of the dams [34]. They have been profusely used in mountain areas where sediment transfer is a problem, although they have collateral, undesirable consequences, such as the incision of the channel immediately downstream of the check dam and their possible collapse during extreme hydrological events [35–39]. Braided rivers and ravines have also been canalized to avoid the flooding of agricultural lands and human settlements, although the containment of stream beds between banks tends to accelerate the velocity of the flood wave and to increase the efficiency of bedload transfer [40]. Besides, during extreme floods the artificial banks can collapse or be overpassed, resulting in long-term damages in the alluvial plain. For these reasons, grey infrastructures are only recommended in exceptional cases, to save lives, settlements, and strategic buildings.

Nevertheless, a combination of grey and green approaches has been implemented in different Mediterranean environments. Generally, grey infrastructures produce immediate effects, however, they are often short-lived (<30–40 years) and they can lose their effectiveness few years after their construction. Contrary, NBS can be a cost-effective long-term solution for hydrological risks and land degradation processes [41], such that shifts from grey infrastructures to green infrastructures and NBS could provide similar effects but without some negative impacts. Boix-Fayos et al. [30] supported the idea that vegetation restoration in abandoned areas and NBS are more sustainable economically with a long-term efficacy, while hydrological control through grey infrastructures serves better for some specific problems and can be used as a short-term solution but with higher economic costs. Likewise, NBS offers many additional benefits than grey infrastructure as the increase in biodiversity or soil carbon sequestration.

Generally, NBS focused on soil and water conservation measures for flood mitigation should include catchment-based interventions at landscape scale and should improve infiltration processes, which slow down runoff velocity (as water passes slowly through the soil), and help in reducing erosion processes [42]. Below, a range of NBS options to control floods in abandoned Mediterranean mountain areas is briefly presented (Fig. 1). In general, these NBS, as suggested by Keesstra et al. [43], should be based on making the landscape less connected, facilitating rainfall to be less transformed into runoff, increasing soil moisture, and thus reducing flood risk, soil erosion, and land degradation.

2.1 *Passive Rewilding*

The main visible consequence of land abandonment in Mediterranean mountain areas is the natural expansion of shrublands and forests. Abandoned agricultural lands are usually occupied by natural revegetation of shrubland and forest cover in a slow-process that can lead around 100 years (rewilding) [44]. Natural revegetation alters the water cycle, the partitioning of rainfall between evapotranspiration, runoff and groundwater flows, increases water infiltration during rainstorm events, and reduces overland flow and sediment yields. Most of the studies in abandoned areas with a natural revegetation process suggested: (1) a decrease in river discharges, (2) a reduction of runoff coefficients and peak flows, and (3) changes in flood hydrographs characteristics, like slower time lags and longer recession limbs [3, 45–47].

However, it is also demonstrated that vegetation recover can present negative effects, as the decrease of runoff volumes especially during dry periods, due to the high evapotranspiration values and water demands of the forest cover, the increase of forest fire risks [43], and the loss of cultural landscapes [13] and biodiversity, although there are differing opinions on the latter [48].

2.2 *Afforestation*

Abandoned farmlands in Mediterranean mountains have also been afforested through large forest hydrological rehabilitation projects, carried out mainly during the second half of the twentieth century. Check dams (considered grey infrastructure) and afforestation programmes (greening-up processes) were used in both semi-arid and humid environments [30, 49]. The idea that abandoned fields and

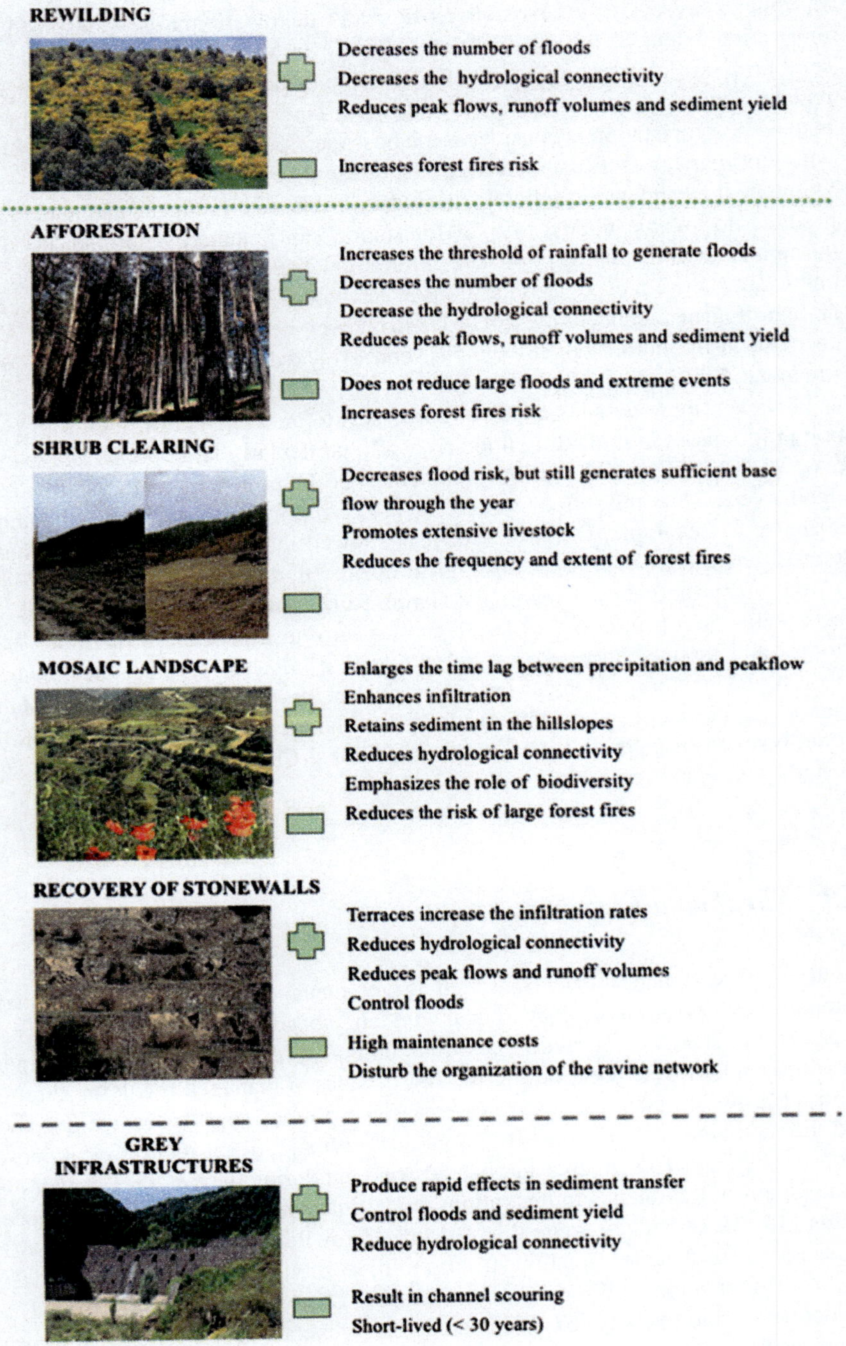

Fig. 1 Nature-based solutions and grey infrastructures to control floods in abandoned Mediterranean mountain areas and grey infrastructures. + means positive effects and – negative effects of the different NBS and grey infrastructures

shrublands were affected by soil erosion and degradation was widespread, thus motivating extensive afforestation programmes. In Mediterranean mountains (both semi-arid and humid environments), afforestation with conifers was used for land reclamation during many years, causing rapid landscape transformations. In most instances, afforestation is considered a landscape change that reduces water supply, whose primary effect is the increase of the threshold of the amount of rainfall needed to initiate flow and generate floods (the number of floods is reduced and many events produce no notable floods) [49]. Afforestation can potentially mitigate flood risk through increasing interception and infiltration rates, and attenuating runoff volumes. Thus, the effects of afforestations at catchment scale were: (1) a clear decrease in hydrological connectivity [50, 51], (2) lower peak flows [49, 52, 53], (3) a decrease in the number of floods [54], and (4) a reduction in sediment supply to streams [55, 56].

However, there remains a lack of consensus to the general efficacy of afforestation programmes in mitigating flood risk [57], particularly large floods and extreme events [31, 58–60]. Thus, afforestation may be helpful in moderating floods for small events, but this effect would be increasingly reduced as rainfall amount increase. For example, Didon-Lescot [61] showed that the effects of afforestation are restricted to less intense hydrological events (floods corresponding to a return period of approximately 5 years) and García-Ruiz et al. [62] considered that it only has notable effects in floods of less than 10-year return period. Furthermore, some studies and models suggest some other negative impacts, as the increase in soil erosion and geomorphological processes during the first years after afforestation (due to the use of heavy machinery) [63], and the increase in forest fires risk, that could be even stronger under the context of Global Change, due to longer drought periods and higher temperatures [64].

2.3 Shrubland Clearing

As already mentioned, scientists, as well as land managers and policy-makers, have taken different positions on how to deal with land abandonment [13]. Some support a passive management (rewilding) while others defend an active management (as afforestation). Likewise, during the last decades, a managed rewilding through shrub clearing has been proposed, leading to the promotion of extensive livestock and the reduction of forest fires [65], but also with significant hydrological consequences (Fig. 2a). This practice has been carried out during the last three decades by some Spanish Regional administrations with financial support of the European Union [13, 65, 66]. They argue that a managed rewilding may be a good NBS to come to a sustainable situation from multiple environmental points of view (i.e. water regulation, soil erosion, biodiversity). Likewise, shrub clearing leads to an increase in biodiversity [67, 68] and can also generate socioeconomic improvements, as the increase in extensive livestock [13, 66]. Alados et al. [68] and Gómez

Fig. 2 Examples of NBS in Mediterranean mountain areas: (**a**) Shrubland clearing landscape in the Leza Valley (Iberian System, Spain) (Photo E. Nadal-Romero); (**b**) Example of mosaic landscape with meadows and shrublands in a south-facing slope in San Juan de Plan (Pyrenees, Spain) (Photo T. Lasanta); (**c**) Meadows in the bottom valley of the Esera River (Pyrenees, Spain) (Photo T. Lasanta); (**d**) Terraced slopes in Munilla (Iberian System, Spain) (Photo D. Lasanta)

et al. [69] also concluded that shrubland clearing is a better strategy than controlled burning to mitigate the effects of shrub encroachment.

Shrub clearing as a NBS can be considered a good water resource management, because of its capacity to decrease flood risk and to generate sufficient base flow through the year [13]. However, there are no catchment scale examples to contrast these results [70]. Nadal-Romero et al. [12] simulated shrub clearing in small plots in the Central Pyrenees, and compared it with natural shrub vegetation. The results showed that runoff coefficient in grassland areas (shrub clearing) are higher than in shrublands, because canopy interception is reduced, increasing overland flow especially during the wet period, whereas soil erosion values do not show changes. Currently, the LIFE project MIDMACC "Mid-mountain adaptation to climate change" funded by the European Commission implements shrub clearing practices in two abandoned and marginal mid-mountain areas in Spain, as a landscape management practice and NBS, and will evaluate the effects of this practice on water resources and hydrological dynamics (https://life-midmacc.eu/es/).

2.4 Recovery of Mosaic Landscapes

Landscape changes moving to a managed mosaic of land uses and land covers should be also considered as a NBS for abandoned lands. The disappearance of mosaic landscapes, after land abandonment, represents a significant loss of socio-cultural and ecological benefits, as the increment of forest fires or the decrease of water [71]. Landscape solutions and the recovery of these landscapes can incorporate high production values (e.g. timber), ecological (e.g. soil quality) and cultural values (e.g. leisure, cultural identity), the support of a large diversity of habitats, and the regulation of hydromorphological processes such as soil conservation, water production and quality, and flood control and mitigation (see Fig. 1).

In that sense, a slope or basin should be considered as a complex hydrological system, in which hydrological processes considered at the plot scale do not allow to understand the global response of the basin [72]. It is well known that runoff dynamics is conditioned by the structure of the landscape (topography and environment). It involves its composition (land uses and covers) and its configuration or spatial distribution of landscape elements, including paths, roads, ditches, ravines, etc. In mosaic landscapes, including *bocage* landscapes (mixed woodland and pasture areas), the spatial distribution of land uses and covers can disconnect (in some cases) sectors of the slope from the ravine or river (water course) (Fig. 2b, c) [73]; likewise, in other cases, the effects of topography, land use and road networks, hedges, ditches, among others, may be combined, to create a complex artificial drainage network, altering the topographic flow pattern [74, 75]. The scientific literature highlights that: (1) in hydrological efficiency the composition of the landscape is as important as the organization of the landscape elements, and (2) the density and spatial organization of the drainage networks determine the hydrological connectivity of a basin [76–79]. For instance, Malek et al. [80] showed, using a modelling approach that enhancing traditional mosaic-like landscapes improve the status of the water resources in the Mediterranean region.

2.5 Recovery of Stonewalls and Terraces

Agricultural terraces occupied important extents in the slopes of Mediterranean mid-mountain areas (Fig. 2d). Generally, terraces increase infiltration rates, reduce hydrological connectivity affecting contributing areas, peak flows, and runoff volumes, and consequently can control floods. Likewise, terraces alter the paths of runoff and sediment transport and erosion processes. Several authors suggest that terraces break up the continuity of water flows, imposing large rainfall thresholds on catchment-scale runoff production [81–84]. However, terrace walls require regular maintenance to continue retaining water and soil [85]. Furthermore, the development of a terraced landscape can lead to severe erosion problems, with the triggering of deep gullies. The reason is that farmers tried to alter the course of natural drainages in the hillslopes in order to concentrate the surface runoff out of the cultivated fields. In some cases, runoff concentration evolved into deep gullies in a few decades [86].

The abandonment of agricultural activities and consequently the lack of maintenance of agricultural terraces triggered a set of negative geomorphological impacts, as soil erosion or small mass movements [87, 88]. In addition, terraced abandonment largely alters runoff production, hydrological processes and connectivity [84]. Given the benefits of agricultural terraces (such as the reduction of flood risk or the enrichment of ground water potential due to the increased permeability of the soil), their maintenance or the restoration of stonewalls should be considered as a priority NBS practice to prevent off-site effects after land abandonment [89–91]. For example, Bellin et al. [82] indicated that terrace maintenance slowed runoff for events with a return period shorter than 8–10 years, and Calsamiglia et al. [92] concluded that terraces conservation encourages the dis-connectivity between the slopes and channels, reducing flood risk. Thus, active maintenance and rehabilitation of stone terraces is probably the most effective approach for controlling off-site effects and negative impacts of land abandonment (such as landslides or small mass movements). Tarolli et al. [93] considered that the reconstruction of the stonewalls should be also accompanied by the simultaneous reconstruction of irrigation and drainage channels and complementary structures of the terrace systems. However, the high costs make this NBS impracticable in most cases because of the large extent of terraced areas. However, despite the highest costs, an example of participatory framework approaches in Cyprus concluded that after land abandonment terraces rehabilitation had the best overall performance followed by afforestation, due to the high environmental benefices together with the local-socio cultural landscape context [94]. During the last decade, some private and public activities (i.e. UNESCO) have promoted the rebuilding of ruined stonewalls in those areas where the connectivity is high, being of interest for hydrological and forestry NBS projects. For example, some of these activities are supported by the LIFE programme (European Commission) and recently two new projects have been funded: "TERRACESCAPE" and "STONEWALLSFORLIFE". The main objective of the TERRACESCAPE project (http://www.lifeterracescape.aegean.gr/en/) is the restoration and re-cultivation of drystone terraces (as a prominent element of the Mediterranean landscape) in the Andros Island (Greece) to demonstrate the economic, cultural, and ecological benefits derived from such elements and to create and adapt green infrastructures to the context of Global Change. The STONEWALLSFORLIFE project (https://www.stonewalls4life.eu) aims at repairing wall terraces and ensuring their long-term maintenance, to protect the territory (Cinque Terre National Park, Italy) and its inhabitants against the effects of extreme climate events (such as floods or landslides).

3 Coping with Floods in the Central Spanish Pyrenees

The occurrence of floods is a common hydrological phenomenon in the Pyrenees, linked to long rainfall events or to short rainstorms. Most of floods concentrate between November and February, although the period between November and May includes important floods related with snowmelt (usually coinciding with rainfall events) [95]. Some floods have been particularly intense, accompanied by mass movements in the slopes, severe erosion periods, high costs in infrastructures and occasionally, casualties (Fig. 3). According to García-Ruiz et al. [96], the calculation of return periods of the most extreme rainfall events confirms that their occurrence shows an erratic spatial and temporal distribution. When precipitation between 150 and 200 mm in 24 h is considered, the relief explains the distribution of the rainfall values recorded, but in the case of precipitation over 200 mm, not any spatial organization explains their occurrence. The most extreme rainfall events can be considered as outliers, with magnitudes much higher than one would expect. For this reason, floods produced during such extreme pluviometric events cannot be controlled through NBS and grey structures, unless if a large reservoir is sufficiently empty to retain a large proportion of the flood. Nevertheless, even if reservoirs can reduce the peakflow downstream of the dams, the most relevant control of floods should be made in the hillslopes. There is where runoff generation and overland flow must be reduced, i.e. before runoff reaches the fluvial channel. This is particularly interesting for the low- and mid-mountain areas, where forests and shrublands can be managed to enhance rainfall interception and infiltration and slow overland flow.

Until the middle of the twentieth century, most of the south-facing slopes in the Central Spanish Pyrenees below 1,600 m a.s.l. were cultivated [11]. The abandonment resulted in the natural revegetation with shrubs and forests and the use of large afforestation programmes as well as hydrological control works that have been performed to regulate floods and control soil erosion and land degradation. During the last 30 years, researchers from the Instituto Pirenaico de Ecología (Spanish

Fig. 3 Some effects of the 19–21 October flood 2012 in the Central Spanish Pyrenees. (**a**) The bridge over the Aragón River at Jaca, dammed by a large mass of trees carried by the floodwaters; (**b**) A family house destroyed after the October flood (Aragón River) (Photos E. Nadal-Romero)

Fig. 4 (**a**) Afforestation carried out in the slopes around the Mediano Reservoir (Cinca River basin) to reduce runoff generation and the transfer of sediment to the reservoir (Photo JM. García-Ruiz); (**b**) A series of check dams in the Escuer Valley, a tributary of the Upper Gállego Valley. This valley was also almost totally afforested in the 1960s since it was affected by intense erosion processes after centuries of cultivation and grazing on steep slopes (Photo JM. García-Ruiz); (**c**) Afforestation carried out in the Ijuez River, a tributary of the Upper Aragón River. The whole basin was also afforested to reduce sediment yield and the magnitude of floods (Photo E. Nadal-Romero); (**d**) Large check dam in the lower course of the Ijuez River. It was constructed in the 1960s using the pillars of an ancient aqueduct (Photo JM. García-Ruiz)

Research Council, IPE-CSIC) have studied the effects of land abandonment and LMPs on water resources, flood generation and flood risk, and soil erosion in the Central Spanish Pyrenees [18]. The results from the Aisa Valley Experimental Station (AVES) with small plots that represent different old and current land uses (from 1991 to 2011) suggested that (1) farmland abandonment leading to dense shrub cover yields the lowest values of runoff generation and sediment yield; (2) cereal cultivation alternating with fallow on steep slopes produces the highest runoff coefficient and sediment yield, particularly in the case of shifting in agricultural practices; and (3) shrub clearing (grazing meadows) produces similar results than those obtained in the abandoned area covered with shrubs [12].

Afforestations have been a common NBS in the Central Pyrenees (Fig. 4a, b). Given the large extent of afforestation different studies were performed to detect changes in the hydrological behaviour of taluses and channels. García Ruiz and Ortigosa [97] reported that channels in afforested basins were those with the highest percentage of plant cover and the lowest average of bare soil and gravels. In the case of taluses located close to the channels, those in afforested basins had a greater plant cover, a greater proportion of no-erosion areas and a lower percentage of severe erosion. Therefore, a reduction in the magnitude and frequency of floods was deduced, together with a loss of energy to transport coarse sediment. Nevertheless, it may be noted that during the first years following the afforestation works, soil erosion and overland flow can be enhanced by the removal of soil, and that the afforestation techniques used are a major factor to explain the hydrological behaviour of afforestation [63], particularly in bench terraces made with machinery.

Other studies performed in some of the main Pyrenean rivers showed that streamflow and sediment yield have been declining in the Central Pyrenees since the middle of the twentieth century [98]. Beguería et al. [99] indicated that discharge clearly decreases from the last 60 years, suggesting that these changes were not attributable to a temperature increase or to a decline in precipitation, but rather to a generalized expansion of shrubs and forests following farmland abandonment. López-Moreno et al. [100] also detected a remarkable decrease in the discharges of Pyrenean rivers, with evidence that climate change alone only explains a small proportion of the observed decrease. Instead, the increase in evapotranspiration and rainfall interception due to the growth of vegetation in abandoned lands should be a major factor for these hydrological changes. Likewise, López-Moreno et al. [101] observed a decrease in the frequency and intensity of floods in most of the Pyrenean rivers, regardless of the evolution of precipitation. The results obtained show a negative trend in flood magnitude since the 1970s, although a change in the precipitation events was not detected. This difference between precipitation and runoff events has been explained as due to the expansion of plant cover in old cultivated fields and grasslands areas, as a consequence of depopulation, farmland abandonment, and the decline of extensive stockbreeding.

To check these general results four small headwater catchments have been monitored since the 1990s, with different history of land uses and land covers [102]. The Arnás, San Salvador, Araguás, and Araguás Afforestation catchments are located in the upper Aragón River basin (Central Pyrenees) between 875 and 1,340 m a.s.l., and all the catchments are close to each other such that they have similar climatic conditions [47]. The San Salvador catchment represents a dense natural forest of *Pinus sylvestris*, with *Fagus sylvatica* and *Quercus faginea*. The Arnás and Araguás Afforestation catchments were both cultivated and abandoned in the late 1950s: the Arnás catchment has been subject to a natural plant colonization process, including a complex vegetation mosaic with shrubs and forest patches, and the Araguás Afforestation catchment was afforested in the 1960s with *P. nigra* and *P. sylvestris*. Finally, the Araguás catchment represents a degraded environment, with a dense network of badlands. Significant hydrological differences were observed among the catchments [102, 103]. Focusing on land abandonment scenarios (Arnás and Araguás Afforestation), the primary effect of plant cover in both abandoned catchments is to raise the threshold for the amount of rain needed to initiate flow: both catchments experience fewer floods compared to the degraded area (Araguás), and many rainfalls produce no notable flow [60, 104]. However, these studies also noted that the number of floods recorded in these areas is twice the

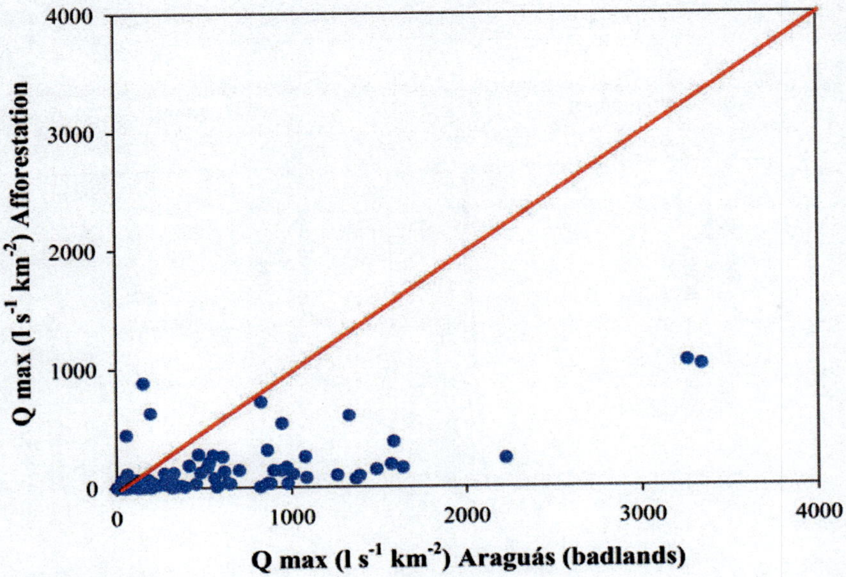

Fig. 5 Peak flows during concurrent floods in the afforested and badlands catchments (Araguás) in the Central Spanish Pyrenees (period 2013–2020). Data recorded by the Pyrenean Institute of Ecology (IPE-CSIC)

number of floods registered in the natural forest catchment (San Salvador). Besides, the abandoned areas are able to generate floods over the entire year, whereas the natural forest generate events mainly in spring.

Comparing the afforested area with the degraded area and the natural forest for the period 2013–2020, the number of floods is twice lower in the afforested area than in the degraded area (163 and 316 floods, respectively), but it is fourfold higher than in the natural forest catchment (163 and 41 floods, respectively). Most peak floods in the afforested catchment (except for extreme events, rainfall intensity >40 mm/h) have commonly recorded one order of magnitude lower than those on the unforested degraded area (Fig. 5). However, flood events lasted longer on the afforested catchment than in the degraded area (Fig. 5), due to a reduced connectivity and an enhanced infiltration in the afforested catchment.

Figure 6 shows the hydrological response in the four catchments during short and intense events recorded during late spring with similar rainfall amount and high rainfall intensity. We can observe that (1) a high peak flow was recorded in the degraded area; (2) in the natural forest a low response was observed, confirming the relevant role of rainfall interception and infiltration into the soil; (3) the time response is faster in the afforested area than in the natural revegetated area; and (4) longer recession times were observed in the afforested and natural revegetated areas, while in the degraded badland areas the rising and decreasing limbs are steeper, with high fast responses. These examples suggested that in afforested

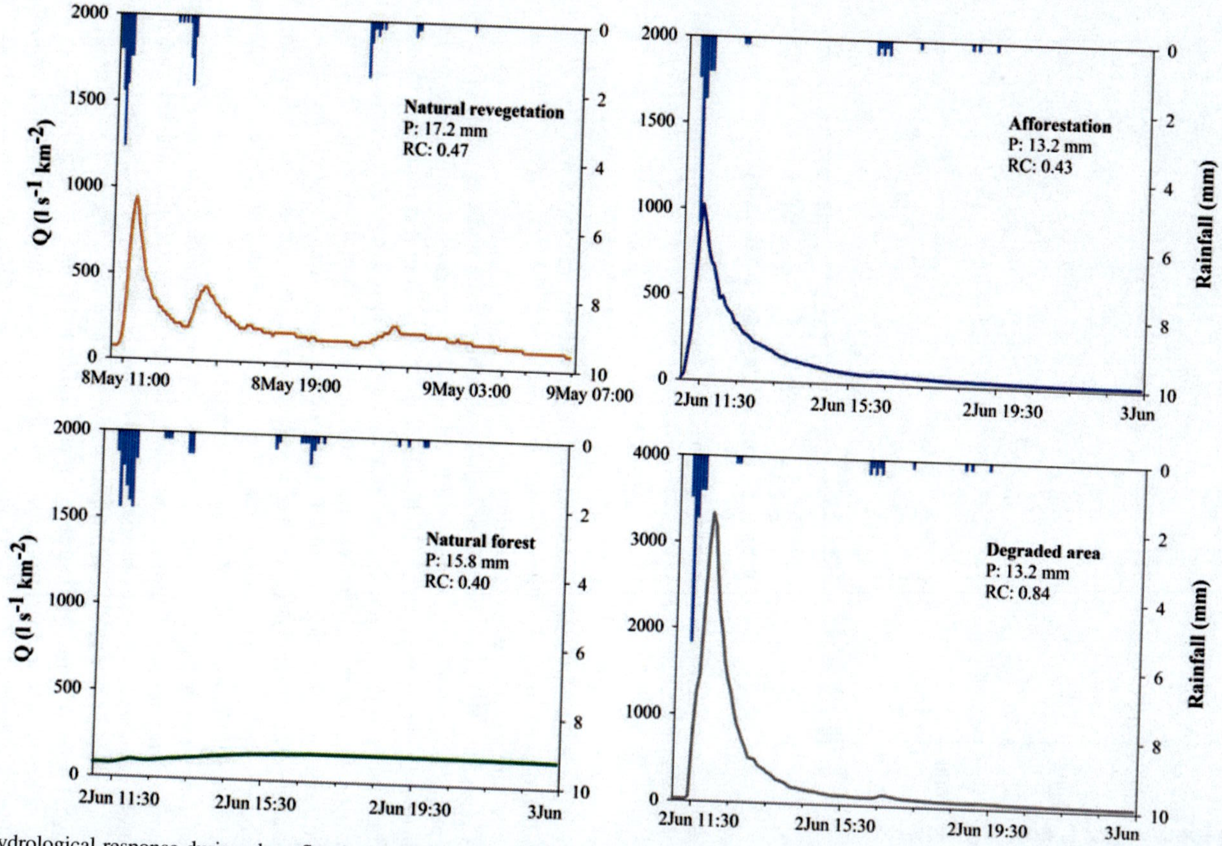

Fig. 6 Hydrological response during short floods recorded in late spring (May and June) in the four experimental catchments (natural revegetation, Arnás; Afforestation, Araguás afforested; Natural forest, San Salvador; and degraded area, Araguás). Note that the Q scale in the degraded area is different (twice the other Y axis). Data recorded by the Pyrenean Institute of Ecology (IPE-CSIC)

areas, flood mitigation effects are greatest for moderate floods, but have less impact during intense rainfall events as observed in Fig. 6.

A declining hydrological connectivity after land abandonment due to the growth of vegetation has been reported in a number of Mediterranean mountain areas [50]. Lana-Renault et al. [47] showed a decrease in connectivity in both abandoned catchments (Arnás and Araguás Afforestation). The largest decrease in the connectivity index was observed in the afforested catchment due to the significant changes from sloping, somewhat terraced fields to tree cover. López-Vicente et al. [51] showed that afforestation in the Araguás afforested catchment promoted lower and more stable connectivity both at hillslope and catchment scales.

Similarly, a decline in connectivity between slopes and the channel was observed in the Ijuez River basin (Central Pyrenees) that was also cultivated with sloping fields up to mid-twentieth century (Fig. 4c, d). The Ijuez River was highly affected by the occurrence of sudden floods with high volumes of sediment transport, mainly bedload, carried out by tributaries and frequent debris flows, resulting in a braided, extremely unstable fluvial channel (Fig. 4c) [56]. For this reason, in the 1950s the basin was purchased by the State Administration and afforested with *P. sylvestris* and *P. nigra* to reduce sediment yield and the transfer of sediments to the Yesa reservoir. A combination of green and grey infrastructures was carried out, with the construction of five small check dams to trap sediment production (Fig. 4d). Afforestation produced a rapid recovery of vegetation, and consequently a significant decrease in the area affected by erosion, declining sediment yield, and the connectivity between hillslope and channels [56]. This resulted in the narrowing and the incision of the channel and a progressive armouring of the bed [105].

Land abandonment and different LMPs based on NBS show also significant hydrological and geomorphological consequences during extreme events. It should not be forgotten that Mediterranean mountain areas are affected by intense rainstorms especially in summer and the beginning of autumn, with rainfall exceeding 200 mm in 24 h [62], and extreme floods are relative common in the area. In addition, an intensification of intense rainfalls due to climate change is also expected [106]. The occurrence of exceptional events can produce catastrophic floods, usually affecting small catchments. For instance, an exceptional hydrological event occurred during 19–21 October 2012 (250 mm in 2 days), in the Aragón River Basin (Fig. 3) [107]. The response was recorded in the four small experimental catchments [108] and the results indicated that: (1) the natural forested catchment showed a slow response and did not react to the most intense rainfall peaks at the beginning of the rainstorm, and only reacted during the final rainfall peaks, with a moderate peak flow; (2) the abandoned farmland catchment registered two moderate peaks at the beginning of the event, and only when the catchment was saturated a high peak flow was recorded; (3) a similar response was observed in the afforested catchment at the beginning of the flood, although unfortunately no data were available during the complete event; and (4) the degraded badland area showed a high fast response, almost immediately, with high peak discharges and high sediment yields coinciding with each rainfall peak.

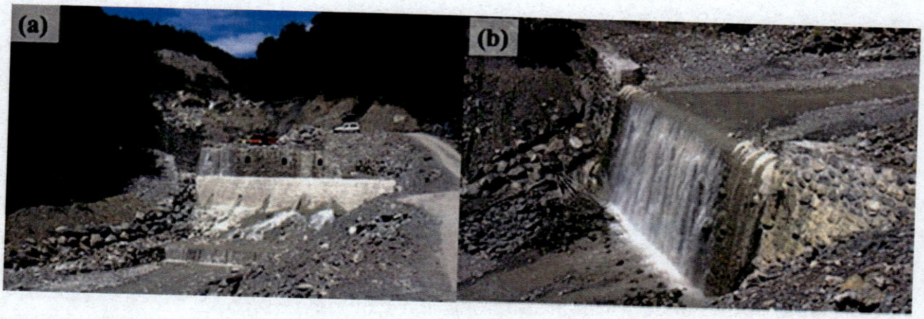

Fig. 7 (**a**) Construction of new check dams in the lower stretch of the Arás ravine (Upper Gállego basin), following the collapse of most of check dams during the extreme flood of August 1996; (**b**) New check dam in the lower course of the Arás ravine (Upper Gállego basin). It was constructed after the collapse of most of check dams during the extreme flood of August 1996 (Photos JM. García-Ruiz)

Finally, it should be also highlighted the consequences of extraordinary events, as the catastrophic flood recorded in the Arás river (Central Pyrenees, August 7, 1996), destroying a campsite situated in the alluvial fan of the torrent and killing 87 people [36]. The Arás catchment was characterized by the presence of a very steep stream channel (up to 20%) in the last 4 km, where the ravine crosses lateral moraines that behave as major sediment sources. As a consequence, the ravine developed a large and active alluvial fan that menaced the road from Saragossa to tourist resorts in the Central Pyrenees. An ambitious programme for reducing the magnitude of floods as well as erosion and sediment transfer from the moraines was designed at the beginning of the twentieth century. A combination of grey and green infrastructures was carried out, including (1) an artificial channel and 40 small check dams that were constructed between 1911 and 1950 in both the alluvial fan and the final stretch of the ravine, respectively; and (2) the afforestation of most of the catchment with *P. sylvestris*. In addition, decades later the forestry service carried out various correction works in the channel and constructed new check dams, as some of the old ones were already clogged (Fig. 7). The extreme rainfall (approx. 250 mm in 1 h and 15 min were indirectly estimated) produced an extreme peak flow of at least 300 $m^3\ s^{-1}$ that caused the collapse of a series of most of check dams in the Arás channel, carrying out around 120,000 t of sediment mainly coming from the collapsed check dams [109]. An event of these characteristics is always possible in the Central Spanish Pyrenees, and land management practices built on NBS should be considered as strategies to minimize and mitigate flood risks in these mountain areas.

A global and highly relevant question still emerges in Mediterranean mountains and in the Central Pyrenees: How should we manage abandoned lands? In that way, we encourage the use of the ecosystem services approach and NBS (as proposed in this chapter) in future public policies to manage land abandonment areas with different objectives as flood mitigation and water resources regulation.

4 Concluding Remarks

During the twentieth century, farmland abandonment has been the most relevant landscape change in Mediterranean mountain areas (as in other mountain areas), causing the expansion of shrublands and forests. The global consequences of this process have been widely discussed in the literature and several positive and negative effects have been identified (i.e. landscape homogenization, increasing forest fire risk and decreasing streamflows and sediment yield). However, the following question is still unresolved: how should we manage abandoned land to mitigate flood risk?

The impacts of land abandonment in flood mitigation in Mediterranean mountain areas depend greatly on how these areas are managed after abandonment (post-land management practices), which, at present is still a controversial topic. Nature-based solutions should consider the use of natural or semi-natural structures to reduce the occurrence of flooding events. The most used NBS in abandoned lands of mountain areas are natural plant recolonization, afforestation, shrub clearing, recovery of mosaic landscapes, and the reconstruction of terrace systems and stonewalls. These practices have been also combined with the implementation of grey infrastructures, as the construction of small check dams and sediment traps.

In general, worldwide land abandonment and the expansion of shrublands and forests (rewilding and afforestation) decrease hydrological and sediment connectivity, peak flows and sediment yield, and decline the frequency and magnitude of most frequent floods in mountain areas. However, some of these practices, as afforestation, have only remarkable effects on small and moderate floods, and have less impact in extreme and extraordinary rainfall events. In the case of extreme rainfall and hydrological events, the role of plant cover seems to be moderate in both slopes and channels, although it also depends on the antecedent conditions of the soil. In the most extreme hydrological events (outliers) not any influence of vegetation to reduce the peakflow has been detected. The construction of some infrastructures in the main and secondary fluvial channels (check dams) reduces sediment transfer, whereas artificial channels can provide a false sensation of safety. Obviously, the presence of large reservoirs tends to change the behaviour and seasonality of large floods, because floods occur when the reservoir is partially empty. In any case, NBS, focused on the reduction of runoff and connectivity, must be applied directly on the slopes, before runoff reaches the fluvial channels. Shrub clearing is another alternative strategy for organizing the landscape to improve grasslands productivity and at the same time controlling sediment yield and runoff generation. This needs a high-resolution knowledge of the landscape heterogeneity in order to decide the location of clearings and to reinforce the declining connectivity between slopes and channels. Finally, the maintenance or rehabilitation of agricultural terraces reduces flood risk because they decrease overland flow and connectivity on hillslopes. For this reason, permanent human intervention is needed to maintain NBS as strategies to flood mitigation in marginal Mediterranean mountains areas.

Acknowledgement This work was funded by the H2020-MSCA-IF-2018 programme (Marie Sklodowska-Curie Actions) of the European Union under REA grant agreement, number 834329-SEDILAND, and the MANMOUNT project (PID2019-105983RB-I00/AEI/10.13039/501100011033) funded by the Spanish National Plan for Research and the European Union ERDF funds.

References

1. MacDonald D, Crabtree JR, Wiesinger G, Dax T, Stamou N, Fleury P, Gutierrez Lazpita J, Gibon A (2000) Agricultural abandonment in mountain areas of Europe: environmental consequences of policy response. J Environ Manag 59(1):47–69. https://doi.org/10.1006/jema.1999.0335
2. Rey Benayas JM, Martins A, Nicolau JM, Schulz J (2007) Abandonment of agricultural land: an overview of drivers and consequences. CAB Rev Persp Agric Vet Sci Nutr Nat Resources 57:2. https://doi.org/10.1079/PAVSNNR20072057
3. García-Ruiz JM, Lana-Renault N (2011) Hydrological and erosive consequences of farmland abandonment in Europe, with special reference to the Mediterranean region. A review. Agric Ecosyst Environ 140(3–4):317–338. https://doi.org/10.1016/j.agee.2011.01.003
4. Lasanta T, Nadal-Romero E, Khorchani M, Romero-Díaz A (2021) Una revisión sobre las tierras abandonadas en España: de los paisajes locales a las estrategias globales de gestión. Geogr Res Lett 47:1–42. https://doi.org/10.18172/cig.4755
5. Campbell JE, Lobell DB, Genova RC, Field CB (2008) The global potential of bioenergy on abandoned agriculture lands. Environ Sci Technol 42:5791–5794. https://doi.org/10.1021/es800052w
6. Pointereau P, Coulon F, Girard P, Lambotte M, Stuczynski T, Sánchez Ortega V, Del Rio A (2008) Analysis of farmland abandonment and the extent and location of agricultural areas that are actually abandoned or are in risk to be abandoned. JRC Scientific and Technical. Reports (EUR 23411). http://www.publicationsjrceceuropaeu/repository/handle/111111111/166
7. Lasanta T, Errea MP, Nadal-Romero E (2017) Traditional agrarian landscape in the Mediterranean mountains. A regional and local factor analysis in the Central Spanish Pyrenees. Land Degrad Dev 28:1626–1640. https://doi.org/10.1002/ldr.2695
8. Tasser E, Walde J, Tappeiner U, Teutsch A, Noggler W (2007) Land-use changes and natural reforestation in the eastern Central Alps. Agric Ecosyst Environ 118:115–129. https://doi.org/10.1016/j.agee.2006.05.004
9. Sluiter R, de Jongs SM (2007) Spatial patterns of Mediterranean land abandonment and related land cover transitions. Landsc Ecol 22(4):559–576. https://doi.org/10.1007/s10980-006-9049-3
10. Peña-Angulo D, Khorchani M, Errea P, Lasanta T, Martínez-Arnáiz M, Nadal-Romero E (2019) Factors explaining the diversity of land cover in abandoned fields in a Mediterranean mountain area. Catena 181:103064. https://doi.org/10.1016/j.catena.2019.05.010
11. Lasanta T (1988) The process of desertion of cultivated areas in the Central Spanish Pyrenees. Pirineos 132:15–36
12. Nadal-Romero E, Lasanta T, García-Ruiz JM (2013) Runoff and sediment yield from land under various uses in a Mediterranean mountain area: long-term results from an experimental station. Earth Surf Process Landf 38:346–355. https://doi.org/10.1002/esp.3281
13. García-Ruiz JM, Lasanta T, Nadal-Romero E, Lana-Renault N, Álvarez-Farizo B (2020) Rewilding and restoring cultural landscapes in Mediterranean mountains: opportunities and challenges. Land Use Policy 99:104850. https://doi.org/10.1016/j.landusepol.2020.104850

14. Viviroli D, Dürr HH, Messerli B, Meybeck M, Weingartner R (2007) Mountains of the world, water towers for humanity: typology, mapping, and global significance. Water Resour Res 43 (7):W07447. https://doi.org/10.1029/2006WR005653
15. Lana-Renault N, Morán-Tejeda E, Moreno de las Heras M, Lorenzo-Lacruz J, López-Moreno JI (2020) Land-use change and impacts. In: Zribi B, Brocca L, Tramblay Y, Molle F (eds) Water resources in the Mediterranean Region. Elsevier, Amsterdam, pp 257–286
16. Bernués A, Ruiz R, Olaizola A, Villalba D, Casasús I (2011) Sustainability of pasture-based livestock farming systems in the European Mediterranean context. Synergies and trade-offs. Livest Sci 139:44–57. https://doi.org/10.1016/j.livsci.2011.03.018
17. Riedel JL, Bernués A, Casasús I (2013) Livestock grazing impacts on herbage and shrub in a Mediterranean Natural Park. Rangel Ecol Manag 66(2):224–233. https://doi.org/10.2111/REM-D-11-00916.1
18. García-Ruiz JM, López-Moreno JI, Lasanta T, Vicente-Serrano SM, González-Sampériz P, Valero-Garcés BL, Sanjuán Y, Beguería S, Nadal-Romero E, Lana-Renault N, Gómez-Villar A (2015) Los efectos geoecológicos del cambio global en el Pirineo Central español: una revisión a distintas escalas espaciales y temporales. Pirineos 170:e012. https://doi.org/10.3989/Pirineos.2015.170005
19. Ortigosa L, García-Ruiz JM, Gil E (1990) Land reclamation by reforestation in the Central Pyrenees. Mt Res Dev 10(3):281–288. https://doi.org/10.2307/3673607
20. Valero Garcés BL, López Moreno JI, García Ruiz JM, Beguería Portugués S (2003) Intensidad de las avenidas y aterramiento de embalses en el Pirineo español. Ería 61:159–167
21. Lasanta T (2019) Active management against shrubland expansion: seeking a balance between conservation and exploitation in the mountains. Geogr Res Lett 45(2):423–440. https://doi.org/10.18172/cig.3726
22. Lana-Renault N, Nadal-Romero E, Cammeraat E, Llorente JA (2020) Critical environmental issues confirm the relevance of abandoned agricultural land. Water 14(4):1119. https://doi.org/10.3390/w12041119
23. Lilli MA, Nerantzaki SD, Riziotis C, Kotronakis M, Efstathiou D, Kontakos D, Lymberakis P, Avramakis M, Tsakirakis A, Protopapadakis K, Nikolaidis P (2020) Vision-based decision-making methodology for riparian forest restoration and flood protection using nature-based solutions. Sustainability 12:3305. https://doi.org/10.3390/su12083305
24. Kumar P, Debele SE, Sahani J et al (2020) Towards an operationalisation of nature-based solutions for natural hazards. Sci Total Environ 731:138855. https://doi.org/10.1016/j.scitotenv.2020.138855
25. Turconi L, Faccini F, Marchese A, Paliaga G, Casazza M, Vojinovic Z, Luino F (2020) Implementation of nature-based solutions for hydro-meteorological risk reduction in small Mediterranean catchments: the case of Portofino Natural Regional Park, Italy. Sustainability 12:1240. https://doi.org/10.3390/su12031240
26. Cohen-Shacham E, Walters G, Janzen C, Maginnis C (2016) Nature-based solutions to address global societal challenges. IUCN Commission on Ecosystem Management (CEM) and IUCN World Commission on Protected Areas (WCPA), Geneva
27. European Commission (2019) Policy topics: nature-based solutions. https://ec.europa.eu/research/environment/. Accessed May 2020
28. Debele SE, Kumar P, Sahani J, Marti-Cardona B, Mickovski SB, Leo LS, Porcù F, Bertini F, Montesi D, Vojinovic Z, di Sabatino S (2019) Nature-based solutions for hydro-meteorological hazards: revised concepts, classification schemes and databases. Environ Res 179:108799. https://doi.org/10.1016/j.envres.2019.108799
29. Krauze K, Wagner I (2019) From classical water-ecosystem theories to nature-based-solutions – contextualizing nature-based solutions for sustainable city. Sci Total Environ 655:697–706. https://doi.org/10.1016/j.scitotenv.2018.11.187
30. Boix-Fayos C, Boerboom LGJ, Janssen R, Martínez-Mena M, Almagro M, Pérez-Cutillas P, Eekhout JPC, Castillo V, de Vente J (2020) Mountain ecosystem services affected by land use

changes and hydrological control works in Mediterranean catchments. Ecosyst Serv 44:101136. https://doi.org/10.1016/j.ecoser.2020.101136
31. Iacob O, Brown I, Rowan J (2017) Natural flood management, land use and climate change trade-offs: the case of Tarland catchment, Scotland. Hydrol Sci J 62(12):1931–1948. https://doi.org/10.1080/02626667.2017.1366657
32. López Moreno JI, Beguería S, García Ruiz JM (2004) Laminación y estacionalidad de avenidas en los embalses pirenaicos. In: Benito G, Díez Herrero A (eds) Riesgos naturales y antrópicos en Geomorfología. SEG and CSIC, Toledo, pp 79–87
33. López Moreno JI, Beguería S, Valero Garcés B, García-Ruiz JM (2003) Intensidad de avenidas y aterramiento de embalses en el Pirineo Central español. Ería 61:159–167
34. García-Ruiz JM, Alatorre LC, Gómez-Villar A, Beguería S (2010) Upstream and downstream effects of check dams in braided rivers, Central Pyrenees. In: Conesa Garcia C, Lenzi MA (eds) Check dams, morphological adjustments and erosion control in torrential stream. Nova Science Publishers, New York, pp 307–322
35. Arnáez-Vadillo J, Gómez-Villar A (1990) Efectos de presas de corrección en la evolución geomorfológica de un torrente de montaña: la morfología del canal. I Reunión Nacional de Geomorfología, Sociedad Española de Geomorfología, Teruel
36. White S, García-Ruiz JM, Martí C, Valero B, Errea MP, Gómez-Villar A (1997) The 1996 Biescas campsite disaster in the Central Spanish Pyrenees, and its temporal and spatial context. Hydrol Process 11:1797–1812. https://doi.org/10.1002/(SICI)1099-1085(199711)11:14<1797::AID-HYP605>3.0.CO;2-7
37. Castillo V, Mosch WM, Conesa-García C, Barberá GG, Navarro-Cano JA, López-Bermúdez F (2007) Effectiveness and geomorphological impacts of check dams for soil erosion control in a semiarid Mediterranean catchment: El Cárcavo (Murcia, Spain). Catena 70:416–427. https://doi.org/10.1016/j.catena.2006.11.009
38. Conesa-García C, García-Lorenzo R (2007) Erosión y diques de retención en la cuenca mediterránea. Fundación Instituto Euromediterráneo del Agua, Murcia
39. Romero-Díaz MA (2007) Los diques de corrección hidrológica. Cuenca de Quípar (Sureste de España). Universidad de Murcia, Murcia
40. Schumm SA (2007) River variability and complexity. Cambridge University Press, Cambridge. 220 p
41. Maes J, Egoh L, Willemen C, Liquete C, Vihervaara P, Schägner JP, Grizzetti B, Drakou EG, Notte AL, Zulian G, Bouraoui F, Paracchini ML, Braat L, Bidoglio G (2012) Mapping ecosystem services for policy support and decision making in the European Union. Ecosyst Serv 1:31–39. https://doi.org/10.1016/j.ecoser.2012.06.004
42. Collentine D, Futter MN (2018) Realising the potential of natural water retention measures in catchment flood management: trade-offs and matching interests. J Flood Risk Manag 11 (1):76–84. https://doi.org/10.1111/jfr3.12269
43. Keesstra S, Nunes J, Novara A, Finger D, Avelar D, Kalantari Z, Cerdà A (2018) The superior effect of nature based solutions in land management for enhancing ecosystem services. Sci Total Environ 610–611:997–1009. https://doi.org/10.1016/j.scitotenv.2017.08.077
44. Molinillo M, Lasanta T, García-Ruiz JM (1997) Managing mountainous degraded landscapes after farmland abandonment in the Central Spanish Pyrenees. Environ Manag 21(4):587–598. https://doi.org/10.1007/s002679900051
45. Keesstra SD, Van Huissteden J, Vanderberghe J, Van Dam O, De Gier J, Pleizier ID (2005) Evolution of the morphology of the river Dragonja (SW Slovenia) due to land-use changes. Geomorphology 69:191–207. https://doi.org/10.1016/j.geomorph.2005.01.004
46. Keesstra SD, Van Dam O, Verstraeten G, Van Huissteden K (2009) Changing sediment dynamics due to natural reforestation in the Dragonja catchment, SW Slovenia. Catena 78 (1):60–71. https://doi.org/10.1016/j.catena.2009.02.021
47. Lana-Renault N, López-Vicente M, Nadal-Romero E, Ojanguren R, Llorente JA, Errea P, Regüés D, Ruiz-Flaño P, Khorchani M, Arnáez J, Pascual N (2018) Catchment based

hydrology under post farmland abandonment scenarios. Geogr Res Lett 44(2):503–534. https://doi.org/10.18172/cig-3475
48. Queiroz C, Beilin R, Folke C, Lindborg R (2014) Farmland abandonment: threat or opportunity for biodiversity conservation? A global review. Front Ecol Environ 12:288–296
49. Cosandey C, Andéassian V, Martin C, Didon-Lescot JF, Lavabre J, Folton N, Mathys N, Richard D (2005) The hydrological impact of the Mediterranean forest: a review of French research. J Hydrol 301(1–4):235–249. https://doi.org/10.1016/j.jhydrol.2004.06.040
50. Quiñorero-Rubio JM, Boix-Fayos C, de Vente J (2013) Development and application of a multi-factorial sediment connectivity index at the catchment scale. Cuad Investig Geogr 39 (2):203–223. https://doi.org/10.18172/cig.1988
51. López-Vicente M, Nadal-Romero E, Cammeraat ELH (2017) Hydrological connectivity does change over 70 years of abandonment and afforestation in the Spanish Pyrenees. Land Degrad Dev 28:1298–1310. https://doi.org/10.1002/ldr-2531
52. Piégay H, Salvador PG (1997) Contemporary flood plain forest evolution along the Middle Ubaye River, Southern Alps. Glob Ecol Biogeogr Lett 6(5):397–406. https://doi.org/10.2307/2997340
53. Buendia C, Bussi G, Tuset J, Vericat D, Sabater S, Palau A, Batalla RJ (2016) Effects of afforestation on runoff and sediment load in an upland Mediterranean catchment. Sci Total Environ 540:144–157. https://doi.org/10.1016/j.scitotenv.2015.07.005
54. Richard D, Mathys N (1999) Historique, contexte technique et scientifique des BVRE de Draix. In: Mathys N (ed) Caractéristiques, données disponibles et principaux résultats acquis au cours des dix ans de suivi. Actes du séminaire: Les bassins versants expérimentaux de Draix, laboratoire d'étude de l'érosion en montagne (Draix, Le Brusquet, Digne, 1997). Cemagref-Editions, coll. Actes de colloques: Cemagref Editions, pp 11–28
55. Piégay H, Walling DE, Landon N, He Q, Liébault F, Petiot R (2004) Contemporary changes in sediment yield in an alpine mountain basin due to afforestation (the upper Drôme in France). Catena 55(2):183–212. https://doi.org/10.1016/S0341-8162(03)00118-8
56. Sanjuán Y, Gómez-Villar A, Nadal-Romero E, Álvarez-Martínez J, Arnáez J, Serrano-Muela MP, Rubiales JM, González-Sampériz P, García-Ruiz JM (2016) Linking land cover changes in the sub-alpine and montane belts to changes in a torrential river. Land Degrad Dev 27:179–189. https://doi.org/10.1992/ldr.2294
57. Dixon SJ, Sear DA, Odoni NA, Sykes T, Lane SN (2016) The effects of river restoration on catchment scale flood risk and flood hydrology. Earth Surf Process Landf 41:997–1008. https://doi.org/10.1002/esp.3919
58. Bathurst JC, Amezaga J, Cisneros F, Gaviño Novillo M, Iroumé A, Lenzi MA, Mintegui Aguirre J, Miranda M, Urciulo A (2010) Forest and floods in Latin America: science, management, policy and the EPIC FORCE Project. Water Int 35(2):114–131. https://doi.org/10.1080/02508061003660714
59. Bathurst JC, Iroumé A, Cisneros F, Fallas J, Iturraspe R, Gaviño Novillo M, Urciuolo A, de Bièvre B, Guerrero Borges V, Coello C, Cisneros P, Gayoso J, Miranda M, Ramírez M (2011) Forest impacts on floods due to extreme rainfall and snowmelt in four Latin American environments 1: field data and analysis. J Hydrol 400(3–4):281–291. https://doi.org/10.1016/j.hydrol.2010.11.044
60. Nadal-Romero E, Cammeraat E, Serrano-Muela MP, Lana-Renault N, Regüés D (2016) Hydrological response of an afforested catchment in a Mediterranean humid mountain area: a comparative study with a natural forest. Hydrol Process 30:2717–2733. https://doi.org/10.1002/hyp.10820
61. Didon-Lescot JF (1996) Forêt et développement durable au Mont-Lozère. Impact d'une plantation de résineux, de sa coupe et de son remplacement, sur l'eau et sur les réserves minérales du sol. Thèse de Doctorat de l'Université d'Orléans, 161 p + bibliographie et annexes
62. García-Ruiz JM, Martí-Bono C, Llorente A, Beguería S (2002) Geomorphological consequences of frequent and infrequent rainfall and hydrological events in Pyrenees mountains of

Spain. Mitig Adapt Strateg Glob Chang 7:303–320. https://doi.org/10.1023/A:1024479630433
63. Ortigosa L (1991) Las repoblaciones forestales en La Rioja: Resultados y efectos geomorfológicos. Geoforma Ediciones, Logroño, 149pp
64. Eekhout JPC, Boix-Fayos C, Pérez-Cutillas P, de Vente J (2020) The impact of reservoir construction and changes in land use and climate on ecosystem services in a large Mediterranean catchment. J Hydrol 590:125208. https://doi.org/10.1016/j.jhydrol.2020.125208
65. Lasanta T, Khorchani M, Pérez-Cabello F, Errea P, Sáenz-Blanco R, Nadal-Romero E (2018) Clearing shrubland and extensive livestock farming: active prevention to control wildfires in the Mediterranean mountains. J Environ Manag 227:256–266. https://doi.org/10.1016/j.jenviman.2018.08.104
66. Lasanta T, Nadal-Romero E, García-Ruiz JM (2019) Clearing shrubland as a strategy to encourage extensive livestock farming in the Mediterranean mountains. Geogr Res Lett 45 (2):487–513. https://doi.org/10.18172/cig.3616
67. Navarro LM, Pereira HM (2015) Rewilding abandoned landscapes in Europe. In: Pereira HM, Navarro LM (eds) Rewilding European landscapes. Springer, Berlin, pp 3–23. https://doi.org/10.1007/978-3-319-12039-3_1
68. Alados CL, Sáiz H, Nuche P, Gartzia M, De Frutos Á, Pueyo Y (2018) Clearing vs. burning for restoring grasslands after shrub encroachment. Geogr Res Lett 45:441–468. https://doi.org/10.18172/cig.3589
69. Gómez D, Aguirre AJ, Lizaur X, Lorda M, Remón JL (2019) Evolution of Argoma shrubland (*Ulex gallii Planch.*) after clearing and burning treatments in Sierra de Aralar and Belate (Navarra). Geogr Res Lett 45(2):469–486. https://doi.org/10.18172/cig.3747
70. Khorchani M, Nadal-Romero E, Tague C, Lasanta T, Zabalza J, Lana-Renault N, Domínguez-Castro F, Choate J (2020) Effects of active and passive land use management after cropland abandonment on water and vegetation dynamics in the Central Spanish Pyrenees. Sci Total Environ 717:137160. https://doi.org/10.1016/j.scitotenv.2020.137160
71. Kozak J, Gimmi U, Houet T, Bollinger J (2017) Current practices and challenges for modelling past and future land use and land cover changes in mountain regions. Reg Environ Chang 17:2187–2191. https://doi.org/10.1007/s10113-017-1217-2
72. Slaymaker O (2006) Towards the identification of scaling relations in drainage basin sediment budgets. Geomorphology 80:8–19. https://doi.org/10.1016/j.geomorph.2005.09.004
73. Mérot P, Gascuel-Oudoux C, Walter C, Zhang X, Molenat J (1999) Influence du réseau de haies des paysages bocagers sur le cheminement de l'eau de surface. Rev Des Sci L'Eau 12 (1):23–31. https://doi.org/10.7202/705342ar
74. Carluer N, Marsily G (2004) Assessment and modelling of the influence of man-made networks on the hydrology of a small watershed: implications for fast flow components, water quality and landscape management. J Hydrol 285:76–95. https://doi.org/10.1016/j.jhydrol.2003.08.008
75. Gascuel-Odoux C, Aurousseau P, Cordier MO, Durand P, Garcia FG, Masson V, Salmon-Monviola J, Tortrat F, Trepos R (2009) A decision-oriented model to evaluate the effect of land use and agricultural management on herbicide contamination in stream water. Environ Model Softw 24:1433–1446. https://doi.org/10.1016/j.envsoft.2009.06.002
76. Bracken LJ, Croke J (2007) The concept of hydrological connectivity and its contribution to understanding runoff-dominated geomorphic systems. Hydrol Process 21:1749–1763. https://doi.org/10.1002/hyp.6313
77. Turnbull L, Wainwright J, Brazier RE (2008) A conceptual framework for understanding semi-arid land degradation: ecohydrological interactions across multiple-space and time scales. Ecohydrology 1(1):23–34. https://doi.org/10.1002/eco.4
78. Lexartza-Artza I, Wainwright J (2009) Hydrological connectivity: linking concepts with practical implications. Catena 79:146–152. https://doi.org/10.1016/j.catena.2009.07.001

79. Wainwright J, Turnbull L, Ibrahim TG, Lexartza-Artza I, Thornton ST, Brazier RE (2011) Linking environmental regimes, space and time: interpretations of structural and functional connectivity. Geomorphology 126:187–204. https://doi.org/10.1016/j.geomorph.2010.07.027
80. Malek Z, Verburg PH, Geijzendorffer IR, Bondeau A, Cramer W (2018) Global change effects on land management in the Mediterranean region. Glob Environ Change 50:238–254. https://doi.org/10.1016/j.gloenvcha.2018.04.007
81. Cammeraat ELH (2004) Scale dependent thresholds in hydrological and erosion response of a semi-arid catchment in southeast Spain. Agric Ecosyst Environ 104(2):317–332. https://doi.org/10.1016/j.agee.2004.01.032
82. Bellin N, van Wesemael B, Meerkerk A, Vanacker V, Barbera GG (2009) Abandonment of soil and water conservation structures in Mediterranean ecosystems. A case study from south east Spain. Catena 76:114–121. https://doi.org/10.1016/j.catena.2008.10.002
83. Lesschen JP, Schoorl JM, Cammeraat LH (2009) Modelling runoff and erosion for a semi-arid catchment using a multi-scale approach based on hydrological connectivity. Geomorphology 109:174–183. https://doi.org/10.1016/j.geomorph.2009.02.030
84. Moreno-de-las-Heras M, Lindenberger F, Latron J, Lana-Renault N, Llorens P, Arnáez J, Romero-Díaz A, Gallart F (2019) Hydro-geomorphological consequences of the abandonment of agricultural terraces in the Mediterranean region: key controlling factors and landscape stability patterns. Geomorphology 333:73–91. https://doi.org/10.1016/j.geomorph.2019.02.014
85. Costra GB, Dal Negro P, Frattini P (2003) Soil slips and debris flows on terraced slopes. Nat Hazards Earth Syst Sci 3:31–42. https://doi.org/10.5194/nhess-3-31-2003
86. Gallart F (2009) Algunos criterios topográficos para identificar el origen antrópico de cárcavas. Geogr Res Lett 35(2):215–221. https://doi.org/10.18172/cig.1219
87. Arnáez J, Lana-Renault N, Lasanta T, Ruiz-Flaño P, Castroviejo J (2015) Effects of farming terraces on hydrological and geomorphological processes. A review. Environ Sci 128:122–134. https://doi.org/10.1016/j.catena.2015.01.021
88. Arnáez J, Lana-Renault N, Ruiz-Flaño P, Pascual N, Lasanta T (2017) Mass soil movement on terraced landscapes of the Mediterranean mountain areas: a case study in the Iberian Range, Spain. Geogr Res Lett 43(1):83–100. https://doi.org/10.18172/cig.3211
89. Lesschen JP, Cammeraat LH, Nieman T (2008) Erosion and terrace failure due to agricultural land abandonment in a semi-arid environment. Earth Surf Process Landf 33:1574–1584. https://doi.org/10.1002/esp.1676
90. Lasanta T, Arnáez J, Ruiz-Flaño P, Lana-Renault Monreal N (2013) Los bancales en las montañas españolas: un paisaje abandonado y un recurso potencial. Bol Asoc Geogr Esp 63:301–322
91. Mekonnen M, Keesstra SD, Baartman JE, Ritsema CJ, Melesse AM (2015) Evaluating sediment storage dams: structural off-site sediment trapping measures in northwest Ethiopia. Geogr Res Lett 41(1):7–22. https://doi.org/10.18172/cig.2643
92. Calsamiglia A, Fortesa J, García-Comendador J, Lucas-Borja ME, Calvo-Cases A, Estrany J (2018) Spatial patterns of sediment connectivity in terraced lands: anthropogenic controls of catchment sensitivity. Land Degrad Dev 29:1198–1210. https://doi.org/10.1002/ldr.2840
93. Tarolli P, Preti F, Romano N (2014) Terraced landscapes: from an old best practice to a potential hazard for soil degradation due to land abandonment. Anthropocene 6:10–25. https://doi.org/10.1016/j.ancene.2014.03.002
94. Zoumides C, Bruggeman A, Giannakis E, Camera C, Djuma H, Eliades M, Charalambous K (2016) Community-based rehabilitation of mountain terraces in Cyprus. Land Degrad Dev 28 (1):95–105. https://doi.org/10.1002/ldr.2586
95. García Ruiz JM, Beguería S, López Moreno JI, Lorente A, Seeger M (2001) Los recursos hídricos superficiales del Pirineo aragonés y su evolución reciente. Geoforma Ediciones, Logroño, 192 p
96. García-Ruiz JM, Arnáez J, White SM, Lorente A, Beguería S (2000) Uncertainty assessment in the prediction of extreme rainfall events: an example from the central Spanish Pyrenees.

Hydrol Process 14(5):887–898. https://doi.org/10.1002/(SICI)1099-1085(20000415)14:5<887::AID-HYP976>3.0.CO;2-0
97. García Ruiz JM, Ortigosa LM (1988) Algunos efectos geomorfológicos de las repoblaciones forestales: cambios en la dinámica de cauces en pequeñas cuencas del Pirineo Central español. Cuat Geomorfol 2:33–41
98. Beguería S, López-Moreno JI, Gómez-Villar A, Rubio V, Lana-Renault N, García-Ruiz JM (2006) Fluvial adjustments to soil erosion and plant cover changes in the Central Spanish Pyrenees. Geogr Ann Ser A Phys Geogr 88(3):177–186. https://doi.org/10.1111/j.1468.0459.2006.00293.x
99. Beguería S, López-Moreno JI, Lorente A, Seeger M, García-Ruiz JM (2003) Assessing the effect of climate oscillations and land-use changes on streamflow in the Central Spanish Pyrenees. Ambio 32(4):283–286
100. López-Moreno JI, Vicente-Serrano SM, Morán-Tejeda E, Zabalza J, Lorenzo-Lacruz J, García-Ruiz JM (2011) Impact of climate evolution and land use changes on water yield in the Ebro basin. Hydrol Earth Syst Sci 15:311–322. https://doi.org/10.5194/hess-15-311-2011
101. López-Moreno JI, Beguería S, García-Ruiz JM (2006) Trends in high flows in the central Spanish Pyrenees: response to climatic factors or to land-use change? Hydrol Sci J 51:1039–1050. https://doi.org/10.1623/hysj.51.6.1039
102. García-Ruiz JM, Lana-Renault N, Beguería S, Lasanta T, Regüés D, Nadal-Romero E, Serrano-Muela P, López-Moreno JI, Alvera B, Martí-Bono C, Alatorre LC (2010) From plot to regional scales: interactions of slope and catchment hydrological and geomorphic processes in the Spanish Pyrenees. Geomorphology 120:248–257. https://doi.org/10.1016/j.geomorph.2010.03.038
103. García-Ruiz JM, Regüés D, Alvera B, Lana-Renault N, Serrano-Muela P, Nadal-Romero E, Navas A, Latron J, Martí-Bono C, Arnáez J (2008) Flood generation and sediment transport in experimental catchments affected by land use changes in the central Pyrenees. J Hydrol 356:245–260. https://doi.org/10.1016/j.hydrol.2008.04.013
104. Lana-Renault N, Latron J, Karssenberg D, Serrano-Muela P, Regüés D (2011) Differences in stream flow in relation to changes in land cover: a comparative study in two sub-Mediterranean mountain catchments. J Hydrol 411(3–4):366–378. https://doi.org/10.1016/j.hydrol.2011.10.020
105. Gómez-Villar A, Sanjuán Y, García-Ruiz JM, Nadal-Romero E, Álvarez-Martínez J, Arnáez J, Serrano Muela MP (2014) Sediment organization and adjustment in a torrential reach of the upper Ijuez River, Central Spanish Pyrenees. Geogr Res Lett 40(1):191–214. https://doi.org/10.18172/cig.2566
106. Allan R et al (2020) Advances in understanding large-scale responses of the water cycle to climate change. Ann N Y Acad Sci 1472:49–75. https://doi.org/10.1111/nyas.14337
107. Serrano-Muela MP, Nadal-Romero E, Lana-Renault N, González-Hidalgo JC, López-Moreno JI, Beguería S, Sanjuán Y, García-Ruiz JM (2015) An exceptional rainfall event in the Central western Pyrenees: spatial patterns in discharge and impact. Land Degrad Dev 26:249–262. https://doi.org/10.1002/ldr.2221
108. Lana-Renault N, Nadal-Romero E, Serrano-Muela MP, Alvera B, Sánchez-Navarrete P, Sanjuán Y, García-Ruiz JM (2014) Comparative analysis of the response of various land covers to an exceptional rainfall event in the Central Spanish Pyrenees, October 2012. Earth Surf Process Landf 39:581–592. https://doi.org/10.1002/esp.3465
109. White S, García-Ruiz JM (1998) Extreme erosional events and their role in Mountain areas of Northern Spain. Ambio 27(4):300–3005

Modelling and Evaluation of the Effect of Afforestation on the Runoff Generation Within the Glinščica River Catchment (Central Slovenia)

Gregor Johnen, Klaudija Sapač, Simon Rusjan, Vesna Zupanc, Andrej Vidmar, and Nejc Bezak

Contents

G. Johnen
Faculty of Science, Radboud University Nijmegen, Nijmegen, Netherlands
e-mail: johnen_gregor@gmx.de

K. Sapač, S. Rusjan, A. Vidmar, and N. Bezak (✉)
Faculty of Civil and Geodetic Engineering, University of Ljubljana, Ljubljana, Slovenia
e-mail: klaudija.sapac@fgg.uni-lj.si; simon.rusjan@fgg.uni-lj.si; andrej.vidmar@fgg.uni-lj.si; nejc.bezak@fgg.uni-lj.si

V. Zupanc
Biotechnical Faculty, University of Ljubljana, Ljubljana, Slovenia
e-mail: vesna.zupanc@bf.uni-lj.si

Carla S. S. Ferreira, Zahra Kalantari, Thomas Hartmann, and Paulo Pereira (eds.),
Nature-Based Solutions for Flood Mitigation: Environmental and Socio-Economic Aspects,
Hdb Env Chem (2022) 107: 215–232, DOI 10.1007/698_2020_649,

Published online: 29 August 2020

Abstract Increases in the frequency of flood events are one of the major risk factors induced by climate change that lead to a higher vulnerability of affected communities. Natural water retention measures such as afforestation on hillslopes and floodplains are increasingly discussed as cost-effective alternatives to hard engineering structures for providing flood protection, particularly when the evaluation also considers beneficial ecosystem services other than flood protection. The present study provides a combined modelling approach and a cost–benefit analysis (CBA) of the effects of afforestation on peak river flows and on selected ecosystem services within the Glinščica River catchment in Slovenia. In order to investigate these effects, the hydrological model HEC-HMS, the hydraulic model HEC-RAS, and the flood damage model KRPAN, which were developed specifically for Slovenia, are used. Three scenarios were evaluated where the main difference was the extent of afforestation. It was found that increasing the amount of tree cover (i.e., ≈15–60%) results in a flood peak reduction ranging from 9 to 13%. Flood extensions were significantly lower for most scenarios leading to reduced economic losses. However, a 100-year CBA only showed positive net present values (NPV) for one of the considered scenarios, where the afforestation was considered only in floodplain areas and the prevailing benefits were those of flood protection measures, which were higher than, for example, biodiversity or recreational benefits. Additional ecosystem co-benefits that are not directly linked with flood protection are considered in case of all three scenarios.

Keywords Afforestation, Cost–benefit analysis, Ecosystem services, Flood hazards, Glinščica River, Land-use change

1 Introduction

Climate change is expected to increase the frequency and intensity of future flood events (e.g., [1]), leading to higher costs of flood damages and increasing the public demand for protective measures. Nature-based solutions such as natural water retention measures (NWRM), which also include afforestation, are increasingly used to mitigate flood risks as a complement to gray infrastructure measures. The influence of afforestation is usually assessed with model-based or observational studies. Evidence that floodplain afforestation leads to peak flow reduction and thus to a lowering of the flood risk is less supported by observations, while "modelled results were found to provide significant evidence that increasing cover reduces flood peaks" [2].

However, NWRMs not only serve to reduce flood risk but also provide additional ecosystem services, including increased biodiversity and recreation opportunities, which can help to manage smaller-scale flooding problems where the high cost of constructing hard defenses cannot be justified, especially for smaller communities [3].

Despite this growing interest in NWRMs, economic appraisals of the flood protection benefit of afforestation measures are rare. Dittrich et al. [4] undertook a detailed cost–benefit analysis of afforestation as a climate change adaptation measure and highlighted the importance of identifying and quantifying additional ecosystem co-benefits of NWRMs. As a result of the limited hydrological and economic appraisals of NWRM, the overall aim of this study is to enhance the knowledge base associated with flood risk and impacts. Given this aim, we applied the hydrological model HEC-HMS, the hydraulic model HEC-RAS, and the flood damage model KRPAN [5] in order to investigate afforestation effects on the runoff generation throughout the calculation of the expected flood damage. In the process of the CBA, we tried to quantify multiple ecosystem services.

2 The Glinščica River Catchment

The Glinščica River catchment area is relatively small with 16.9 km^2 located in the western part of Ljubljana, Slovenia. The catchment is located in a temperate continental climate and has torrential characteristics. The mean annual precipitation in the area is around 1,500 mm, while snow falls regularly in winter. Moreover, floods are most often generated by either summer thunderstorms or by spring and autumn prolonged rainfall events. Since part of this catchment also covers the urban part of the Ljubljana City, the population density is relatively high given the rest of Slovenia. Moreover, the investigated area is easily accessible and its tourism well-developed, and especially local people tend to use it for recreational activities such as hiking, running, or cycling. The location of the catchment on the map of Slovenia can be found in previously published studies [6, 7]. The catchment area was selected as a case study to investigate the effects of afforestation, as the hydrological and hydraulic models were already used in the previous study by Bezak et al. [6] and just needed to be adjusted in order to be able to model the effects of afforestation. The Glinščica River is part of a larger catchment area of the Gradaščica River, which flows into the Ljubljanica River. The topography of the Glinščica River catchment area consists of hilly areas in the east and west with elevation ranging from 210 to 590 m a.s.l. It has a diverse relief with hilly headwaters as well as flat plain areas [6]. According to the earlier studies by Bezak et al. [6] and Šraj et al. [7], the Glinščica River catchment area is one of the hydrological test areas in Slovenia. The lowlands of the Gradaščica River, formerly natural floodplain areas, mainly consist of agricultural fields and meadows (Fig. 1). They have been partially urbanized in recent decades, increasing the vulnerability to floods. A flood in October 2014 with a 10–50-year return period, when extensive urban areas and more than 1,000 homes were flooded, is acknowledged as the major recent flood event [6]. Moreover, a large flood also occurred in 2010 with a return period of 50–150 years. Because the river is heavily regulated (i.e., concrete trapezoidal channel) in the lowlands, a primary goal

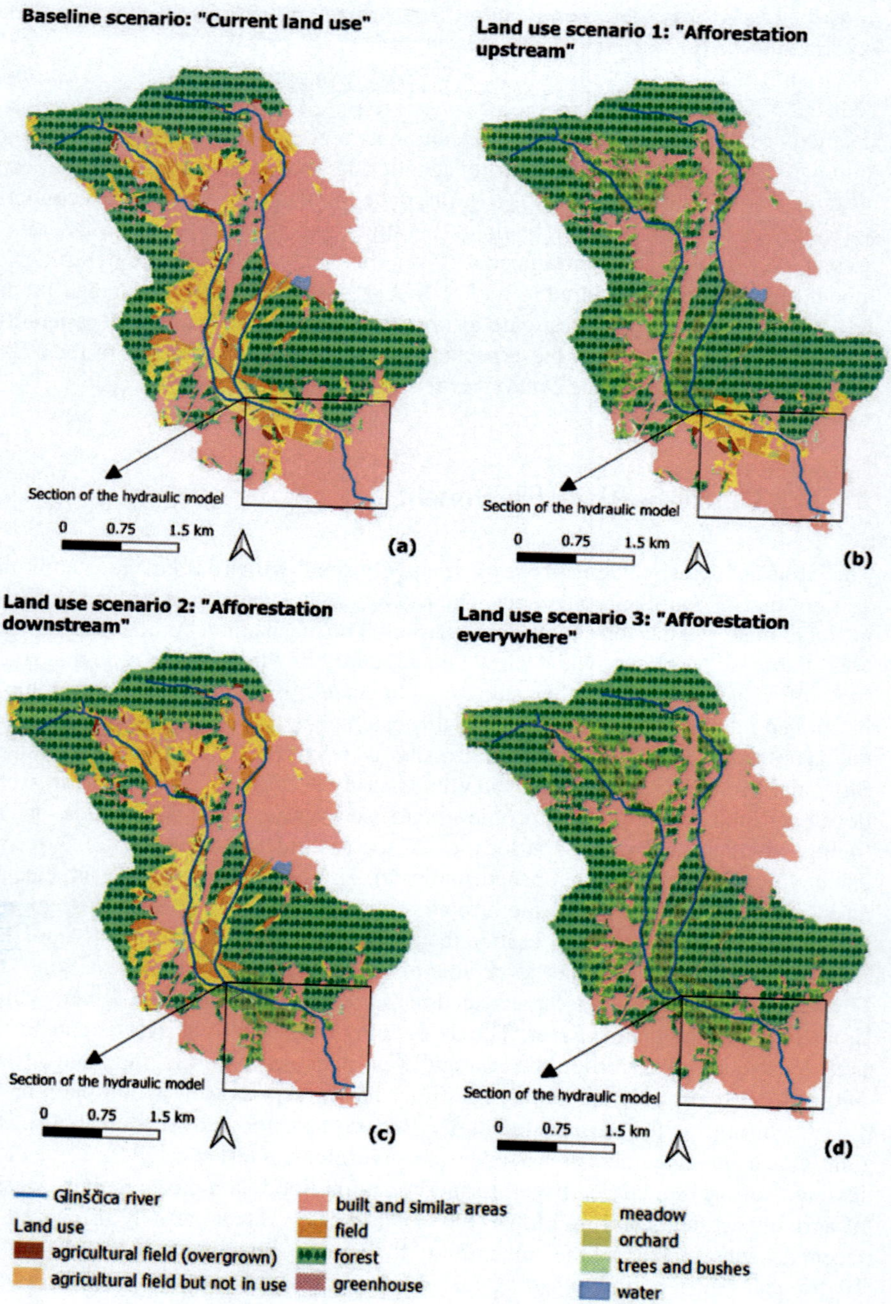

Fig. 1 Current (**a**) and adopted land-use scenarios (**b**), (**c**), and (**d**)

of the study is to examine how new forests could help alleviate the flooding in the City of Ljubljana.

3 Model Simulations

This study, like the one in Bezak et al. [6], presents a combined hydrological and hydraulic model of the Glinščica River catchment. While the entire catchment was considered in the hydrological model, the hydraulic model was used only for the downstream part, which is also indicated in Fig. 1. By using a calibrated and validated hydrological model, the inputs (i.e., hydrographs) for the hydraulic model were determined. In order to best detect the possible effects of afforestation, three different hypothetical afforestation (i.e., mixed forest) scenarios were chosen.

In the scope of this study, the following four scenarios were considered (Fig. 1):

- Scenario "Current land use" where hydrological and hydraulic models represent the current situation.
- Scenario "Afforestation upstream" where afforestation in the hydrological model is considered in the upper part of the catchment (afforestation area is 244 ha).
- Scenario "Afforestation downstream" where afforestation in the hydrological and hydraulic model is considered only in the lower part of the catchment (afforestation area is 77 ha).
- Scenario "Afforestation everywhere" where afforestation in the hydrological and hydraulic model is considered in all parts of the catchment (afforestation area is 321 ha).

The scenario "Afforestation upstream" represents the afforestation of 244 ha of former meadows, agricultural fields, and greenhouses upstream of the hydraulic model, the scenario "Afforestation everywhere" represents a complete cover of woodland within all floodplains (321 ha), and the scenario "Afforestation downstream" represents the afforestation of the floodplains only within the hydraulic modelling section (77 ha). Scenario "Afforestation upstream" was only incorporated within the hydrological model where the effect of afforestation was conceptually modelled, while the "Afforestation everywhere" and "Afforestation downstream" scenarios could also be displayed by a change in the roughness values of the 2D floodplain area within the HEC-RAS hydraulic model (Fig. 1).

The same design hyetographs (i.e., Huff curves) as of Dolšak et al. [8] and shown in Bezak et al. [6] were used in order to compute the design hydrographs, which then served as inputs into the hydraulic model. A detailed description of the calibration and validation of the hydrological model is provided by Šraj et al. [9]. However, adjustments were made regarding the assigned loss method which was changed to Soil Conservation Service (SCS)-Curve Number (CN) as this allowed parameter alteration for the lag time and the initial abstraction (i.e., afforestation modelling). Increasing the tree cover is advocated as an affective measure of increasing the water storage capacity of catchment areas, which is reflected/expressed in the model with the increased lag time and the initial abstraction. This leads to a decrease in discharge and subsequently lowers the hydrographs which will be the input data of the hydraulic model [10].

The lag time is one of the parameters used in the hydrological model and is denoted as the time difference between the center of the unit rainfall event and the runoff peak. In our case, the empirical formula for the lag time from the SCS was used [10].

To determine the initial abstraction, the SCS developed an empirical relationship between the maximum potential retention and the initial abstraction [10]. Similarly, as done in Bezak et al. [6], we combined Huff curves that describe internal rainfall distribution and intensity-duration-frequency (IDF) curves in order to define the design rainfall events. As stated by Šraj et al. [9] and Bezak et al. [6], the catchment time of concentration is around 6 h. Thus, the design rainfall duration of 6 h was used. Similarly, as in Bezak et al. [6], IDF curves for the Ljubljana-Bežigrad rainfall station were used in order to define the design rainfall event. Return periods of 2, 10, and 25 years were used in this study. Results of the hydrological model were used as inputs to the hydraulic model (i.e., hydrographs). The same hydraulic model settings as those explained by Bezak et al. [6] were used. The section of the hydraulic modelling is indicated in Fig. 1. A detailed description of the hydraulic model is given by Bezak et al. [6].

In order to display floodplain afforestation in the hydraulic model, Manning's roughness coefficient (n) was used to represent the energy lost in the water flowing across the floodplain. There are several methods available for calculating Manning's n, and separate values are required for the river channel and the floodplain. After the import of the land-use dataset (i.e., detailed land-use map provided by the Slovenian Ministry of Agriculture, Forestry and Food) into HEC-RAS was completed, Manning's values were chosen according to USACE [11].

Table 1 gives an overview of the different land-use classes and their corresponding Manning's values for the areas outside the main channel (i.e.,

Table 1 Land use with correspondent roughness values before (i.e., scenario "Current land use") and after implementing the afforestation scenarios (scenarios "Afforestation upstream," "Afforestation downstream," and "Afforestation everywhere")

Land use	Manning's n	Adapted manning's n
Field	0.05	0.2
Greenhouse	0.07	0.2
Orchard	0.04	0.04
Meadow	0.035	0.2
Agricultural field (overgrown)	0.05	0.2
Trees and bushes	0.15	0.15
Agricultural field but not in use	0.04	0.2
Forest	0.2	0.2
Built and similar areas	0.045	0.045
Water	0.01	0.01

floodplains). Implementing this computation approach enabled us to trade-off detailed spatial information with relative simplicity and speed while preserving the key real-world hydraulic floodplain water movement dynamics. Thus, hydrological and hydraulic models were used to compute the floodplain water depth and the extent for different scenarios and for 2-, 10-, and 25-year return periods.

The water depths in the floodplain areas for different return periods and scenarios were subsequently exported from HEC-RAS to ArcGIS in order to simplify it to fewer polygons including the attribute of the flooding depth. These polygons were the input to the flood damage model KRPAN, which gave us the expected economic damage for each scenario and return period [5]. The flood damage model can calculate damage in different sectors such as infrastructure, cultural heritage, residential areas, etc. [5]. The model uses data from the census and market values and applies functions (i.e., depth-damage curves) to estimate the damage based on the floodplain water depth [5]. Therefore, the main input is the polygon flood map with information about water depth [5]. The result of the model is the flood damage for different sectors [5].

4 Cost–Benefit Analysis

The chosen time frame of the anticipated CBA was 100 years, which could be regarded as a reasonable lifetime of such a scenario. Costs and benefits are calculated with the most recent available prices of elements included in the CBA and explained in the following paragraphs. The discount rate applied, which takes into consideration how much someone prefers benefits now rather than in the future, was 4% because similar values were adopted in other studies [4]. It is known from previous studies such as Dittrich et al. [4] that ecosystem services (ES), such as biodiversity, usually represent also a significant portion of the beneficial character of afforestation measures. Due to temporal and financial constraints of the present study, the only option to obtain values for different ES was to use a benefit transfer (BT). We used and transferred the results of existing studies to generate and determine monetary values for our case study. As stated in the extensive review by Müller et al. [12], a basic requirement is the similarity of background conditions. However, since similar studies with respect to scale, perspective, and dimension are rare, we tried to focus on studies of temperate European forests. In cases where values could not be determined without extensive data collection, we used the study of Dittrich et al. [4], given that the scale, dimension, and perspective of this study were the most similar we could find in comparison with our research area. They were then transferred to Euro. In total, our study included benefits of flood protection measures, costs of afforestation measures, as well as benefits for biodiversity, carbon, recreation, and drinking water as part of ES. A result of the CBA is the net present value (NPV), which determines if a specific measure has a positive or a negative value. The NPV of the different afforestation scenarios was calculated by using a discount rate of 4% and a series of future costs (negative values) and ecosystem service benefits (positive values). Assuming n is the number of cash flows in the list of values, Eq. 1 displays the formula of the NPV. Furthermore, the internal rate of return (IRR) indicates the rate for which the NPV equals zero.

$$\mathrm{NPV} = \sum_{i=1}^{n} \frac{\mathrm{values}_i}{(1+\mathrm{rate})^i} \tag{1}$$

4.1 Cost of Afforestation

To calculate the costs of afforestation, the price of the area to be afforested as well as the opportunity costs need to be determined. Opportunity costs stem from the potential returns from agricultural crops harvested in the fields to be afforested [13]. Based on the current land use, the prices were identified using the online portal [14]. Based on the database (i.e., previous sales of agricultural land in the area), an average price of 60,000 €/ha was set related to the previous land use. Furthermore, construction costs need to be quantified. It was assessed that 3,500 trees can on average be planted on 1 ha. Without protection through fences, the costs would be 1 €/tree which results in total construction costs of 3,500 €/ha as implemented within the CBA.

4.2 Benefits of Flood Protection Measures

After the hydrological and hydraulic models computed the various scenarios, the results in terms of flood depth and flood extent were exported from HEC-RAS and used as an input for the flood damage model to calculate the expected flood damage and the benefits of measures subsequently. Benefits of flood protection measures were obtained using the recently developed flood damage model KRPAN [5]. For each of the sectors (environment, cultural heritage, economic activity), the model uses a simple equation (Eq. 2) in order to assess the expected damage (ED) due to the different scenarios used in our study. To calculate the benefits, the calculations are carried out with and without the implementation of a flood risk management scheme to obtain a comparison: the damage avoided under the scenario is equal to the benefits of the scenario. As the output per scheme is one value per scenario, we statistically generated the yearly benefit considering a period of 100 years that was selected for the CBA. This was done based on the results of the flood frequency approach. Thus, based on the measured annual maximum peak discharge values from the nearby discharge gauging station, Dvor station on the Gradaščica River, we fitted the generalized extreme value (GEV) distribution to the data where the parameters were estimated using the L-moments method as defined by Hosking and Wallis [15]. Based on the fitted distribution, we generated 1,000 × 100 data points, which represent 100 years of annual maximum data. For each data point, the return period was determined based on which damage could be calculated. With such a procedure, we estimated how many flood events with different return periods are to be expected in the next 100 years. For example, we could have 2 events with a return period above 100 years, 8 events with a return period between 50 and 100 years, 10 events with a return period between 2 and 50 years, and 80 events with a return period below 2 years (i.e., 100 in total because we are using the annual maxima approach). The 1,000 random realizations were used in order to capture some of the natural variability. The median values of the 1,000 realizations were used.

$$\mathrm{ED} = S * D * E * \mathrm{Vu} * \mathrm{Va} \tag{2}$$

where S = strength of the event (i.e., water depth or velocity), D = dimension (number or size of the exposed elements), E = exposure (probability that an individual sector element will be present in a given area at a given time), Vu = vulnerability (structural damage of the individual element), and Va = economic value (of the individual element)

4.3 Biodiversity

As one of the core elements of ecosystems, biodiversity influences multiple aspects of ES, which makes them difficult to evaluate [13]. Nevertheless, Müller et al. [12] saw it as necessary to quantify even single components if the valuation should contribute to forest management and planning. After Dittrich et al. [4], biodiversity services to be valued are divided into use and nonuse values. Dittrich et al. [4] provide a range of values for different woodlands. The use of values (like recreation itself) will be also included within the valuation of recreation. Therefore, to avoid double counting, nonuse values are transferred from Dittrich et al. [4]. According to Dittrich et al. [4], "Non-use values are existence value (the benefit people receive from just knowing that wildlife exists even though they never see it) and bequest value (the benefit people derive from knowing that wildlife will be protected and preserved for the benefit of future generations)." Their central estimation of 281.05 €/ha/year was used in this study.

4.4 Carbon

Concerning the impact of forests on atmospheric carbon dioxide concentration, the main aspects to be valued are storage and sequestration. Bernal et al. [16] developed a database of different types of forest restoration actions worldwide across different latitudes that indicates what removal rates apply to each subnational unit in the world which can be accessed via [17]. The afforested area per scenario was multiplied with the relevant carbon price of 15.41 €/tCO_2 set out by [18] in their central scenario and by per hectare carbon sequestration rates in tons (based on the removal rate database for Slovenia with restoration type "other broadleaf"), which was 11.73 tCO_2/ha/year.

4.5 Recreation

Recreational benefits stem from direct contact between people and natural resources. They are to a huge extent dependent on local characteristics and societal perceptions about nature as well as on the size of the afforested landscape [12]. Recreational values of small forests may be marginal and initially increase with size. However, at a certain size, it is expected that aesthetic values reach a maximum. There is evidence that mosaic landscapes, which present a combination of forested areas and permanent meadows, offer a different kind of aesthetic values. The area of the anticipated afforestation is already accessible and touristically developed [19]. Furthermore, based on the questionnaire of Japelj et al. [20], the needs and well-being of the population are already sufficiently addressed with urban forests in place leading to reduced surplus benefits through the anticipated afforestation measures. As, however, forest landscapes are still considered attractive landscapes whose benefit for the local population should be taken into account, we decided to choose a particularly low value of 50 €/ha/year from Dittrich et al. [4], which adopted this value after Tinch et al. [21] for rural woodlands. Thus, we aim to sufficiently account for the uncertainties regarding this explicit ecosystem co-benefit.

4.6 Water Quality

Afforestation might possibly impact the surface and subsurface water quality. In Ljubljana, most of the water supplied to the City of Ljubljana stems from the Ljubljansko polje aquifer system which is characterized by a strong interconnection between surface and groundwater and thus an increased vulnerability to chemical pollution [22, 23]. As the computations show, afforestation in the Glinščica catchment will lead to a more balanced hydrological regime with reduced peak flows, flooding extents, and by an increased control in the groundwater recharge and watershed protection. Through that it is reasonable to expect that the amount of sediment and nutrient input into the river and the aquifer will be lowered [22]. To quantify these benefits related to surface water and groundwater, quality is challenging, as there are no relevant studies in Slovenia. Therefore, we took the mean value within the range of most values for forest ecosystems, which resulted from the systematic review of Müller et al. [12]. Accordingly, the price for the improvement of drinking water quality was set to 118 €/ha/year.

5 Results and Discussion

Significant peak flow reductions could be achieved by conceptually implementing the measures within the hydrological model. Peak flow reductions were 14% for the 2-year return period, 10% for the 10-year return period, and 9.5% for the 25-year return period tending to decrease for events with a larger return period. This trend was not detected comparing the differences in peak times. For both the 2- and 10-year return periods, the differences compared to the current state scenario were 10 min. A 5-min difference was detected for the 25-year return period. This reduction in peak flow also influenced the flooding extent as can be seen in Fig. 2 (yellow inundation boundary). The smaller input hydrograph, which was modelled within the hydrological model, did as such directly translate to a smaller inundation extent.

However, changing the hydraulic roughness of the floodplains (i.e., scenario "Afforestation downstream") within the hydraulic model had a similar effect to that in the scenario "Afforestation upstream" (Fig. 2, orange and yellow boundaries) although comprising approximately 245 ha less afforested surface. Thus, the effect of the afforestation is relatively small. The main reason is that in one case, the scenario is implemented only in the hydrological model, and in the other case, the afforestation is also included in the hydraulic model where obviously the changed roughness coefficients have a large impact on the floodplain water movement. Additionally, this might be due to the uncertainty in predicting the maximum flooding extent as explained by Bezak et al. [6]. Comparing the outflow hydrographs for the 10-year return period, scenario supports this explanation as one can see that the hydrograph for the scenario "Afforestation downstream" is only insignificantly smaller than that of the scenario "Current land use" (Fig. 3). While peak flows were not influenced by changing the hydraulic roughness of the 2D flow, the extent of the inundated area could be narrowed by almost 25% (Fig. 2). The 2D flow areas within the hydraulic model allowed for a more detailed assessment of the effects of floodplain afforestation on flood depth and water velocity. However, maximum water depths within some parts of the floodplains were low (below 1 cm), and strong variations by a factor of 10 were found with regard to the water velocities within the floodplains. Therefore, we could not detect a clear trend here.

In view of the economic damage, the smaller extent of the inundated area also manifested in a lower cumulative economic damage caused by the event. Only small damages were calculated for the 2- and 10-year return periods, as mainly natural areas were flooded with a small flooding depth as mentioned before.

Therefore, the smaller flooding extent in case of 2- and 10-year return periods could also not significantly contribute to the flood protection measure benefits ranging between a maximum of 8,000 € and 18,000 €, respectively. Damage costs increase with the return period; for the 25-year return period, a damage of 610,752 € was calculated as the expected flood damage under the current land-use scenario ("Current land use"). In relation to the costs of the 25-year return period with the initial land use, significant reductions of approximately 78% and 80% could be

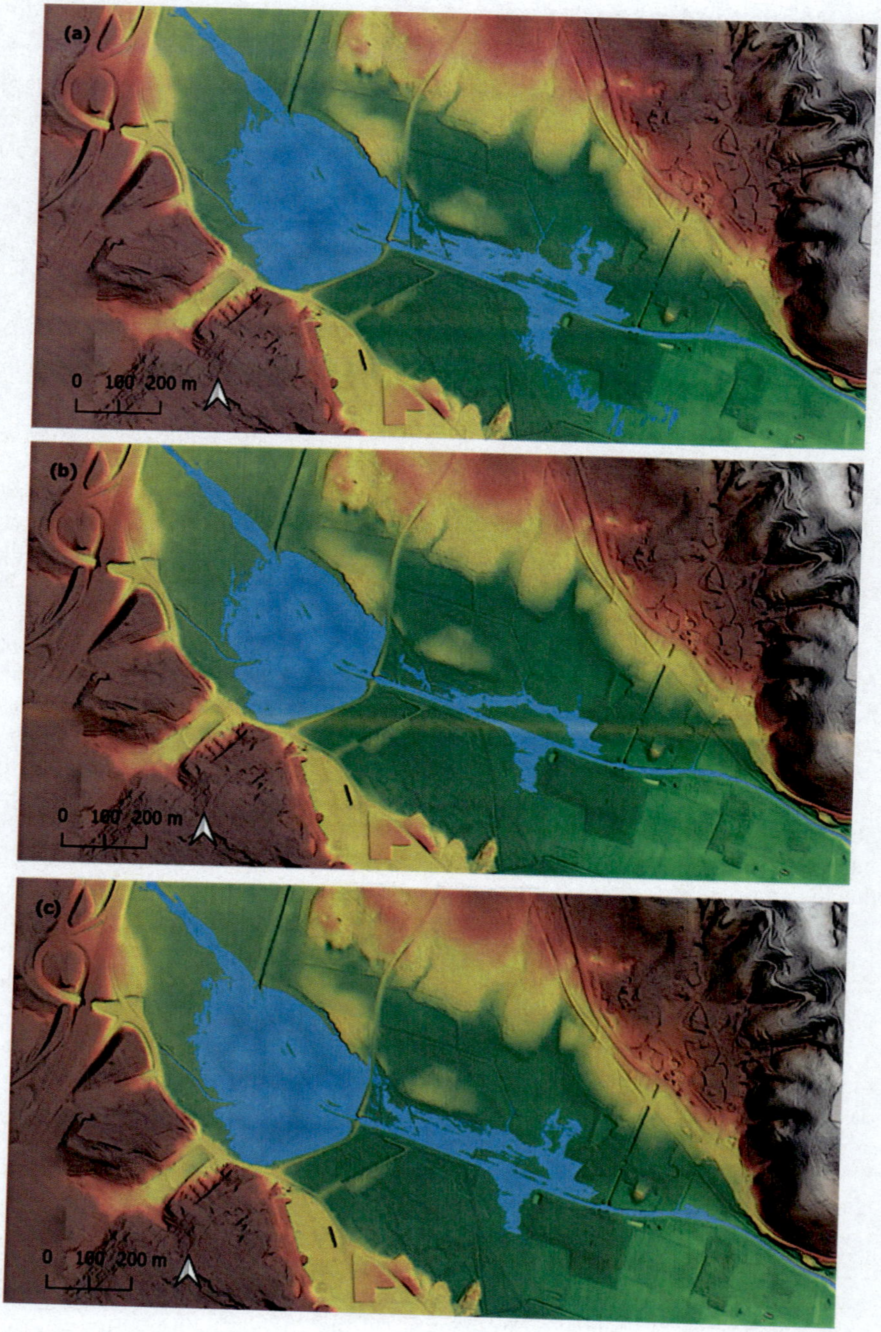

Fig. 2 Comparison between the inundation boundaries for the scenarios "Current land use" (**a**), "Afforestation upstream" (**b**), and "Afforestation downstream" (**c**) for the 10-year return period

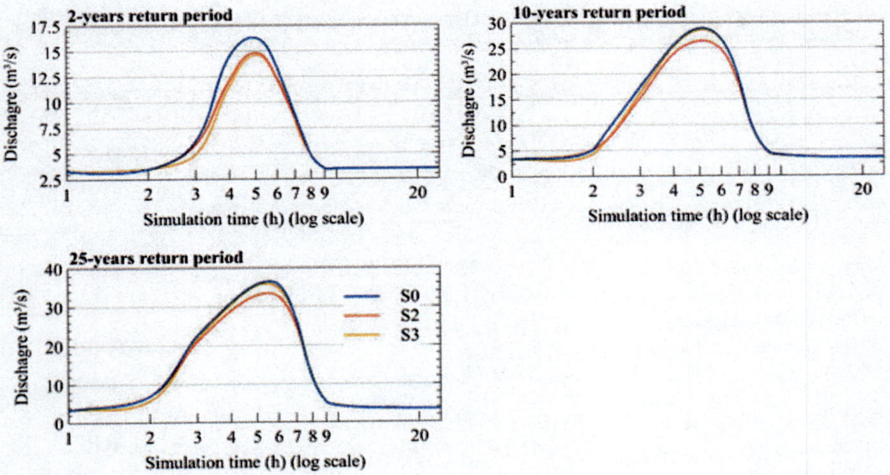

Fig. 3 Comparison of the outflow hydrographs of the hydraulic model for different return periods considering the different scenarios: S0 ("Current land use"), S3 ("Afforestation everywhere"), and S2 ("Afforestation downstream")

achieved for scenarios "Afforestation upstream" and "Afforestation everywhere," respectively. Even though the largest reductions could always be achieved for scenario "Afforestation everywhere," which comprised the largest hypothetical afforestation area of 321.17 ha, it should be mentioned that especially for the 25-year return period, a flood damage reduction of 65% could be achieved with the scenario "Afforestation downstream." The benefits of the flood protection measure represent 396,744 € when afforesting an area which is 244.50 ha smaller than in scenario "Afforestation everywhere." This, an almost similar cost reduction, despite this huge difference in the afforested area, might indicate that Manning's n roughness is a more sensitive parameter with regard to displaying effects of afforestation than the CN, which was used in the hydrological model. In addition, Hall et al. [24] found that Manning's n roughness coefficient has the dominant impact on uncertainty in the hydraulic model prediction. Therefore, results should be interpreted carefully. Based on the previous findings, the overall benefits of the scenario were the largest for the scenario "Afforestation downstream" as the costs of afforestation were low with 268,334 € for only 76.67 ha of afforested area. A positive NPV of 4,184,035.48 € could be found for the scenario "Afforestation downstream" only (Table 2).

The scenarios "Afforestation everywhere" and "Afforestation upstream" show negative NPV for a CBA period of 100 years. This means that only the scenario "Afforestation downstream" is economically sustainable and would be worth implementing from the economic point of view when taking into account flood protection measure benefits plus other ecosystem service co-benefits. The main reason for the negative NPV values lies in the fact that large areas would need to be afforested in case of "Afforestation everywhere" and "Afforestation upstream"

Table 2 Net present values (NPVs), internal rate of return (IRR), and benefit-cost ratio (B/C) of the different scenarios over a period of 100 years

	"Afforestation upstream"	"Afforestation downstream"	"Afforestation everywhere"
Costs	€ 15,525,924	€ 4,868,351	€ 20,394,275
NPV	€ −2,836,497.85	€ 4,184,035.48	€ −6,124,130.82
IRR	3.28%	7.72%	2.59%
B/C	0.84	1.89	0.69

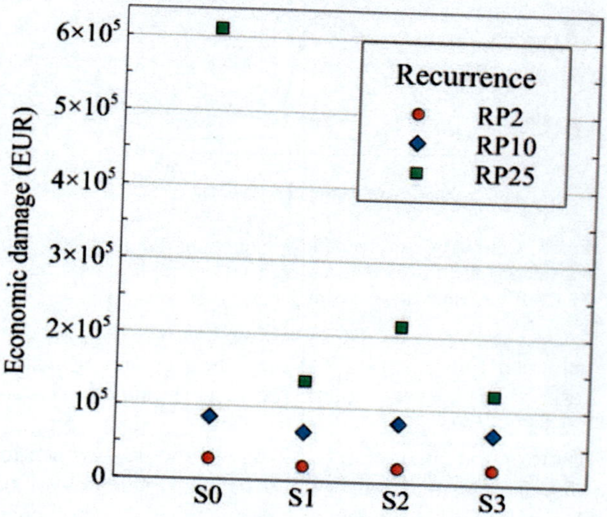

Fig. 4 Economic damage in EUR for different afforestation scenarios under changing probability of flooding (RP: return period). S0 ("Current land use"), S1 ("Afforestation upstream"), S2 ("Afforestation downstream"), S3 ("Afforestation everywhere")

scenarios. Consequently, costs of land acquisition are high, and obviously flood damage is smaller than these costs and maintenance costs.

During the CBA we found that the benefits of flood protection measures exceed the costs of planting and protection in case of afforestation of larger areas. Furthermore, considering also ecosystem service benefits, it was found that the overall benefits are dominated by benefits of flood protection measures. One possible reason could be that the functions, in order to extrapolate the yearly benefits of flood protection measures, were fitted based on the three values calculated for the different return periods. Since the benefits (damage avoided under the scenario) were by far the largest for the 25-year return period (Fig. 4), the function therefore implies that afforestation becomes more effective under higher flows. On the contrary, several studies found that the opposite is the case, which would produce negative net benefits for larger scenarios [2, 4].

With respect to the considered ecosystem services, the values vary significantly based on their correlation with the afforested surface, which also undermines the uncertainty of the underlying data for ecosystem services. Since most of the values were transferred and estimated based on the size of the afforested area, the largest ecosystem service benefits were achieved for scenario "Afforestation everywhere."

Due to temporal constraints of the present study, it was impossible to apply the valuation function transfer based on a broad set of indicators, which would allow for the possibility to adapt the valuation function to changing background conditions as recommended by Müller et al. [12]. However, despite the large uncertainties that underlie the current estimation of ecosystem co-benefits other than flood protection (which might be slightly overestimated in this study), they deliver relevant NPV benefits. This implies that the consideration of multiple ecosystem benefits in addition to benefits of flood protection measures will make the implementation of NWRM in general more likely.

6 Conclusion

This study aimed to provide a better understanding of hydrological and hydraulic dynamics and the economic evaluation of NWRM within the Glinščica River catchment in Slovenia. The findings of the initial modelling approach suggest a considerable scope for using floodplain afforestation as an aid to flood control. It was found that reductions of up to 25% in the extent of the inundated area could be achieved by parameterizing afforestation measures within both hydrological and hydraulic models. The hydrological model computed smaller input hydrographs for the hydraulic model. The hydraulic computation revealed no significant peak flow; nevertheless it revealed an overall reduction of the inundated area by 10%, mostly in the vulnerable urbanized parts. For the flood damage, this translated to significant cost reductions of, e.g., 396,744 € compared to the current land-use scenario for the 25-year return period. This is significant as scenario "Afforestation downstream" is characterized by a much smaller afforested area just within the section of the hydraulic model, compared to scenarios "Afforestation everywhere" and "Afforestation upstream." Consequently, positive net present values could be found only for scenario "Afforestation downstream"; even though the NPV benefits were dominated by benefits of flood protection measures, we underline the importance to also value other ecosystem co-benefits of the NWRM in order to make their implementation economically more worthwhile.

As such, although it is very unlikely that floodplain afforestation on its own would be able to provide a complete protection for the downstream City of Ljubljana, it could make a valuable contribution alongside existing flood defenses in order to tackle the potential increased risk of flooding. Additionally, the provision and the valuation of ecosystem services in afforested areas of former meadows or fields can deliver crucial information for informed decision-making and for sustainable investment choices. Especially for small-scale flooding problems, it could be a decisive point in order to decide upon the implementation. Future research must, therefore, focus on adequately displayed afforestation measures within a hydraulic model software, minimizing the presently large uncertainties in site- and time-specific ecosystem services.

Acknowledgments The study was conducted as part of a short-term scientific mission (STSM) within the LAND4FLOOD: Natural Flood Retention on Private Land (CA16209) Cost Action. The study was also conducted as part of the research program P2-0180 financed by the Slovenian Research Agency (ARRS).

References

1. Blöschl G, Hall J, Viglione A, Perdigão RA, Parajka J, Merz B et al (2019) Changing climate both increases and decreases European river floods. Nature 573(7772):108–111
2. Stratford C, Miller J, House A, Old G, Acreman M, Duenas-Lopez MA, et al (2017) Do trees in UK-relevant river catchments influence fluvial flood peaks? A systematic review
3. Nisbet T (2004) Interactions between floodplain woodland and the freshwater environment. Forest Res Ann Rep Accounts 2003(4):32–39
4. Dittrich R, Ball T, Wreford A, Moran D, Spray CJ (2018) A cost-benefit analysis of afforestation as a climate change adaptation measure to reduce flood risk. J Flood Risk Manage 12: e12482
5. Vidmar A, Zabret K, Sapač K, Pergar P, Kryžanowski A (2019) Development of an application for estimating the benefits of structural and non-structural measures for flood risk reduction. In: Biondić D, Holjević D, Vizner M (eds.) Croatianwaters in environmental and nature protection: proceedings of 7th Croatian water conference with international participation in Opatija, 30 May-1 June 2019. Grafički zavod Hrvatske, Zagreb, pp 615–624
6. Bezak N, Šraj M, Rusjan S, Mikoš M (2018) Impact of the rainfall duration and temporal rainfall distribution defined using the huff curves on the hydraulic flood modelling results. Geoscienes 8(2):69
7. Šraj M, Bezak N, Rusjan S, Mikoš M (2016b) Review of hydrological studies contributing to the advancement of hydrological sciences in Slovenia. Acta Hydrotech 29(50):47–71
8. Dolšak D, Bezak N, Šraj M (2016) Temporal characteristics of rainfall events under three climate types in Slovenia. J Hydrol 541:1395–1405
9. Šraj M, Dirnbek L, Brilly M (2010) The influence of effective rainfall on modeled runoff hydrograph. J Hydrol Hydromech 58(1):3–14
10. Feldman AD (2000) Hydrologic modeling system HEC-HMS: technical reference manual. US Army Corps of Engineers, Hydrologic Engineering Center
11. USACE (2016) HEC-RAS river analysis system. Hydraulic reference manual, version 5.0. US Army Corps of Engineers Institute of Water Resources, Hydrologic Engineering Center, Davis
12. Müller A, Knoke T, Olschewski R (2019) Can existing estimates for ecosystem service values inform forest management? Forests 10(2):132
13. Masiero M, Pettenella D, Boscolo M, Barua SK, Animon I, Matta JR (2019) Valuing forest ecosystem services: a training manual for planners and project developers. Forestry working paper no. 11. Rome, FAO. 216 pp. Licence: CC BY-NC-SA 3.0 IGO
14. Prostor (2019) Spatial portal of the Republic of Slovenia. http://prostor3.gov.si/ETN-JV/. Accessed Dec 2019
15. Hosking JRM, Wallis JR (1997) Regional frequency analysis: an approach based on L-moments. Cambridge University Press. https://doi.org/10.1017/CBO9780511529443
16. Bernal B, Murray LT, Pearson TR (2018) Global carbon dioxide removal rates from forest landscape restoration activities. Carbon Balance Manag 13(1):1–13
17. IUCN (2018) Global emissions and removals databases. https://infoflr.org/what-flr/global-emissions-and-removals-databases. Accessed Dec 2019
18. OECD (2016) Effective carbon rates: pricing CO2 through Taxes and emissions trading systems. OECD Publishing, Paris

19. Ljubljana Tourism (LT) (2019) The path of remembrance and comradeship. https://www.visitljubljana.com/en/visitors/explore/things-to-do/active-holidays/article/the-path-of-remembrance-and-comradeship/. Accessed Dec 2019
20. Japelj A, Mavsar R, Hodges D, Kovač M, Juvančič L (2016) Latent preferences of residents regarding an urban forest recreation setting in Ljubljana, Slovenia. Forest Policy Econ 71:71–79
21. Tinch R, Thomson C, Dickie I, Leslie R (2010) The economic contribution of the public forest estate in England. In: Report prepared for the Forestry Commission, London (London: eftec)
22. Broadmeadow S, Nisbet TR (2004) The effects of riparian forest management on the freshwater environment: a literature review of best management practice
23. Ogrinc N, Tamše S, Zavadlav S, Vrzel J, Jin L (2019) Evaluation of geochemical processes and nitrate pollution sources at the Ljubljansko Polje aquifer (Slovenia): a stable isotope perspective. Sci Total Environ 646:1588–1600
24. Hall JW, Tarantola S, Bates PD, Horritt MS (2005) Distributed sensitivity analysis of flood inundation model calibration. J Hydraul Eng 131(2):117–126

Long-Term Impacts of Land Use Change Upon the Natural Flood Storage Reservoirs Along the North Bulgarian Black Sea Coast

Iliyan Kotsev, Bogdan Prodanov, and Radoslava Bekova

Contents

Abstract The Black Sea is a micro-tidal basin, where significant storm-related marine inundations are characterized by comparably long return periods and moderate extents. Thus, floods resulting from heavy rainfalls, rapid snowmelts, and the related extreme runoff in the watersheds are far more common. River inundations are among the properties of the hydrological regime on the Bulgarian coast, dictated by specifics of the humid subtropical climate. Nevertheless, extreme fluvial flood

I. Kotsev (✉) and B. Prodanov
Department "Coastal Zone Dynamics", Professor Fridtjof Nansen Institute of Oceanology – Bulgarian Academy of Sciences, Varna, Republic of Bulgaria

R. Bekova
Department "Biology and Ecology of the Sea", Professor Fridtjof Nansen Institute of Oceanology – Bulgarian Academy of Sciences, Varna, Republic of Bulgaria

Carla S. S. Ferreira, Zahra Kalantari, Thomas Hartmann, and Paulo Pereira (eds.), *Nature-Based Solutions for Flood Mitigation: Environmental and Socio-Economic Aspects*, Hdb Env Chem (2022) 107: 233–262, DOI 10.1007/698_2021_765,

Published online: 6 May 2021

extents come primarily as a consequence of land use changes, unbalanced real estate development, erroneous spatial planning, poor land administration, and the lack of proper integrated coastal zone management (ICZM). Accordingly, floods pose a severe hazard to coastal communities and economic activities. The estuaries of the Black Sea tributaries once successfully maintained endemic hygrophilous forests known as *longozes*. Nowadays, the areals of these riparian woods are severely decreased and fragmented. Longozes used to play a vital role as peak outflow regulators and flood storage reservoirs. Baltata and Longoza localities were picked as case studies to analyze the magnitude of landscape change, affecting the endemic forests' contemporary flood retention capabilities. The study herein consists of two core components – literature review for periodization of significant land use changes, backed by GIS-aided assessment of the landscape change at these two case study sites. The literature review reveals records on outflow peaks in periods of rapid snowmelts and torrential rainfalls, which may be attributed to impaired retention capabilities of the longoz landscapes. Likewise, an intensified frequency of the recent flood events is observed. These findings are in a good correlation with results of the GIS-aided change detection interpretations. The comparative analyses of historical topographic maps revealed significant decrease and fragmentation of the longozes at both investigated localities. The GIS analyses also demonstrate ubiquitous drainage of adjacent wetlands for conversion into agricultural, residential, and recreational areas.

Keywords Batova River, Floodplain forests, GIS-aided comparative analyses, Kamchia River, Landscape transformations

1 Introduction

Longozes represent endemic hygrophilous (primarily riparian and floodplain) forests native to the eastern part of the Balkan Peninsula. In coastal environments, they occur along slow-flowing river downstream sectors, floodplains, estuaries, and brackish lagoons [1]. Apart from the Bulgarian seaside, they are also preserved at a few sites along the Black Sea and Aegean shores of European Turkey [2], as well as in the region of Western Trace in Northern Greece, between the Nestos and Evros river deltas [3, 4]. Their presence in the Danube River delta (Romania) is quite possible, yet a subject of scientific debates [1]. Being nicknamed the "temperate mangroves" of the East Balkans, the formation and existence of these alluvial forests is dependent from the humid subtropical climate with mild, wet winters and periodic high-water fluvial inundations occurring twice yearly, once in early spring and again in late autumn [1].

A recently recognized important ecosystem service that longoz forests are able to provide along with the coastal wetlands is their property as peak runoff regulators and natural flood storage reservoirs. Nowadays, however, due to widespread land conversion, longoz forests in Bulgaria are characterized by severely decreased and fragmented areals, and as a consequence, they are enlisted in the country's Red Data Book as a sensitive habitat type [1]. Nevertheless, thanks to the excellent flood retention properties, since very recently, the environmental management notion about longozes goes beyond the realm of nature conservation and habitats preservation agenda. These endemic forests are no longer seen by spatial planners as merely a critically endangered habitat type [1] requiring preservation in compliance with Council Directive 92/43/EEC (i.e., the Habitats Directive) [5, 6], but also as a nature-based solution in coastal flood risk management [7]. It should be pointed out that since the Black Sea is a micro-tidal basin, where significant storm-related marine inundations are with comparably long return periods and of moderate spatial extents, floods of non-marine (particularly fluvial) origin represent the principal threat of this kind, which has to be mitigated in compliance with Directive 2007/60/EC (i.e., the Floods Directive) [8]. Finally, flood risk management in coastal areas, along with nature conservation and the preservation of sensitive habitats, represents an essential component of the ICZM. Thereby, the objectives of the study herein can be summarized in the following:

- To investigate the spatio-temporal pattern of the longoz forests' areals along the North Bulgarian Black Sea coast by reviewing the available literature and by applying a GIS-aided comparative-historical analysis for assessment of the magnitude of human-driven landscape change;
- To lay the fundament for an assessment of the longoz forests' impaired flood retention capabilities in light of their role as a nature-based solution for flood hazard mitigation.

2 Study Area

2.1 *Coastal Geology and Geomorphology*

The North Bulgarian Black Sea coast stretches between Cape Sivriburun (marking the state border with Romania) to the north and Cape Emine (marking the easternmost tip of the Balkan Mountains) to the south. It encompasses the coastal sectors of five major geological structures, namely the Moesian (Danubian) Platform, the Batova Depression, the Varna-Beloslav Grabben, the Kamchia Depression, and the easternmost part of the Balkan Young-Folded Alpine Zone [9]. In geotectonic sense, the Batova Depression is a tectonically predisposed erosional valley (a fault) formed between the Vranino-Balchishki (Dobrudzhean Plateau) and the Frangenski (Franga Plateau) coastal morphotectonic blocks [10]. Respectively, the Kamchia Depression serves as a transitional zone that separates the Moesian Platform from the easternmost part of the Balkan Young-folded Alpine Zone [11]. In terms of geomorphology, the littoral sectors of both depressions represent depositional lowlands of the delta-estuarine type, formed in the floodplains of the Batova and Kamchia rivers [12].

2.2 *Climate on the Bulgarian Coast*

While some of the publications (mostly by Bulgarian scholars, e.g., [13–16]) describe the climate along the Bulgarian Black Sea coast (or parts of it) as transitional to continental-Mediterranean or even as typical for the continental part of the Northeast Mediterranean, it seems to be much properly categorized by foreign authors who apply the Köppen-Geiger classification scheme (e.g., [17–19]). Pursuant to the above classification, the climate of the Bulgarian Black Sea coast is humid subtropical, which better describes the wetter transitional seasons and the warm, but noticeably cooler summer months in comparison to the Mediterranean region. There is a gradual increase of the mean annual precipitation quantities in the coastal sector between Cape Kaliakra in Maritime Dobrudzha to the north and Cape Emine in the East Balkan Mountains to the south, i.e., from 412 mm/y to 593 mm/y. This increase goes along with a gradual shift of the rainfall maxima from one in the late spring-early summer season into two occurring during the late autumn-early winter and the early-spring seasons [15, 20].

In terms of climate requirements, a limiting factor for the occurrence of longozes is precisely the cited mild and humid winters. This is the reason why in the northern part of Bulgaria these forests used to dominate at the landscape scale at sites close to the Black Sea, i.e., in the valleys and floodplains of the rivers Batova, Kranevska (Ekrenska or Chaltika Dere), Pasha Dere, Kamchia, Fandakliyska, Perperi Dere, Dvoynitsa, and Vaya. Some evergreen and thermophilic species are also present in the forest communities – a coenological property that sets longozes apart from the remaining riparian and floodplain forests in Bulgaria [1, 5].

2.3 *Hydrological Properties*

Among the principal ecological factors determining the spatial distribution of the longoz forests along the Bulgarian coast is moisture availability. A general hydrological peculiarity of the region is the gradual increase of water availability in north-to-south direction. This is attributed to the specifics of the coastal climate, lithologic composition, topography, and afforestation of the catchment basins. The cited factors altogether consequently determine the properties of the surface runoff, ground water discharge, occurrence of major Black Sea tributaries, and presence of freshwater and brackish coastal lakes. Consecutively, the banks of the aforementioned water bodies found along the shore are the one providing the necessary ecological conditions for the existence of longoz forests [1, 3, 21, 22].

A significant number of rivers and ravine-like creeks flow into the Black Sea basin, whose total catchment area within the Bulgarian sector exceeds 16,900 km^2 [23–26]. Among them, the Batova and Kamchia rivers are of utmost interest. The Kamchia River is the longest Bulgarian river that empties into the Black Sea, having a total length of approximately 245 km and storing considerable ground water volumes within its fluvial terraces [23–25]. The specifics of the Kamchia River runoff, analogous to Batova River, are associated with early-spring and late-autumn inundations, which facilitate the mass development of riparian forests of the endemic longoz type [1, 3, 4, 5, 22, 27].

2.4 Floristic Characteristics of the Longoz Forest Communities

As being woods of the hygrophilous type, longozes comprise specific forest ecosystems, where edificators altogether demonstrate excellent water retention capabilities at the landscape scale. It is among the natural properties of these endemic communities that are being of scientific interest to various interdisciplinary studies. There are three main specifics that set the longozes apart from the rest of the riparian and floodplain forests on the Balkan Peninsula:

- polydominance and great diversity of arboreal species, adding up to more than forty kinds;
- vast abundance of climbing plants with winding stems (lianas), giving these woods a rainforest-like appearance;
- fluvial inundations occurring twice a year, related to the early-spring snowmelts upstream, the early-spring and late autumn-winter precipitation maxima, as well as the strong coastal wave activity in result of the prevailing east and northeast winds, which hinders the river outflow into the sea during the cold months [22, 27, 47].

The most representative longoz forest in Europe (and in Bulgaria, respectively) is found along the Kamchia River estuary. It has been designated as a biosphere reserve, and a NATURA 2000 protected site in compliance with Council Directive 92/43/EEC (Habitats Directive) [7, 27–30]. Main associations forming these hygrophilous woods are these of the field elm (*Ulmus minor* Mill.) and the common ash (*Fraxinus oxycarpa* Willd.). Other common tree species that participate in them are the European white elm (*Ulmus laevis* Pall.), field maple (*Acer campestre* L.), common oak (*Quercus robur* L.), pedunculate oak (*Quercus pedunculiflora* C. Koch), white poplar (*Populus alba* L.), black poplar (*Populus nigra* L.), black elder (*Sambucus nigra* L.), black alder (*Alnus glutinosa* (L.) Gaerth), mahaleb cherry (*Prunus mahaleb* L.), bird cherry (*Prunus padus* L.), common pear (*Pyrus communis* L.), European crab apple (*Malus sylvestris* (L.) Mill.), various species of willow (*Salix* spp.), etc. The shrub layer consists of Tatarian maple (*Acer tataricum*

L. 1753), common hawthorn (*Crataegus monogyna* Jacq.), small-flowered black hawthorn (*Crataegus pentagyna* Waldst. and Kit. ex Willd.), small-leaved hawthorn (*Crataegus microphylla* K.Koch), common dogwood (*Cornus sanguinea* L.), blackthorn (*Prunus spinosa* L.), wild privet (*Ligustrum vulgare* L.), European spindle (*Euonymus europaeus* L.), broad-leaved spindle (*Euonymus latifolius* (L.) Mill.), *Euonymus verrucosus*, common hazel (*Corylus avellana* L.), etc. [1, 29]. Typical climbing plants with winding stems for these forests are species that are widespread in the country, e.g., old man's beard (*Clematis vitalba* L.), ivy (*Hedera helix* L.), wild grapevine (*Vitis vinifera* L. subsp. *Sylvestris* (C.C. Gmel.) Hegi), common hop (*Humulus lupulus* L.), hedge bindweed (*Calystegia sepium* (L.) R.Br.), black bryony (*Dioscorea communis* (L.) Caddick and Wilkin), as well as the Submediterranean species high catbrier (*Smilax excelsa* L.) and the rare for the Bulgarian flora Ponto-Mediterranean element Greek silkvine (*Periploca graeca* L.) [1, 21, 22, 27]. There is a typical spring synusium of the flowers Bithynian squill (*Scilla bithynica* Boiss.) and summer snowflake (*Leucojum aestivum* L.), the latter being a valuable medicine plant [1].

The total number of protected plants found in the longoz forests well exceeds twenty species and includes European Fritillaria (*Fritillaria pontica* Wahl.), *F. stribrnyi* Velen., Bulgarian galium (*Galium* bulgaricum Velen.), wild parsnip (*Pastinaca umbrosa* Steven ex. DC), Sibthorp primrose (*Primula vulgaris* subsp. *Sibthorpii* (Hoffmanns.) W.W.Sm. and Forrest), Bithynian squill (*Scilla bithynica* Boiss.), bastard stone-parsley (*Sison amomum* L.), summer snowflake (*Leucojum aestivum* L.), etc. [1, 22, 27, 28, 30].

2.5 *Baltata and Longoza Localities as Case Studies*

Baltata and Longoza localities (Fig. 1) are the two most representative sites with longoz forests along the North Bulgarian Black Sea coast [1, 3, 5, 22, 27]. Therefore, these two areas were selected as case studies for analysis of the human-induced landscape transformations and the accompanying impacts upon the endemic forest areals.

Baltata Locality (Fig. 2) occupies the downstream sector, floodplain and estuary of the Batova River. It is a protected area since 1962, designated in order to preserve Europe's northernmost longoz forest as a follow-up to the expert advice by the British researchers Guy Mannford and Eric Hosking [27]. The core conservation area has been established as Baltata Managed Reserve (Fig. 3) [28]. The reserve's buffer zone is a separate protected area with the status of a protected locality, called Blatno Kokiche [29] (translated as summer snowflake in English, which is the name of the medicine plant *Leucojum aestivum* that grows abundantly in the area). Both the managed reserve and the protected locality are included in BG 0000102 Dolinata na reka Batova (or Batova River Valley) NATURA 2000 protected site, designated under the EU's Habitats Directive [7]. Main conservation goal of the protected site's

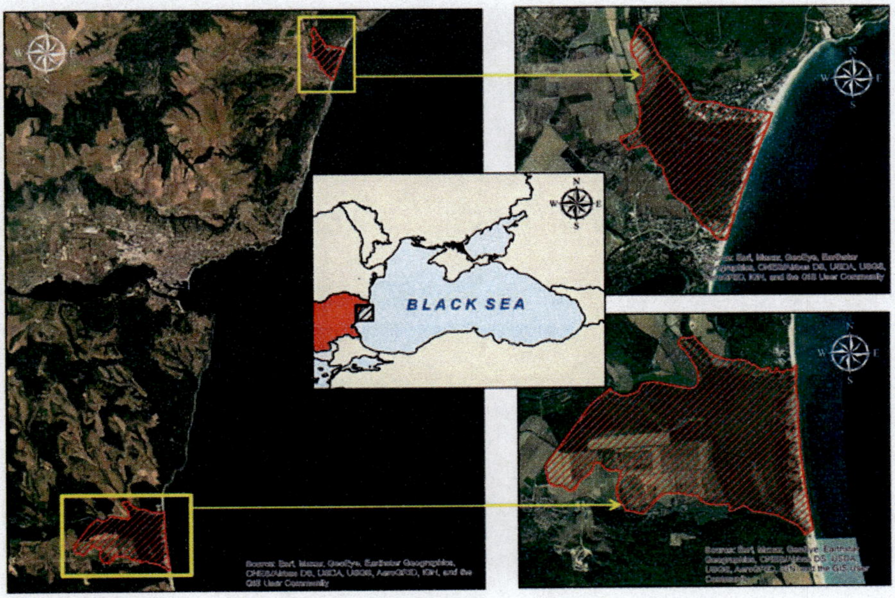

Fig. 1 Location of the two pilot case study sites on the North Bulgarian Black Sea coast: a. Baltata Locality; b. Longoza Locality. Satellite imagery provided by the World Imagery basemap service of ESRI for ArcGIS Desktop

designation is to guarantee the preservation of the remnants of the once vast longoz forest formed in the floodplain of the Batova River [32].

Similarly, Longoza Locality (Fig. 4) encompasses the downstream sector, floodplain and mouth of the estuarine type of the Kamchia River. It is a nature reserve called Kamchia since 1951 (Fig. 5), designated as a biosphere reserve from 1977 until 2017 [27–29]. The reserve's buffer zone, similar to the spatial zoning at Baltata, has also been established as a separate protected locality called Longoza, named after the area itself. Both the biosphere reserve and the protected locality are nowadays included in the spatial extent of BG 0000116 Kamchia NATURA 2000 protected site, designated under the EU's Habitats Directive [7]. Main nature conservation goal of the protected site's designation is to preserve the remnants of the longoz forest formed in the river downstream sector and floodplain of the Kamchia River [32]. In terms of nature conservation, an intriguing fact is that Longoza Locality is probably among the oldest protected areas in present-day Bulgaria. It used to have the status of a protected forest from the sixteenth century onwards, while the country was a constituent part of the Ottoman Empire, until the declaration of the semi-independent Principality of Bulgaria in 1878, when, unfortunately, the forest lost its preservation status and became subject to an extensive exploitation for timber by national and foreign companies [3].

Fig. 2 A winter image over the longoz forest at the Baltata Locality, formed in the floodplain of the Batova River. The estuary visible to the north (upper part of the photograph) represents the old river mouth of the Batova. The estuary located to the south (lower part of the photograph) is the new river mouth, artificially translocated there in order to clear the marshy terrain for the construction of the Albena International Maritime Resort (a few of the hotels are discernible in the upper central section of the image)

3 Materials and Methods

The study herein consists of two main parts:

Part 1. literature review for analysis of the human-driven landscape-scale transformations that occurred at Baltata and Longoza localities from historical perspective, followed by logically grouping of the land conversion events into actual historical periods;

Part 2. GIS-aided assessment and interpretation of the spatio-temporal land use changes at the two case study sites, based on the analysis of historical and archive cartographic resources.

In order to make correct implications about the overall spatio-temporal change at the landscape scale at Baltata and Longoza localities, it is necessary to review the main stages in the anthropogenic transformations of the two areas and to subsequently aggregate them into historical periods. In Part 1, the literature review entails thorough analysis of the available publications and gray literature upon the matter, including non-indexed scientific papers in Bulgarian, research articles published prior to the 1990s, information materials about protected areas within the spatial extent of the Regional Inspectorate for Environment – Varna [28], NATURA 2000

Fig. 3 A field photograph of Baltata Managed Reserve, established in the downstream sector of the Batova River. The marshy terrains within the banks of the slow-flowing river provide excellent conditions for the development of polydominant hygrophilous forest communities of the endemic longoz type. Longozes are characterized by vast abundance of climbing plants with winding stems (lianas), giving these woods a rainforest-like appearance

dossiers of the two relevant protected sites (BG 0000102 Dolinata na reka Batova and BG 0000116 Kamchia) freely available from the national geoportal (http://natura2000.moew.government.bg/) [32], as well as data obtainable from the national online register of the protected areas and NATURA 2000 sites to the Bulgarian Executive Environment Agency (http://eea.government.bg/zpo/en/index.jsp [29]).

In Part 2, initial entry data for analysis in GIS was a set of archive topographic maps covering the Bulgarian Black Sea coast, as well as two up-to-date cadastral layers, of Baltata and Longoza localities, respectively. The cadastral data was obtained from the Geodesy, Cartography and Cadastre Agency of the Republic of Bulgaria. The topographic dataset comprised maps produced at various scales and relevant to four different periods: the late nineteenth century; the 1930s, 1970s, and 1980s of the twentieth century. The cited historical topographic maps were obtained in a digital format as a web map service (WMS) for geo-software, freely distributed by an independent Bulgarian website for GIS professionals (https://kade.si) [52]. In the first step of the GIS procedures, datasets on historical land cover/land use at Baltata and Longoza localities were created by on-screen digitizing of the areas of interests from the archive topographic maps. As a product of the digitizing activities, two pairs of land cover/land use layers synchronous to the above-mentioned historical periods were created. In addition, LANDSAT imagery relevant to the 1980s was

Fig. 4 The longoz forest at the Longoza Locality, formed in the floodplain of the Kamchia River. Naturally, the fluvial floodplain is occupied by this dense endemic hygrophilous forest, as it is visible on the oblique aerial image. Furthermore, part of the river estuary is discernible in the central part of the photograph

used for further cross-check and data refinement of the 1970s–1980s land cover/land use digital layers.

Crosstabulations carried out in GIS environment represent the core technique of the change detection analyses carried out in relation to the study discussed herein. Accordingly, the resultant two historical GIS datasets on land cover/land use (relevant for the periods late nineteenth century–1930s and 1970s–1980s) were crosstabulated versus each other and versus the two up-to-date cadastral layers. These crosstabulation procedures made it possible to trace and quantify the rats of human-driven landscape transformations at Batova and Longoza localities that occurred historically in result of land conversion and land use changes.

4 Results

4.1 *Part 1: Literature Review. Land Use Dynamics Along the North Bulgarian Coast. Main Land Use and Land Conversion Periods*

The issue concerning detailed historical periodizations of the main stages in land conversion with associated land use dynamics and overall landscape change, for the

Fig. 5 A field photograph of the Kamchia River downstream sector, established as UNESCO's Kamchia Biosphere Reserve for the period 1977–2017. The Kamchia River represents the longest and the most water-abounding Bulgarian tributary emptying into the Black Sea, having a total length of approximately 245 km

area of present-day Bulgaria, presents certain problems due to the peculiarities of the Ottoman and pre-Ottoman periods [33, 34]. A general feature of the aforementioned epochs is the scarcity of easily available sources on these topics. Another issue is related to the historical fate of Baltata Locality (Fig. 6) over the pace of the early twentieth century [35]. As being a constituent part of Maritime Dobrudzha, it was annexed by Romania for the period from 1913 (shortly after the Second Balkan War) until 1940 (when it was returned to Bulgaria following the Treaty of Craiova). Therefore, the historical periods described in the present chapter of the study reflect to a great extent the availability of narrative and cartographic data with regard to the study area.

4.1.1 Ottoman Period

The preserved narrative and cartographic data concerning the areas of Baltata and Longoza during the Ottoman period is very scarce [34]. Nevertheless, there are some quite interesting written facts available about the status of the Longoza Locality, communicated by Dimitrov [3]. It used to be a protected forest, where felling of trees for timber and grazing were strictly prohibited, while hunting was a privilege

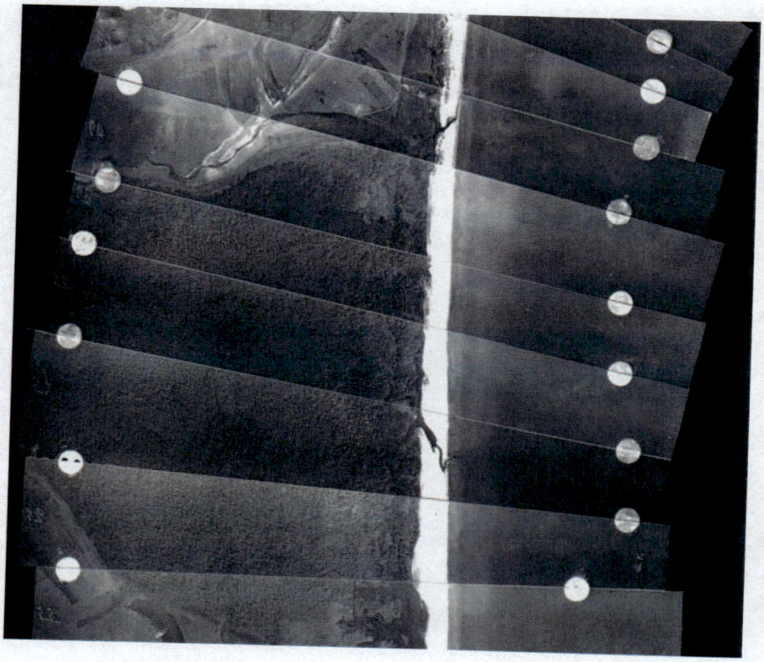

Fig. 6 An orthorectified aerial photograph series over Baltata Locality, as of the beginning of the twentieth century. From the year 1913 until 1940, Baltata used to be an isolated coastal area at the Bulgarian–Romanian border, almost completely deprived of any human activity. Photo source: Archive of the Bulgarian Academy of Sciences

allowed to noble Ottomans only [3]. Based on this fact, it may be assumed that the Longoza Locality is probably among the earliest protected areas in present-day Bulgaria.

4.1.2 Period from the Liberation of Bulgaria Until the Second World War

This period is marked by certain specifics of the land use dynamics and the concomitant processes of anthropogenic transformation, dictated primarily by the turbulent political situation and the associated socio-economic changes on the Balkans that inevitably affected the young Bulgarian Principality (an independent from the Ottoman Empire kingdom since 1908) [35]. As a consequence of the refugee waves from Macedonia and Thrace following the Ilinden–Preobrazhenie Uprising of 1903 and the Second Balkan War of 1913, new settlements and quarters emerged, while the previously existing increased significantly their spatial extents [36–38]. This was accompanied by reclamation of terrains for construction and farming [39]. The participation of Bulgaria in three consecutive wars during the period 1912–1918 [35] (which, according to some Bulgarian climatologists, coincided with several successive harsh winters on the Balkans [14]) and its involvement in the Second World War had a devastating impact upon the forest landscapes found along the seaside. Besides, shortly after the liberation from the Ottoman rule, the Longoza locality lost its status as a protected forest. The wood resources of the most representative European longoz forest were extensively exploited for timber by national and foreign companies, which in turn led to significant reduction of its spatial extent [3].

4.1.3 Socialist Period (1944–1989)

The epoch of Totalitarianism is associated with drastic changes in the political and socio-economic life in Bulgaria, which inevitably reflected upon the land use dynamics and led to a widespread land conversion along the Black Sea coast [40, 41]. Although generally perceived as a dark period in Bulgaria's recent history, the Socialist period had both good and bad aspects regarding the land administration, spatial planning, environmental protection, and nature conservation. On the one hand, the Socialist period marks great advancements in the fields of cadastral mapping and regulation, construction of critically important infrastructure, expert-based planning of the national economy with the associated land use and land administration activities, development of the international tourist sector, initiation of measures for flood hazard mitigation, as well as the establishment and preservation of newly designated protected areas. On the other hand, these actions were often at the expense of property rights of the Bulgarian citizens, sometimes irrational exploitation of the natural resources and lack of flexibility in terms of planning, neglecting of the landscape's properties and carrying capacity, blind copying of Soviet models, and so forth [41–46].

A characteristic feature of this period is the collectivization of all private farmland and the creation of state-owned agricultural cooperatives that later evolved into agro-industrial complexes, following the Soviet model [41, 44, 46]. This process was accompanied by reclamation of new fertile areas, converting entire coastal regions (e.g., the Batova and the Lower Kamchia river valleys) into highly intensive agricultural landscapes. Accompanying effects of the large-scale intensive agriculture were aggravation of the soil erosion, soil and water pollution issues, etc. [45].

The extensive exploitation of the coastal wood resources for timber continued during the Socialist period, but also mass afforestation campaigns were carried out as well, including the creation of shelterbelt forests in the rural areas. Lamentably, far from all forestry activities were planned or undertaken with the required level of botanical expertise. This in turn resulted in the formation of secondary, less valuable forest communities and, however, led to the ill-considered widespread creation of artificial, often inappropriate or unstable in long-term perspective forest patches with non-native tree species [20, 21, 31].

Thanks to the excellent climate conditions and the consistent long-term policy of the Bulgarian communist governments, oriented towards fast development and expansion of the tourist sector, the country became a well-known Balkan destination for maritime summer recreation alongside the Socialist Yugoslav republics of Croatia and Montenegro [43, 46]. Numerous resorts and recreational areas were established in the study area during this period, e.g., Albena International Maritime Resort and Kranevo Village near the town of Balchik in Maritime Dobrudzha, as well as Kamchia Recreational Area and Shkorpilovtsi Village in the Kamchia coastal area. These processes of landscape transformation, land conversion, and the associated land use changes were naturally accompanied by construction of new infrastructure, e.g., roads, water and sewage pipelines, water-treatment plants, etc. However, it should be pointed out that the greater share of the accommodation facilities for maritime tourism in Bulgaria used to be concentrated in small recreational areas with campsites, caravan parks, and bungalows. In addition, all coastal settlements, e.g., Kavarna, Balchik, Varna, Shkorpilovtsi, Byala, and Obzor, used to accommodate a great share of the visitors in private, family-run guest houses and apartments. Therefore, the human pressure upon the coastal nature as a straight consequence of tourism was rather small compared to the disgraceful situation along the Bulgarian Black Sea coast nowadays [42, 43, 46].

Nonetheless, although carefully planned in general, yet the fast-growing international recreation led to certain negative human-induced disturbance exerted upon the longoz landscapes. A common feature of the Black Sea coast is that the most important nature conservation sites are also the most attractive ones when it comes to tourism [20, 27, 45, 46]. Thus, a few of the resorts and the recreational facilities were built in the vicinity of environmentally sensitive areas, e.g., Albena Maritime Resort and Kranevo contiguous to Baltata Locality, as well as Kamchia Recreational Area and Shkorpilovtsi close to Longoza Locality [43].

Among the good features of the Socialist period is the significant progress of the nature conservation initiatives, followed by the designation of multiple new protected areas along the North Bulgarian Black Sea coast [45]. Among these were Kamchia Nature (i.e., Strict) Reserve (established in 1951, an UNESCO's biosphere reserve from 1977 until 2017), Baltata Reserve (declared in 1962, recategorized into a managed reserve in 1999), and Kamchiyski Pyasatsi (Kamchia Sands) Protected Locality (initially established in 1980 as a buffer zone of Kamchia Reserve) [27–29].

However, the environmental issues at the cited protected areas seriously aggravated as a consequence of land conversion and the associated land use shifts. For instance, Baltata Reserve was temporarily effaced as a protected area in the period 1974–1978 due to a state decision to construct Albena International Maritime Resort contiguous to the Batova beach. Although reestablished in 1978, the nature reserve's spatial extent was severely reduced and as a consequence of the contiguous tourist infrastructure for mass recreation, it represents one of Bulgaria's coastal geoecological hotspots nowadays [27]. At Longoza Locality, the drainage and follow-up conversion into agricultural land of the Oryahovo Wetlands, which used to act as natural flood storage reservoir during the high-water inundations of the Kamchia River, led to irreversible alterations of the moisture regime of the endemic longoz forest at the homonymous locality [27]. The cited environmental issues aggravated additionally after the construction of four large reservoirs in the river's catchment basin, created in order to meet the growing demands for water of the expanding coastal economy, agriculture, tourism, and growing population. Batova River, although having less water quantities, had a similar fate. The captation of its karst springs upstream led to severe deterioration of the state of the longoz forest communities at Baltata Reserve. As a consequence, most of the representative hygrophilous longoz vegetation began to perish and is currently being replaced by other, more tolerant but less valuable floristic communities [5, 27]. Another aspect of the negative alterations caused by the decreased river outflow was the shortage of the sediment supply transported and deposited by the Kamchia and Batova rivers, which in turn led to imbalance in the coastal morphodynamic system, end eventually resulting in intensification of the coastal erosion processes at both the Batova beach and Kamchiyski Pyasatsi Protected Locality [48].

4.1.4 Contemporary Period (1989–Present Day)

This period is perhaps the most detrimental for the Bulgarian Black Sea coastal nature [20, 27, 49, 50]. After the crash of the totalitarian regime in the country, the majority of the recreational facilities were privatized and consequently densely overbuilt. Unlike the common tendency in Europe, oriented towards nature-based and environmentally friendly forms of recreation, as well as the widespread increase of the campsites throughout the continent, most of these grounds along the Bulgarian Black Sea coast were purposely effaced. Later, the terrains were sold and subsequently converted into large resorts, without taking into consideration the carrying capacity of the coastal landscape [42]. Certain protected areas were wiped out from the register of the Bulgarian protected areas (i.e., Kamchia sands) and were planned for building [27–29]. Although sounding irrational for an EU member state, most of the coastal settlements and resorts in Bulgaria still lack any general spatial development plans. Nowadays, the negative consequences of these actions are obvious. Bulgaria lost its fame as a preferred Balkan destination for summer recreation, with the lack of pristine nature near the resorts, pollution, and over construction of the coast being among the main reasons for the tourist outflow.

Currently, the Bulgarian agriculture is in unprecedented decline [41]. After the termination of the state-owned agricultural cooperatives, the land was returned to the private owners and their inheritors. Unfortunately, due to the general lack of interest in farming initiatives by the local population, nowadays the abandoned agricultural landscapes are a common feature of the coast [42]. Nevertheless, from landscape ecological point of view, these former agricultural areas and abandonment processes are rather intriguing since they facilitate the monitoring of the natural landscape dynamics and the restoration abilities of longozes and coastal wetlands.

Among the good features of the contemporary period is the launch of the NATURA 2000 project in the country and along the Bulgarian Black Sea coast in particular [30, 32]. Accordingly, as being among the most important nature conservation areas on both national and international level, Baltata and Longoza localities were included in the spatial extents of BG 0000102 Dolinata na reka Batova (Batova

River Valley) and BG 0000116 Kamchia NATURA 2000 protected sites, designated in compliance with the EU Habitats Directive [7, 30, 32]. Nonetheless, the questions concerning habitat mapping, spatial planning, land administration, and land use regulations at these sites of community importance remain open.

4.2 *Part 2: GIS-Aided Assessment and Interpretation of the Spatio-Temporal Land Use Changes at Baltata and Longoza Localities*

4.2.1 Baltata Locality

Baltata Locality spreads over an area of roughly 528 ha. As of the late nineteenth century, 97.7% of it used to be covered by longozes, with the remainder 2.3% being the beach–dune complex developed at the land–sea interface in front of the forest and stretching between the river mouths of Batova to the north and Kranevska (Ekrenska) to the south (Fig. 7).

As of the late 1970s–early 1980s, the extent of the longoz forests has decreased from 97.7% to barely 46.4% due to land clearing for agricultural and resort-building demands (Fig. 8). In this period, the river mouth of Batova was artificially translocated to the south via a system of water canals. Subsequently, Albena Maritime Resort was built on the former marshy terrain, meanwhile becoming one of the greatest coastal consumers of freshwater, supplied by the karst streams feeding Batova River and on which the longoz ecosystems at Baltata Locality highly depend on. The aggregate landscape transformation at the case study site (late nineteenth century vs. late 1970s–early 1980s) adds up to 52.4%, while the transformed area of the longoz forests is 53.6%.

The final time frame subject to analysis is the period early 1980s–2019 (Fig. 9). Overall, the landscape transformation at Baltata Locality as of 2019 in comparison to the early 1980s is much lower – 5.8%. Further transformation of the longoz forests (around 0.1%) is demonstrated by the crosstabulation results, which, however, fall within the statistical error. There is an overall increase of the longoz forests extent at Baltata Locality, which is attributed to the abandonment of the former camp site near the southeastern tip of the pilot study site.

4.2.2 Longoza Locality

The second pilot study site has an area of approximately 2,500 ha. As of the late nineteenth century, 88.9% of it used to be covered by longoz forests, while the contiguous beach–dune complex, known as Kamchia-Shkorpilovtsi beach, comprised 2.3% of the site's aggregate area (Fig. 10). The remainder used to represent wetlands (the so-called azmatsi, representing old river beds of Kamchia) – 7.2%, and the river watercourse – 1.6%.

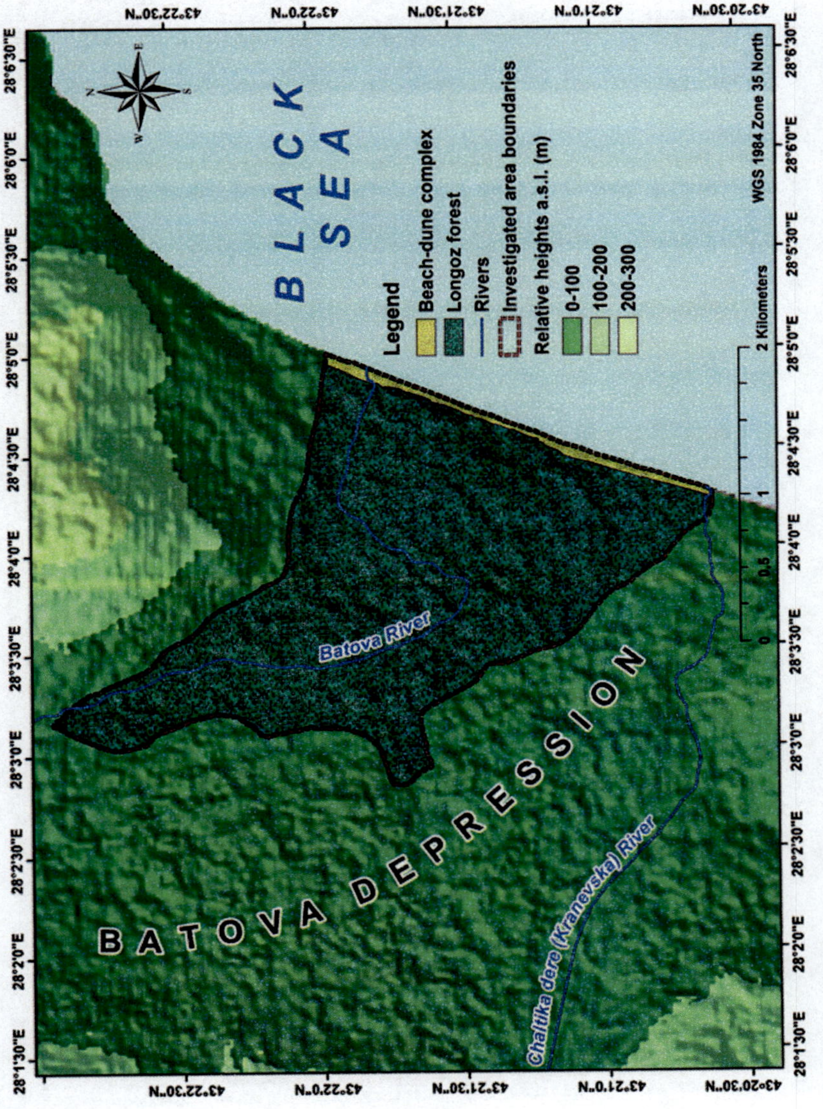

Fig. 7 Map of Baltata Locality as of the late nineteenth century, reconstructed from archive topographic maps compiled at scale 1: 126,000. The endemic longoz forest used to dominate the area's landscape pattern

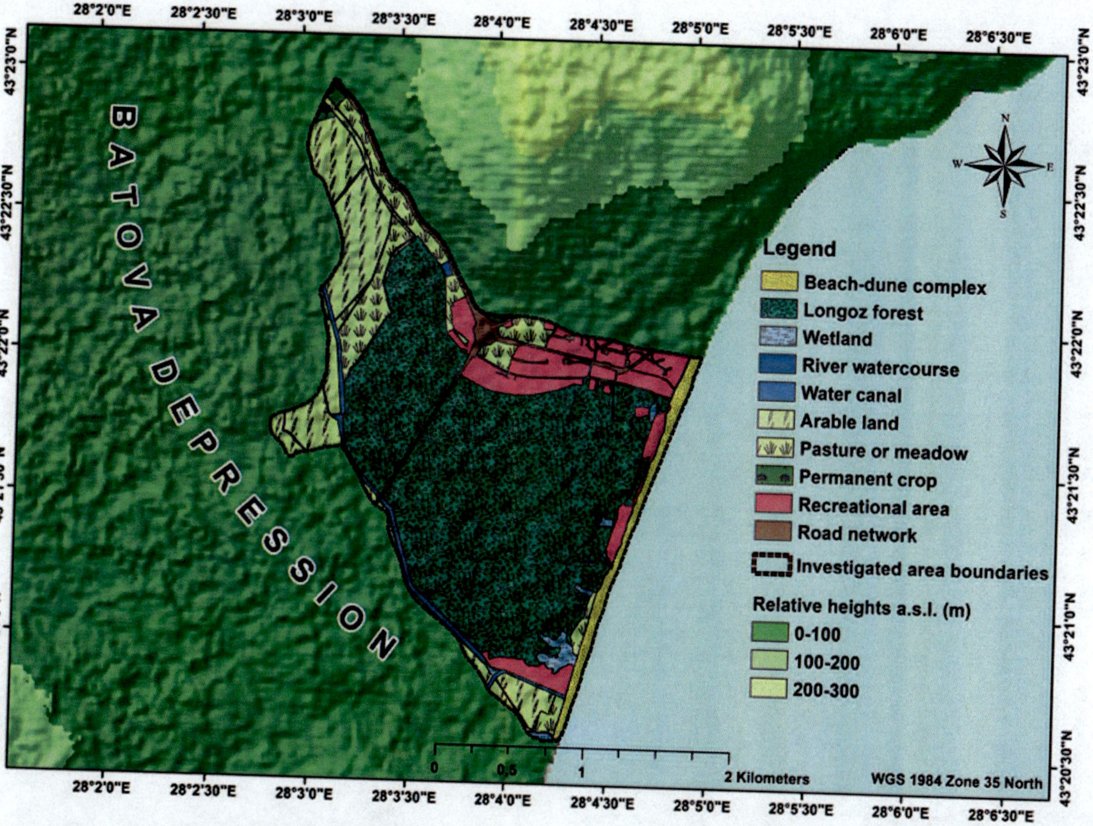

Fig. 8 Map of Baltata Locality as of the late 1970s–early 1980s of the twentieth century, reconstructed from archive topographic maps compiled at scale 1: 5,000 and 1: 50,000, refined using LANDSAT imagery. Various anthropogenic land cover/land use classes are already present in result of land conversion during the Socialist Period. These new land cover/land use types have emerged at the expense of the once dominant longoz landscape

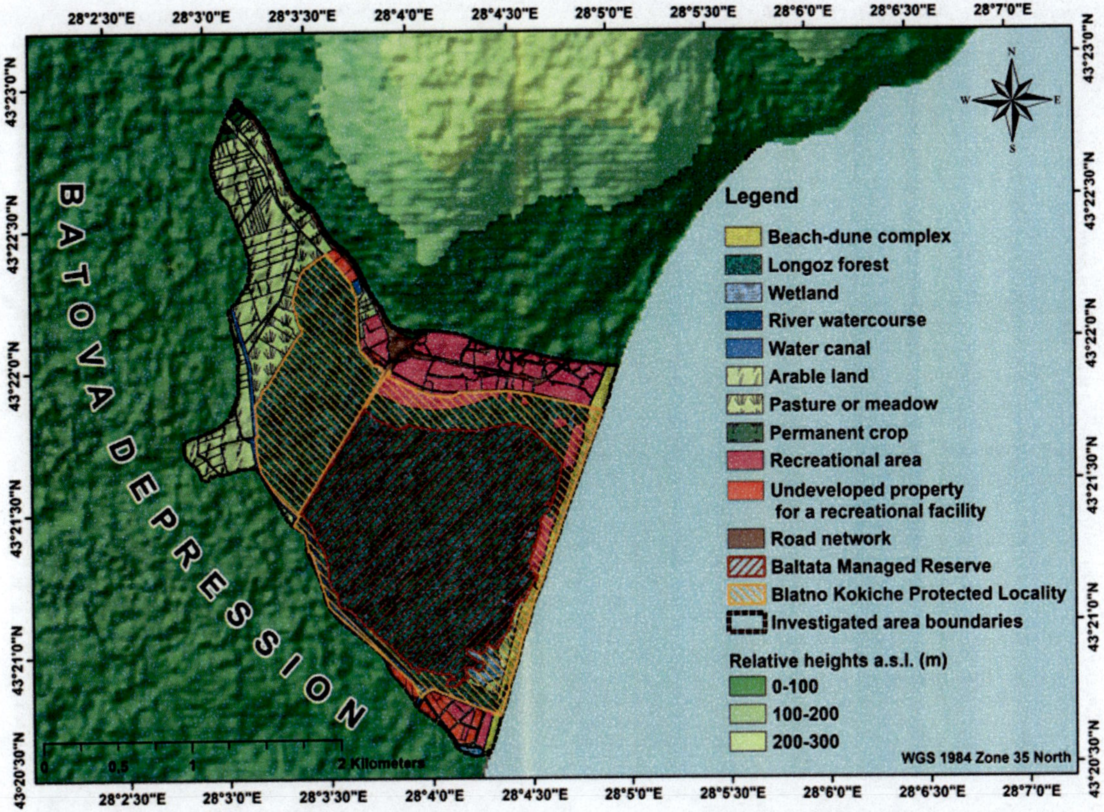

Fig. 9 Map of Baltata Locality as of 2019, compiled on the basis of cadastral data. All remnants of the once vast longoz forest are nowadays preserved as protected areas (illustrated with yellow and red crosshatches on the map)

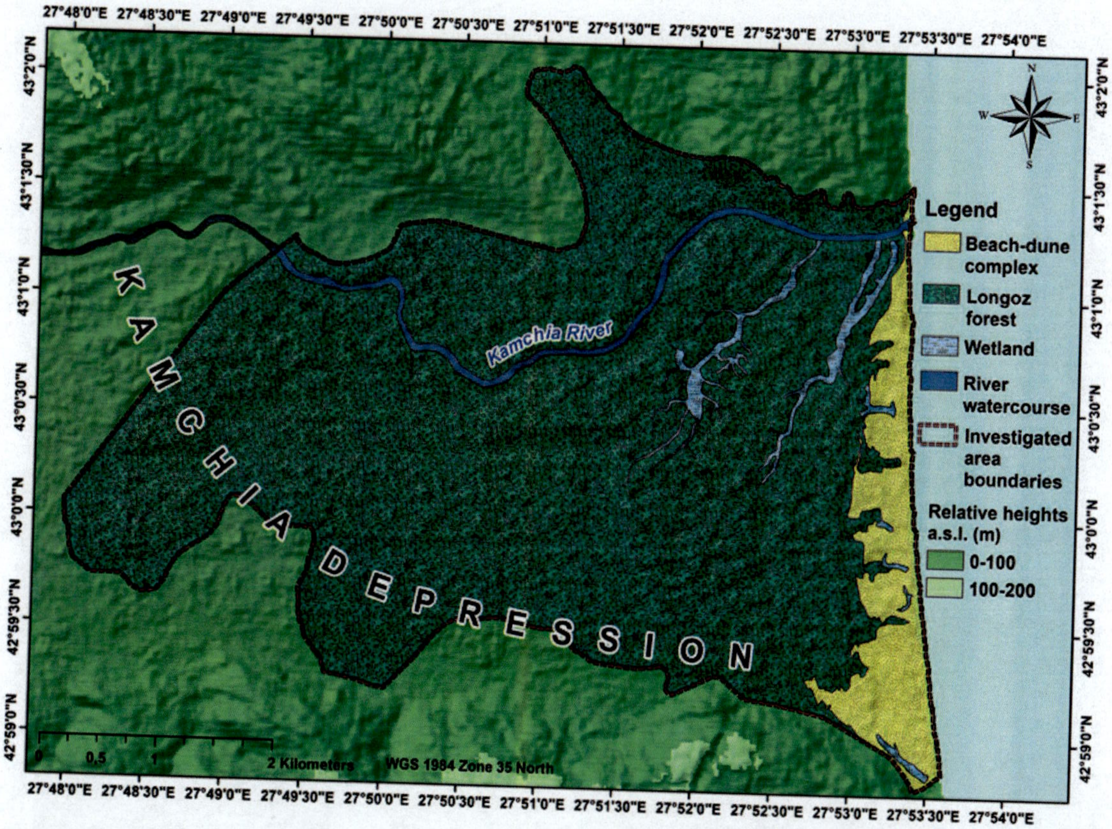

Fig. 10 Map of Longoza Locality as of the late nineteenth century, reconstructed from archive topographic maps compiled at scale 1: 126,000. Analogous to the Baltata Locality, greatest portion of the area represents longozes occupying the Kamchia River floodplain and its downstream sector

As of the late 1970s–early 1980s, the relative area of the longoz forests at the case study site dropped drastically, from almost 89% to barely 49.1%. Among the greatest anthropogenic alterations of the Kamchia River's hydrologic regime was the construction of four large reservoirs in its catchment basin, as well as the conversion of the Oryahovo Wetlands into agricultural land. As of comparably smaller environmental impact, but nevertheless incompatible with the nature conservation value of the area, may be assessed the establishment of the Kamchia Recreational Area. The total landscape transformation in comparison to the late nineteenth century is 46.7%, while the transformed area of the longoz forests is 47.6% (Fig. 11).

The last compared period (early 1980s–2019) reveals further moderate landscape transformations at Longoza Locality, estimated at 0.9%. The relative area of the longoz forests is estimated at 49%, which demonstrates an insignificant decrease in comparison to the precedent period (Fig. 12).

5 Discussion

The analyses of the crosstabulation results reveal that the aggregate landscape transformation at Baltata Locality is approximately 37%, while at Longoza Locality these values are even higher – nearly 46% when comparing the late nineteenth century versus the contemporary states of the two pilot study sites as of the year 2019. Most impacted by these anthropogenic-driven landscape-scale alterations are indeed the longoz forest ecosystems.

One of the drastic landscape-scale transformations at Baltata Locality is related to the artificial translocation of the Batova's river mouth to the south of the floodplain longoz forest via a system of water canals, where the Kranevska (Chaltika Dere) River flows into the Black Sea. As discussed in the preceding sections of the text, the reason for this action, which occurred over the Socialist Period (1970s of the twentieth century), was to clear the low-lying marshy terrain for subsequent construction of the Albena International Maritime Resort. Nevertheless, field observations confirm that the old estuary still exists to this date, while the area of the old river bed gets periodically drowned during peak outflow periods or torrential rains. The latter represents a typical feature of the coastal humid subtropical climate during the early-summer season [17, 18]. Accordingly, floods with fluvial and pluvial genesis represent a peculiar characteristic of the cited coastal resort nowadays as a consequence of this ill-planned and inexpertly executed land conversion project.

The Oryahovo Wetlands, which used to serve as a natural runoff regulator at the Longoza Locality in the past, share a similar faith in terms of land conversion, although transformed primarily into arable land for intensive agriculture. Similarly to the area of the Albena International Maritime Resort, frequent inundations are observed nowadays there, especially in the events of peak water outflows within the floodplain of the Kamchia River. Thus, it may be concluded that both case study sites, hosting hygrophilous forests of the endemic longoz type and selected for the

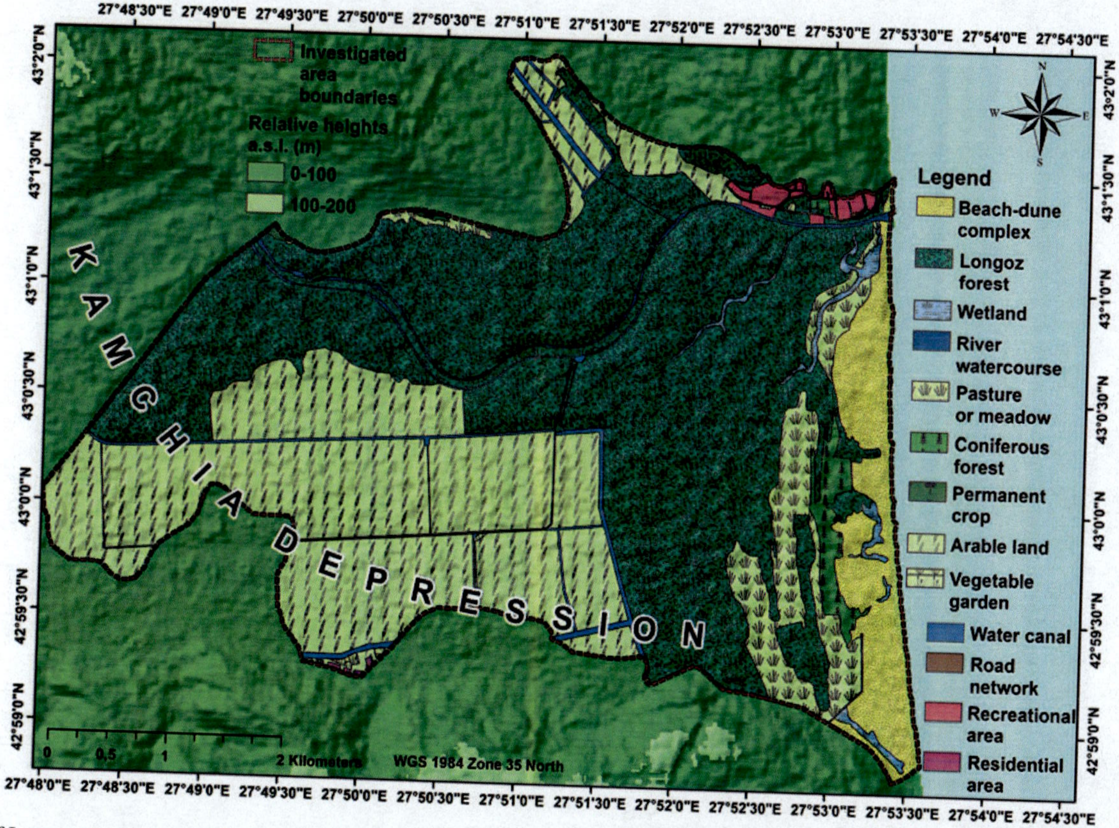

Fig. 11 Map of Longoza Locality as of the late 1970s– early 1980s of the twentieth century, reconstructed from archive topographic maps compiled at scale 1: 5,000 and 1: 50,000, refined using LANDSAT imagery. The once vast and uniform floodplain longoz landscapes have undergone severe reductions and fragmentation in result of land use changes, among which the land conversion into irrigated agricultural terrains prevails noticeably

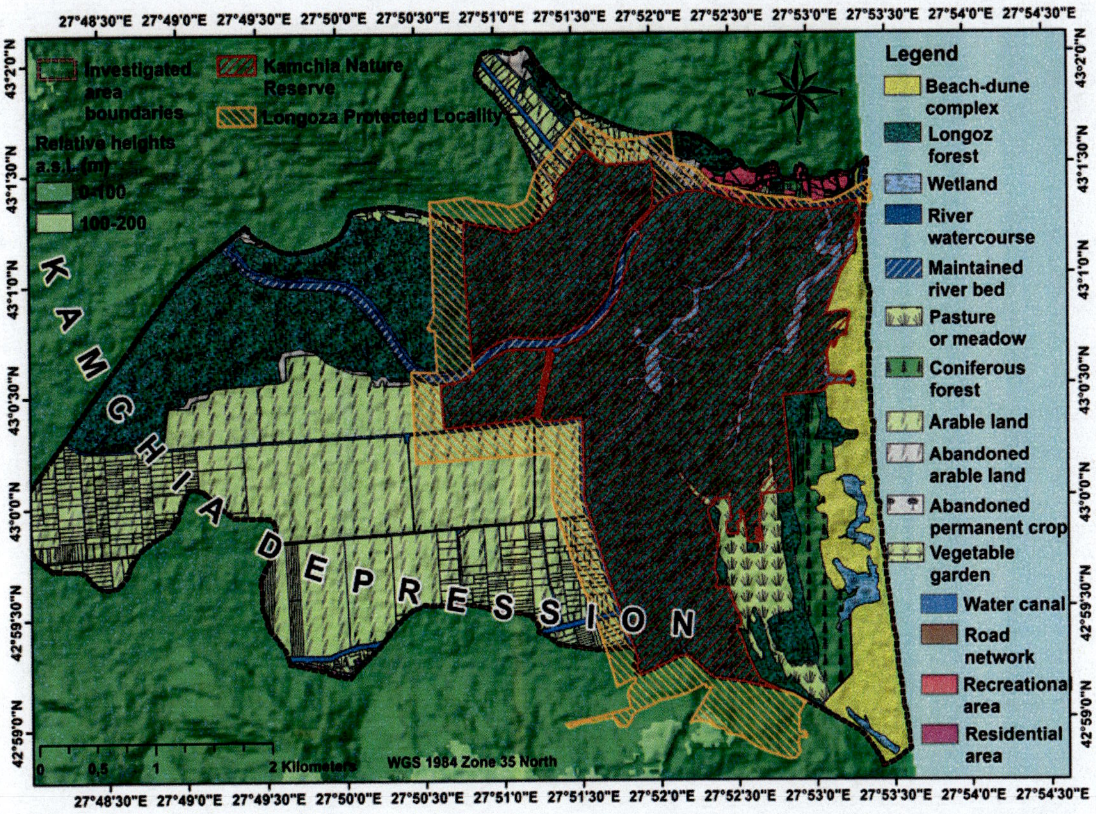

Fig. 12 Map of Longoza Locality as of 2019, compiled on the basis of cadastral data. Further landscape transformation in result of resort and residential construction takes place at insignificant scales in the area's periphery, affecting predominantly former agricultural terrains. A vast portion of the Longoza Locality has been established as protected areas nowadays (illustrated with yellow and red crosshatches on the map)

survey presented herein, share a lot of similarities. These parallels are not simply with regard to the array of mutual physiographic, botanical, ecological, habitat, and nature conservation properties, but also concern the pattern of land use change and the related spatio-temporal transformations at the landscape scale at Baltata and Longoza localities. Nevertheless, while the cited peculiarities of the present-day inundation regime at Baltata Locality represent a spatial management issue, in the latter case discussed for Longoza Locality it may actually facilitate flood risk management. This can be achieved by using the agricultural terrains of the former Oryahovo Wetlands as a temporal flood storage reservoir in the light of nature-based solutions.

The areas deprived of vegetation cover that used to be occupied by longoz forest patches in the past are definitely highly flood-prone nowadays. The analysis of past floods within the catchment basins of Batova and Kamchia rivers [51] reveals an intensified frequency of these extreme events after the year 1990. Thus, 9 floods have been registered versus just 1 for the preceding Socialist period that spanned between 1945 and 1990 in the catchment basin of the Batova River. Moreover, a total of 28 floods have occurred after the year 1990 in the catchment basin of the Kamchia River versus just 3 registered for the Socialist period and just 2 registered for the pre-Socialist period. Finally, both localities, Baltata and Longoza, are classified as extremely susceptible areas to floods with return periods 20, 100, and 1,000 years [51]. Meanwhile, rather intensified rates of resort construction are taking place at these two localities, with all implications resulting from it.

6 Conclusions

Results of the present study confirm that both pilot study sites, namely Baltata and Longoza localities, have undergone significant levels of anthropogenic-driven landscape transformation in result of land conversion and land use changes. Nevertheless, the available scientific publications about both localities that are cited herein are with quite different scopes. At present, there are very few (if any) papers attempting to link human-induced transformations of the longoz landscape pattern in result of land use change to their impaired capacity as flood storage reservoirs and the resulting fluvial flood hazard. In this regard, our study represents a pioneer attempt. Longozes are recognized internationally as a critically endangered, endemic type of polydominant hygrophilous forests, found exclusively in the eastern part of the Balkan Peninsula, and as such they are an integral part of Europe's fragile natural heritage [1, 2, 22, 27, 28]. Accordingly, their long-term preservation and restoration, apart from being of continental-scale interest in the realm of nature conservation and habitats preservation, will also ensure an improved, nature-based flood risk resilience of the coastal areas concerned. The analyses conducted with focus on the longoz forests at both investigated locations in the present paper are of significantly decreased spatial extents nowadays in comparison to the late nineteenth century. The most significant landscape transformations and land conversions occurred during the

Socialist period in Bulgaria, which was marked by extensive tourist, infrastructural, and agricultural development, often running counter to the landscape's carrying capacity. Nevertheless, alterations of the landscape pattern (i.e., by construction of recreational facilities and tourist infrastructure) continue nowadays as well and take place mainly on former agricultural land, which represented longoz forest patches and wetlands prior to the beginning of the twentieth century. These transformations and decreased areals of the longoz forests imply for impaired flood storage capabilities of the Batova and Kamchia catchments nowadays. The aforementioned statement is backed by the data on past flood events and modeled flood extents that comprise relevant parts of the preliminary flood risk assessment in the Black Sea river basin district for water management [51]. The cited findings of the present study are in a good correlation with the recent extreme floods of non-marine (particularly fluvial) origin along the North Bulgarian Black Sea coast and dictate the necessity about the initiation of landscape-scale restorations of the endemic longoz forests in Bulgaria in the light of nature-based solution for flood hazard mitigation in Bulgaria.

Acknowledgments The coastal surveys discussed herein were supported financially by the National Science Fund to the Ministry of Education and Science of the Republic of Bulgaria under project entitled "UAV-based mapping and monitoring of depositional river mouth sectors along the North Bulgarian Black Sea Coast" (National Science Fund, Project № KP-06-COST-12/ August 6, 2019).

References

1. Dimitrov M, Tzonev R (2015) Riparian and lowland mixed woodlands and longoses. In: Biserkov V et al (eds.) Red data book of the Republic of Bulgaria, vol III (Natural Habitats). Bulgarian Academy of Sciences and Ministry of Environment and Water, Sofia, pp 281–284
2. Ozyavuz M, Ataseven Y, Sısman EE (2011) Biosphere reserve approach: a case study Igneada longoz (flooded) Forests National Park (Kirklareli). J Environ Prot Ecol 12(3A):1383–1385
3. Dimitrov B (1992) Vliyanie na rechnite razlivaniya na Kamchiya warhu longoznite ekosistemi. Godishnik na Sofiyski universitet "Sv. Kliment Ohridski", Geologo-Geografski fakultet, Kniga 2: Geografiya, (The influence of river Kamchia overflows on the longoz ecosystems), (**Димитров Б** (1992) **Влиянието на речните разливания на Камчия върху лонгозните екосистеми. Годишник на Софийски Университет „Св. Климент Охридски", Геолог-о-Географски Факултет, Книга** 2: **География**). Yearbook of Sofia University "St. Kliment Ohridski", Book No2: Geography 84:55–63 (in Bulgarian)
4. Dimitrov B (1994) Izsledvane na izmeneniyata na rechnite razlivaniya na Kamchiya posredstvom dendrohronologichen analiz. Godishnik na Sofiyski universitet "Sv. Kliment Ohridski", Geologo-Geografski fakultet, Kniga 2: Geografiya, (Investigation of the flood changes of the river Kamchia by applying dendrochronological analysis), (**Димитров Б** (1994) **Изследване на измененията на речните разливания на Камчия посредством дендрохронологичен анализ. Годишник на Софийски Университет „Св. Климент Охридски", Геолого-Географски Факултет, Книга** 2: **География**). Yearbook of Sofia University "St. Kliment Ohridski", Book No2 (Geography), 85: 131–137 (in Bulgarian)

5. Tashev AN, Vitkova AA, Alexandrova AV (2018) Floristic composition and current state of forest natural habitats in Natura 2000 protected site "Kamchia" (BG0000116). Acta Zool Bulg Suppl 11:69–74
6. Flood Risk Management Plan of the Black Sea Basin Directorate for the Period 2016–2021 (2016) https://www.bsbd.org/UserFiles/File/PURN/BSBD%20PURN%202016-2021.pdf. Accessed 1 Nov 2020
7. Council Directive 92/43/EEC of 21 May 1992 on the conservation of natural habitats and of wild fauna and flora (Habitats Directive) (1992) https://eur-lex.europa.eu/legal-content/EN/TXT/?uri=celex%3A31992L0043. Accessed 1 Nov 2020
8. Directive 2007/60/EC of the European Parliament and of the Council of 23 October 2007 on the assessment and management of flood risks (Floods Directive) (2007) https://eur-lex.europa.eu/legal-content/EN/ALL/?uri=CELEX%3A32007L0060. Accessed 1 Nov 2020
9. Stefanov P (2002) Relef. Morfografska harakteristika. In: Kopralev I, Yordanova, M, Mladenov CH (red) Geografiya na Balgaria (Relief. Morphographic characteristics. Geography of Bulgaria), (**Стефанов П, Релеф. Морфографска характеристика. В: Копралев И, Йорданова М, Младенов Ч (ред) География на България**). ForKom, Sofia, pp 29–44. (in Bulgarian)
10. Keremedchiev ST (2005) Morfostrukturni i geodinamichni predpostavki za savremennoto morfolozhko razvitie na Balgorskoto Chernomorsko kraybrezhie. Trudove na Instituta po okeanologiya 5:181–204 (Morphostructural and geodynamic prerequisites for the up-to-date morphological development of the Bulgarian Black Sea coastal zone. Proceedings of the Institute of Oceanology - BAS, 5: 181–208), (**Керемедчиев, Ст** (2005) **Морфоструктурни и геодинамични предпоставки за съвременното морфоложко развитие на Българското Черноморско крайбрежие. Трудове на Института по Океанология**, 5: 181–204). (in Bulgarian)
11. Popov VI, Mishev K (1974) Geomorfologiya na Balgarskoto Chernomorsko krayvrezhie i shelf. Izdatelstvo na Balgarska akademiya na naukite, Sofia (Geomorphology of the Bulgarian Black Sea coast and shelf. Academic Press of the Bulgarian Academy of Sciences, Sofia), (**Попов Вл, Мишев К** (1974) **Геоморфология на Българското Черноморско крайбрежие и шелф. Издателство на Българска академия на науките, София**). (in Bulgarian)
12. Peychev V (2004) Geolozhki stroezh i geomorfologia na Balgarskoto Chernomorsko krayvrezhie. V: Morpfodinamichni i litodinamichni procesi v bregovata zona (Geologic structure and geomorphology of the Bulgarian Black Sea coast. In: Morphodynamical and lithodynamical processes in the coastal zone), (**Пейчев В** (2004) **Геоложки строеж и геоморфология на Българското Черноморско крайбрежие. В Морфодинамични и литодинамични процеси в бреговата зона**). Slavena, Varna, pp 29–69. (in Bulgarian)
13. Velev St (1990) Klimatichno rayonirane na Balgariya. V: Klimat na Balgariya, (Climate regionalization of Bulgaria. In: Climate of Bulgaria), (**Велев Ст** (1990) **Климатично райониране на България. В: Климат на България**). Marin Drinov Academic Press, Sofia, p 87-88 (in Bulgarian)
14. Velev St (1997) Klimat. V: Geografiya na Balgariya, (Climate. In: Geography of Bulgaria), (**Велев Ст** (1997) **Климат. В: География на България**). Marin Drinov Academic Press, Sofia, pp 108–130. (in Bulgarian)
15. Penin R (2007) Klimat na Balgariya. V: Prirodna geografiya na Balgariya (Climate of Bulgaria. In Physical Geography of Bulgaria) (**Пенин Р** (2007) **Климат на България. В: Природна география на България**). Bulvest 2000, Sofia, pp 45–70. (in Bulgarian)
16. Harding AE, Palutikoff J, Holt T (2009) The climate system. In: Woodward JC (ed) The physical geography of the Mediterranean. Oxford University Press, Oxford, pp 69–88
17. Kotteck M, Grieser J, Beck CH, Rudolf BR, Rubel FR (2006) World map of the Köppen-Geiger climate classification updated. Meteorol Z 15(3):259–263
18. Peel MC, Finlayson BL, McMahon TA (2007) Updated map of the Köppen-Geiger climate classification. Hydrol Earth Syst Sci 11:1633–1644

19. Rubel FR, Kottek M (2010) Observed and projected climate shifts 1901-2100 depicted by world maps of the Köppen-Geiger climate classification. Meteorol Z 19(2):135–141
20. Penin R (2007) Balgarsko Chernomorsko kraybrezhie. V: Prirodna geografiya na Balgariya (Bulgarian Black Sea coast. In: Physical Geography of Bulgaria), (**Пенин Р (2007б) Българско Черноморско крайбрежие. В: Природна география на България**). Bulvest 2000, Sofia, pp 223–230. (in Bulgarian)
21. Kitanov B (1978) Rastitelnost. V: Cherno more (sbornik), (Vegetation. In: Black Sea (miscellany)), (**Китанов Б** (1978) **Растителност. В: Черно море (сборник)**). Georgi Bakalov, Varna, pp 202–218. (in Bulgarian)
22. Asenov A (2006) Biogeografski rayon na Chernomorskoto kraybrezhie. V: Biogeografiya na Balgaria (Biogeographic region of the Black Sea coast. In: Biogeography of Bulgaria), (**Асенов А** (2006) **Биогеографски район на Черноморското крайбрежие. В: Биогеография на България**). AN-DI Andrian Tassev, Sofia, pp 426–446. (in Bulgarian)
23. Ivanov K, Marinov IV, Panayotov T, Petkov AI (1961) Hidrografska harakteristika na rekite. Chernomorski basein V: Hidrologiya na Balgariya (Hydrographic characteristic of the rivers. Black Sea (drainage) basin. In: Hydrology of Bulgaria), (**Иванов К, Маринов Ив, Панайотов Т, Петков Ал** (1961) **Хидрографска характеристика на реките. Черноморски басейн. В: Хидрология на България**). Nauka i Izkustvo State Press, Sofia, pp 72–85. (in Bulgarian)
24. Dimitrov P, Solakov D, Peychev V, Dimitrov D (2003) The source provinces of the Black Sea. Trudove na Instituta po okeanologiya – BAN (Proceedings of the Institute of Oceanology – BAS), (**Ттрудове на Института по океанология – БАН**) 4:29-35
25. Hristova N (2012) Chernomorski ottochen basein. V: Rechni vodi na Balgariya (Black Sea catchment basin. In: Riverine waters of Bulgaria), (**Христова Н** (2012) **Черноморски отточен басейн. В: Речни води на България**). Tip-Top Press, Sofia, pp 153–193. (in Bulgarian)
26. Ivanova EI, Nedkov RD, Ivanova IB, Radeva KL (2012) Morpho-hydrographic analysis of the Black Sea catchment area in Bulgaria. Procedia Environ Sci 14:143–153
27. Georgiev G (2004) Chernomorski biogeografski rayon. V: Natsionalnite i prirodnite parkove i rezervatite v Balgariya (Black Sea biogeographic region. In: National and nature parks and reserves in Bulgaria), (**Георгиев Г** (2004) **Черноморски биогеографски район. В: Националните и природните паркове и резерватите в България**). Gea-Libris Press, Sofia, pp 217–271. (in Bulgarian)
28. Zashtiteni teritorii v obhvata na RIOSV-Varna (Protected areas within the spatial extent of the Regional Inspectorate for Environment – Varna), (**Защитени територии в обхвата на РИОС-В-Варна**), (2004). Published by the Regional Inspectorate for Environment – Varna, 56 p. (in Bulgarian)
29. Register of the protected areas and NATURA 2000 sites in Bulgaria, Executive Environment Agency (2014). http://eea.government.bg/zpo/en/index.jsp. Accessed 1 Nov 2020
30. Biodiversity Act of the Republic of Bulgaria (2002). https://lex.bg/laws/ldoc/2135456926. Accessed 1 Nov 2020. (in Bulgarian)
31. Filipova-Marinova M, Pavlov D, Kotsev I (2017) Vegetation cover. In: Kotsev I, Stanchev HR (eds) Sensitivity mapping and analysis of the Bulgarian Black Sea coastal zone. Pulsio, Sofia, pp 59–71. https://doi.org/10.7546/IO.BAS.2018.2
32. Information system for protected sites of the Bulgarian NATURA 2000 network (2013) http://www.natura2000.moew.government.bg. Accessed 31 Oct 2020. in Bulgarian
33. Angelov D (1978) Chernomorieto prez IV-XV vek. V: Cherno more (sbornik), (The Black Sea and the Black Sea coast in the period 4th – 15th century A.D. In: Black Sea (miscellany)), (**Ангелов Д** (1978) **Черноморието през IV-XV век. В: Черно море (сборник)**). Georgi Bakalov, Varna, pp 452–485. (in Bulgarian)
34. Paskaleva V (1978) Cherno more i Chernomorieto prez epohata na Osmanskoto vladichestvo v balgarskite zemi. V: Cherno more (sbornik), (The Black Sea and the Black Sea coast during the epoch of the Ottoman rule in the Bulgarian lands. In: Black Sea (miscellany)), (**Паскалева В** (1978) **Черно море и Черноморието през епохата на Османското владичество в**

българските земи. В: Черно море (сборник)). Georgi Bakalov, Varna, pp 486–507. (in Bulgarian)
35. Pavlov Vl (1978) Cherno more v politikata na kraybrezhnite darzhavi i na Velikite sili (1878–1918). V: Cherno more (sbornik), (The Black Sea in the political agenda of the littoral states and the Great Powers (1878–1918). In: Black Sea (miscellany)), (Павлов Вл (1978) Черно море в политиката на крайбрежните държави и на Великите сили. В: Черно море (сборник)). Georgi Bakalov, Varna, p 531–568 (in Bulgarian)
36. Mladenov CH (2002) Dinamika na naselenieto za perioda 1880–2001 g. In: Kopralev I, Yordanova, M, Mladenov Ch (red) Geografiya na Balgaria, (Population dynamics during the period 1880–2001. In: Kopralev I, Yordanova, M, Mladenov Ch (eds) Geography of Bulgaria), (Младенов Ч (2002а) Динамика на населението за периода 1880–2001 г. В: Копралев И, Йорданова М, Младенов Ч (ред) География на България). ForKom, Sofia, pp 443–445. (in Bulgarian)
37. Mladenov Ch (2002) Migratsii na naselenieto v Balgaria. In: Kopralev I, Yordanova, M, Mladenov Ch (red) Geografiya na Balgaria, (Migration movements in Bulgaria. In: Kopralev I, Yordanova, M, Mladenov Ch (eds) Geography of Bulgaria), (Младенов Ч (2002б) Миграции на населението в България. В: Копралев И, Йорданова М, Младенов Ч (ред) География на България). ForKom, Sofia, pp 457–467. (in Bulgarian)
38. Ninov Z (2002) Gradsko i selsko naselenie. In: Kopralev I, Yordanova, M, Mladenov Ch (red) Geografiya na Balgaria, (Urban and rural population. In: Kopralev I, Yordanova, M, Mladenov Ch (eds) Geography of Bulgaria), (Нинов З (2002а) Градско и селско население. В: Копралев И, Йорданова М, Младенов Ч (ред) География на България). ForKom, Sofia, pp 468–470. (in Bulgarian)
39. Ninov Z (2002) Zarazhdane i razvitie na selishtata i selishtnata mrezha. In: Kopralev I, Yordanova, M, Mladenov CH (red) Geografiya na Balgaria, (Formation and development of the settlements and the settlement network. In: Kopralev I, Yordanova, M, Mladenov Ch (eds) Geography of Bulgaria), (Нинов З (2002б) Зараждане и развитие на селищата и селищната мрежа. В: Копралев И, Йорданова М, Младенов Ч (ред) География на България). ForKom, Sofia, pp 517–520. (in Bulgarian)
40. Dimitrov E (2002) Urbanizatsiya. In: Kopralev I, Yordanova, M, Mladenov CH (red) Geografiya na Balgaria, (Urbanization. In: Kopralev I, Yordanova, M, Mladenov Ch (eds) Geography of Bulgaria), (Димитров Е (2002) Урбанизация. В: Копралев И, Йорданова М, Младенов Ч (ред) География на България). ForKom, Sofia, pp 525–529. (in Bulgarian)
41. Ilieva M (2002) Obshta harakteristika na natsionalnoto stopanstvo. In: Kopralev I, Yordanova, M, Mladenov CH (red) Geografiya na Balgaria, (General characteristics of the national economy. In: Kopralev I, Yordanova, M, Mladenov Ch (eds) Geography of Bulgaria), (Илиева М (2002) Обща характеристика на националното стопанство. В: Копралев И, Йорданова М, Младенов Ч (ред) География на България). ForKom, Sofia, pp 537–544. (in Bulgarian)
42. Ilieva M, Ninov Z, Kolev B, Terziyska E, Dimitrov E, Rukova P, Kopralev I, Grozeva M (2002) Sektorna i otraslova struktura na stopanstvoto. In: Kopralev I, Yordanova, M, Mladenov Ch (red) Geografiya na Balgaria, (Sectoral and branch structure of the national economy. In: Kopralev I, Yordanova, M, Mladenov Ch (eds) Geography of Bulgaria), (Илиева М, Нинов З, Колев Б, Терзийска Е, Димитров Е, Рукова П, Копралев И, Грозева М (2002) Секторна и отраслова структура на стопанството. В: Копралев И, Йорданова М, Младенов Ч (ред) География на България). ForKom, Sofia, pp 545–668. (in Bulgarian)
43. Georgiev A (1978) Cherno more i razvitieto na turizma. V: Cherno more (sbornik), (The Black Sea and the tourist development. In: Black Sea (miscellany)), (Георгиев А (1978) Черно море и развитието на туризма. В: Черно море (сборник)). Georgi Bakalov, Varna, pp 365–390. (in Bulgarian)
44. Donchev D (1978) Balgarskoto Chernomorie po patya na sotsialisticheskite preobrazuvaniya. V: Cherno more (sbornik), (The Bulgarian Black Sea coast on the path to the Socialist conversions. In: Black Sea (miscellany)), (Дончев Д (1978) Българското Черноморие по

пътя на социалистическите преобразувания. В: Черно море (сборник)). Georgi Bakalov, Varna, pp 580–588. (in Bulgarian)

45. Tsvetkov L, Velev Vl (1978) Zashtita na prirodata. V: Cherno more (sbornik), (Nature conservation and environmental preservation. In: Black Sea (miscellany)), (**Цветков Л, Велев Вл** (1978) **Защита на природата. В: Черно море (сборник)**). Georgi Bakalov, Varna, pp 295–318. (in Bulgarian)
46. Daneva M, Mishev K (1979) Balgarskoto Chernomorsko kraybrezhie. Izdatelstvo na Balgarskata akademiya na naukite, Sofia (the Bulgarian Black Sea coast. Academic Press of the Bulgarian Academy of sciences, Sofia), (**Данева М, Мишев К** (1979) **Българското Черноморско крайбрежие. Издателство на Българската академия на науките, София**). (in Bulgarian)
47. Antsyferov S, Belberov ZDR, Massel S (1990) Dynamical processes in coastal regions. Academic Press of the Bulgarian Academy of Sciences, Sofia
48. Dachev V, Nikolov HR (1977) Integralni izmeneniya na bregovata liniya pri akumulativnite uchastatsi mezhdu Cherni nos i kurortniya kompleks "Albena" (Integral coastline changes at the accumulative sectors between Cape Cherni nos and Albena Maritime Resort), (**Дачев В, Николов Хр** (1977 **Интегрални изменения на бреговата линия при акумулативните участъци между Черни нос и курортния комплекс „Албена"**). Okeanologiya 2: 57–64. (in Bulgarian)
49. Velikov V, Drenovski IV, Mihaylov M, Vlaskov Vl (2002) Prirodni predpostavki i antropogenni faktori za narusheniya na okolnata sreda. V: Kopralev I, Yordanova, M, Mladenov Ch (red) Geografiya na Balgaria, (Natural prerequisites and anthropogenic factors for environmental degradation. In: Kopralev I, Yordanova, M, Mladenov Ch (eds) Geography of Bulgaria), (**Великов В, Дреновски Ив, Михайлов М, Власков Вл** (2002) **Природни предпоставки и антропогенни фактори за нарушения на околната среда. В: Копралев И, Йорданова М, Младенов Ч (ред) География на България**). ForKom, Sofia, p 411–416 (in Bulgarian)
50. Petrov P (2002) Problemi po sahranyavaneto i ratsionalnoto izpolzvane na prirodnoto nasledstvo. V: Kopralev I, Yordanova, M, Mladenov Ch (red) Geografiya na Balgaria, (Problems of preservation and rational use of the natural heritage. In: Kopralev I, Yordanova, M, Mladenov Ch (eds) Geography of Bulgaria), (**Петров П** (2002) **Проблеми по съхраняването и рационалното използване на природното наследство. В: Копралев И, Йорданова М, Младенов Ч (ред) География на България**). ForKom, Sofia, p 436–437 (in Bulgarian)
51. Predvaritelna otsenka na riska ot navodneniya v Chernomorski rayon za baseynovo upravlenie na vodite (2012) Preliminary flood risk assessment in the Black Sea river basin district for water management. (**Предварителна оценка на риска от наводнения в Черноморски район за басейново управление на водите**). https://www.bsbd.org/UserFiles/File/Copy%20of%20BSBD_PFRA_v2.pdf. Accessed 1 Nov 2020
52. Topographic and other maps as a free WMS service. https://kade.si. Accessed 31 Oct 2020

Hydrological Aspects of Nature-Based Solutions in Flood Mitigation in the Danube River Basin in Croatia: Green vs. Grey Approach

Kristina Potočki, Damir Bekić, Ognjen Bonacci, and Tin Kulić

Contents

Abstract Flood protection based around "grey" infrastructure measures, such as dikes and dams, comprise traditional flood alleviation schemes. Besides their advantages, reflected in well-established design principles and construction techniques, they cannot fully respond to the challenge of increasing trends of flood magnitudes driven by impacts of climate and land use changes. The application of nature-based solutions for flood mitigation, such as providing "space for water" and inclusion of

---

K. Potočki (✉), D. Bekić, and T. Kulić
Department of Hydroscience and Engineering, Faculty of Civil Engineering, University of Zagreb, Zagreb, Croatia
e-mail: kristina.potocki@grad.unizg.hr

O. Bonacci
Faculty of Civil Engineering, Architecture and Geodesy, University of Split, Split, Croatia

Carla S. S. Ferreira, Zahra Kalantari, Thomas Hartmann, and Paulo Pereira (eds.),
Nature-Based Solutions for Flood Mitigation: Environmental and Socio-Economic Aspects,
Hdb Env Chem (2022) 107: 263–288, DOI 10.1007/698_2021_770,

Published online: 30 May 2021

natural retention areas, provides more flexible and effective solution that may also enhance local biodiversity. This chapter presents a short overview of large flood risk management schemes in the Republic of Croatia with the focus on schemes that utilize nature-based solution aspects on larger scale. The contribution of nature-based solutions was studied through a comparison of longitudinal distribution of peak flows in the two 20-year periods along the major schemes in the Danube River basin in Croatia. The lower sections of the Sava, Mur, Drava and Danube Rivers host exceptional examples of large natural retention areas, such as Lonjsko polje, Kopački rit and Drava floodplains. In addition to flood alleviation impact, the conservation of biodiversity in the Lonjsko polje and Kopački rit wetland areas is also emphasized.

Keywords Croatia, Danube River Basin, Flood mitigation, Nature-based solutions, Wetlands

1 Introduction

Riverine floods are extreme events causing the largest monetary losses in the EU and projections show their five-fold increase by 2050 [1, 2]. The traditional paradigm of flood management focused on flood protection with engineering infrastructure was abandoned in the 90s, and a new paradigm based on flood risk management (FRM) has been implemented through the Water directive and Flood directive at the EU level. Concepts such as *living with the floods* and *making room for the rivers* are incorporated in this new approach. Technical measures based on conventional engineering structures ("grey" infrastructure), such as dikes and dams, are still the foundation of the flood protection systems. Besides all their advantages, reflected in the well-established design and modelling principles, "grey" infrastructure objects are not flexible enough and cannot respond fully to the challenges of climate change and land use changes that magnify the flooding mechanism [3]. Therefore, the introduction of nature-based solutions (NBS) is that "use, or mimic, natural processes to contribute to the improved management of water" [4] has increased in recent years as a complementary FRM approach to the mitigation of floods [5]. Similar to the nature-based solution (NBS) concept, the Natural Water Retention Measures (NWRM) concept aims to restore or enhance natural storage capacities of wetlands, rivers and floodplains by increasing water retention and groundwater storage with different multifunctional agricultural, forestry, urban and hydromorphological changes [6]. Agricultural and forestry measures (various cropping, tilling, sowing and planting methods, land use conversion, reconstruction of the existing driveways, etc.) are conducted in order to increase the soil infiltration and retention capacity, reduce potential soil erosion, improve water quality and soil fertility. Urban measures (permeable surfaces, bioretention basins, green roofs, etc.) are implemented with the objective of reducing the hazardous impact of urban floods

Fig. 1 Territory of the Republic of Croatia divided into two major basins—The Danube River basin (green) and The Adriatic Sea basin (blue). Large rivers in the Danube River basin are presented (Sava, Drava-Mur and Danube) [11]

by increasing the infiltration rates in such areas, as well as surface and soil retention capacity. Hydromorphological measures of floodplain restoration and management, especially if applied on the large scale of the river basin, can significantly contribute to flood attenuation while also improving water quality and enhancing habitat biodiversity and landscape aesthetic values [7, 8]. The NBS approach to flood risk management through retention of water in floodplains and wetlands can be less costly than structural engineering measures if benefits from the natural ecosystem services are considered [9, 10].

The area of the Republic of Croatia is 56,566 km^2 and is divided between the Danube River basin and the Adriatic Sea basin (Fig. 1) [11]. The territory of Croatia spreads over three different climatic, hydrological and biogeographical regions (the Mediterranean, mountainous and Pannonian region), with specific conditions and unique hydromorphological features and habitat types [12]. According to the flood events inventory, Croatian territory is mostly exposed to riverine and pluvial floods [11]. About 52% of the country's area (29,772 km^2) is exposed to flooding, over 64% of which occurs in the Danube River basin [11]. Over the last two decades, many floods have been recorded in Croatia. The most recent catastrophic flood with human fatalities and flooded settlements was in the year 2014, within the Sava River Basin, caused by extreme precipitation combined with highly saturated soil [13, 14].

Large lowland floodplains with significant capacity are located in the Danube River basin and are used for water retention during floods [10, 11]. Areas of alluvial wetlands and large lowland forests cover part of the Drava and especially the Sava River basin. The largest alluvial wetlands and Ramsar sites are Lonjsko polje in the Sava River basin and Kopački rit in the Drava-Danube confluence. They present examples of NBS for flood alleviation with their significant water retention capacity (up to one billion m^3 each), but also provide additional ecosystem services, such us supporting biodiversity and eco-tourism [10, 15]. In particular, Lonjsko polje, located in the Central Sava Basin flood protection system, represents a unique example of the large scale NBS for flood mitigation, with complementary interaction of "grey" (artificial relief canals, sluices, weir and embankments) and "green" (system of lowland near-natural retention areas) flood protection measures, and is recognized as a world-wide example of good practice [10, 16, 17].

The main objective of this chapter is to present a short overview of the approaches to flood risk management in Croatia and provide examples of flood mitigation projects that have included aspects of NBS on the large scale. Emphasis is given to the utilization of the natural and semi-natural flood inundation areas in the Danube River Basin. An overview of the existing floodplain retention areas and protected areas, as well as their hydrological capabilities for flood alleviation, is provided. Finally, a detailed description of the two largest floodplains that can serve as NBS examples of good practice is given: Lonjsko polje and Kopački rit which are also natural protected areas due to their relevance for biodiversity conservation in wetland habitats.

2 Flood Risk Management in Croatia

As defined in the Floods Directive [18], flood risk is a combination of the probability of a flood event and the potential adverse consequences for human health, environment, cultural heritage and economic activities associated with a flood event. The Flood Risk Management Plan (FRMP) provides a set of measures in the areas of prevention, protection and preparedness for flooding. The FRMP should focus on the reduction of potential adverse consequences of flooding, and, if considered appropriate, on non-structural initiatives as well as on the reduction of the probability of flooding. With the goal of providing more space for rivers, the FRMP should prioritize the maintenance and/or restoration of floodplains.

The Floods Directive was transposed into Croatian legislation through the Water Act [19], with supporting sub-acts. As a consequence, preliminary flood risk assessments including flood hazard and flood risk maps, as well as FRMPs are published by Croatian waters[1] every 6 years. The application of the Floods Directive is coordinated with the Water Framework Directive [20] focusing on opportunities for improving efficiency and achieving common synergies of environmental objectives by both directives. All major Croatian Rivers are transboundary, hence the flood risk management objectives are shared with other countries (Slovenia, Hungary, Serbia, Bosnia and Herzegovina), as defined in multi- and bi-lateral agreements and implemented in common projects.

2.1 Legal Documents and Projects Supporting NBS Implementation for Flood Risk Management

The NBS approach to flood risk mitigation is supported by different EU and national policies and legislations that enable meeting of the main goals established within the (1) Water Framework Directive, (2) Floods Directive, (3) Habitats Directive, (4) Birds Directive, (5) Europe 2020 strategy – resource efficient Europe, (6) Blueprint to safeguard Europe's waters, (7) The Biodiversity Strategy including the Green Infrastructure strategy and (8) Climate Change Adaptation Strategy. National legislation related to water management and FRM in Croatia is harmonized with the EU legislation through the Water Act [19] which stipulates adoption of the river basin management plan (RBMP) as the basic instrument for the management of flood risks. The current RBMP adopted for the period 2016–2021 emphasizes the choice and implementation of solutions and measures to reduce flood risk through an effective combination of "grey" and "green" infrastructure, such as the preservation of natural retentions, wetlands and wide inundation areas along the watercourses [11]. NBS in the Republic of Croatia have also recently been introduced in the national climate adaptation strategy [21]. The strategy recommends the implementation of measures for flood adaptation through the development of "green" infrastructure. Measures are focused on restoration measures for the existing natural watercourses and on controlled flooding of the natural lowland areas for reducing the peak discharges during flood events. Such measures need to include the protection of the areas under Natura 2000 and the identified Ramsar sites. Although NBS have been recently introduced in the national legislation, they have not yet been fully recognized by all stakeholders and practitioners as crucial FRM measures which simultaneously enable adaptation to a changing climate and enhance biodiversity. Additionally, several national and international transboundary water management projects are being implemented, including some measures of NBS for the mitigation of flood risk (and drought risk and/or improving and attaining good ecological status of water bodies). Examples of significant water management projects including NBS

[1] Croatian waters is the legal entity for water management in Republic of Croatia.

Table 1 Examples of national and international projects using NBS approach for flood risk management (FRM) implemented on Croatian territory

Project name/ project acronym	Description of main objective related to the NBS approach for FRM	Project area (River Basin)	Implementation period	Website
Wading toward Integrated Basin management/ IBM - CENTRAL POSAVINA	National project that addressed, investigated, and conducted the measures for protecting the biological and landscape diversity of the Lonjsko Polje Nature Park. NBS approach included maintaining traditional land use, managing the wetlands and controlled flooding of the Lonjsko Polje	Sava River basin – Lonjsko Polje Nature Park (Danube River basin)	2006–2008	https://ec.europa.eu/environment/life/project/Projects/index.cfm?fuseaction=search.dspPage&n_proj_id=2962
DRAVA LIFE – Integrated River management/ DRAVA LIFE	National project with the objective of preserving existing and creating new water bodies and flooding areas within the already existing floodplains. NBS approach includes river restoration measures for increasing the number of natural and dynamic riverine habitats	Drava River in Croatia (Danube River basin)	2015–2020	https://www.drava-life.hr/en/project/
Integrated cross-border monitoring and Management Systems for Flood Risks, environmental and	International project with the objective of reducing the flood risk through monitoring and	Sava River; natural floodplains in the region of the Spačva-Morović; forests in the border area of Croatia	2017–2020	https://www.interreg-croatia-serbia2014-2020.eu/project/forret/

(continued)

Table 1 (continued)

Project name/ project acronym	Description of main objective related to the NBS approach for FRM	Project area (River Basin)	Implementation period	Website
biodiversity protection and forestry through transboundary Forest retentions and other measures/FORRET	"green infrastructure" measures. NBS approach includes FRM with water retention in the forest floodplains	and Serbia (Danube River basin)		
Reducing the flood risk through floodplain restoration along the Danube River and tributaries/ DANUBE FLOODPLAIN	International project with the objective of improving transnational water management and flood risk prevention while maximizing benefits for biodiversity conservation. NBS approach includes restoration of water storage capacity of floodplains and development of best practices on using "green infrastructure" for sustainable flood risk management	Danube River and tributaries (Danube River basin)	2018–2020	http://www.interreg-danube.eu/approved-projects/danube-floodplain
Project for the improvement of non-structural flood risk management measures in the Republic of Croatia/VEPAR	National project with the objective of increasing the effectiveness of the implementation of non-structural measures for FRM, such as data collection and analysis, the	Croatian territory (Danube River Basin and Adriatic Sea Basin)	2019–2023	https://www.voda.hr/hr/novosti/projekt-vepar-modernija-preciznija-sigurnija-rjesenja-za-smanjenje-rizika-od-poplava

(continued)

Table 1 (continued)

Project name/ project acronym	Description of main objective related to the NBS approach for FRM	Project area (River Basin)	Implementation period	Website
	establishment of flood forecasting and early warning systems but also encouraging the implementation of "green infrastructure" measures			

for flood risk management implemented on Croatian territory are presented in Table 1.

3 Hydrological Aspects of Retention Areas in the Danube River Basin in Croatia

This section gives an overview of the main sub-basins, water bodies, alluvial floodplains, protected areas in the Danube River basin and shows characteristic discharges along selected hydrological stations corresponding to average and high flow conditions. Effectiveness of natural floodplains and wetlands on flood mitigation is presented with hydrological analysis of changes in peak discharges and in hydrographs for selected flood events along river sections in the Danube River basin. Flood events are selected for two periods: 1951–1970 and 2000–2019, representing periods before and after construction of some large "grey" infrastructure (dams, reservoirs, flood protection system) in the Danube River Basin.

3.1 *Hydrological Characteristics of the Danube River Basin*

The Danube River basin covers an area of 35,132 km^2 (62% of the country territory), and spreads across the Pannonian valley in the north and a mountainous area in the south (Fig. 2). The entire basin area has a diverse geological and pedological composition dominated by surface runoff with numerous rivers and streams. The Danube River basin contains 80% of the national watercourses network and contributes to the total national water resources with 84 × 10^6 m^3, which is estimated as

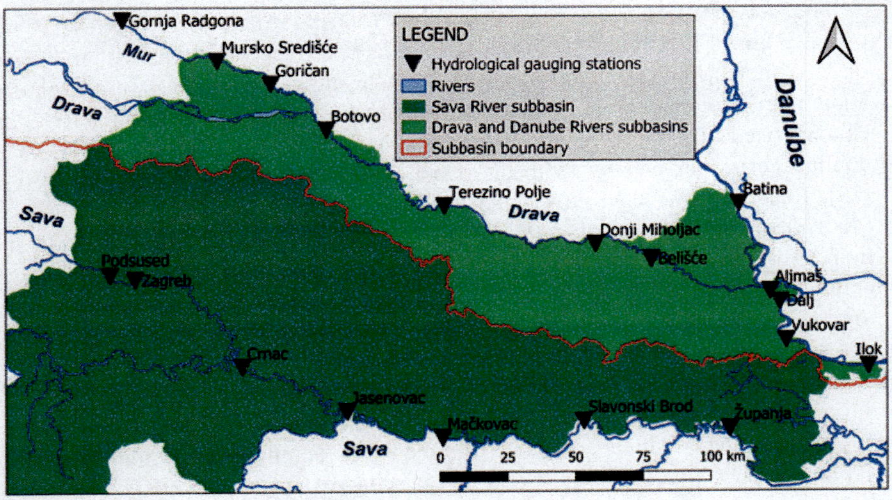

Fig. 2 Danube River basin in Croatia, including the Sava River and Drava and Danube Rivers sub-basins. Selected hydrological gauging stations along Sava, Drava, Mur and Danube are also presented

an average annual value for the period 1960–1990. The largest sub-basins in the Danube River Basin are the Sava River Basin and the Drava River Basin with its main tributary Mur. The section of the Danube River has the length of 138 km and is located on the eastern county border [22].

The average annual precipitation in the lowlands ranges from 900–1,200 mm in the west to 600–700 mm in the east, with the highest monthly precipitation in June (95.3 mm) and the lowest in February (35.1 mm). In the mountainous area, the average annual precipitation is significantly higher (up to 3,500 mm), with the highest monthly precipitation in November (168.7 mm), and the lowest in January (95.5 mm) [12].

The Sava River basin covers 46% of Croatian territory (25,770 km^2). The Sava River is the largest river in Croatia, with the length of 510 km [22]. According to the hydrological regime, high flows occur in late fall or early winter (November/December), caused by intensive rainfall, and in the early spring (March/April), driven by snowmelt [23]. Both, decreasing and increasing trends in monthly, seasonal and annual water levels and discharges related to climate and anthropogenic factors, have been detected by Bonacci and Ljubenkov [24], Trninić and Bošnjak [25], Potočki et al. [26], Orešić et al. [27]. The upper course of the Sava River, upstream from hydrological station Zagreb (see Fig. 2), exhibits longitudinal slopes from ~2‰ to ~0.05‰. In the downstream sections the Sava River exhibits lowland hydromorphological characteristics with wide alluvial floodplains [28]. Regarding the land cover/land use, the basin is covered mainly by forest and seminatural areas (54.7%), as well as agriculture fields (42.4%), while artificial surfaces cover 2.2%.

The Drava River basin covers 12% of the Croatian territory which represents area of 7,015 km^2. The Drava River is the second largest river in Croatia, with the length of 323 km [22]. The main left tributary, the Mur, flows into the Drava in Croatia at about river kilometre 235. Channel morphology and typology of the Drava River are characterized with straight type in the upstream part while downstream the Mur confluence is characterized with the transitional braided and the meandering type [29]. Regarding the land cover/land use, the basin is covered mainly by forests (45.8%) and agriculture fields (28.7%), while natural grasslands and sparse vegetation cover 9.0% and 3.9%, respectively. The hydrological regime of the Drava River exhibits maximum water levels in late spring and summer (May, June and July) and minimum water levels during winter period (January and February) [30]. Hydrological and hydromorphological changes in discharges, water levels and suspended sediments upstream of the confluence with the Mur River are heavily modified by three dams and reservoirs built for the hydropower plants [31]. The downstream section, near the confluence with the Danube River, is influenced by the backwaters of the Danube River. Water levels and discharges for characteristic minimum, average and maximum values show a decreasing trend in all downstream parts of the river course from the dams [30].

The hydrological characteristics of Sava, Drava-Mur and Danube rivers are presented in form of the changes in characteristic discharges along the course of the river. Characteristic discharges are defined by discharge frequency analysis for the period 2000–2019, by defining discharges with 1% (Q1% or first percentile) and 50% (Q50% or 50th percentile) duration for the selected 20 hydrological gauging stations and they are presented together with the longitudinal cross section of the river (Fig. 3). Characteristic discharge values present the information about median discharges (Q50%) and high flow discharges (Q1%) in selected gauging station. Frequency analysis is conducted for seven gauging stations along the Drava-Mur Rivers, seven stations on the Sava River and five stations along the Danube River. The hydrological gauging stations have been selected according to two criteria. The first criterion is data availability, where only stations with more than 90% of available daily discharge data during the period 2000-2019 were selected. The second criterion is the spatial coverage of the river with hydrological stations, the intention being to select the most upstream station in Croatia and stations that are approximately equidistant from each other. The hydrological stations without discharge data were excluded from the analysis. The locations of hydrological stations are shown on the Danube River basin map (Fig. 2).

Characteristic discharges for the period 2000–2019 along the Sava River for seven analysed hydrological gauging stations (Podsused, Zagreb, Crnac, Jasenovac, Mačkovac, Slavonski Brod, Županja) show values of median discharges (Q50%) between 200 and 980 m^3/s, while flood discharges (Q1%) range between 1,400 m^3/s upstream from Zagreb and the 3,300 m^3/s downstream in Županja (Fig. 3a). The range of characteristic discharges (Q1% and Q50%) for the period 2000–2019 along the Drava-Mur River is shown for three hydrological gauging stations on the Mur River: Gornja Radgona, Mursko Središće and Goričan, and for four on the Drava River: Botovo, Terezino polje and Donji Miholjac. Median discharge (Q50%) is

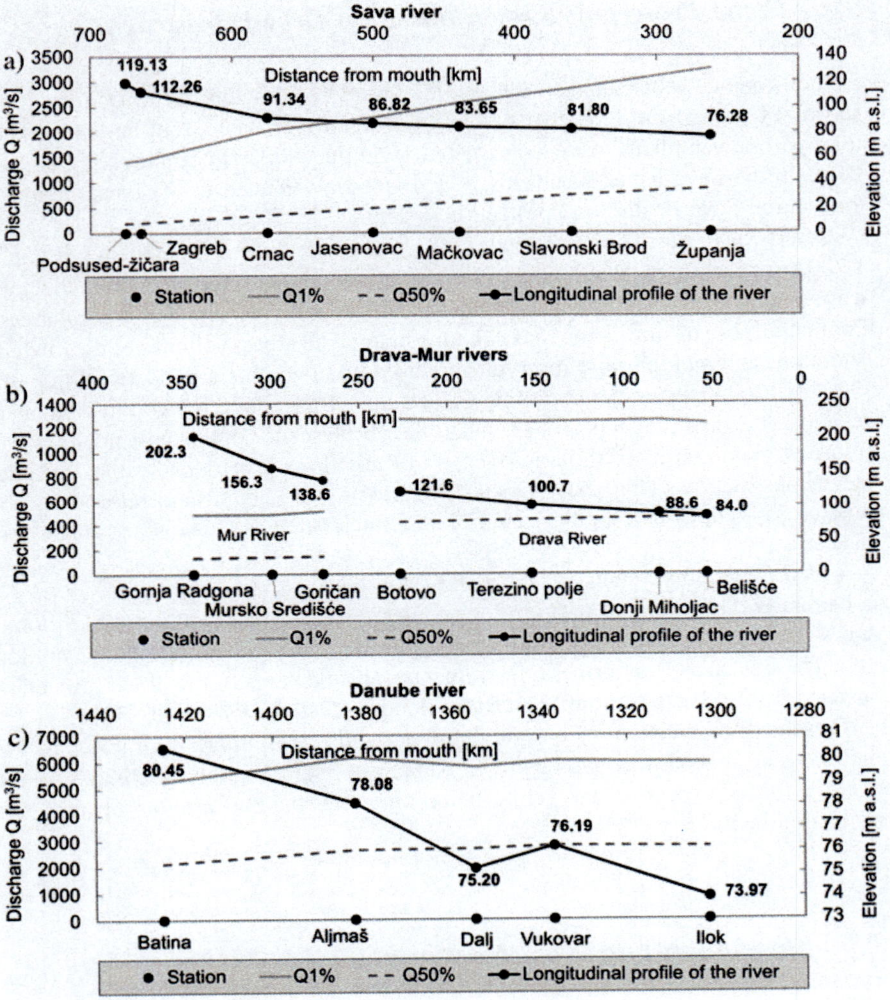

Fig. 3 Characteristic discharges with 1% (Q1%) and 50% (Q50%) duration in the period 2000–2019 at selected hydrological gauging stations on (**a**) Sava River, (**b**) Drava and Mur Rivers and (**c**) Danube River. Longitudinal profiles are also presented for each river

around 100 m^3/s along the Mur River section and between 400 and 500 m^3/s along the Drava River section. Average flood discharges (Q1%) are around 500 m^3/s on the Mur River and between 1,200 and 1,300 m^3/s on the Drava River section downstream from the confluence with the Mur River (Fig. 3b). Characteristic discharges (Q1% and Q50%) along the five hydrological gauging stations Batina, Aljmaš, Dalj, Vukovar and Ilok for the period 2000–2019 are represented for the Danube River. The range of the median discharges (Q50%) is from 2,100 to 2,800 m^3/s, while flood discharges (Q1%) are in the range from 5,200 to 6,100 m^3/s (Fig. 3c).

3.2 *Flood Protection System Based on Grey Infrastructure*

Several major water management facilities were constructed in the Danube River Basin in Croatia from 1972 to 1990. In the Drava River Basin, four large reservoirs for hydropower plants were constructed (Formin, Varaždin, Čakovec, Dubrava) while in the Sava River Basin this period is mainly related to the construction of the Central Sava Basin flood protection system.

Central Sava Basin flood protection system, besides the constructed dikes (around 1,200 km), uses three lateral relief channels (around 530 km) for evacuating the flood wave volume surplus upstream of the protected areas (urban and agricultural areas). For the purpose of storing the flood volume in lowland areas, volumes of the lowland retentions and multipurpose reservoirs are used (total volume around 1.6 billion m^3) [32, 33] The biggest issues regarding Central Sava Basin flood protection system are related to the insufficient height as well as the obsolescence of the existing dikes. So, future works in terms of the improvement of this flood protection system consider the reconstruction of the existing dikes (increasing the height and reinforcing the dike cross section) and the flow control structures (such as sluice gates and weirs) [34, 35].

Flood protection system on Drava and Danube Rivers sub-basins is based on the constructed dikes (around 650 km long), lateral relief canals (around 60 km long) and storage volume of the existing reservoirs (total volume around 164.52 million m^3). The issues, as well as the improvement of the elements of this flood protection system, are similar to Central Sava Basin flood protection system [36, 37].

Presented information about "grey" infrastructure are important in terms of the hydrological analysis that will be presented in Sect. 3.4, regarding the comparison of historical flood events for the period before and after the construction of the flood protection facilities.

3.3 *Floodplains and Protected Areas*

Large water retention areas in Croatia are mainly located in the floodplains of large rivers in the Danube River basin along the Drava-Mur, Sava and Danube River sections. The extent of floodplains is over 2,800 km. The retention capacity of existing floodplains in the Danube River basin is estimated by Schwartz [10] to be between 2,640 and 5,900 million m^3 by combining data on the size of the retention area and the maximum average flood depth. Large wetlands, located mainly in the floodplains of the Danube River basin, are also extremely important for the conservation of biodiversity. Numerous carp ponds along the Sava, Drava and Danube Rivers form complexes of semi-natural wetland habitats important for the nesting

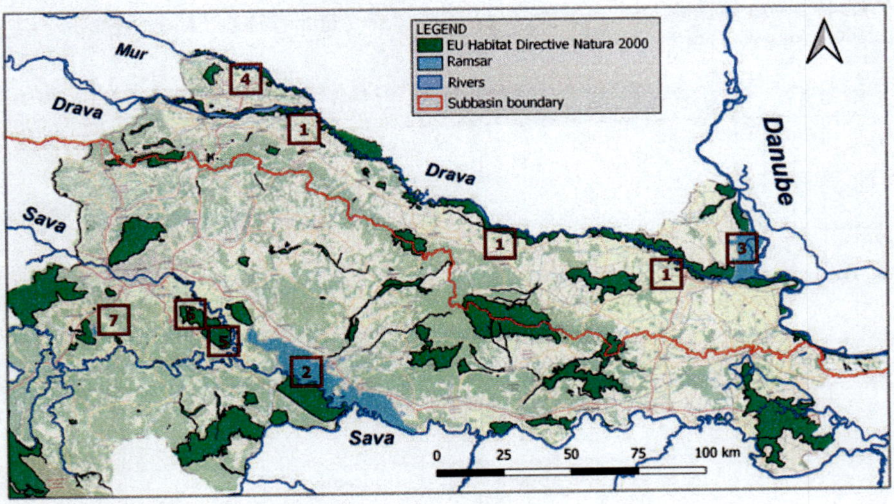

Fig. 4 Protected areas in the Danube River basin in Croatia (source of data www.bioportal.hr) labelled on topographic map with numbers 1–7. The description of the protected areas is provided in Table 2

and migration of wetland birds. Wetland complexes also host moist grasslands, shrubs and water-dependent forests, where a number of protected and indigenous species can be found, such as kockavica (*Fritillaria Meleagris*), willows and German tamarisk (*Salici-Myricarietum*). The majority of forest stands (95%) have a natural and indigenous composition of species, which is rare and extremely valuable worldwide. The large forests of pedunculate oak, hornbeam and ash complexes are located in the Sava River floodplains (Lonjsko polje and Spačva) while very valuable white willow marsh forests are located along the Drava and Danube Rivers (Kopački rit Nature Park) [38]. The majority of these areas are protected through status of nature parks, national parks, Ramsar sites and Natura 2000 sites. Seven protected areas designated as Natura 2000 sites cover 28% (2,558.07 km^2) of the Danube River basin, with three of them being Ramsar sites (Fig. 4, Table 2) [39]. The two largest wetlands are Lonjsko polje in the Sava River Basin and Kopački rit in the Drava-Danube confluence, and they account for 35% of the total volume capacity of floodplains [10].

The effectiveness of four largest natural floodplains and wetland areas in the Danube River basin on flood alleviation is presented in the coming sections for:

- Floodplains along the Drava and Mur Rivers (Drava-Mur Rivers).
- Lonjsko polje (Sava River).
- Special Zoological Reserve Kopački rit (Drava-Danube Rivers).
- The Mur River in the area of Međimurje county (Mur River).

Table 2 Natura 2000 protected areas and Ramsar sites in the Danube River basin in Croatia (source of data www.bioportal.hr)

Nr.	ID	River	Site	Category of designation	Area (ha)	Date of designation
1	466	Drava-Mur	Floodplains along the Drava and Mur Rivers	Regional park	87,449	2011-02-26
2	377	Sava	Lonjsko Polje and Mokro Polje	Ramsar	51,173	1990-03-28
3	327	Danube-Drava	Special zoological reserve Kopački rit	Ramsar	23,143	1977-01-02
4	439	Mur	The Mur River in the area of Međimurje county	Nature reserve	14,438	2001-04-18
5	461	Odra	Odransko Polje	Nature reserve	9,399	2006-07-25
6	456	Odra	Turopoljski lug and wet meadows along the Odra River	Nature reserve	3,344	2003-05-21
7	332	Kupa	Special ornithological reserve Crna Mlaka	Ramsar	694	1980-07-23

Special attention will be given to the description of two Ramsar sites of Kopački rit and Lonjsko polje which are the largest natural water retention areas in Croatia, and examples of efficient incorporation of protected wetlands for flood risk mitigation.

3.4 *Effectiveness of Floodplains and Wetland for Flood Alleviation*

The effectiveness of floodplains and wetlands on flood alleviation in the Danube River Basin was assessed through a progression of peak discharges and hydrograph shapes along the river sections. For this purpose, a total of five highest flood events along the rivers Mur, Drava, Sava and Danube were analysed in the two historical periods: before (1951–1970) and after (2000–2019) the construction of large water management systems in the Danube River Basin in Croatia (see Sect. 3.2). The patterns of peak discharges and hydrograph shapes for five events are presented for seven gauging stations along the Drava-Mur Rivers, for seven stations along the Sava River, and for five stations along the Danube River (see Fig. 2). The daily discharge data from the national hydrological network were provided by Croatian waters.

A longitudinal progression of average peak discharges from five events (dashed line) shows that the Mur and Drava floodplains have stronger effect on peak discharge alleviation when compared to the Danube River (Fig. 5). The Sava River average peak alleviation is evident up to Jasenovac. Such a pattern has not changed significantly after the period of construction of large schemes (after 2000).

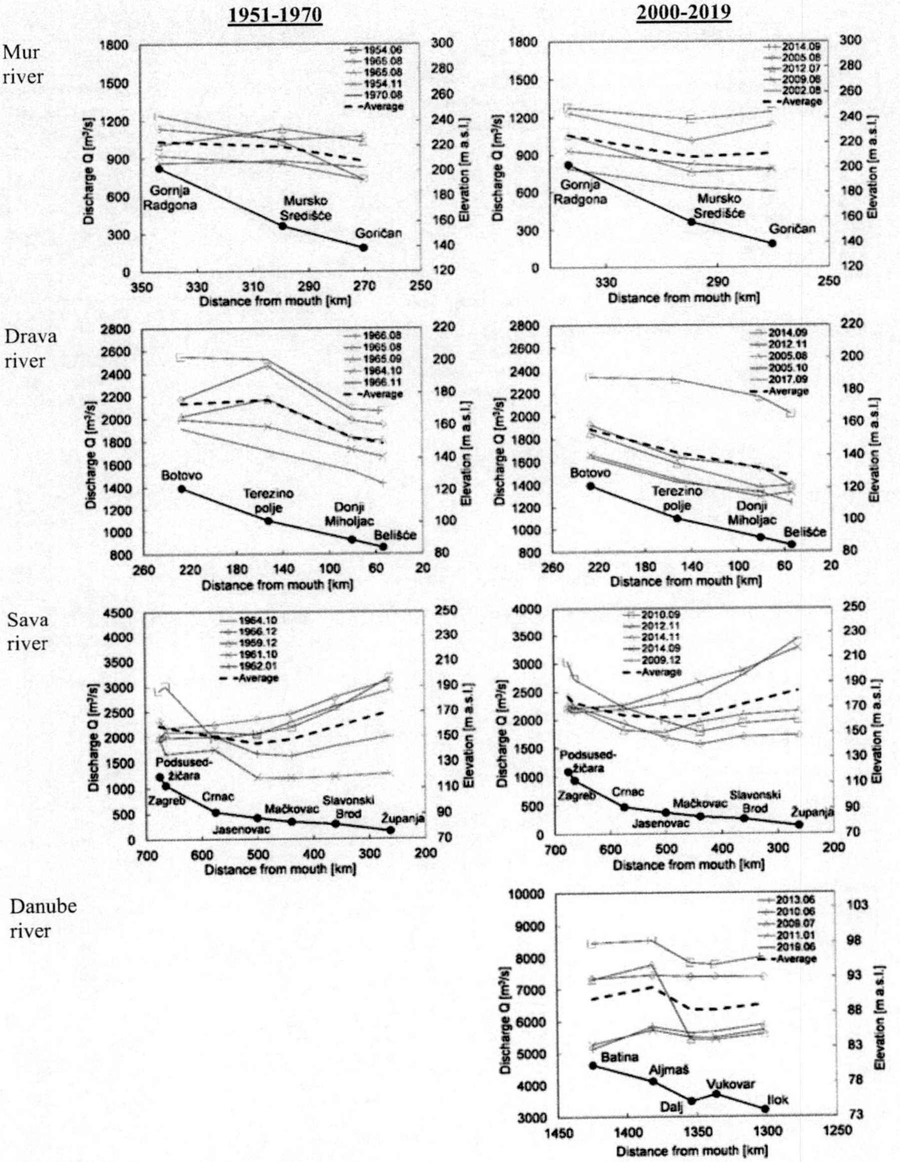

Fig. 5 Comparison of longitudinal distribution of peak discharges of five largest flood events in the two analysed periods (1951–1970 and 2000–2019) at the selected gauging stations along the river. Locations of hydrological stations (black circles) are shown on river longitudinal profile (black solid line)

Average peak discharges from five events (dashed line) in period 2000–2019 along the Danube River section show decrease by approx. 800 m^3/s after the Aljmaš station

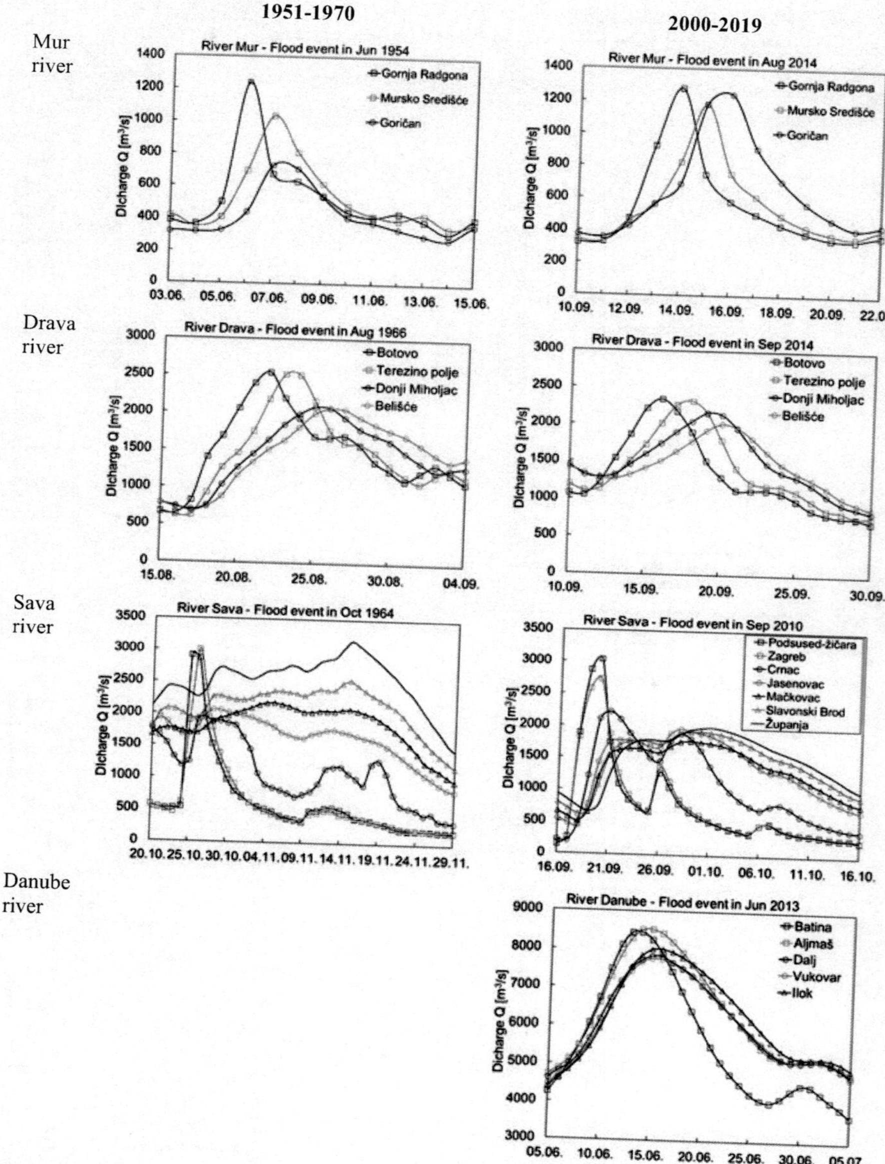

Fig. 6 Comparison of hydrographs in the two analysed periods 1951–1970 and 2000–2019 for the highest flood event at the selected gauging stations along the river (presented in Fig. 2)

(located downstream the River Drava confluence and *Kopački rit* floodplain area) (Fig. 5). The contribution of this Danube section was significant for the June 2013 flood event, during which significant damages were recorded in the upstream parts of the Hungarian and Austrian Danube [40]. Namely, the observed attenuation of peak **flows by approx. 1000 m³/s and a delay of several days (Fig.** 6), achieved by the floodplain areas located upstream from Aljmaš station, together with the Kopački rit wetland, ensured that no damage was recorded in the Danube section in Croatia [10].

A longitudinal progression of average peak discharges of five highest flood events, recorded for both analysed periods (1951–1970 and 2000–2019), showed a decrease in the peak discharge of approx. 300–400 m^3/s along the Sava River section. The peak discharges were reduced in the river section between the Podsused and Mačkovac hydrological stations (located downstream from the floodplains of Lonjsko polje retention areas) (see Figs. 5 and 6). This river section stretches along with the Central Sava Basin flood protection system and shows the contribution of the Sava River floodplains to flood protection of the central Sava River Basin area, both in the period of pre-construction and post-construction of the system. Contribution of these semi-regulated floodplains to the flood protection in the current state is especially visible when comparing highest flood events from October 1964 (disastrous flood event that initiated the design and construction of the Central Sava Basin system) and flood from September 2010. The flood wave from September 2010 was significantly reduced in its peak discharge by over 1,000 m^3/s, and the main settlements in the area were successfully protected.

4 Wetlands as NBS for Flood Risk Mitigation

Examples of two wetlands with significant flood retention capabilities are presented below: the semi-natural retention area of Lonjsko polje in the Sava River Basin, and the natural retention area of Kopački rit at the Drava-Danube confluence. A short description of the sites is given regarding the importance to natural alleviation of floods and to supporting of biodiversity.

4.1 *Lonjsko Polje and Central Sava Basin Flood Protection System*

The Lonjsko polje, a Nature park and Ramsar site, has a wider area ("green") incorporated into the Central Sava Basin flood protection system ("grey"). It presents a unique example on the European level of efficient application of NBS at the large scale for flood protection. The central Sava Basin flood protection system stretches along the Sava River and protects the central Sava River basin with a system of lowland near-natural retention areas, artificial relief canals, sluices, weir and embankments (Fig. 7).

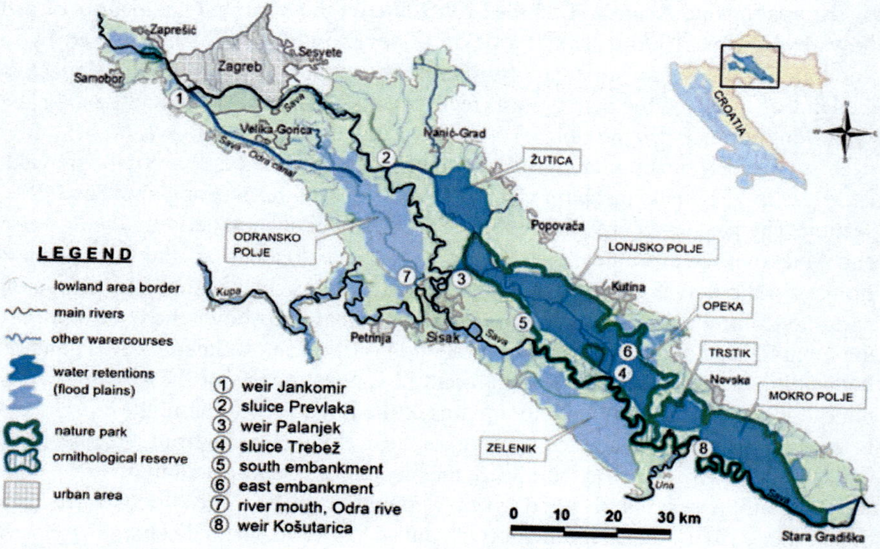

Fig. 7 Map with the main elements (1–8) of the Central Sava Basin flood protection system, including Lonjsko polje Nature Park (adjusted from [41])

A catastrophic flood recorded in the city of Zagreb in 1964 initiated the design and construction of the Central Sava Basin flood protection system [24]. The initial design for the Central Sava Basin flood protection system was proposed by the UN programme [42] (UNDO 1972) and it was based on using the existing floodplains of the Kupa and the Sava River for the retention and attenuation of flood waves. In the initial design of the system, existing dikes along larger watercourses were exploited, and 58,800 ha of flood storages with their total designed capacity of 1.8 billion m^3 were planned (Lonjsko polje, Mokro polje, Zelenik, Kupčina). Additionally, three artificial relief canals were planned. Capacity of the system was designed to provide protection from the 100-year flood, and for the urban areas from the 1,000-year flood. Such approach, based on mimicking the natural flooding process with incorporation of floodplains, was unique at that time [41].

After a lengthy period of construction, during the 1970s and 1980s, main elements of Lonjsko polje retention area were constructed, as well as a smaller section of the artificial relief canal with accompanying dikes and two sluices, Prevlaka and Trebež [41]. In the 1990s there was a delay in the construction and financing due to the war and a partially constructed system was used for the flood defense. Although only 40% of the initially planned works were carried out by then, flood protection effects were great. The initial project, based on the artificial relief canals and many flow control structures, was redesigned at the beginning of the 2000s to adjust for the newly emerged principles of integrated and sustainable water management. This new principle needed to consider the influence of built hydrotechnical structures on environment and biodiversity. A newly developed solution was founded on the functionality of the partially built system and on the expansion of existing floodplain retention areas. The important change from the initial design was related to the less strictly controlled flow process aiming to maintain the natural water regime. One part of this solution was achieved by utilizing existing river networks for water transport and abandoning solution with the costly artificial relief canal. Since designed flood storages still did not have fully constructed dikes, existing floodplain of Lonjsko polje could be enlarged by approx. 7,000 ha. Previously mentioned changes, together with enlargement of the retention areas and free inflow in Mokro polje retention area, resulted in lowered maximal water depth and flow conditions similar to the near-natural state [41].

The main upstream elements of the system reduce the peak discharge of 3,600 m^3/s (1000-year flood) from the Sava River near Zagreb urban area by 1,000 m^3/s. Additionally, discharge of 500 m^3/s from the Sava River can enter the Lonjsko polje and Mokro polje areas where water can be retained for several weeks before it is gradually discharged back to the Sava River [43, 44]. The system is partly completed (see Sect. 3.2.) but has proved to be very effective in the recent years in protecting important towns, such as Zagreb and Sisak, as well as large agricultural areas from flooding with a capacity to redirect 1,500 m^3/s from the Sava River. The effectiveness of flood risk reduction was sustained by hydraulic modelling and field monitoring [16]. With the retention capacity for a flood volume of 1.6 billion m^3 over the extent of 112,000 ha, this is the largest floodplain ecosystem in the Danube River Basin and an important nutrient sink for the Upper and Central Sava Basin [33, 45].

Flood protection based on water retention in lowland areas enables the maintenance of ecologically favourable conditions on floodplains. Due to its exceptional natural values, part of the Central Sava Basin protection system has been declared a Nature Park. The Lonjsko Polje Nature Park covers the largest retention area of the Central Sava basin system with over 50,000 ha in extent and with a retention capacity of 600 million m^3. Lonjsko Polje is one of the rare preserved complex wetlands in Europe [10, 17]. It is a highly representative example of an extensive river floodplain covered by a mixture of alluvial forests, wet grasslands, watercourses, oxbows and other wetland habitats (Figs. 8 and 9). Seminatural forests are managed in such a way that they contain very rich biodiversity, including several rare and threatened species at the European or even the global level. Due to its ecological characteristics, Lonjsko polje plays a very important role in the natural functioning of the Sava River as well as of the whole Danube River basin [46].

The appearance of bird species, primaeval forest structures and large undisturbed woodland complexes indicates the well-preserved nature of the recent riparian forest ecosystem in the Lonjsko Polje Nature Park. The combination of habitats and their regular flooding as well as high groundwater levels provide conditions for rich biodiversity [38]. About 250 bird species are recorded in the Nature park representing more than two-thirds of those found in Croatia [47]. More than 110 different animal species [48] and more than 550 different plant species have been recorded [49, 50]. Part of the area is an internationally protected Ramsar site since 1993 which is on the list of internationally Important Bird Areas since 1989

Fig. 8 Landscape in Žutica retention (left) and Lonjsko polje retention area, part of Ramsar site (right), both within the Central Sava basin flood protection system (photo by Maja Pintar & Vedrana Ričković)

Fig. 9 Traditional grazing of livestock in the Lonjsko polje supports site-specific plant and animal species (photo by Vedrana Ričković)

[51]. The wetland is of cultural and ecological importance. Vegetation in retention area is being sustained by the traditional farming and traditional grazing of livestock (Fig. 9) [38]. Retention areas in Lonjsko polje are important for the water purification processes from streams and in maintaining high water quality standards [46].

4.2 *Kopački Rit*

Kopački rit is an inland delta at the confluence of the Danube and Drava Rivers, which constitutes the last remaining large natural floodplain area along the entire upper reach of the Croatian Danube. The area is a geomorphological depression of triangular shape used for flood retention since the nineteenth century. Today, with its floodable area of approximately 20,000 ha and with water depths of up to 5 m, it can store water up to 1.0 billion m^3, effectively protecting the city of Osijek and buffering the flood wave downstream the Danube [10]. Kopački rit is also one of the largest (around 23,000 ha area) and the best-preserved riparian wetland in Croatia. It was protected by law in 1967, along with the classification as a nationally valuable area, as well as being included in the List of Wetlands of International Importance under the Ramsar Convention, in 1993 (see Table 2) [52].

Floods in the Osijek urban area are mostly caused by the Danube River backwaters, and Kopački rit retention capacity together with the dikes contributes to its protection. Hydrological analysis of Kopački rit determined the main directions of water into inundation area during the flooding. The area is mainly flooded from the Danube River, and only partly from the Drava River (Fig. 10) [53]. Located within the floodplain of the Danube and Drava Rivers, Kopački rit can be divided into two distinctive areas: (1) central floodplain area that extends from the River Danube to the dikes on both sides of the river (1–15 km wide) and (2) former floodplain that extends from the east and west dikes right to the geological borders of the floodplain and forms a buffer zone around the central floodplain. Hydrology and natural features of the area were modified by construction of artificial drainage system in the nineteenth and twentieth century, while the western parts were converted into arable land or reclaimed for construction. Central part of floodplain is the most important part of Kopački rit concerning its wetland functions and values. The area is flooded over a month per year in the higher parts of the former floodplain area, and up to 3 months in the lower parts, usually from March to May. Water starts to enter the floodplain when the Danube's water level at Apatin gauging station reaches 300 cm. During the year, the water level recorded on this station fluctuates 5–7 m on average, while the maximum-recorded fluctuations have been 9.40 m [52, 53].

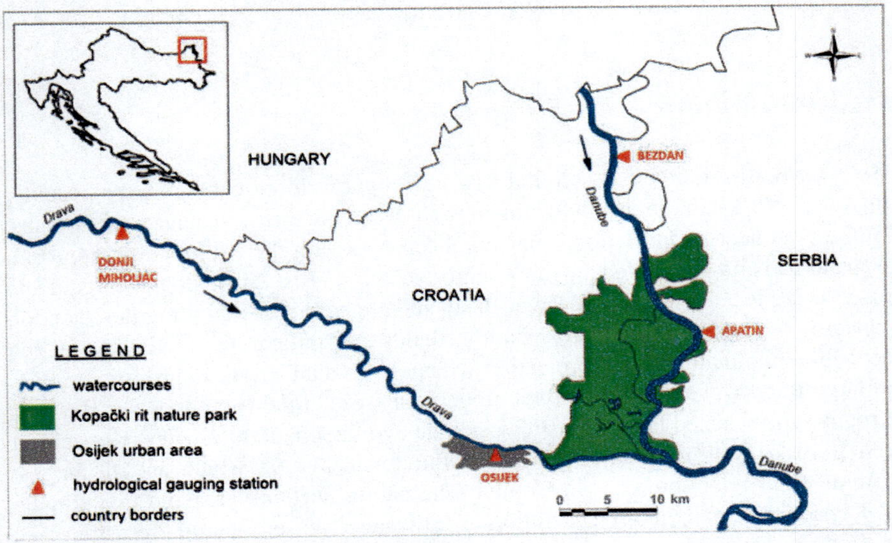

Fig. 10 Kopački rit Nature Park retention area located at the confluence of the Drava and the Danube Rivers (adjusted from [52])

Fig. 11 Water retention areas in Kopački rit (photo by Maja Pintar)

Kopački rit Nature Park landscape consists of lakes, marshes, wet grasslands, reed beds, riverine forests, numerous channels, oxbow lakes and fishponds (see Fig. 11). The interaction of water and land generates high biological diversity with more than 2000 species registered in Kopački rit Nature Park (over 460 vascular plants, 300 birds, 55 mammals, 55 fish, 11 amphibians and 10 reptile species) [53]. The main hydrological values of Kopački rit wetlands are related to flow regulation and flood control, bio-chemical/physical purification of waters, groundwater recharge, as well as sedimentation and nutrient retention capacity. Water quality within the recent floodplain of Kopački rit depends on the Danube River's water quality, while in the former floodplain area it is affected by point and non-point sources of pollution (animal farms, settlements, runoff pollution from arable land) [54].

5 Conclusion

Nature-based solutions as a relatively new approach in Croatia were recently included in the national legal documents and in the flood risk management policies. But, the pioneering large flood protection schemes with the nature-based approach were already implemented in the country in the 1970s. Most of the large scale projects are located in Danube River Basin and are based on exploiting the retention capacity of existing floodplains and wetlands for mitigating flood waves. The contribution of natural flood protection schemes was analysed through a progression of large flood waves along the three floodplain areas in the Danube River Basin in Croatia. The selected floodplain areas include two famous Ramsar sites (Kopački rit and Lonjsko polje) and the Drava-Mur floodplain system which are the largest natural water retention areas in Croatia. The natural wetland area and inland delta of Kopački rit can store up to 1 billion m^3 of water. The semi-natural retention area of Lonjsko Polje Nature Park together with the Central Sava Basin flood control system, with capacity of 1.6 billion m^3, is one of the most prominent flood control and retention systems in Europe. The analysis of flood wave progression was provided for five highest flood events in the two periods before (1951–1970) and after (2000–2019) the construction of major water management facilities. The analysis showed that all floodplain areas contribute to flood hazards mitigation in downstream sections for both pre- and post-construction periods. For the longitudinal distribution of average peak flows it is evident that the large floodplain systems, such as Drava-Mur, may have larger spatial consistency in the alleviation of major flood waves. The large semi-natural flood protection scheme in the Sava River (Lonjsko Polje and Central Sava Basin) proved significant reduction of peak flows (decrease of 1,500 m^3/s on the Sava River during several weeks) for the largest flood event in the post-construction period (September 2010 flood). Besides flood alleviation, the floodplain areas are mimicking near-natural conditions and thus support preservation of biodiversity and provision of additional ecosystem services. They are positive examples of efficient incorporation of protected wetlands for the mitigation of flood risk while still providing outstanding biological and landscape diversity. The presented examples can serve as a good practice for the implementation of nature-based solutions for flood mitigation projects, both in Croatia and internationally. The benefits of this "green" nature-based solution approach in comparison to the "grey" engineering approach lie in recognizing and considering additional ecosystem services that are provided by wetlands: sediment and nutrient retention and export, groundwater replenishment, water purification, climate change mitigation & adaptation, and recreation & tourism [46].

Acknowledgments We are grateful for the hydrological time series and data provided by Croatian Waters. This work has been supported by the Croatian Science Foundation under the project number DOK-2020-01.

References

1. European Environment Agency (EEA). Economic losses from climate-related extremes. https://www.eea.europa.eu/data-and-maps/indicators/direct-losses-from-weather-disasters-3/assessment-1. Accessed 10 Sep 2018
2. Jongman, B., Hochrainer-Stigler, S., Feyen, L., Aerts, J. C., Mechler, R., Botzen, W. W., ., Bouwer, L. M., Pflug, G., Rojas, R., & Ward, P. J., 2014. Increasing stress on disaster-risk finance due to large floods. Nat Clim Chang, 4(4), 264–268
3. Bonacci O (2016) River–the bloodstream of landscape and catchment. Acta Hydrotech 29 (50):1–12
4. WWAP (United Nations World Water Assessment Programme)/UN-Water (2018) The United Nations world water development report 2018: nature-based solutions for water. UNESCO, Paris
5. Koseoglu N, Moran D (2014) Review of current knowledge. Demystifying natural water retention measures (NWRM). Foundation of Water Research, Marlow, Bucks pp 3–15
6. Bekić D, Halkijević I, Gilja G, Lončar G, Potočki K, Carević D (2019) Examples of trends in water management systems under influence of modern technologies. Građevinar 71(10):833–842

7. Bonacci O (2016) Natural water retention measures. Hrvatske vode: časopis za vodno gospodarstvo 24(96), 161–169. (in Croatian)
8. European NWRM platform. http://nwrm.eu/. Accessed 31 Jul 2020
9. Reguero BG, Beck MW, Bresch DN, Calil J, Meliane I (2018) Comparing the cost effectiveness of nature-based and coastal adaptation: a case study from the Gulf coast of the United States. PLoS One 13(4):e0192132
10. Schwarz U, Pokrajac S, Bockmühl K, Stolpe G (2018) Nature-based solutions for flood risk prevention in South-Eastern Europe. BfN-Skripten 511. Bundesamt für Naturschutz. https://www.bfn.de/fileadmin/BfN/service/Dokumente/skripten/Skript511.pdf. Accessed 15 Jun 2020
11. Croatian River Basin Management Plan (CRBMP), 2016–2021 http://www.voda.hr/sites/default/files/plan_upravljanja_vodnim_podrucjima_2016._-_2021.pdf. Accessed 10 Aug 2020
12. Zaninović K, Gajić-Čapka M, Perčec Tadić M, Vučetić M, Milković J, Bajić A, Cindrić K, Cvitan L, Katušin Z, Kaučić D, Likso T, Lončar E, Lončar Ž, Mihajlović D, Pandžić K, Patarčić M, Srnec L, Vučetić V (2008) Climate atlas of Croatia: 1961-1990. 1971–2000. Državni hidrometeorološki zavod, Zagreb
13. Kuspilić N, Oskoruš D, Vujnović T (2014) Simple truth – extreme hydrological event. Građevinar 66(7):653–661. (in Croatian)
14. Đuroković Z (2014) Provedba mjera obrane od poplava u republici hrvatskoj. Konferencija Lokalna uprava u zaštiti od poplava. Zagreb (4.11.2014)
15. International Sava River Basin Commission (ISRBC) (2010) Sava River Basin analysis summary. http://www.savacommission.org/dms/docs/dokumenti/documents_publications/publications/sava_river_basin_analysis_-_summary/sava_booklet_eng.pdf. Accessed 5 Jul 2020
16. Baptist MJ, Haasnoot M, Cornelissen P, Icke J, Van Der Wedden G, De Vriend HJ, Gugic G (2006) Flood detention, nature development and water quality along the lowland river Sava, Croatia. In: Living Rivers: trends and challenges in science and management. Springer, Dordrecht, pp 243–257
17. UN (2018) The United Nations world water development report 2018: natural-based solutions for water. Paris. https://reliefweb.int/sites/reliefweb.int/files/resources/261424e.pdf. Accessed 20 Jun 2020
18. European Community (2007) Directive 2007/60/EC of the European Parliament and of the Council of 23 October 2007 on the assessment and management of flood risks. Official Journal of the European Community, L288/27
19. Official Gazette of the Republic of Croatia, 66/19. Water Act. https://www.zakon.hr/z/124/Zakon-o-vodama
20. European Community (2000) Directive 2000/60/EC of the European parliament and of the council establishing a framework for community action in the field of water policy. Official Journal of the European Community, L327
21. Official Gazette of the Republic of Croatia, 127/19. Climate Change Adaptation Strategy in the Republic of Croatia until 2040 with an outlook to 2070
22. Hrvatske vode/Croatian waters (2009) Strategija upravljanja vodama. Croatian Water Management Strategy
23. Čanjevac I (2012) Promjene i tipologija režima protoka rijeka u Hrvatskoj. PhD thesis, University of Zagreb
24. Bonacci O, Ljubenkov I (2008) Changes in flow conveyance and implication for flood protection, Sava River, Zagreb. Hydrol Process Int J 22(8):1189–1196
25. Trninić D, Bošnjak T (2009) Characteristic discharges of the Sava river at Zagreb. Hrvatske vode 17(69/70):257–268
26. Potočki K, Kuspilić N, Oskoruš D (2013) Wavelet analysis of monthly discharge and suspended sediment load on the river Sava. In: Thirteenth international symposium on water management and hydraulic engineering-proceedings, Slovak University of Technology, Bratislava, pp 615–623
27. Orešić D, Čanjevac I, Maradin M (2017) Changes in discharge regimes in the middle course of the Sava River in the 1931-2010 period. Prace Geograficzne 151

28. Babić-Mladenović M, Bekić D, Grošelj S, Kupusović T, Mikoš M, Oskoruš D (2013) Towards practical guidance for sustainable sediment management using the Sava River basin as a showcase, estimation of sediment balance for the Sava River, International Sava River Basin Commission
29. Schwarz U (2007) Pilot study: hydromorphological survey and mapping of the Drava and Mura Rivers. FLUVIUS, Vienna
30. Tadić L, Brleković T (2019) Hydrological characteristics of the Drava River in Croatia. In: The Drava River. Springer, Cham, pp 79–90
31. Bonacci O, Oskoruš D (2010) The changes of the lower Drava River water level, discharge and suspended sediment regime. Environ Earth Sci 59(8):1661–1670
32. Hrvatske vode/Croatian waters (2019) Prethodna procjena rizika od poplava 2018
33. Brundić D, Barbalić V, Omerbegović V, Schneider-Jacoby M, Tušić Z (2001) Alluvial wetlands preservation in Croatia – The experience of the Central Sava Basin flood control system. In: Nijland HJ, Cals MJR (eds.) River restoration in Europe, practical approaches, proceedings of the conference on river restoration 2000 - July 17, 2000, RI ZA rapport nr.:2001.023, pp 109–118
34. Hrvatske vode/Croatian waters (2014c) Provedbeni plan obrane od poplava branjenog područja, Sektor C – Gornja Sava, Branjeno područje 14: Središnji dio područja malog sliva Zagrebačko Prisavlje
35. Hrvatske vode/Croatian waters (2014d) Provedbeni plan obrane od poplava branjenog područja, Sektor D – Srednja i Donja Sava, Branjeno područje 10: Područje maloga sliva Banovina
36. Hrvatske vode/Croatian waters (2014) Provedbeni plan obrane od poplava branjenog područja, Sektor A – Mura i Gornja Drava, Branjeno područje 33: Međudržavne rijeke Drava i Mura na područjima malih slivova Plitvica-Bednja, Trnava i Bistra
37. Hrvatske vode/Croatian waters (2014) Provedbeni plan obrane od poplava branjenog područja, Sektor B – Dunjav i Donja Drava, Branjeno područje 34: Međudržavne rijeke Drava i Dunav na područjima malih slivova Baranja, Vuka, Karašica-Vučica i županijski kanal
38. Hrvatske vode/Croatian waters (2013) Prethodna procjena rizika od poplava Republika Hrvatska: vodno područje rijeke Dunav i jadransko vodno područje
39. Lieb GK, Sulzer W (2019) Land use in the Drava Basin: past and present. In: The Drava River. Springer, Cham, pp 27–43
40. The International Commission for the Protection of the Danube River (ICPDR) (2014) Countries of the Danube River Basin. http://www.icpdr.org/main/danubebasin/countries-danube-river-basin. Accessed 27 Sep 2020
41. Petričec M, Filipović M, Kratofil L, Šurlan S, Tusić Z (2004) Toward integrated water management in the middle Sava Basin. In: Third European conference on river restoration, Zagreb, Croatia, pp 17–21
42. UNDO (1972) Study for regulation and management of the Sava River in Yugoslavia. United Nations Development Office, Consortium Polytechna-Hydroprojekt-Carlo Lotti & C. Prag-Roma
43. Schwarz U (2016) Sava white book. The river Sava: threats and restoration potential. Radolfzell/Wien: EuroNatur/Riverwatch. Accessed 15 Jul 2020
44. VPB (2012) Sustav obrane od poplava Srednjeg Posavlja. Aktualizacija rješenja. Vodoprivredno-projektni Biro (VPB). Zagreb
45. DPRP - Danube Pollution Reduction Programme (PCU UNDP/GEF) (1999) Evaluation of Wetlands and Floodplain Areas in the Danube River Basin, Final Report, May 1999 prepared by WWW Danube Carpathian Programme and WWF-Auen-Institut
46. Secretariat of the Convention on Wetlands (n.d.) Ramsar sites information service. https://rsis.ramsar.org/. Accessed 12 Oct 2020
47. Mužinić J (1996) The preservation of ornithofauna of wetlands in Croatia (in Croatian). Hrvatska vodoprivreda 5:38–49

48. Schneider-Jacoby M (1990) The distribution of and threats facing the typical waterplant species in the region of the proposed Lonjsko Polje Nature Park. Acta Bot Croat 49:125–136
49. Gugić G (2012) Croatia: the floodplain ecosystem of the Central Sava River basin. In: protected landscapes and wild biodiversity. IUCN, pp 19–26
50. Nikolić T (2013) Flora Croatica database. Department of Botany and Botanical Garden, Faculty of Science, University of Zagreb, p 1
51. Karadžić B, Jarić S, Pavlović P, Mitrović M (2015) Aquatic and wetland vegetation along the Sava River. In: The Sava River. Springer, Berlin, pp 249–316
52. Tadić L, Bonacci O, Dadić T (2016) Analysis of the Drava and Danube rivers floods in Osijek (Croatia) and possibility of their coincidence. Environ Earth Sci 75(18). https://doi.org/10.1007/s12665-016-6052-0
53. Tadić L, Bonacci O, Dadić T (2014) Dynamics of the Kopački Rit (Croatia) wetland floodplain water regime. Environ Earth Sci 71(8):3559–3570
54. Elektroprojekt Plc (2003) Park prirode Kopački rit plan upravljanja "Prijedlog Plana upravljanja Parkom prirode Kopački rit", Ministarstvo zaštite okoliša, prostornog uređenja i graditeljstva

Assessment of NBS Impact on Pluvial Flood Regulation Within Urban Areas: A Case Study in Coimbra, Portugal

Luís Valença Pinto, Paulo Pereira, Milan Gazdic, António Ferreira, and Carla S. S. Ferreira

Contents

L. V. Pinto
Research Centre for Natural Resources, Environment and Society (CERNAS), Polytechnic Institute of Coimbra, Coimbra Agrarian Technical School, Coimbra, Portugal

Environmental Management Laboratory, Mykolas Romeris University, Vilnius, Lithuania
e-mail: lmpinto@esac.pt

P. Pereira
Environmental Management Laboratory, Mykolas Romeris University, Vilnius, Lithuania

M. Gazdic
Department of Forestry, University of Belgrade, Belgrade, Republic of Serbia

A. Ferreira
Research Centre for Natural Resources, Environment and Society (CERNAS), Polytechnic Institute of Coimbra, Coimbra Agrarian Technical School, Coimbra, Portugal
e-mail: aferreira@esac.pt

C. S. S. Ferreira (✉)
Research Centre for Natural Resources, Environment and Society (CERNAS), Polytechnic Institute of Coimbra, Coimbra Agrarian Technical School, Coimbra, Portugal

Department of Physical Geography, Stockholm University, and Bolin Centre for Climate Research, Stockholm, Sweden

Navarino Environmental Observatory, Messinia, Greece

Carla S. S. Ferreira, Zahra Kalantari, Thomas Hartmann, and Paulo Pereira (eds.),
Nature-Based Solutions for Flood Mitigation: Environmental and Socio-Economic Aspects,
Hdb Env Chem (2022) 107: 289–312, DOI 10.1007/698_2021_769,

Published online: 6 May 2021

Abstract The majority of the world population is living in urban areas. As cities expand, soil sealing increases the vulnerability of urban areas to pluvial floods, and the consequent impacts on social and economic domains. Flood mitigation typically relies on grey infrastructures, but the implementation of Nature-Based Solutions (NBS) can be critical to cope with increasing flood hazard driven by urbanization and climate change. By mimicking natural hydrological processes, NBS enhance water retention, infiltration and evapotranspiration through greening, leading to lower runoff and flood hazard. The effectiveness of NBS on flood mitigation is affected by several factors including the type of NBS and the biophysical characteristics of the area. Nevertheless, a relatively limited number of studies have monitored the impact of NBS, and thus the lack of knowledge is still a barrier to the widespread implementation of this approach. This chapter assesses the impact of a Green Infrastructure (GI) located in Coimbra (Portugal), which performs as a NBS for runoff management and flood hazard mitigation. The study applies the widely used Curve Number method to estimate runoff within the *Quinta de São Jerónimo* study site, driven by rainfall events of 2-, 5-, 10- and 20-years recurrence, based on Intensity–Duration–Frequency precipitation curves. The results show that the implemented NBS can retain runoff produced by 20-years flood, decreasing the flood peak and flood hazard in downstream urban areas. This efficiency is achieved by combining blue, green and grey elements, and proved useful to enhance urban resilience. Furthermore, the green and blue elements of the NBS provide additional ecosystem services, including environmental, social and economic benefits (co-benefits), relevant for human well-being in urban areas.

Keywords Co-benefits, Green infrastructure, Nature-based solutions, Pluvial floods, Runoff management, Urban areas

1 Introduction

Urban areas encompass over half of the world's population [1] and are expected to embrace 70% of the population by 2050 [2]. Urbanization enhances soil sealing with impervious materials (e.g. concrete, asphalt or buildings). In 2006, sealed soils covered 2.3% of the European Union [3]. Sealing is one of the main problems associated with sustainable urban development [4], given, for example, the potential impacts on the hydrological cycle [5]. Expanding impervious surfaces reduce evapotranspiration (although few studies show small increases [6, 7]), decrease infiltration rates, and increase stormwater runoff, thus enhancing the susceptibility to floods [5, 8–10].

To mitigate flood hazard driven by urbanization, hydrologic flows are generally shifted to a complex series of drains, pipes, and other grey infrastructures, designed to facilitate the centralized collection of stormwater and quickly divert it away from the urban areas [11]. These traditional systems often produce unintended consequences, such as changes in the hydrological behaviour and increase pollutant concentrations [12, 13], which affect the urban water quality [14]. Nevertheless, even with these drainage systems, high-intensity rains may trigger low-grade flooding of streets, homes, and basements, causing economic losses, adverse physical and mental problems, and amplification of social inequalities [15]. Changes in precipitation patterns associated with more extreme events (e.g. intensity and frequency of rainfall) driven by climate change, coupled with urbanization trends, will exacerbate cities' vulnerability to flooding [16]. Since grey infrastructures are typically dimensioned for specific volumes of water, often not considering realistic urbanization rates or the impact of climate change, additional solutions are required to enhance urban adaptation and resilience [17].

Over the last decades, urban water drainage management options changed substantially, moving from an approach primarily focused on grey infrastructures to a multifunctional one, based on engineered green/ecological systems which mimic the natural hydrological cycle [18]. This nature-based solutions (NBS) approach aims to restore pre-development flow-regimes within urban catchments and address the degradation of urban water quality [19].

Green Infrastructures (GIs), defined as "*a strategically planned network of natural and semi-natural areas with other environmental features designed and managed to deliver a wide range of ecosystem services*" [20], are at the very heart of NBS approach [21]. GI aims to increase the cover of permeable surfaces to maximize infiltration and water storage capacity of the soil, retain surface runoff near its source, and slow water transfer downslope. This will delay flood peaks and alleviate urban drainage systems [14, 22, 23]. In this context, GI can be understood as an operationalization of NBS [24]. Urban GI includes diverse types of green and blue spaces, such as public parks, community gardens, bioswales, dry ponds and wetlands [25–27]. In the literature and practice, however, a wide range of terms referred to similar GI applications have been applied, such as Sustainable Drainage Systems, Low Impact Development, and Sponge Cities [21, 28, 29]. These NBS range from solutions with low human intervention to solutions involving the creation of new ecosystems [30], as well as solutions considering a combination of green and grey infrastructures (hybrid solutions) [24, 30, 31].

NBS for stormwater management have been studied extensively by engineers and urban planners [15], and have become popular in several countries to mitigate urban floods [32]. Numerical hydrologic and hydrodynamic models have been widely used to select and design stormwater management strategies, such as the Storm Water Management Model [33], and the Model for Urban Stormwater Improvement Conceptualization (MUSIC) [34]. However, these useful tools for planning purposes often lack the details needed to consider site-specific aspects [35]. Field studies have shown that NBS performance can be highly dependent on their design, implementation aspects, and local biophysical aspects, including the intensity and duration of rainfall events [36].

NBS proved to be effective in managing runoff [37] and efficient in substituting grey infrastructures such as dikes or levees [30, 38, 39]. They are effective and flexible strategies to tackle climate change and enhance urban resilience [40–42], and often less cost-effective when compared to grey options [43]. Although literature provides evidence on the positive impacts of NBS on water management, most studies are based on qualitative assessments [44]. Thus, the lack of evidence-based knowledge of NBS effectiveness, developed upon monitoring data from implemented solutions, represents one of the major barriers for a wider implementation of this approach [14]. Nevertheless, NBS is an effective way to increase the greening in urban environments and to provide a wide range of ecosystem services (co-benefits are driven by several ecological, social and economic functions), relevant to promote the well-being of residents [27, 45, 46]. These co-benefits must be taken into consideration when assessing NBS effectiveness [21].

This chapter aims to assess the impact of an NBS on stormwater regulation and mitigation of pluvial floods in urban areas. The NBS investigated includes green and blue elements, coupled with grey elements, designed and implemented as a mandatory requirement for the approval of an extensive urbanization project implemented in Coimbra, Portugal, where pluvial floods are recurrent. The effectiveness of the NBS on flood mitigation is based on the comparison of runoff estimates for several recurrent floods (2, 5, 10 and 20 years) and the water retention capacity of the NBS, using widely accepted methods. In addition, this study explores the co-benefits provided by the NBS, in order to provide a holistic evaluation of the NBS approach used by local authorities to enhance urban resilience.

2 Flood Management in Coimbra and Green Infrastructures

2.1 Location and Characterization of the Urban Areas

Coimbra is the largest city in the Portuguese Centre region (Fig. 1a). The municipality of Coimbra (319 km^2) accommodates a population of 143,397 inhabitants [47]. The urban perimeter (Fig. 1b), including all urban and urbanizable spaces, covers 16% of the municipality surface area and comprises over 64% of its population [48]. Coimbra's urban consolidated area (Fig. 1b), designated as city core and considering stabilized urban soils and infrastructures (Regulating Decree 9/2009), however, extends over 13 km^2 and settle 44,534 inhabitants [49].

The origins of Coimbra city date back to the pre-Roman period, and until nowadays, it records a significant urbanization trend, driven by a massive increase in the population (Fig. 1c), which lead to extensive surface sealing. In 2018, the urban land use covered 22% of the municipality, while agriculture, forest and water

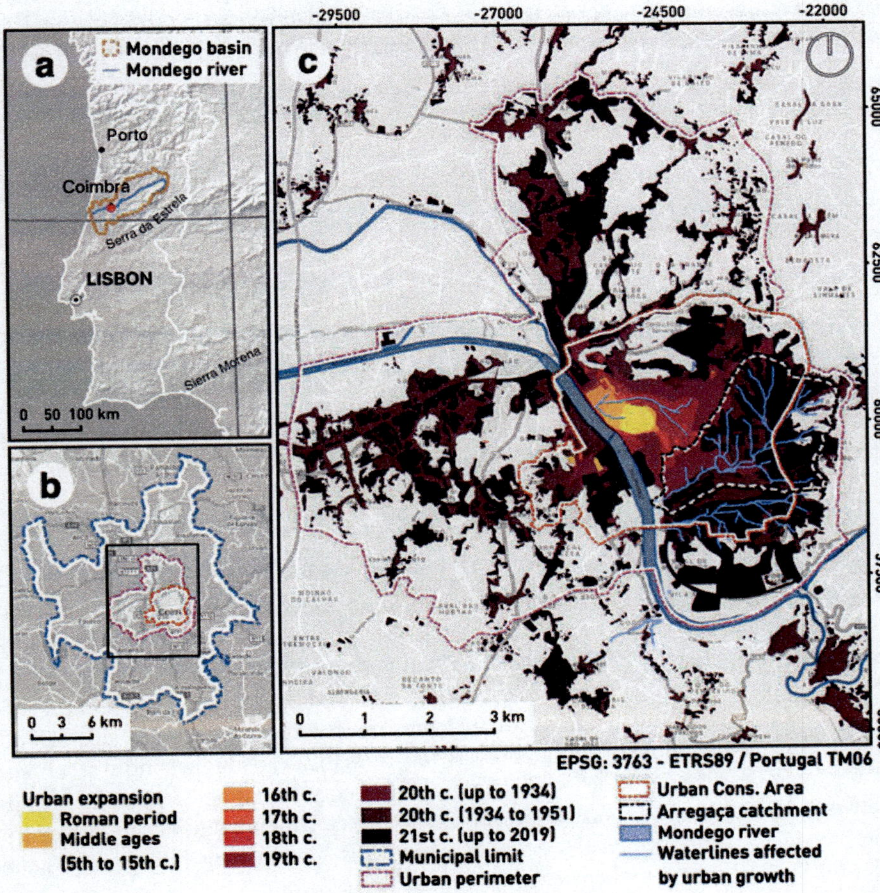

Fig. 1 Location of Coimbra in the central region of Portugal (**a**), extent of the urban perimeter and urban consolidated areas within Coimbra municipality (**b**), and expansion of the urban areas since the Roman period (**c**)

occupied 32%, 39% and 1%, respectively [50]. In the urban perimeter of Coimbra several GI extend over 2,567 ha, including a wide variety of GI from which arable land and forests are dominant (Table 1). Although the extent of GI has been decreasing over the last 15 years (from 53.4% in 2006 to 51.1% in 2018), as a result of urbanization, the green urban areas, and sports and leisure facilities were expanded from 5.0% to 5.3% and 1.1% to 1.2% from 2006 to 2018, respectively [50]. This increase was driven by an effort performed by local authorities to achieve a greener and more sustainable city. According with these aims, the approval of urbanization projects over the last years required the inclusion of GI elements.

Table 1 Changes in the area (ha) occupied by all types of GI (based on Urban Atlas land use classes) and their surface cover within the urban perimeter of Coimbra city (in % of the total urban area), between 2006 and 2018 [50]

GI types	Land use				
	2006		2018		
	(ha)	(%)	(ha)	(%)	% of change
Green urban areas	249.6	5.0	265.3	5.3	6.3
Sports and leisure facilities	55.6	1.1	58.1	1.2	4.4
Arable land (annual crops)	1,450.8	28.9	809.1	16.1	−8.0
Permanent crops			31.9	0.6	
Pastures			212.1	4.2	
Herbaceous vegetation associations			281.1	5.6	
Forests	793.5	15.8	776.7	15.5	−2.1
Water	132.7	2.6	132.7	2.6	0.0
Total	2,682.3	53.4	2,566.9	51.1	−4.3

2.2 Water Management and Floods

Coimbra expanded from the margins of the Mondego river (227 km), which drains the second largest basin (approximately 6,645 km^2) entirely in the Portuguese territory (Fig. 1a) [51]. Coimbra has a Mediterranean hot summer climate (Csa, according to Köppen-Geiger classification), with average annual temperature of 16°C and average annual rainfall of 922 mm, recorded between 1941 and 2000. The average annual flow of Mondego was 108 m^3/s [52]. The highest flow recorded in Coimbra reached 2,457 m^3/s, in January 1962, corresponding to a return period between 25 (2,131 m^3/s) and 100 years (2,756 m^3/s), which led to severe floods in the city [53]. Coimbra and the Mondego lowlands have a long history of floods [51, 54, 55], triggered by heavy winter rainfalls and favoured by the large size and marked orography of the river basin. These characteristics lead to peak flows reached in a few hours after extreme rainfall onsets [52].

At the end of the eighteenth century, the Mondego river was largely artificialized, namely in the section crossing Coimbra city and in the downslope alluvial plain [51, 52], to reduce the impacts of the river floods. Despite the intervention, the measures implemented were not sufficient to mitigate floods, and during the twentieth century several management plans based on grey infrastructures were implemented. The most extensive measures were the three dams constructed upstream Coimbra city, a weir bridge in the river stretch crossing the city, and five large dikes with one being located immediately upstream the city. Despite these infrastructures, periodic floods still affect Coimbra and settlements placed in the river floodplain, leading to major economic losses in urban infrastructures and agriculture fields [52].

Although riverside floods are quite relevant given their magnitude, pluvial floods across the city have been more frequent and intense over the last decades, due to progressive soil sealing and increasing frequency of short but intense rainfalls. Pluvial floods have been increasingly noticed due to overflowing of the grey

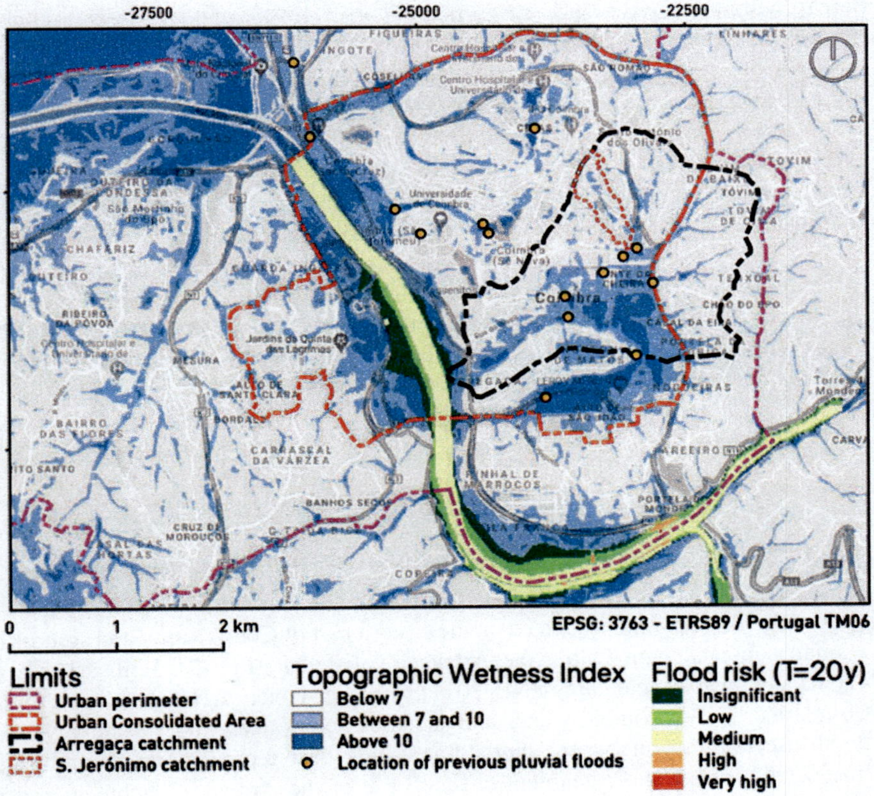

Fig. 2 Location of major pluvial floods recorded since 2006 in the urban perimeter and urban consolidated area of Coimbra, and identification of the flood-prone areas estimated using the Topographic Wetness Index, and the flood risk areas for the 20 years return period identified in the Directive 2007/60/EC

stormwater drainage systems, and/or lack of maintenance of the urban drainage systems (e.g. gutters bridged with litter and sediments). Since 2006, at least 10 large pluvial flood episodes were recorded in the city, with major constrains for vehicular traffic within the main roads and avenues, inundation of private and commercial buildings, and causing occasional shallow landslides. Most of these floods were observed in winter, but also during spring and late summer. These floods tend to affect specific areas of the city, usually located in flood-prone areas (Fig. 2).

The flood-prone areas identified in Fig. 2 were assessed using the Topographic Wetness Index (TWI), calculated in QGIS (3.14) using the SAGA algorithm, based on a Digital Elevation Model (DEM) with 10 m resolution provided by the Portuguese Directorate General for Territory. This method has been widely used as a proxy to identify flood-prone areas [56]. It was applied to identify flood susceptible areas within the city, since the official flood risk maps prepared to fulfill the European Floods Directive (Directive 2007/60/EC) only identify critical areas near the Mondego river, associated to fluvial floods (Fig. 2).

Since water management approaches based on grey infrastructures are not sufficient to prevent floods, local authorities have been implementing additional measures based on NBS over the last decades. Thus, several GI have been implemented or adapted to perform as NBS for flood mitigation. The NBS approach used include the installation of (1) alluvial woods in all the areas susceptible to 20-year return floods, (2) an urban park in the area adjacent to the flood-prone urban perimeter, (3) conservation of the vegetation on Mondego river margins (still under implementation) and (4) GI for recent urbanization projects [57]. *Quinta de São Jerónimo* GI is one example of the latter strategy, comprising a small infrastructure developed to fulfill legal criteria for the implementation of a new urbanization project.

3 Case Study of *Quinta de São Jerónimo* GI

3.1 Location and Biophysical Characterization

Quinta de São Jerónimo is located on the eastern part of Coimbra city and comprises a small sub-catchment within the Arregaça catchment (Fig. 2). With an area of 420 ha and 20,900 inhabitants (INE, 2011), Arregaça covers an important part of the Coimbra urban consolidated area and includes some areas under relatively high flood susceptibility, and where pluvial floods have been recorded over the last years (Fig. 2). *São Jerónimo* catchment covers 3.8% of the Arregaça catchment and is placed in a narrow and steep valley, with slopes up to 45%, ranging from 164 m a.s.l. in the northern part to 69 m a.s.l. in the southeast area. *São Jerónimo* catchment is not prone to local floods but rather contributes to downslope floods in the urban area. One of the most recurring flood sites identified over the last years is located immediately downslope *São Jerónimo* catchment (Fig. 2).

São Jerónimo catchment was subject to a strong urbanization in 1999, driven by the implementation of *Quinta de São Jerónimo* project. This urbanization project, involving the construction of 21 individual housing lots, 30 collective housing lots and 6 lots for private equipment, led to the extent of the impervious surface in *São Jerónimo* catchment from 37.4% in 1995 to 65.2% in 2018, at the expense of forest areas (Fig. 3). This urbanization project, developed as a residential area for high social strata (which became the most expensive residential area in Coimbra), also included the implementation of *Quinta de São Jerónimo* GI (mandatory for the approval of the urbanization project). This GI extends over 5.6 ha and comprises extensive green areas, walking routes, a tennis club with sports fields, a lake, a swimming pool, an amphitheatre, a bar, an old chapel with an atrium, a few management infrastructures and a parking area. Although it is a public garden, it has a condominium function and is managed by owners and residents of *Quinta de São Jerónimo*, through a cooperation agreement for the management of green spaces and collective use.

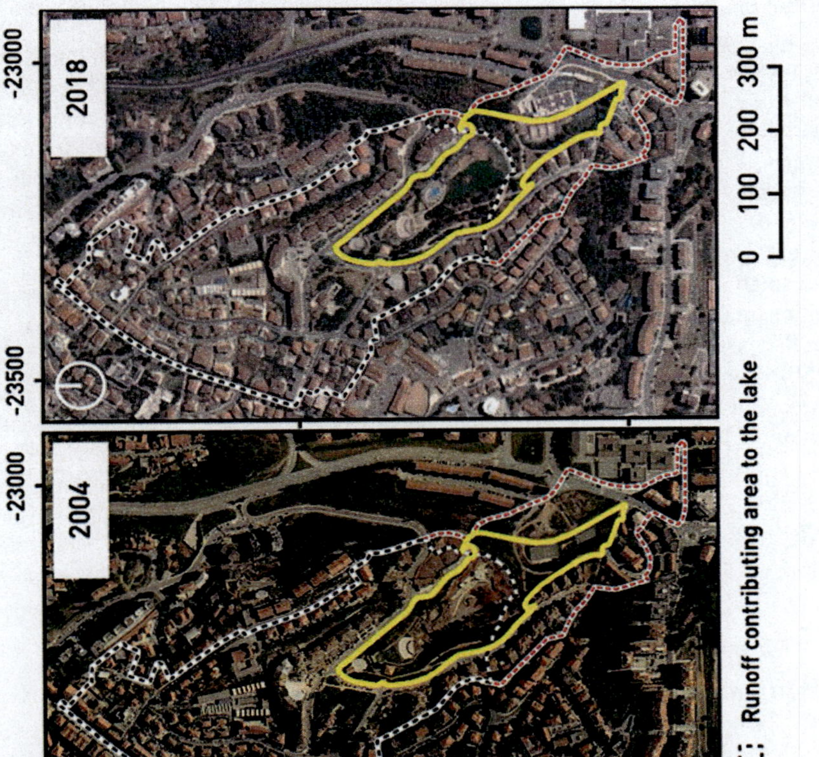

Fig. 3 Land use changes in *São Jerónimo* catchment driven by the implementation of *Quinta de São Jerónimo* urbanization project, which includes the *Quinta de São Jerónimo* green infrastructure, for the years 1995, 2004 and 2018

Quinta de São Jerónimo GI, although designed to provide an attractive and beautiful landscape, was also conceived to retain stormwater runoff and slow down its transfer to downslope areas. Thus, it has been claimed by municipal authorities as an NBS for flood mitigation. However, the water management system within this GI combines natural water storage principles with a grey engineered infrastructure, being classified as a hybrid NBS [24, 31]. Stormwater runoff from the catchment is collected and piped to the GI which includes ~2.4 ha of green areas, a small retention basin in the amphitheatre area with a water storage capacity of 75 m^3, a lake with ~0.3 ha, and a sequence of five settling ponds with a total capacity of 24 m^3 located upslope the lake to retain sediments and pollutants (Fig. 4).

The small retention basin receives stormwater runoff generated from the 700 m^2 amphitheatre (Fig. 5a) sealed surface and the surrounding area, and slows its release to the first settling pond by reduced discharge controlled through a small outlet (Fig. 5b). The first settling pond receives stormwater runoff from *Quinta de São Jerónimo* and transfers the runoff through the sequence of ponds until the lake. The bottom of the lake was sealed with concrete, and a spillway structure was installed to provide a slow release of incoming stormwater runoff to the downslope Arregaça drainage system (Fig. 4). The lake structure and the spillway system provide an additional storage capacity apart from the usual water level.

3.2 *The Role of* Quinta de São Jerónimo *GI on Flood Mitigation*

3.2.1 Methodology

Field surveys were performed to develop a topographic assessment of the lake and the surrounding area, in order to calculate the water storage capacity at typical water level, at the spillway level (when runoff is piped into the urban drainage system), and the maximum water storage capacity considering the flooding of part of the green area (Fig. 4).

Within *São Jerónimo* catchment, an artificial drainage system was installed to convey and pipe surface runoff from sealed surfaces. Although field surveys were developed to investigate the real contributing area of the catchment supplying runoff to *Quinta de São Jerónimo* GI, the lack of detailed information about the subsurface drainage system (despite the contacts established with local water authorities) was a major constrain for the study. Thus, the estimates of the stormwater runoff to *Quinta de São Jerónimo* GI considered the contribution of all the topographic catchment upslope the lake. Since runoff measurements are not performed in the study site, runoff estimates were based on Curve Number (CN) method developed by the Soil Conservation Service [58]:

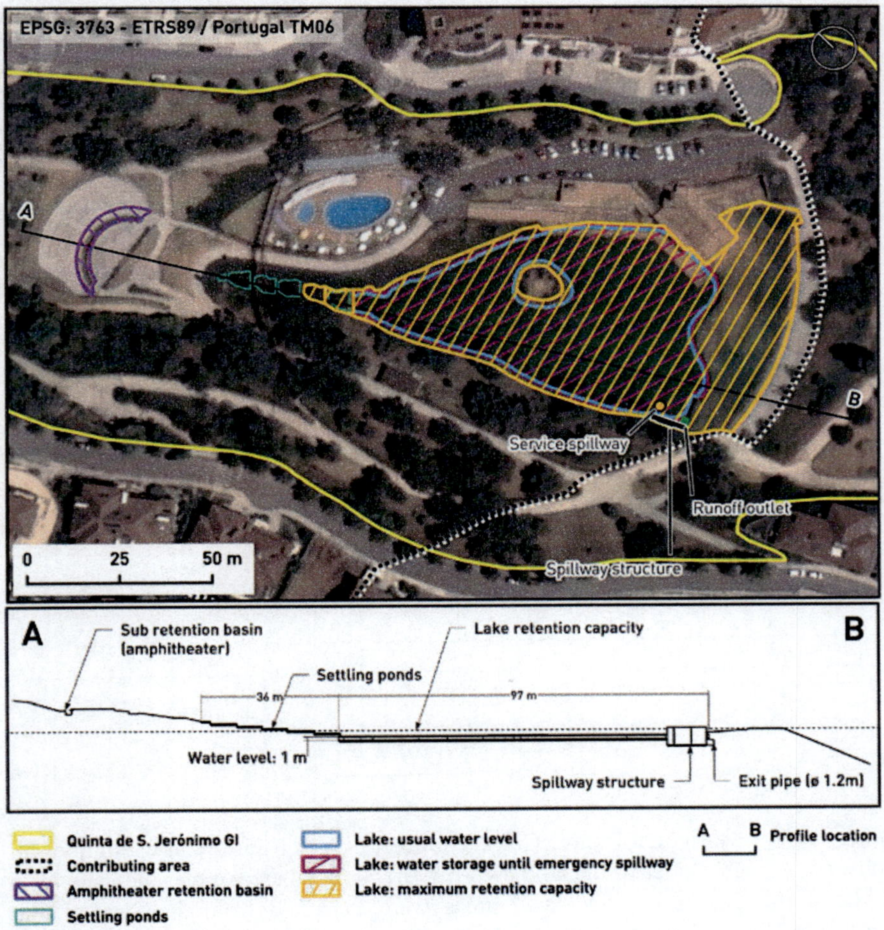

Fig. 4 Detailed view of Quinta de São Jerónimo GI, and the stormwater management system including a retention basin, a sequence of five ponds and a lake with typical water level and maximum water storage capacity, controlled by the spillway. The A-B profile of the GI provides a lateral view with details on the spatial relationship between all the water management devices

$$Q = \frac{\left[P - 0.2 \times \left(\frac{1000}{CN} - 10\right)\right]^2}{\left[P + 0.8 \times \left(\frac{1000}{CN} - 10\right)\right]} \quad (1)$$

where Q = runoff (mm), P = rainfall (mm), CN = Runoff Curve Number.

Since the topographic catchment includes several land-uses, a weighted Curve Number was calculated as follows:

Fig. 5 View of the Amphitheatre in the foreground, with a small water retention volume (a), and the reduced outlet connecting to the settling ponds (b)

Table 2 Runoff Curve Numbers for different land-uses and hydrological soil groups (A: soils with low runoff potential; B: soils with moderate infiltration rates when thoroughly wetted; C: soils with slow infiltration rates when thoroughly wetted; and D: soils with high runoff potential) [42]

Land cover	Hydrologic soil group			
	A	B	C	D
Impervious surface	98	98	98	98
Forested pervious area	30	55	70	77
Non-forested pervious area	49	69	79	84
Open water[a]	n/a	n/a	n/a	n/a

[a]Areas of open water are not included in the calculation of stormwater runoff

$$CNw = \frac{\Sigma(CNi \times Ai)}{At} \quad (2)$$

where CNw = weighted Runoff Curve Number, CNi = Runoff Curve Number for the land use i, Ai = area of the land use i (m^2), At = Total area of the study site (m^2).

Land use types and associated areas were extracted from the Urban Atlas [50]. The CN values were obtained from Table 2, based on the Soil Conservation Service values [58] and adapted from Tsegaya et al. [42]. The hydrological soil group considered for *São Jerónimo* topographic catchment was C, due to the relatively fine-textured soils, their slow infiltration rate and the shallow soil depth assessed during field visits.

The rainfall (P) used in Eq. (1) to estimate catchment runoff was based on rainfall intensity [59], calculated from the Intensity–Duration–Frequency (IDF) curves of Coimbra (Table 3), using Eqs. 3, 4 and 5. P and Q (from Eq. 1) were calculated for the return periods of 2, 5, 10 and 20 years. Stormwater runoff (mm) estimates were then converted into volume (m^3) by multiplying for the topographic contributing area.

$$P = h = t \times I \quad (3)$$

Table 3 IDF curves developed for Coimbra, for durations between 5 to 30 min [60] and 30 min to 6 h [61], for different return periods

	Return period (years)							
	2		5		10		20	
Duration	a	b	a	b	a	b	a	b
5–30 min	202.72	−0.577	259.26	−0.562	290.68	−0.549	317.74	−0.538
30 min–6 h	280.69	−0.653	374.38	−0.647	436.65	−0.644	496.49	−0.643

$$I = \frac{h}{t} \tag{4}$$

$$I = at^b \tag{5}$$

where I = rainfall intensity (mm/min), h = height of rainfall (mm), t = duration of rainfall (min), a and b = parameters from the Intensity–Duration–Frequency curves.

This methodology was also applied to estimate the surface runoff from Arregaça catchment, to understand the magnitude of *São Jerónimo* runoff within the larger urban catchment. In this case, the calculation of CN was performed considering the hydrological group B instead of C, given the higher soil permeability in Arregaça than *São Jerónimo* catchment.

3.2.2 Water Storage Capacity of the Lake

The spillway determines the water level and the storage capacity of the lake, and provides a controlled release of flows into the downslope drainage system of Arregaça catchment, during large rainfall events. The spillway structure is made of concrete and comprises a service spillway, an auxiliary spillway and an emergency spillway, associated with three distinct water levels in the lake, triggered by storm events, which produce increasing runoff excess (Fig. 6). The service spillway controls the normal water level. The auxiliary spillway comprises a lateral grid, placed 0.22 m above the service spillway, and provides an additional water storage capacity, besides which the runoff discharges to the downslope drainage system. The emergency spillway, comprising a larger upper grid in the overall spillway structure, is activated when the water exceeds 0.62 m above the normal water level in the lake. The maximum water level capacity of the retention basin is reached at 1.66 m above the normal water level. Under the highest water levels, the three types of spillways are functioning simultaneously, but all the runoff discharge is controlled by a single exit pipe.

The lake usually accommodates 2,995 m^3 of water. Thus, the storage capacity to retain additional runoff during the storms is provided by the spillway structure and the local topographic settings. The 0.22 m between normal water level (controlled by the service spillway) and the bottom of the auxiliary spillway provides an additional

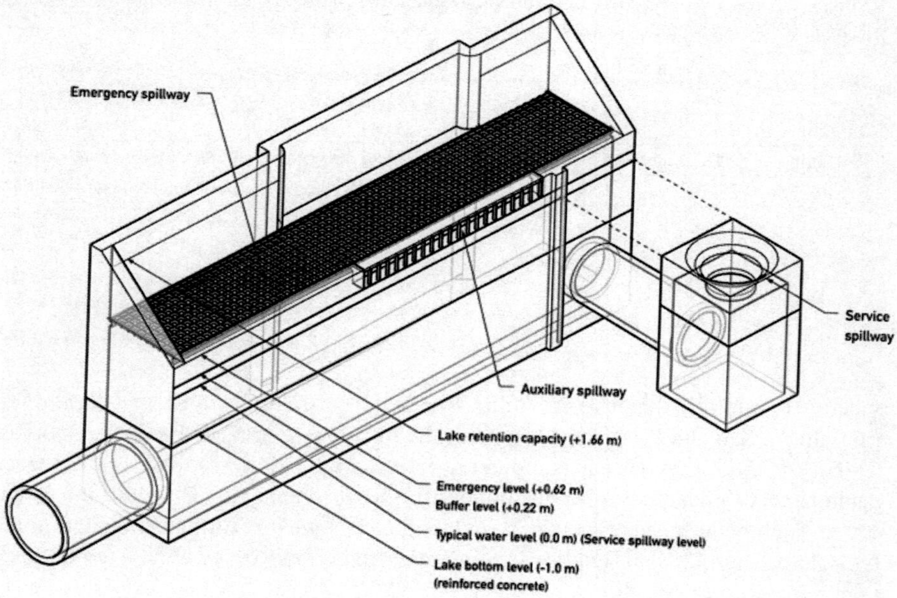

Fig. 6 Schematic representation of the spillway structure installed in *Quinta de São Jerónimo* GI lake, controlling the water storage capacity within the NBS

water storage of 667 m^3, before the auxiliary spillway is activated. This volume represents 22% of the total water storage capacity. After reaching the auxiliary spillway, which increases runoff discharge into the downslope drainage system, an extra storage capacity of 1,240 m^3 is provided (up to 0.62 m above normal water level), just before reaching the emergency spillway. Both volumes of water (1,907 m^3) are kept inside the normal lake boundaries. After surpassing the 0.62 m water level, where the emergency spillway provides extra runoff discharge volume, an additional capacity of 4,120 m^3 is provided through the water volume that overflows to a grass-covered embankment located in the south part of the lake (Table 4). A total retention volume of 6,027 m^3 is ensured (not including normal water volume), after which runoff will flow to *Quinta de São Jerónimo* GI downslope area and, if not infiltrated and/or retained, will contribute to downslope urban floods. If the GIs water storage capacity includes the capacity provided by the upslope retention basin located in the amphitheatre (75 m^3), the total storage of the blue infrastructure is 6,102 m^3, which represents 2 times the normal volume of the water in the lake.

3.2.3 Performance of the Blue Structures to Mitigate Downslope Floods

The performance of the NBS to mitigate downslope floods was based on comparing the water storage capacity and the potential stormwater runoff generated in the

Table 4 Typical water storage capacity of the lake and additional storage capacities affected by the spillway structure (see details on Fig. 6)

Water level	Water storage capacity (m^3)	Description
Typical water level (service spillway level)	*2,995*	Typical volume stored in the lake (maintained by the service spillway)
Buffer level (+0.22 m)	667	Retention capacity provided until the water level reaches the auxiliary spillway, located 22 cm above the service spillway. Water volume is kept inside the lake borders
Emergency level (+0.62 m)	1,240	Retention capacity provided before the water level reaches the emergency spillway, located 62 cm above the service spillway. Auxiliary spillway device in use. Water volume is kept within the lake
Lake retention capacity (+1.66 m)	4,120	Maximum retention capacity provided when the water level reaches 1.66 m above normal water level. All three spillway components in use. This water storage considers the overflow of the lake and flooding of the grass embankment
Total retention volume	***6,027***	Represents the maximum storage capacity of the lake, excluding the typical volume stored in the lake

Table 5 Land cover types and weighted CN values for *São Jerónimo* and *Arregaça* topographic catchments

	São Jerónimo			Arregaça		
	Area			Area		
Land cover type	(m^2)	(%)	CNw	(m^2)	%	CNw
Impervious areas	103,342	66.7	89.5	2,279,822	54.2	81.2
Forest pervious areas	36,927	23.8		708,857	16.9	
Non-forest pervious areas	14,762	9.5		1,215,175	28.9	
Total area	158,157	–	–	4,203,854	–	–

contributing topographic catchment, estimated from the CN method (Table 5). The stormwater runoff results for the different rainfall durations and return periods analysed are presented in Table 6. Comparing the runoff estimated for *São Jerónimo* topographic catchment (assuming that all the runoff reaches the blue infrastructures of *Quinta de São Jerónimo* GI) with the total storage capacity of the GI (6,102 m^3), it is possible to understand that this GI can accommodate runoff from rainfall events up to 60 min, associated with return periods up to 20 years. However, if only the capacity of the blue structure is considered (amphitheatre and lake), without letting part of the green area (grass embankment) to be overflowed (1,315 m^3), the GI would cope only with runoff from rainfall events up to 10 min, associated with return periods of 2 years, and events up to 5 min and return periods of 5 years.

The high runoff volume stored in the GI (0.62 m above normal water level) is not effectively retained in the NBS but rather partially released at control rate by the spillway, which slows the water outflow into the downslope drainage system.

Table 6 Rainfall intensity and runoff volume estimated for *São Jerónimo* and *Arregaça* topographic catchments, for rainfalls of different duration and return periods

	Rainfall intensity (mm/h)				Runoff (m^3)							
					São Jerónimo				Arregaça			
Rainfall duration (min)	2 year	5 year	10 year	20 year	2 year	5 year	10 year	20 year	2 year	5 year	10 year	20 year
5	80.1	104.9	120.1	133.7	845	1,160	1,354	1,527	19,004	27,188	32,286	36,861
10	53.7	71.1	82.1	92.1	1,191	1,635	1,918	2,174	28,005	39,744	47,281	54,109
15	42.5	56.6	65.7	74.0	1,447	1,990	2,342	2,662	34,762	49,185	58,606	67,193
30	30.5	41.5	48.9	55.7	2,156	3,006	3,577	4,110	53,632	76,437	91,833	106,202
60	19.4	26.5	31.3	35.7	2,795	3,895	4,635	5,321	70,780	100,391	120,392	138,934

Although runoff generated in *São Jerónimo* catchment represents only 4% of the Arregaça catchment runoff, *Quinta de São Jerónimo* GI provides a relevant storage capacity and delay in the peak discharge, which may alleviate the flood risk downslope.

The results showed that combining blue and green infrastructures was relevant to maximize the runoff storage capacity of GI, and that NBS can provide a relevant complement to runoff management with conventional grey infrastructures, maximizing the mitigation of downslope pluvial floods. These findings support the increasing evidence that incorporating GI in urban design can alleviate flood risk due to their effectiveness in managing urban floods, reducing peak flow rates, and controlling the total volume of stormwater runoff [14, 62]. Furthermore, this case study demonstrates the relevance of GI to manage stormwater near its origin, as reported by previous authors [42].

Even though *Quinta de São Jerónimo* GI was fully operational in 2006, storm events recorded during that year in June and October (both with rainfall equivalent to 60 min duration and return periods of 20 years) led to floods in the urban area placed immediately downslope (Fig. 2). Thus, albeit *Quinta de São Jerónimo* GI can support water management in Arregaça, additional NBS measures are required to mitigate runoff within the extensive urban area of Arregaça catchment. The current water management system in Arregaça, mainly depending on grey infrastructures, has proved insufficient to prevent floods and NBS can provide an important complement to enhance urban resilience.

3.3 *Co-benefits of* Quinta de São Jerónimo

As stressed by some authors, the evaluation of NBS should not focus only on water management aspects, but also include additional benefits provided to the society [11, 21]. Similar to other NBS, *Quinta de São Jerónimo* GI supports local stormwater management but also provides multiple secondary benefits (co-benefits) far beyond that of flood protection, relevant for people and the environment, through direct and indirect use of ecosystem services delivered by the green and blue components.

Quinta de São Jerónimo GI has a green area of 13,452 m^2, with a wide variety of trees, shrubs and herbaceous species, and a blue component including a lake of approximately 3,000 m^2, and some springs and water tanks. These green and blue areas provide habitat for several plants (e.g. at least 25 different trees) and animals (e.g. small birds, ducks and fishes), some of them with high conservation value (e.g. *Quercus rubra* and *Quercus ilex*). Besides the relevant ecological benefit, improving biodiversity and ecological resilience, this GI provides some food items since it includes an edible garden with a few fruit trees (e.g. oranges and lemons) and aromatic plants. Several studies highlight the impact of GI on improving biodiversity, namely through the provision of wildlife habitat [63], but also timber and food items [46]. Few authors argue that urban gardens can decrease the overall urban

Fig. 7 Overview of *Quinta de São Jerónimo* green infrastructure (GI); (**a**) view to the south part, with the retention lake (south-centre part of the GI) and tank (in the northern part), the amphitheatre (in the centre) and edible gardens; (**b**) example of a tree with slab providing botanical information; (**c**) view to the north of the GI, showing few deposition ponds in the foreground, the amphitheatre on the midground, and the upper limit of the GI with the old chapel and fountains

footprint, and decrease the reliance of urban dwellers on external provision services [64].

The impact of *Quinta de São Jerónimo GI* on water regulation is beyond that of stormwater volume storage. It includes water evapotranspiration and infiltration by the green areas, and a small contribution for water quality regulation driven by reduced erosion (favoured by vegetation cover), filtration of contaminants through the soil and sediment retention in the tanks and lakes. The relevance of green areas, namely woody vegetation, rainfall interception, increased evapotranspiration, and infiltration in urban areas, has been widely identified [42, 65]. *Quinta de São Jerónimo* GI offers additional regulating ecosystem services such as temperature regulation through shading and evaporative cooling, which mitigates heat-island effect and reduces the energy used in buildings [11]. It also provides airborne particulate filtration and improves air quality [66], noise reduction [67], biological carbon capture and storage [68], and thus climate change mitigation [14]. These co-benefits can occur even if not considered or maximized in the original design of the GI [14]. However, some authors argue that the magnitude of GI benefits on regulation of ecosystem services and biodiversity is affected by the connectivity between green and blue spaces and should be assessed at a larger scale such as regional and national [27].

Quinta de São Jerónimo GI plays a major role in cultural services, allowing the residents to reconnect to nature and improve their well-being [64]. This GI promotes a healthier lifestyle by supporting physical activities, such as walking and sports practices, enhanced by the presence of multi-sport infrastructures, including tennis field and swimming pool [67]. *Quinta de São Jerónimo* GI has a high aesthetic value (Fig. 7a) and provides education and recreation opportunities. This GI includes a wide variety of trees, with several of them placed nearby the walking routes, providing botanical information through slabs with the species common and Latin names and their origin (Fig. 7b). It also comprised an aromatic plant zone with a wide variety of species, identified with high education value slabs.

Furthermore, *Quinta de São Jerónimo* GI supports social networks, improving social benefits such as cohesion and entertainment. This is enhanced by available supporting infrastructures, including bar and restaurant, and an amphitheatre (Fig. 7c) where some cultural events are organized (e.g. music festivals). In contrast, grey infrastructure lacks involvement and engagement with community initiatives [14]. This GI also includes a small heritage chapel and a viewpoint for part of Coimbra city. Recreational settings are used by residents living in close proximity and visitors that come to *Quinta de São Jerónimo* GI for relaxation and socialization purposes. These cultural services have been widely reported in other GI implemented in urban areas [14, 69]. Green spaces reduce stress, anxiety, depression, and increase the level of happiness and life satisfaction [68].

Additionally, *Quinta de São Jerónimo* GI provides economic benefits by supporting the local economy by promoting the bar, restaurant, swimming pool, and sports fields. The maintenance of GI and existent infrastructures provides work opportunities in the private sector, called by previous researchers as collar jobs [70].

Although this chapter does not aim to perform an economic valuation of the investigated GI, some authors stress the relevance of cost-benefit analysis to assess GI projects developed for water management purposes [11]. These analyses are commonly restricted to the cost of measures to increase safety and reduce expected damages. Thus, grey options typically appear as the only economically viable strategy for flood mitigation [11]. However, Vincent et al. [71] demonstrated that GI's economic feasibility is substantially improved if multiple benefits are considered. The monetary valuation of co-benefits would help decision-makers when choosing among different solutions [72]. However, the costs and benefits of GI change when green and blue infrastructures are combined with grey solutions [11], such as the *Quinta de São Jerónimo* GI. A mix of green, blue and grey infrastructures have been identified as the best strategy to enhance urban resilience since they complement each other to provide several benefits in limited urban spaces [35], and green components have higher adaptability and resistibility to deal with the uncertain future [17].

4 Final Considerations

Coimbra is a city historically vulnerable to floods. Over the last years, however, increasing urbanization and frequency of short but intense rainfalls have led to a relatively higher number of pluvial floods, raising concerns about the insufficiency of the water management system, largely based on grey infrastructures. These problems raised awareness among local authorities, which started to consider NBS approach to mitigate flood hazard. Some NBS were already implemented across the city, and it became mandatory that large urbanization projects include Green Infrastructures to get the approval from the authorities.

Quinta de São Jerónimo GI is an example of NBS implemented to mitigate the impacts of an urbanization project, involving the construction of 57 lots of individual and collective houses and private equipment. The NBS includes blue and green elements, such as ponds, a lake and grassed areas, integrated with a grey infrastructure (spillway) which controls the runoff storage capacity of the semi-natural elements. Based on a simple methodology used worldwide to estimate runoff generated within the *São Jerónimo* topographic catchment (CN method), and the calculation of the water storage capacity of the NBS from the topographic characteristics, this study demonstrates the effectiveness of the NBS to mitigate floods. The relatively small scale NBS has the capacity to cope with runoff driven by rainfalls with recurrence up to 20 years, providing runoff storage near to its source (sealed surfaces within urban development), and a slow release of runoff which delays the peak flow into downslope urban areas. These findings demonstrate that incorporating GI in urban design can be an important strategy to manage urban floods and alleviate flood risk.

The investigated GI comprises an appropriate strategy to cope with runoff from the relatively small urban area, which is important to mitigate downslope floods, frequently recorded in nearby urban areas. This NBS, however, is not enough to prevent downslope floods in urban areas of the Arregaça catchment, as noticed with the 2006 urban floods. These floods were triggered by runoff provided from an extensive urban area, with only 4% being supplied by *São Jerónimo* sub-catchment. Therefore, a network of NBS should be considered to complement the current urban drainage system, and effectively mitigate floods and enhance urban resilience in large cities. This is especially important under climate change context, where extreme precipitation events are expected to be more frequent and severe.

The implementation of NBS in urban areas also provides additional ecosystem services, including regulation, provisioning and cultural services, particularly relevant in urban areas given the limited access to green areas, triggered by the limited available space in the cities. Thus, planning and implementing NBS for stormwater management should also consider the additional co-benefits, important for the environment and human well-being.

The strategy of the authorities to include GI as a mandatory element for new urbanization projects is interesting to support the implementation of NBS. However, it may lead to ad-hoc planning strategies, and less than optimal outcomes regarding flood mitigation. Despite there is an interest and an effort to implement NBS, previous studies show that the lack of a coherent approach can hinder the effectiveness of implemented NBS, or even its proper implementation [29, 38].

Acknowledgment This research was supported by the LAND4FLOOD Cost Action, through the Short-Term Scientific Mission of Milan Gazdic, the ERA-LEARN 2020 UrbanGaia project (grant agreement No 645782), and the Portuguese Foundation for Science and Technology, through the post-doctoral and doctoral grants SFRH/BPD/120093/2016 and SFRH/BD/149710/2019. The authors would like to thank Barbara Frigione for the support with the field work, and José Vilhena from Coimbra city council for providing technical information regarding *Quinta de São Jerónimo* urbanization project.

References

1. UN DESA (2018) Revision 2018 world urbanization prospects. United Nations Department of Economics and Social Affairs. https://www.un.org/development/desa/publications/2018-revision-of-world-urbanization-prospects.html. Accessed 2 Jul 2020
2. United Nations, Department of Economic and Social Affairs, Population Division (2014) World urbanization prospects: the 2014 revision, highlights (ST/ESA/SER.A/352)
3. Prokop G, Jobstmann H, Schönbauer A (2011) Report on best practices for limiting soil sealing and mitigating its effects. Study contracted by the European Commission, DG Environment. Technical Report-2011–50. www.ec.europa.eu/environment/archives/soil/pdf/sealing/Soil%20sealing%20-%20Final%20Report. Accessed 10 Jul 2020
4. Artmann M (2016) Urban gray vs. urban green vs. soil protection — development of a systemic solution to soil sealing management on the example of Germany. Environ Imp Assess Rev 59:27–42
5. Sun N, Limburg KE, Hong B (2019) Chapter 6 – The urban hydrological system. In: Hall MHP, Balogh SB (eds.) Understanding urban ecology. https://doi.org/10.1007/978-3-030-11259-2_6
6. Wang Z, Xie P, Lai C, Chen X, Wu X, Zeng Z, Li J (2017) Spatiotemporal variability of reference evapotranspiration and contributing climatic factors in China during 1961–2013. J Hydrol 544:97–108
7. Strohbach MW, Doring AO, Mock M, Sedrez M, Mumm O, Schneider AK, Weber S, Schroder B (2019) The "hidden urbanization": trends of impervious surface in low-density housing developments and resulting impacts on the water balance. Front Environ Sci 7:29
8. Ferreira CSS, Walsh RPD, Nunes JPC, Steenhuis TS, Nunes M, de Lima JLMP, Coelho COA, Ferreira AJD (2016) Impact of urban development on streamflow regime of a Portuguese peri-urban Mediterranean catchment. J Soils Sediments 16:2580–2593
9. Kalantari Z, Ferreira CSS, Walsh RPD, Ferreira AJD, Destouni G (2017) Urbanization development under climate change: hydrological responses in a peri-urban Mediterranean catchment. Land Degrad Dev 28(7):2207–2221
10. Abebe Y, Kabir G, Tesfamariam S (2018) Assessing urban areas vulnerability to pluvial flooding using GIS applications and Bayesian belief network model. J Clean Prod 174:1629–1641
11. Alves A, Gersonius B, Kapelan Z, Vojinovic Z, Sanchez A (2019) Assessing the co-benefits of green-blue-grey infrastructure for sustainable urban flood risk management. J Environ Manag 239:244–254
12. Ferreira CSS, Walsh RPD, Costa ML, Coelho COA, Ferreira AJD (2016) Dynamics of surface water quality driven by distinct urbanization patterns and storms in a Portuguese peri-urban catchment. J Soils Sediments 16:2606–2621
13. Ferreira CSS, Walsh RPD, Steenhuis TS, Ferreira AJD (2018) Effect of Peri-urban development and lithology on streamflow in a Mediterranean catchment. Landsc Degradat Dev 29:1141–1153
14. Li L, Collins AM, Cheshmehzangi A, Shun FK (2020) Identifying enablers and barriers to the implementation of the Green Infrastructure for urban flood management: A comparative analysis of the UK and China. UFUG: 126770
15. Venkataramanan V, Packman A, Peters D, Lopez D, McCuskey DJ, McDonald RI, Miller WM, Young SL (2019) A systematic review of the human health and social wellbeing outcomes of green infrastructure for stormwater and flood management. J Environ Manag 246:868–880
16. Sota CD, Ruffato-Ferreira VJ, Ruiz-García L, Alvarez S (2019) Urban green infrastructure as a strategy of climate change mitigation. A case study in northern Spain. UFUG 40:145–151
17. Dong X, Guo H, Zeng X (2017) Enhancing future resilience in urban drainage system: green versus grey infrastructure. Water Res 124:280–289
18. European Commission (2015) HORIZON 2020 – work programme 2016–2017: 12. Climate Action, Environment, Resource Efficiency and Raw Materials. 2016 1349 of 9 March 2016. Accessed via European Commission Decision C (23-05-2018). http://ec.europa.eu/research/

participants/data/ref/h2020/wp/2016_2017/main/h2020-wp1617-leit-space_v1.1_en.pdf. Accessed 12 Jul 2020
19. Kalantari Z, Ferreira CSS, Page J, Romain G, Olsson J, Destouni G (2019) Meeting sustainable development challenges in growing cities: coupled social-ecological systems modeling of land use and water changes. J Environ Manag 245:471–480. https://doi.org/10.1002/ldr.3264
20. European Commission (2013) Green Infrastructure (GI)—Enhancing Europes Natural Capital. Com, 2013. 249 final. Communication from the Commission to the European Parliament, the Council, the European Economic and Social Committee, and the Committee of the Regions, Brussels (23-05-2018). http://ec.europa.eu/environment/nature/ecosystems/docs/green_infrastructures/1_EN_ACT_part1_v5.pdf. Accessed 12 Jul 2020
21. Jato-Espino D, Sañudo-Fontaneda LA, Andrés-Valeri VC (2018) Green infrastructure: cost-effective nature-based solutions for safeguarding the environment and protecting human health and well-being. In: Hussain C (ed) Handbook of environmental materials management. Springer, Cham. https://doi.org/10.1007/978-3-319-58538-3_46-1
22. Alçada e Silva C (2017) Alterações climáticas, precipitação e água em zonas urbanas. Dissertation, University of Coimbra
23. Gunnel K, Mulligan M, Francis RA, Hole DG (2019) Evaluating natural infrastructure for flood management within the watersheds of selected global cities. Sci Total Environ 670:411–424
24. Kumar P, Debele SA, Sahani J, Aragão L, Barisani F, Basu B, Bucchignani E, Charizopoulos N, Di Sabatino S, Domeneghetti A, Sorolla EA, Finér L, Gallotti G, Juch S, Leo LS, Loupis M, Mickovski SB, Panga D, Pavlova I, Pilla F, Löchner PA, Renaud FG, Rutzinger M, Sarkar Basu A, Aminur Rahman Shah M, Soini K, Stefanopoulou M, Toth E, Ukonmaanaho L, Vranic S, Zieher T (2020) Towards an operationalisation of nature-based solutions for natural hazards. Sci Total Environ 731:138855
25. Haase D (2015) Reflections about blue ecosystem services in cities. Swage 5:77–83
26. Koc CB, Osmond P, Peters A (2016) A green infrastructure typology matrix to support urban microclimate studies. PRO 169:183–190
27. Ronchi S, Arcidiacono A, Pogliani L (2020) Integrating green infrastructure into spatial planning regulations to improve the performance of urban ecosystems. Insights from an Italian case study. Sustain Cities Soc 53:101907
28. Ballard BW, Kellagher R, Martin P, Jefferies C, Bray R, Shaffer P (2007) The SUDS manual. CIRIA- Construction Industry Research and Information Association, London
29. Kuller M, Bach PM, Roberts S, Browne D, Deletic A (2019) A planning-support tool for spatial suitability assessment of green urban stormwater infrastructure. Sci Total Environ 686:856–868
30. Martín EG, Costa MM, Máñez KS (2020) An operationalized classification of nature based solutions for water-related hazards: from theory to practice. Ecol Econ 167:106460
31. Grimm N, Cook EM, Hale RL, Iwaniec DM (2016) A broader framing of ecosystem services in cities: benefits and challenges of built, natural, or hybrid system function. In: Seto KC-Y, Solecki W, Griffith C (eds) The Routledge handbook of urbanization and global environmental change1st edn. Routledge, Taylor & Francis Group, New York
32. Alexander K, Hettiarachchi S, Ou Y, Sharma A (2019) Can integrated green spaces and storage facilities absorb the increased risk of flooding due to climate change in developed urban environments? J Hydrol 579:124201
33. Rossman LA (2015) Storm water management model user's manual version 5.1. EPA, Cincinnati
34. Wong THF, Fletcher TD, Duncan HP, Coleman JR, Jenkins GA, Siriwardena L (2002) Model for urban stormwater improvement conceptualisation (MUSIC) (version 1.00). CRC for Catchment Hydrology, Melbourne
35. Alves A, Vojinovic Z, Kapelan Z, Sanchez A, Gersonius B (2020) Exploring trade-offs among the multiple benefits of green-blue-grey infrastructure for urban flood mitigation. Sci Total Environ 703:134980

36. Massoudieh A, Maghrebi M, Kamrani B, Nietch C, Tryby M, Aflaki S, Panguluri S (2017) A flexible modeling framework for hydraulic and water quality performance assessment of stormwater green infrastructure. Environ Model Softw 92:57–73
37. Sorensen J, Emilsson T (2019) Evaluating flood risk reduction by urban blue-green infrastructure using insurance data. J Water Res Plan Manag:145–142. https://doi.org/10.1061/(ASCE)WR.1943-5452.0001037
38. Schäffler A, Swilling M (2013) Valuing green infrastructure in an urban environment under pressure – the Johannesburg case. Ecol Econ 86:246–257
39. Liquete C, Udias A, Conte G, Grizzetti B, Masi F (2016) Integrated valuation of a nature-based solution for water pollution control. Highlighting hidden benefits. Ecol Ser 22:392–401
40. Demuzere M, Orru K, Heidrich O, Olazabal E, Geneletti D, Orru H, Bhave AG, Mittal N, Feliu E, Faehnle M (2014) Mitigating and adapting to climate change: multifunctional and multi-scale assessment of green urban infrastructure. J Environ Manag 146:107–115
41. Stessens P, Khan AZ, Huysmans M, Canters F (2017) Analysing urban green space accessibility and quality: a GIS-based model as spatial decision support for urban ecosystem services in Brussels. Ecol Ser 28:328–340
42. Tsegaye S, Singleton TL, Koeser AK, Lamb DS, Landry SM, Lu S, Barber JB, Hilbert DR, Hamilton KO, Northop RJ, Ghebremichael K (2019) Transitioning from gray to green (G2G) – a green infrastructure planning tool for the urban forest. UFUG 40:204–214
43. Kalantari Z, Ferreira CSS, Deal B, Destouni G (2019) Nature-based solutions for meeting environmental and socio-economic challenges in land management and development (Editorial). Land Degradation and Development, pp 1–4
44. Raymond CM, Berry P, Breil, M, Nita, MR, Kabisch, N, de Bel, M, Enzi, V, Frantzeskaki, N, Geneletti, D, Cardinaletti, M, Lovinger, L, Basnou, C, Monteiro, A, Robrecht, H, Sgrigna, G, Munari, L, Calfapietra, C (2017) An impact evaluation framework to support planning and evaluation of nature-based solutions projects. Report prepared by the EKLIPSE expert working group on nature-based solutions to promote climate resilience in urban areas. Centre for Ecology & Hydrology, Wallingford
45. Kabisch N, Strohbach M, Haase D, Kronenberg J (2016) Urban green space availability in European cities. Ecol Indic 70:586–596
46. Leitão IA, Ferreira CSS, Ferreira AJD (2019) Assessing long-term changes in potential ecosystem services of a peri-urbanizing Mediterranean catchment. Sci Total Environ 660:993–1003
47. INE (2011) General population census per NUTS. https://www.ine.pt/xportal/xmain?xpid=INE&xpgid=ine_publicacoes. Accessed 12 May 2020
48. INE (2020) Resident population per NUTS (estimates). https://www.ine.pt/xportal/xmain?xpid=INE&xpgid=ine_publicacoes. Accessed 12 May 2020
49. INE (2011) General population census per statistical unit. https://www.ine.pt/xportal/xmain?xpid=INE&xpgid=ine_publicacoes. Accessed 12 May 2020
50. European Union, Copernicus Land Monitoring Service (2020) European Environment Agency (EEA) https://land.copernicus.eu/local/urban-atlas/urban-atlas-2018. Accessed 20 Apr 2020
51. Lourenço L, Velez F, Cunha PP, Lima IP, Tavares A (2017) Flood risk in the lower Mondego. Guidebook of the study trip No 3. In: IV international congress on risks. 23rd–26th May. RISCOS, Chã do Freixo, Portugal
52. Cardielos JP, Lobo R, Peixoto P, Mota E, Duxbury N, Caiado P (2016) Mondego O Surdo Murmúrio do Rio. In: A água como património: experiências de requalificação das cidades com água e das paisagens fluviais. Coimbra University Press. pp 95–112
53. APA (2011) Plano de Gestão de Bacia Hidrográfica – Região Hidrográfica do Vouga, Mondego e Liz (RH4). Lisboa
54. Martins AF (1940) O esforço do Homem na Bacia do Mondego. Ensaio Geográfico. Coimbra
55. Tavares AO, Barros L, Santos PP, Zêzere JL (2013) Desastres naturais de origem hidrogeomorfológica no Baixo Mondego no Período 1961-2010. Territorium 20:65–76

56. García-Rivero AE, Olivera J, Sallinas E, Yuli RA, Bulege W (2017) Use of hydrogeomorphic indexes in SAGA-GIS for the characterization of flooded areas in Madre de Dios, Peru. Int J Appl Eng Res 12(19):9078–9086
57. APA (2016) Plano de Gestão dos Riscos de Inundações - Região Hidrográfica 4 - Vouga, Mondego e Lis - Zonas Críticas: Coimbra, Estuário do Mondego, Águeda, Ria de Aveiro e Pombal. APA, Lisboa
58. United States Department of Agriculture (USDA) (1986) Urban hydrology for small watersheds, 2nd ed. TR-55, Washington
59. Ramke H-G (2018) Collection of surface runoff and drainage of landfill top cover systems. In: Cossu R, Stegmann R (eds) Solid waste landfilling, concepts, processes, technologies. Elsevier, Amsterdam, pp 374–416
60. Matos MR, Silva MH (1986) Estudos de precipitação com aplicação no projeto de sistemas de drenagem pluvial. Curvas intensidade-duração frequência da precipitação em Portugal. In: Encontro Nacional de Saneamento Básico/86. Laboratório Nacional de Engenharia Civil, Lisboa
61. Brandão C, Pinto da Costa J, Rodrigues R (2001) Análise de fenómenos extremos precipitações intensas em Portugal Continental. INAG, Lisboa
62. Zhang T, Xiao Y, Liang D, Tang H, Yuan S, Luan B (2020) Rainfall runoff and dissolved pollutant transport processes over idealized urban catchments. Front Earth Sci 16
63. Goddard MA, Dougill AJ, Benton TG (2010) Scaling up from gardens: biodiversity conservation in urban environments. Trends Ecol Evol 25(2):90–98
64. Calderón-Contreras R, Quiroz-Rosas LE (2017) Analysing scale, quality and diversity of green infrastructure and the provision of urban ecosystem services: a case from Mexico City. Ecol Ser 23:127–137
65. Ferreira CSS, Walsh RPD, Shakesby RA, Keizer JJ, Soares D, González-Pelayo O, Coelho COA, Ferreira AJD (2016) Differences in overland flow, hydrophobicity and soil moisture dynamics between Mediterranean woodland types in a peri-urban catchment in Portugal. J Hydrol 533:473–485
66. Setala H, Viipola V, Rantalainen AL, Pennanen A, Yli-Pelkonen V (2013) Does urban vegetation mitigate air pollution in northern conditions? Environ Pollut 183:104–112
67. Chen S, Wang Y, Ni Z, Zhang X, Xia B (2020) Benefits of the ecosystem services provided by urban green infrastructures: differences between perception and measurements. UFUG, p 126774
68. Navarrete-Hernandez P, Laffan K (2019) A greener urban environment: designing green infrastructure interventions to promote citizens' subjective wellbeing. Landsc Urban Plann 191:103618
69. Zuniga-Teran AA, Gerlak AK, Mayer B, Evans TP, Lansey KE (2020) Urban resilience and green infrastructure systems: towards a multidimensional evaluation. Curr Opin Environ Sustain 44:42–47
70. King A, Shackleton M (2020) Maintenance of public and private urban green infrastructure provides significant employment in Eastern Cape towns, South Africa. UFUG 126740
71. Vincent SU, Radhakrishnan M, Hayde L, Pathiran A (2017) Enhancing the economic value of large investments in sustainable drainage systems (SuDS) through inclusion of ecosystems services benefits. Water 9:841
72. Chenoweth J, Anderson AR, Kumar P, Hunt WF, Chimbwandira SJ, Moore TLC (2018) The interrelationship of green infrastructure and natural capital. Landsc Use Pollut 75:137–144

Identification of Surface Runoff Source Areas as a Tool for Projections of NBS in Water Management

Jiří Jakubínský, Vilém Pechanec, Ondřej Cudlín, Jan Purkyt, and Pavel Cudlín

Contents

Abstract The location of surface runoff source areas is one of the most important information for the conservation of water resources, their sustainable management as well as mitigation of the frequency of extreme hydrological phenomena. The location of the surface runoff source areas in the landscape affects not only the hydrological and sediment regime of the watercourse itself, but also some physico-chemical properties of the water. The exact parameters of the hydrological and sediment regime are the result of a number of variables – mostly soil properties, morphometric parameters of the terrain, and the level of anthropogenic influence, but

J. Jakubínský (✉), O. Cudlín, J. Purkyt, and P. Cudlín
Global Change Research Institute CAS, Brno, Czech Republic
e-mail: jakubinsky.j@czechglobe.cz; cudlin.o@czechglobe.cz; purkyt.j@czechglobe.cz; cudlin.p@czechglobe.cz

V. Pechanec
Department of Geoinformatics, Palacký University Olomouc, Olomouc, Czech Republic
e-mail: vilem.pechanec@upol.cz

Carla S. S. Ferreira, Zahra Kalantari, Thomas Hartmann, and Paulo Pereira (eds.),
Nature-Based Solutions for Flood Mitigation: Environmental and Socio-Economic Aspects,
Hdb Env Chem (2022) 107: 313–338, DOI 10.1007/698_2021_775,

Published online: 30 May 2021

also the type of surface cover. This research deals with the relationship between types of habitats (in terms of their qualitative properties – e.g. naturalness of the habitat, diversity of structures, sensitivity of the habitat to external interventions, etc.) and their impact on runoff processes in the landscape; i.e., the extent to which the habitat type can affect the soil water retention and infiltration capacity, and thus runoff processes. The creation of habitats with an identified positive effect on the hydrological regime (mitigation the frequency of drought and flash floods), or the creation of conditions suitable for the natural formation of these habitats, can be considered as a good example of nature-based solutions for water management. Within a study area in the Czech Republic, a medium-sized watercourse catchment with forest-agricultural landscape, a newly developed water retention model LOREP was applied. This model takes into account a multiple-flow regime, providing more accurate results than previous models. The analysis revealed that there are several types of natural or close-to-nature habitats able to retain a significant amount of rainwater, even in soils with limited retention capacity. A possible increase in the area of these habitats may indirectly contribute to the mitigation of hydrological extremes and the increase of surface water quality.

Keywords Environmental modelling, Habitat structure, Soil water retention, Surface runoff, Water resources

1 Introduction

The hydrometeorological conditions of a given place primarily contribute to the formation of surface runoff. Precipitation and evapotranspiration are spatially and temporally variable factors, which usually represent one of the fundamental impulses leading to the creation, change or extinction of surface flow in the landscape. These factors represent the key variables influencing the formation of extreme hydrological situations – not only flood events, but also drought episodes, which the Central European region is increasingly facing [1–3]. While the long-term flow regime is mainly shaped by the climatic conditions of the river basin (e.g., the prevailing climate conditions), sudden changes in the hydrological characteristics of the stream are the result of the current meteorological situation and direct anthropogenic influences Apart from the extreme weather conditions there is also a number of other factors influencing the surface runoff formation, such as the soil saturation driven by antecedent precipitation, and other characteristics related to the physical–geographical properties of the landscape. The extent of surface runoff within the river basin depends on the rate of infiltration of rainwater, namely determined by the soil water retention capacity and partly also the degree of rainfall interception by vegetation [4].

The relationship between the hydrological regime of watercourses, the natural conditions of a given river basin, and the impact of human intervention in the landscape have been discussed by a number of authors, in order to quantify the influence of individual factors on runoff processes [5–9]. Since vegetation (and especially forest cover) directly affects the precipitation–runoff processes and also acts as a protective cover for soil surface, it represents an important segment of the landscape, affecting the current and long-term water balance of the area. The relationship between the ecological status and structure of vegetation cover and the hydrological regime has been analysed by many authors [10–13]. The general conclusion is that forest stands and soil properties significantly modify the culmination and shape of the hydrograph of the precipitation–runoff episodes (the response time is delayed and the peak flood is lower); in addition to the physical geographical conditions of the river basin itself, these mainly include the composition of the forest, its age and health status of individual trees or the management practices used [10].

Compared to forest stands, agricultural land contributes in a greater extent to the overall impact on surface runoff characteristics. The gradual transformation of the original, highly diversified natural landscapes into the current large monotonous blocks of arable land represents the most significant change that has affected the rural landscape [14]. The great land cover and land-use change called as "green revolution" can be described as a turning point in global land management, which refers to, among other things, a fundamental change in the hydrological conditions of the affected environment (and indirectly also the water regime of most streams). An example of a European country where a very significant change in the land-use took place during the twentieth century is the Czech Republic. The change of land-use is also connected with the drainage of large areas, which was carried out due to efforts to cultivate the land using heavy machinery and to establish the collective agricultural systems [15]. According to Štěrba et al. [16], in the second half of the last century, more than 600,000 ha of land were drained in the Czech Republic (especially wetlands and waterlogged soils) in which the case study presented in this chapter was carried out. Land reclamation (i.e. change in soil utilization) affected practically the entire territory of the republic, especially the headwater areas, in which the original waterlogged soil caused difficulties during the conversion from non-agricultural to agricultural land [1]. Similar issues related to land-use change and its impacts on the hydrological regime have also been addressed in other European countries [17–20]. The main negative consequence of land-use change related to the land drainage and its cultivation was the hydromorphological degradation of the river network, the disruption of which was subsequently reflected in the hydrological regime. The increase in the rate of water runoff and the imbalance of flows during the year led to many changes in the morphological and ecological parameters of the river environment (especially related to the channel incision). The secondary – although not less significant – impact of agricultural activity on the river network is the deterioration of the biochemical quality of water, due to the flushing of harmful substances from cultivated land.

It is also necessary to mention that, although runoff processes are influenced nowadays by a number of anthropogenically conditioned factors, some extreme water levels cannot be eliminated even by the best management of water resources or landscape planning. Although the overall ecological status of the river basin has a significant impact on the occurrence of floods, the main stimulus remains atmospheric precipitation and its parameters in agricultural catchments. According to Pithart et al. [21], it is obvious that, with increasing precipitation, natural conditions and level of anthropogenic influence decreases; for example, while after a 20 mm precipitation a river basin may capture up to 75% of the rainwater under optimal conditions, with a total precipitation of 100 mm or more, the captured amount may be less than 10% of the precipitation (based on measurements made in a medium-sized river basin in an agricultural landscape). The ecological status of the river basin has a significant effect on the formation of water runoff for precipitation amounts approximately up to a recurrence interval of 10–20 years; in the case of catastrophic precipitation events with a longer recurrence interval, the influence of the landscape significantly decreases [21].

Another factor that significantly affects the runoff generation process whose effect is not possible to completely eliminate is the presence of urban areas and impermeable surfaces. In addition to the fact that built-up areas reduce the water retention capacity of the landscape, urban areas reduce water evaporation [22] and also accelerate the water runoff flowing downslope [8, 23]. The most common cause of the accelerated runoff of water is the straightening of riverbeds, channel deepening and especially the concrete revetment of river banks. These features are typical for watercourses affected by the so-called urban stream syndrome [7, 24–26].

The most widespread measures to mitigate the impacts of hydrological extremes – especially flood situations – in the past were in general technical adjustments directly in the actual riverbeds, i.e., straightening, channel fortification, and incisions. These measures often had another equally important goal, namely to improve the soil permeability and thus increase its agricultural potential. However, in connection with the increasingly frequent impacts of climate and environmental change, the negative aspects of the implemented technical measures began to appear. Drained and compacted soil, which also leads to loss of connectivity between the watercourse and groundwater, began to be extremely prone to drying up at a time of gradually rising air temperatures and the unbalanced distribution of precipitation over time [27]. Only in the period with a significantly increased number of these hydrological extremes (after a period of numerous flash floods and the subsequent long period of drought) the need to implement other types of measures that could contribute to alleviating these hydrological extremes begin to be demanded in most parts of Europe. Appropriate solutions could include the so-called nature-based solutions, which are defined by the European Commission [28] as measures inspired by, supported by or copied from nature. Within the water management sector, these are mainly "natural flood management" and "natural water retention measures". According to Hartmann et al. [29], these are targeted interventions to improve water retention by plants (interception) and their evapotranspiration as well as the infiltration of water into the soil, and to support the formation of ponds and wetlands; the measures are intended to restore the connectivity of rivers with their surroundings. Although these measures are important in the entire area of the river basin, the vast majority of these interventions are usually concentrated in the area of river landscapes, as narrow strips lining watercourses, which have the potential to accumulate not only rainwater but also water from the riverbed at elevated water levels, and thus significantly support the flood protection function of current landscapes. In order to ensure the positive effect of the mentioned measures (i.e. to reduce the frequency of hydrological extremes), it is necessary to implement them in areas that are as large as possible, which may require the cooperation of private landowners. However, the landowners are usually not interested in this type of measures, as they very often reduce crop yields, and some solutions also present certain technical complications in terms of land management [30]. One of the tools to stimulate the interest of landowners in the implementation of water retention measures could be various subsidy titles provided by the governments [31]. A certain solution that would help a society to realize the importance of the discussed measures is to express the financial value of individual ecosystem services provided by the landscape (for example, flood control, water storage, soil erosion prevention, and climate stabilization) in optimal conditions in terms of the impact on the hydrological regime [32].

Proposals for nature-based solutions to alleviate hydrological extremes assume a relatively detailed knowledge of local runoff processes and the main variables that affect the hydrological regime of the area. The use of various environmental models seems to be very effective in identification of surface runoff generation areas, especially erosion–accumulation modelling techniques, based on the outputs of which it is possible to identify localities that are prone to surface runoff generation and especially localities with some potential to experience the mentioned extreme phenomena. Most of the existing hydrological models use a digital terrain model (DTM) and information on selected soil characteristics that affect water transport within the soil profile (e.g. infiltration capacity and soil water retention capacity) as background data. These data are supplemented by information concerning the land-use; in particular, the land cover type, the crop rotation and applied sowing procedures in the case of agricultural land. Since the mentioned nature-based solutions that can be implemented in the catchment area also include, in general, a change in the nature of land cover, information on the impact of individual habitat types on runoff processes is one of the key features that should be reflected in landscape management. Thus, this study analyses the influence of different vegetation structures on local hydrological regime.

2 Identification of Runoff Source Areas Using Environmental Modelling Techniques

2.1 Characteristics of the Study Area

The study area is the Dřevnice River basin, located in the eastern part of the Czech Republic, near the state border with Slovakia (see Fig. 1). The basin has an area of 438.2 km^{2}and especially the headwater area is characterized by a relatively rugged

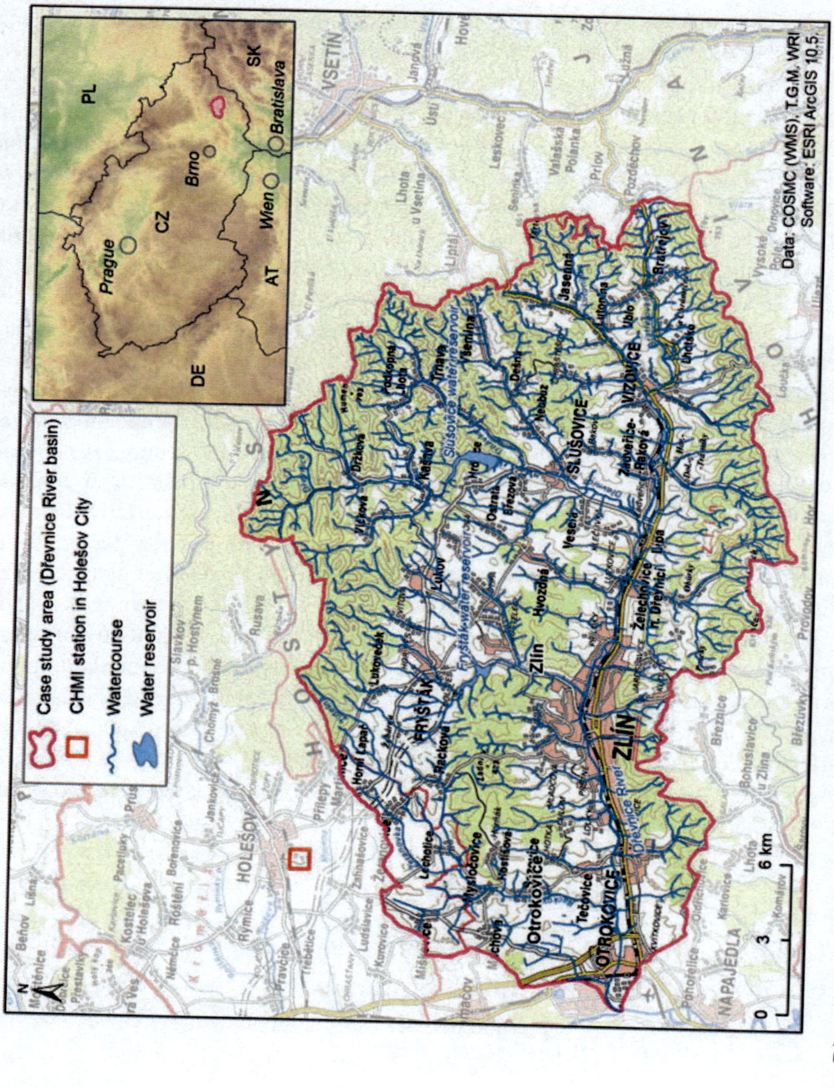

Fig. 1 Dřevnice River basin located in the eastern part of the Czech Republic

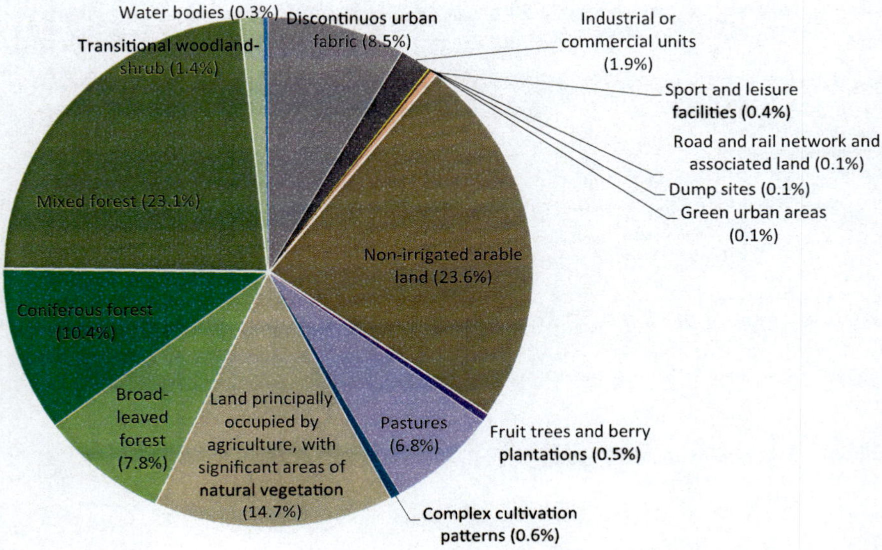

Fig. 2 Prevailing land-use categories in the Dřevnice River basin (based on the CORINE Land Cover 2012 dataset)

relief. In terms of land-use, it is a predominantly forest–agricultural to agricultural landscape, which is divided by a relatively dense network of rural settlements. Non-irrigated arable land (23.6%) and mixed forest (23.1%) comprise the dominant type of land-use; urban areas cover almost 10% of the river basin. The Dřevnice River basin also includes a relatively large urban area, concentrated around the city of Zlín (located in southern edge of the river basin). Detailed information on the type of land-use in the Dřevnice River basin is presented in Fig. 2. An overview of the 15 most represented types of habitats is given in Fig. 3. The plain consists almost exclusively in the floodplain of the Dřevnice River and several of its major tributaries, while a large part of this flat area with fertile soil is covered by urban or industrial areas and transport infrastructures. The area is severely affected by soil erosion through water, which reaches a value of almost 30 t ha^{-1} $year^{-1}$ in on non-forested steep slopes [33].

The average annual air temperature in the region ranges from 7 to 10°C, and the total average annual rainfall is 600–1,000 mm [34], depending on the specific location within the river basin. In the headwaters, cambium soils are the predominant soil type, whereas in downslope areas and in the western part of the river basin, the Luvizems and brown soils are dominant. The Dřevnice river floodplain and floodplains of some larger tributaries are formed by gley alluvial soils (the soils occasionally fully saturated with water). More than two-thirds of the area consist of highly rugged terrain (with an average terrain slope of more than 5°), which together with a specific geological subsoil (Flysch formation) significantly affect the hydrological regime of local watercourses (especially in terms of accelerating runoff). The

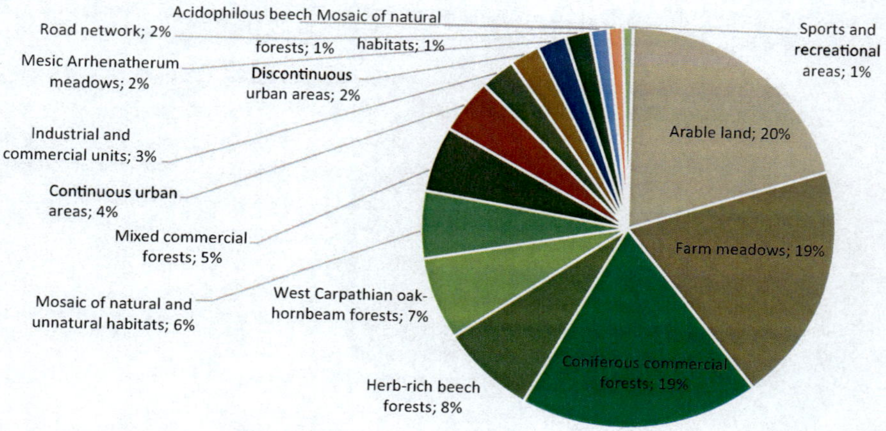

Fig. 3 Fifteen most frequent habitats (both natural and unnatural) in the Dřevnice River basin (using data from [37])

Dřevnice River originates in the highlands located in the north-eastern part of the basin at 560 m above sea level (a. s. l.) and empties into the Morava River (177 m a. s. l.) near Otrokovice city, 42.3 km from the spring area. The average long-term discharge at the mouth to the Morava River is 3.15 $m^3\ s^{-1}$ [35]. There are two dams located in the basin. One is the Slušovice water reservoir, situated directly on the Dřevnice River with a total retention volume of 9.95 million m^3, the main purpose of which is flood protection and ensuring minimum ecological flows in dry periods. The second dam is the Fryšták water reservoir on the Fryštácký Brook, which has a volume of 2.95 million m^3 and was implemented for the same use as the Slušovice dam. The Dřevnice River basin is relatively prone to the occurrence of floods, either by summer floods caused by long-lasting rains or flash floods, which typically affect especially smaller watercourses and tributaries of the Dřevnice River. The largest floods recently recorded in the study area were observed in 1987, 1997, 2006, and 2010 [36], with residential houses always being flooded and property damages caused. Due to frequently recurring floods, technical flood protection measures were implemented in various parts of the river basin, especially within the Zlín City (see the example in Fig. 4).

2.2 *Modelling of the Surface Runoff Source Areas*

In order to analyse the water retention capacity of soils in the study area, a recently developed LOREP model [38] was applied. The LOREP model is a tool for the identification and spatial localization of areas with low water retention capacity, allowing to work with a structured catalogue of non-technical measures (e.g., implementation of grass belts in arable land, and change in sowing procedures) to

Fig. 4 Flood dike delimiting the polder area (on the right side), as an inundation area of the Dřevnice River on the outskirts of Zlín City, Czech Republic

increase and support the water retention function of the landscape. One of the main advantages of the LOREP model is the ability to capture multiple flows in sloping terrain, which many other hydrological models do not allow (they work only with single flow direction – i.e. runoff from one pixel to a neighbouring pixel with lower altitude). The model is based on the use of GIS technology and available hydrological equations. The whole model consists of four consecutive steps: (1) the determination of the volume of territorially specified direct surface runoff (runoff caused by and directly following a rainfall or snowmelt event), (2) the spatial delineation of hydrological zones of the basin (i.e. spatial distribution of sites with higher ability to retain rainwater), and (3) the localization and determination of causes of low water retention capacity.

The procedure for the computation of territorial-specific surface runoff is based on a combination of specific functions in GIS, hydrological equations of the runoff curve number method and spatially distributed unit hydrographs. The LOREP model was developed in Python and designed for ArcGIS Pro. The input data are expressed as a grid of pixels in agreement with the rules of raster representation in ArcGIS. The spatial resolution is chosen to be high enough to allow identification of the influence of linear features on the surface runoff volume.

The modelling technique itself consists of several consecutive steps:

Table 1 Curve number (CN) categories reflecting antecedent precipitation index for 5 preceding days (API5) (according to [44]). Three categories of CN number are commonly determined for the purpose of refining the antecedent precipitation index, based on prevailing conditions

CN number category	API5 – out of growing season [mm]	API5 – during growing season [mm]
I	<13	<36
II	13–28	36–53
III	>28	>53

1. Creation of the necessary GIS layers, capturing the current land-use and hydrological characteristics of soils (an overview of all data sources used is given in Table 1),
2. Reclassification of input data into individual categories according to the runoff curve number (CN) methodology,
3. Determination of CN for individual landscape patches (homogeneous in terms of land-use),
4. Modification of CN after taking into account the antecedent precipitation index – API5 (volume of rainwater from the previous 5 consecutive days preceding to the day with the analysed precipitation event),
5. Calculation of current soil loss values (IaA),
6. Calculation of runoff height and soil water retention capacity.

The CN method is a recognized and globally used approach developed by the American Soil Conservation Service (SCS) in 1954 and documented for the first time in the National Engineering Handbook [39]. The general starting point of the methodology is the assumption that the ratio of the runoff volume to the total of torrential precipitation is equal to the ratio of the volume of water retained during runoff to the potential volume that can be retained. The potential volume of water retention depends on the soil type, surface cover, cultivation method, and previous conditions of soil moisture and vegetation [40]. Water drains can occur only after the initial loss, which is caused by interception, infiltration, and surface retention. The standard CN curve method is used to determine the runoff curve numbers according to U.S. Soil Conservation Service [41]. Individual CN curve values are available for each land-use category based on land-use type, cultivation method, hydrological conditions, and the hydrological groups of soils. Precisely defined categories are described in detail in the methodology itself [41].

The last distinguishing factor is the hydrological group of the soil: A, B, C or D. Based on CN numbers, the runoff height and the volume of water drained from the area can be estimated. With knowledge of the CN, the maximum water retention *A* [mm] can be calculated from the following equation:

$$\mathrm{A} = 25.4 \times \frac{1,000}{\mathrm{CN} - 10} \tag{1}$$

Therefore, if the torrential rain exceeds 20% of the calculated maximum water retention, the height of the drained water H_R can be calculated as (for $H_P > 0.2A$):

$$H_R = \frac{(H_P - 0.2A)^2}{H_P + 0.8A} \tag{2}$$

where "H_R" is the runoff height (mm), "H_P" is the total of the causal precipitation measured in 24 h (mm), and "A" is the maximum water retention (mm).

2.3 *Identification of Runoff Source Areas in the Dřevnice River Basin, Using the LOREP Model*

Using the above modelling approach, the runoff height and the related water retention ability of the landscape for the entire study area were computed. Our method involved the modelling of the hydrological response of the Dřevnice river basin to a causal precipitation of a total of 42 mm (during approximately 6 h of precipitation on 13 May 2020), evenly distributed over the entire area of the river basin. The previous precipitation conditions of the given locality were considered in the calculation –API5 tooks the value of 51.10 mm. This is based on the effect of the total precipitation in the previous days on the soil's ability to retain further precipitation. A five-day total referred to as antecedent precipitation index (API) is commonly used [42], and was divided into three categories in this study according to the precipitation volume (see Table 1 for CN values verified for the Czech Republic). CN values were based on the spatial intersection between the soil hydrological groups and the agricultural and forest areas. By distinguishing the hydrological conditions into good, medium, and poor (respecting the methodology of SCS [41]), three variants of the layer were created. The CN value was assigned to the records according to the conversion table [43]. Since 1971, a specification of the CN values according to antecedent moisture conditions has appeared in the National Engineering Handbook (SCS 1971). Clarification of the initial losses preserves the calculated CN numbers of the mean hydrological conditions and the maximum retention resulting from them. The current value of initial losses is specified on the basis of the total precipitation in previous days. Instead of a fixed 20% of the maximum possible water retention, as in the standard SCS methodology [41], the actual value of the initial losses is calculated using the total precipitation of the last 5 days. Adjusting the initial losses has a direct effect on the change in the runoff height and outflow volume.

The meteorological station nearest to the study area (operated by the Czech Hydrometeorological Institute) is located in the town of Holešov, at a distance of about 5 km from the boundary of the basin at the closest point. The values of API5 in the growing period before the selected precipitation event (as of 13 May 2020) were taken from this station. The input data concerning total precipitation in the area of interest correspond to real conditions, as these are data related to a selected precipitation event that occurred during Spring 2020. The selected precipitation represents an above-average precipitation event in this region – an average maximum amount in individual precipitation episodes is 38.5 mm for the period 1961–2019 and average monthly total precipitation is 52.3 mm for the same period [34]. In addition to meteorological data, information on terrain properties and surface cover is very important when solving the problem of rainfall–runoff processes; this especially includes morphometric (e.g. terrain fragmentation, slope) and morphological features (soil structure, soil depth and its particle size distribution, and other properties of the soil-forming substrate). In general, the key information includes the infiltration capacity of the soil and its maximum water retention capacity, which are characteristics that are significantly affected by anthropogenic activities, whether through the compaction of the topsoil by heavy agricultural machinery or its complete covering with artificial materials (construction of buildings, road networks and other related infrastructure) [45].

The second value affected by the total precipitation in previous days is the indicator of the initial loss. Mishra et al. [46] investigated on data from the U.S. Department of Agriculture Agricultural Research Service (USDA-ARS) Water Database. This database contains data on total precipitation and direct runoff in small river basins in the USA. They worked with many models, of which the most accurate results were achieved by the modified Mishra–Singh model (MMS) determined by the following equations:

$$IAa = \frac{\lambda + A2}{\mathrm{CN} - A + M} \quad (3)$$

$$M = \frac{((\mathrm{API5} - 0.2\mathrm{A}) \times A)}{(\mathrm{API5} + 0.8\mathrm{A})} \quad (4)$$

where IAa is the current value of the initial losses (mm), λ is a coefficient of individual partial losses (cm), A is the potential water retention derived from CN numbers (mm), M is the soil moisture from antecedent precipitation (%), and API5 is the total precipitation in the previous 5 days (mm).

The fact that linear landscape features (such as lines of trees, indicating the riparian vegetation along the streams) can be a part of land-use analysis and represents one of the key benefits of the model. This is possible because high-resolution raster data (with a pixel size of 5 m) was used, and because the modelling focused on the hydrology of small basins (it was not necessary to work with extremely large datasets). List of all the datasets and their sources used for modelling is given in Table 2.

Freely accessible images of the Sentinel 2 satellite, captured on 18 April 2020, were used as input data to create a layer with spectral reflectance information. On the Sentinel portal [47], the image with the least extensive coverage of clouds over the study area was selected from the products available. The images were uploaded to

Table 2 List of data sources used in the case study area

GIS layer	Data source	Data publisher	Scale	Type of classification
Land-use	Open street maps (OSM)	OSM individual contributors	1: 2,116	Each polygon has its own attribute
Hydrological soil groups (HSG)	Map of HSG	Research Institute for Soil and Water Conservation	1: 5,000	Four soil groups – A, B, C, D; according to the infiltration rate
Typological forest units	Regional forest development plan; forest typological map	Institute of Forest Management (IFM)	1: 10,000	Typological classification system of IFM
Spectral reflectance	Sentinel 2 imagery	European space Agency	10–60 m/pixel	12 spectral bands
Aerial images	World imagery	Digital globe	1 m/pixel	Raster data
River basin boundary	Hydrological division – river basin of IV. Stream order	T.G.M. Water Research Institute	1: 10,000	Vector data
Habitat map	Habitat mapping layer	Nature Conservation Agency of the Czech Republic	1: 10,000	Vector data, 165 habitat types

Table 3 NDMI intervals based on moisture conditions (according to [48])

Moisture conditions	Level of moisture conditions	NDMI interval
Dry	1	−1.0–0.0
Moderate	2	0.0–0.2
Wet	3	0.2–1.0

SNAP software, in which the resolution of the bands had to be unified. After unifying the band resolution to 10 m and creating a connection of the images into a mosaic, the Normalized Difference Moisture Index (NDMI) was calculated according to the following equation:

$$\mathrm{NDMI}_{S2} = \frac{(\mathrm{Band}_{8A} - \mathrm{Band}_{11})}{(\mathrm{Band}_{8A} + \mathrm{Band}_{11})} \tag{5}$$

After calculating the NDMI values, the output raster was exported to the ESRI ArcGIS Pro software. This software ensured the conversion of the raster into polygons and the classification of pixel polygons into three levels of the current moisture conditions (Table 3).

The identified runoff source areas were compared with land-use categories and also with the layer of habitats occurring in the river basin (Fig. 3). For this purpose, a habitat mapping layer according to the Habitat Valuation Method – HVM [49] was used. This layer contains all habitat types occurring in the Czech Republic, including habitats significantly affected by man or completely unnatural habitats.

3 Results of the Soil Water Retention Modelling in the Case Study Area

3.1 *Surface Runoff Heights*

Detailed information about the land-use types in the Dřevnice river basin was grouped into a total of nine categories according to their water permeability, and it was found that more than 70% of the area, it can be assumed that the permeability of the soil is not completely limited, although it is anthropogenically influenced (e.g. arable land, orchards, gardens, urban greenery and parks). The remaining 20% of the area, including natural or close-to-nature forests or scrublands, have a completely natural infiltration capacity. An overview of individual categories of land-use differentiated according to their water retention ability is given in Fig. 5. The study area was divided into two categories – "permeable" and "impermeable", while the following categories of land cover were considered impermeable according to the CORINE Land Cover database [50]: continuous urban fabric, industrial or commercial units, road and rail network and associated land, and construction sites. In Fig. 3 these areas are included in the category of "built-up and impermeable surface", which covers approximately 10% of the river basin. All other land-use categories were considered as permeable. Thus, permeable areas are represented in the river basin in a vast majority, although these are unnatural or nature-distant sites in many cases.

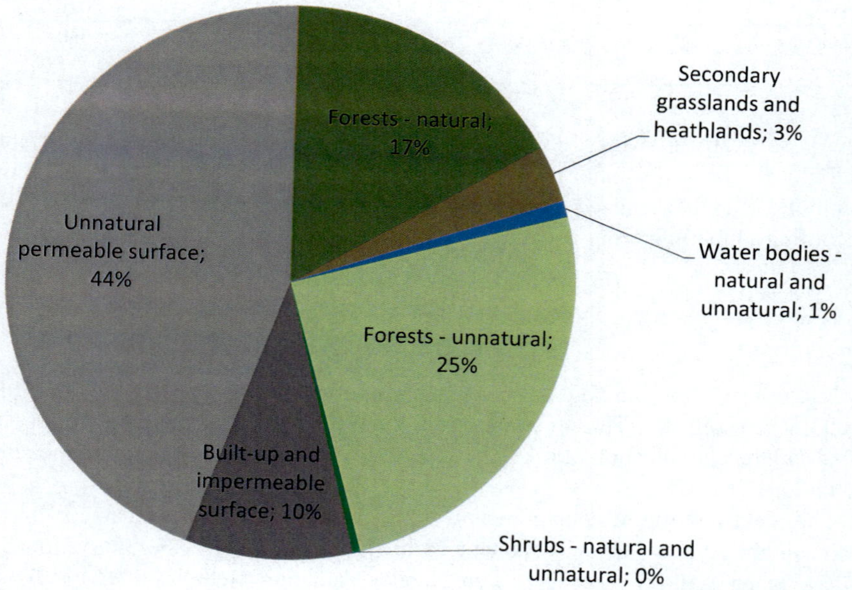

Fig. 5 Individual land-use types merged according to their water permeability in the Dřevnice River basin

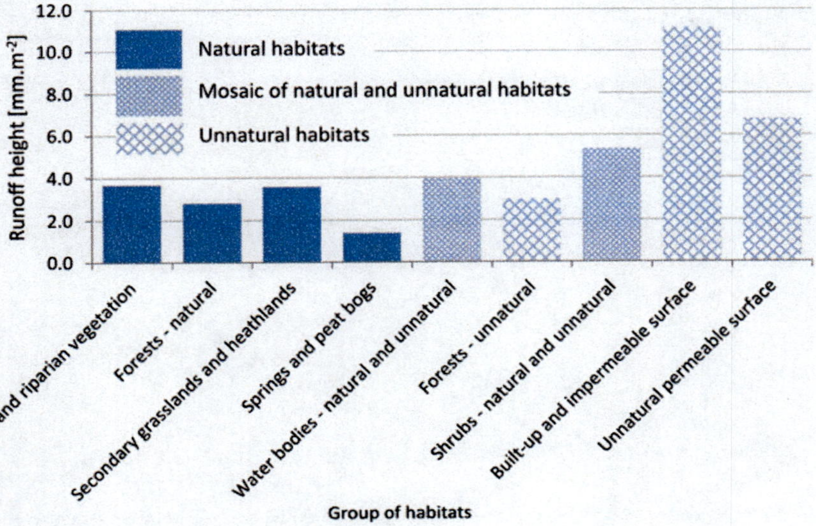

Fig. 6 Runoff heights for the nine land cover categories, driven by a causal precipitation of 42 mm (homogeneous throughout the basin)

Within the above-mentioned nine land-use categories, the surface runoff height was calculated in the next step of this study; i.e., the volume of rainwater draining in the form of surface runoff after the described causal precipitation of 42 mm, and API5 = 51.10 mm (see Fig. 6). The results confirmed that the largest amount of water drains from the built-up and impermeable surfaces. In these localities, on average, up to 11.12 mm of water drains after a causal precipitation, which means that approximately 239 m^3 of water drains in this form in the Dřevnice river basin (approximately 22% of the total surface runoff volume after the investigated causal precipitation). Another significant amount of water runoff is generated under an unnatural permeable surface (e.g., green urban areas, arable land, and pastures), from which about 6.79 mm of drains after causal precipitation. Due to its large area, the presence of unnatural permeable surfaces is a key element influencing runoff processes; in the entire river basin, almost 300 m^3 of water flows out of these areas – i.e., about 28% of the total runoff generated in the basin. Considering natural or near-to-nature land-use categories, the highest values of water runoff after causal precipitation were recorded in wetlands and riparian vegetation (3.68 mm) and secondary grasslands and heathlands (3.58 mm). Lower values were shown by natural and close-to-nature forest stands (2.82 mm) and springs with peat bogs (1.39 mm). In total, about 24% of the total runoff from the above-mentioned causal precipitation drains from the areas covered by natural or near-to-nature landscape in the Dřevnice River basin.

In terms of the spatial distribution of surface runoff within the Dřevnice River basin, it is evident that increased runoff values may occur, especially in the areas with a more developed floodplain, such as in Zlín and Otrokovice city, as well as in

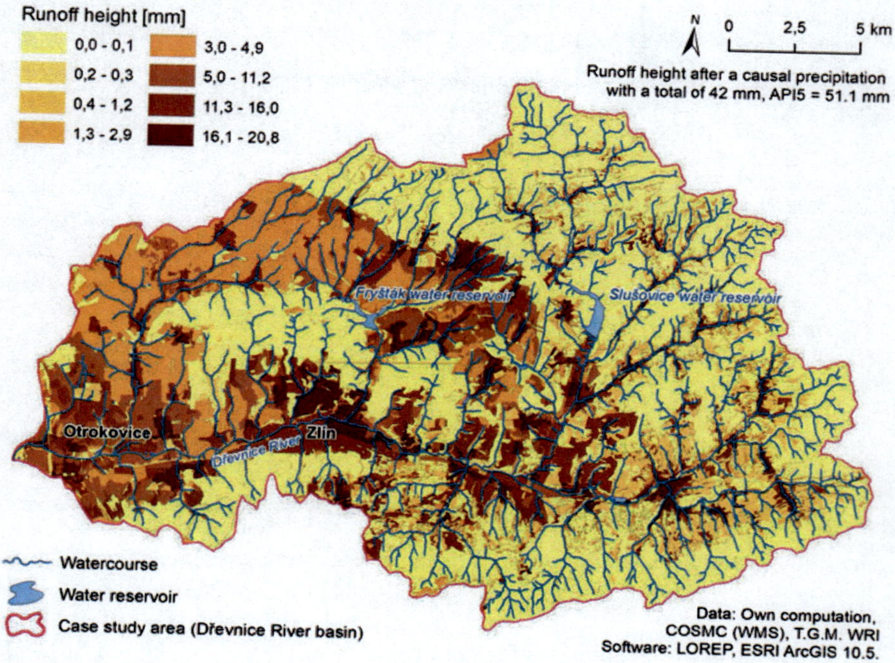

Fig. 7 Spatial distribution of runoff height in the Dřevnice River basin after a causal precipitation of 42 mm and an antecedent precipitation index (API5) of 51.1 mm

the Racková and Fryštácký brook basins (both right-hand tributaries of the Dřevnice river in the middle and lower part of the basin). In these sites, larger units with a relatively homogeneous land-use occur more often; at the same time, this is a significantly less fragmented landscape structure compared to the rest of the river basin. Although the lower slope gradient in this area should make a positive contribution to the lower runoff heights (as verified for example by [51]), intensive land-use dominated by non-irrigated arable land and discontinuous urban fabric has the opposite effect, and the runoff intensity is therefore relatively high. Based on Fig. 7, an obvious difference in runoff processes can be seen in the eastern and western part of the river basin, while for the western part, the conditions that have already been mentioned above apply (i.e., homogeneous blocks with a uniform land-use on less sloping land, which are, however, intensively utilized). In the eastern part of the basin, we can observe different processes. The considerable rugged relief and the presence of relatively deep valleys also determine the nature of land-use: individual sites with homogeneous land-use are significantly smaller, and the runoff height is therefore very variable. The lower runoff values are rather obtained in the headwater parts of the sub-basins present here, which are often forested and less inclined (the slope ranges usually from 5 to 10°). The outlined difference between the eastern and western half of the river basin is also evident when evaluating the density of the river network, which is affected by terrain; while in the western part it

Table 4 Average soil water retention after a causal precipitation of 42 mm, computed for the nine groups of habitats/land cover categories

Group of habitats	Average soil water retention [mm]	% of a causal precipitation
Wetlands and riparian vegetation	38.4	91.5
Forests – natural	37.1	88.3
Secondary grasslands and heathlands	37.2	88.5
Springs and peat bogs	**39.1**	**93.1**
Water bodies – natural and unnatural	38.9	92.7
Forests – unnatural	37.3	88.9
Shrubs – natural and unnatural	36.2	86.1
Built-up and impermeable surface	**34.8**	**82.8**
Unnatural permeable surface	36.7	87.3

Extreme values are marked in bold

ranges between 0.2 and 1.4 km km^{-2}, in the hilly eastern part, it acquires values between 3.0 and 3.5 km km^{-2} [35].

For the nine defined categories of land-use, which differ in terms of their soil permeability and degree of anthropogenic influence, the average volume of captured water after a causal precipitation of 42.0 mm (homogeneous in the whole river basin) was further calculated (see Table 4). The analysis confirmed the working hypothesis, namely that natural or close-to-nature habitat types (land-use categories) will be able to retain the largest amount of rainwater and, conversely, the lowest water retention capacity will be provided by significantly anthropogenically affected habitats. It should be emphasized that habitats classified as "built-up and impermeable surface", in addition to continuous urban areas, include habitats of "rocks and quarries" or habitats of "landfills and construction sites", which maintain a certain water retention function (up to 83%); i.e., the water is still infiltrated and does not drain in the form of surface runoff.

We can summarize that after the causal precipitation of 42 mm, more than 18.4 million m^3 of rainfall fall in the study area, and approximately 88% of this volume was retained in the river basin. Immediately after the precipitation, approximately 2.1 million m^3 of water drained out of the basin in the form of surface runoff, which is an amount corresponding to about two-thirds of the operating volume of the Fryšták water reservoir, located in the central part of the Dřevnice river basin. The average value of water retention in the entire analysed river basin (on all types of land cover) is 36.71 mm m^{-2}. In order to identify specific sites where surface runoff is primarily formed, areas from which more than 20% of a causal precipitation flowing downslope as surface runoff were selected from the runoff height model (Fig. 8). As can be seen from the map in Fig. 8, the sites understood as the "surface runoff source areas" include larger cities in the river basin (especially the city of Zlín), but there is also a relatively large area of arable land in the central part of the basin, east of the Fryšták water reservoir, with high potential for surface runoff formation.

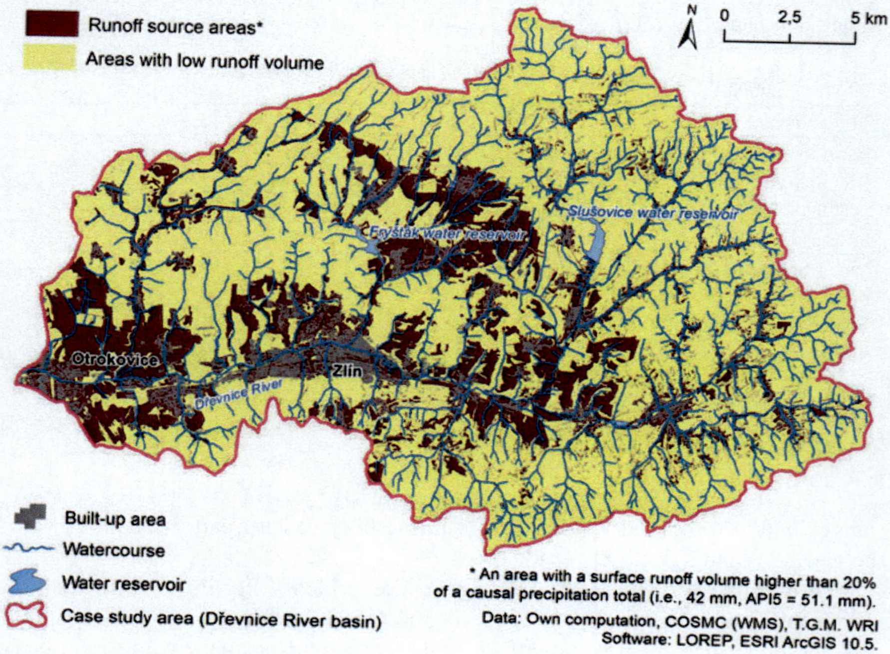

Fig. 8 Spatial distribution of surface runoff source areas within the Dřevnice River basin

3.2 Links between Soil Water Retention and Habitat Types

In order to obtain more detailed information about the water retention function of the landscape, the dependencies between individual habitats (natural, anthropogenically influenced and completely artificial habitats were taken into account) and their soil water retention capacity were analysed in addition to the relationship between land-use and soil water retention. A more detailed look at the soil water retention within the habitats occurring in the area of interest (see Fig. 9) shows a relatively large variability between individual types of habitats. High water retention values, similar to the amount of water fell during the analysed precipitation episode (i.e., 42.0 mm), have been reported in several habitats, but in many cases, vegetation in close proximity to watercourses (e.g., "*Petasites* fringes of montane brooks", "muddy river banks", and "reed vegetation of brooks") occur on permanently wet soils or muddy substrates, and low values of surface runoff are therefore a completely natural phenomenon due to permanent waterlogging. Among the habitats in which a large water retention capacity was found, and at the same time in which the vegetation was not directly conditioned by the presence of watercourses, we can especially highlight "Pollonian oak-hornbeam forests" (with computed soil water retention of 41.05 mm m^{-2}), "Broad-leaved dry grasslands without significant occurrence of orchids and with Juniperus communis" (39.94 mm m^{-2}) and "Dry

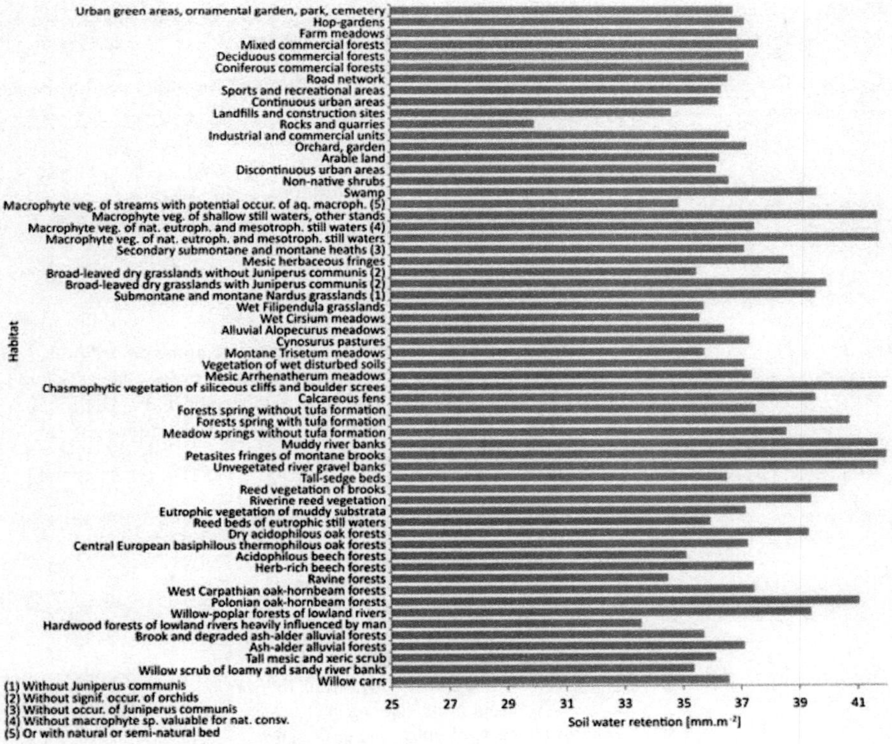

Fig. 9 Soil water retention capacity computed for all habitat types observed in the Dřevnice River basin, following a causal precipitation of 42 mm

acidophilous oak forests“ (39.32 mm m^{-2}). These habitats can therefore be considered to be very effective in terms of their ability to retain rainwater and improve the water retention function of the landscape. On the contrary, habitats that are significantly anthropogenically influenced are characterized by the most limited water retention function; e.g., “landfills and construction sites” (34.60 mm m^{-2}) and “hardwood forests of lowland rivers heavily influenced by man” (33.57 mm m^{-2}). Low values were also recorded in natural or near-to-nature habitats, but these were again cases in which soil water retention was not possible or was significantly limited for natural reasons, such as “vegetation of exposed fishpond bottoms” (31.29 mm m^{-2}) or “ravine forests” (34.48 mm m^{-2}). Overall, it can be stated that the “mixed commercial forests” (37.60 mm m^{-2}) and “coniferous commercial forests” (37.28 mm m^{-2}) have the highest water capacity from a set of unnatural habitats or habitats most significantly affected by humans.

All analysed habitats occurring on naturally very poorly permeable soils (i.e. a total of 28 habitats) were divided into three categories for easier interpretation, depending on their maximum water retention potential. The first category included habitats with the lowest ability to eliminate surface runoff and increase the water

Table 5 Habitats classified into three categories according to their water retention potential on poorly permeable soils (hydrological group "D")

Category	Max. soil water retention [mm]	Habitats
I	25.99	Willow scrub of loamy and sandy river banks Tall-sedge beds Vegetation of exposed fishpond bottoms
II	38.62	Hardwood forests of lowland rivers heavily influenced by man Ravine forests Eutrophic vegetation of muddy substrata Forests spring with tufa formation Forests spring without tufa formation Broad-leaved dry grasslands without significant occurrence of orchids and with *Juniperus communis* Macroph. Vegetation of natural eutrophic and mesotrophic still waters without macroph. Species valuable for nature conservation
III	41.09	Tall mesic and xeric scrub Ash-alder alluvial forests Brook and degraded ash-alder alluvial forests Polonian oak-hornbeam forests Carpathian oak-hornbeam forests Mesic Arrhenatherum meadows Herb-rich beech forests Acidophilous beech forests Central European basiphilous thermophilous oak forests Dry acidophilus oak forests Reed beds of eutrophic still waters Meadow springs without tufa formation Vegetation of wet disturbed soils Cynosurus pastures Wet Cirsium meadows Wet Filipendula grasslands Broad-leaved dry grasslands without significant occurrence of orchids and without Juniperus communis Mesic herbaceous fringes

retention function of the landscape (a maximum of 62% of the volume of causal precipitation was captured), but these were mostly relatively specific habitats (see Table 5) that were dependent on the presence of a watercourse, and their potential enlargement is possible only in isolated cases (e.g., during complex restorations of watercourses and their floodplains). The second category included more habitats (only examples of the most widespread habitats, covering the largest area are given in Table 5) that had a "moderate" ability to retain rainwater (up to 92% of the precipitation volume). The third category included habitats with the greatest potential to perform a water retention function (in exceptional cases, they can retain up to 98% of the precipitation volume). In the given overview, only the most represented habitats within the study area are mentioned. It should be noted that the stated maximum values of soil water retention in individual habitats are valid for causal precipitation, taking into account the antecedent precipitation of 51.1 mm in 5 days.

4 Discussion and Conclusions

The results of the analysis of soil water retention and related surface runoff source areas are consistent with the established hypotheses, concerning the potential influence of the habitat type on the volume of surface runoff, if we neglect the influence of terrain properties. Although the main variables shaping runoff conditions are mainly the soil properties (infiltration capacity), terrain roughness (steepness of slopes), and prevailing land-use category, it has been proven that individual habitat types also have a verifiable effect on the hydrological regime of the landscape. Based on data from the forest-agricultural river basin of a medium-sized watercourse located in the Czech Republic, it was verified that the natural and near-to-natural habitats are characterized by a higher ability to accumulate water and can thus make a positive contribution to mitigating the effects of climate change, manifested in Europe by more frequent flash floods [52].

However, it must be taken into account that the spatial distribution of individual habitat types is also the result of local conditions, including soil properties; thus, the primary variables determining the character of runoff processes in the basin remain the relief and soil. At the same time, it is important to mention that the modelling technique used do not take into account the rainfall interception by vegetation canopy or the effect of evapotranspiration, and so the reported final values of water retention and surface runoff for different habitats will actually be lower, especially in forest habitats. The interception loss can be relatively high for certain types of habitats [53]. Kermavnar and Vilhar [54] state, based on a case study from Slovenia, that the highest values of interception are reached by a mixed forest (18.0% of causal precipitation) and the intermediate level of interception rate (7.1% of causal precipitation) was found for the floodplain hardwood forest. Černý et al. [55] state that, in the Czech Republic, the interception loss of the spruce forests canopy can be up to 30% of the causal precipitation (the exact interception loss depends on the total precipitation – e.g., interception of 30% can be valid for gross precipitation up to 15 mm).

The results from our study, therefore, relate exclusively to the influence of the character of individual habitats on the formation of surface runoff (in terms of their species composition, representation of herbaceous, shrub and tree vegetation floor, and average surface cover due to vegetation).

Data on the potential impact of individual habitats on surface runoff coefficient, and thus the water retention capacity of the given habitats, can represent a relatively important source of information when designing restoration projects for terrestrial parts of river landscapes, but also in other localities outside the river landscape, aimed at mitigating the frequency of hydrological extremes.

Within the study area, localities with an increased potential to generate surface runoff (i.e., performing as runoff source areas) were identified, which were located mainly in the built-up area. The highest runoff values (more than 38% of the initial precipitation), and thus the minimum water capacity, were identified in continuous urban areas located on sloping land; i.e., especially larger settlements and the outskirts of cities. Therefore, it can be stated that the source areas of surface runoff in the Dřevnice River basin are located out of the headwater areas of the main watercourses, which drain larger parts of the river basin. This fact is relatively favourable in terms of groundwater quality and quantity, as in the spring areas in the eastern part of the river basin there is a significant infiltration of rainwater, which positively contributes to groundwater replenishment and the overall balance of discharges in local watercourses. The higher water retention capacity of the landscape in the headwater areas of local watercourses also has a positive impact on reducing the level of flood risk in municipalities located in narrow valleys where these streams flow. The retention of higher amounts of water and its subsequent release reduces the likelihood of potential flash floods, because these occur after intense rainfall in an environment that is unable to retain rainwater and to distribute water runoff over a longer time interval.

For more effective flood protection of larger towns in the lower part of the river basin, it is appropriate to implement certain measures in the floodplains along the middle part of the watercourses and its headwater areas, which would increase the water retention capacity of the landscape under the conditions of ongoing climate and environmental change. Based on our results that identified the runoff source areas located especially in the central part of the river basin, mainly occupied by agricultural areas, selected structural measures can have potential for reducing surface runoff volume – e.g. the creation of retention ponds, flood polders and fills, changes in sowing practices or grassing of river banks. In the headwater area, non-structural measures such as change in infrastructure policy or land-use management by planning tools become important factors, because there is no immediate surface runoff in these localities, but a higher ability of the landscape to retain rainwater can have a positive effect on ensuring a balanced discharge in watercourses and eliminating the occurrence of possible extreme hydrological situations. As the model outputs confirmed that the water retention capacity is significantly higher in several selected habitat types (e.g., ash-alder alluvial forests or Polonian oak-hornbeam forests), it is appropriate to support the expansion of these habitats where conditions allow for river and floodplain restoration plans. The vast majority of the above-mentioned measures can be classified as nature-based solutions, which are characterized by lower financial demands of implementation and at the same time these measures increase the ecological stability and support the quality of ecosystem services provided. In general, we can summarize that in the current intensively utilized agricultural landscapes the surface runoff source areas are very often concentrated in the middle and lower part of the Dřevnice River basin, within wider floodplains whose naturally higher water retention capacity is degraded by the presence of artificial and impermeable surfaces. Therefore, the measures that are feasible especially in the middle river basin zone (e.g. the construction of flood polders or changes in sowing practices in the case of an agricultural landscape) should be implemented in order to eliminate the risks associated with the formation of surface runoff (soil erosion and possible occurrence of flash floods in particular).

Acknowledgment The authors of this chapter are grateful to BSc. Kateřina Machová for her assistance in processing of background data for modelling and cooperation during review of the literature. The authors also thank two reviewers for their valuable comments and recommendations that helped to significantly improve the quality of this chapter. This research was funded by the Ministry of Education, Youth and Sports of the Czech Republic within the project "LAND4FLOOD: Natural Flood Retention on Private Land", Reg. No. LTC18025.

References

1. Brázdil R, Trnka M, Řezníčková V, Balek J, Bartošová L, Bičík I, Cudlín P, Čermák P, Dobrovolný P, Dubrovský M, Farda A, Hanel M, Hladík J, Hlavinka P, Janský B, Ježík P, Klem K, Kocum J, Kolář T, Kotyza O, Krkoška Lorencová E, Macků J, Mikšovský J, Možný M, Muzikář R, Novotný I, Pártl A, Pařil P, Pokorný R, Rybníček M, Semerádová D, Soukalová E, Stachoň Z, Štěpánek P, Štych P, Treml P, Urban O, Vačkář D, Valášek H, Vizina A, Vlnas R, Vopravil J, Zahradníček P, Žalud Z (2015) Sucho v českých zemích: minulost, současnost, budoucnost. Centrum výzkumu globální změny Akademie věd České republiky, v. v. i., Brno. 401 p. (in Czech)
2. Blauhut V, Stahl K, Stagge JH, Tallaksen LM, De Stefano L, Vogt J (2016) Estimating drought risk across Europe from reported drought impacts, drought indices, and vulnerability factors. Hydrol Earth Syst Sci 20(7):2779
3. Spinoni J, Vogt JV, Naumann G, Barbosa P, Dosio A (2018) Will drought events become more frequent and severe in Europe? Int J Climatol 38(4):1718–1736
4. Pan R, Martinez A, Brito T, Seidel EP (2018) Processes of soil infiltration and water retention and strategies to increase their capacity. J Exp Agr Int 20(2):1–14
5. Brown AG, Quine TA (1999) Fluvial processes and environmental change. Wiley, Chichester. 413 p
6. Kantor P, Krečmer V, Šach F, Švihla V, Černohous V (2003) Lesy a povodně. Praha: Ministerstvo životního prostředí ČR. 48 p. (in Czech)
7. Walsh CJ, Leonard AW, Ladson AR, Fletcher TD (2004) Urban stormwater and the ecology of streams. Cooperative Research Centre for Freshwater Ecology, Cooperative Research Centre for Catchment Hydrology, Canberra, 44 p
8. Zeigert SJ, Hubbart JA (2019) Quantifying relationships between watershed characteristics and hydroecological indices of Missouri streams. Sci Total Environ 654:3005–3015
9. Bogan E, Doina STAN, Vărvăruc D (2015) The impact of anthropogenic activities on components of the natural environment of the Titu plain. GEOREVIEW: scientific annals of Stefan cel Mare University of Suceava. Geogr Ser 24(1):54–64
10. Tabacchi E, Lambs L, Guilloy H, Planty-Tabacchi AM, Muller E, Decamps H (2000) Impacts of riparian vegetation on hydrological processes. Hydrol Process 14(16–17):2959–2976
11. Cramer VA, Hobbs RJ (2002) Ecological consequences of altered hydrological regimes in fragmented ecosystems in southern Australia: impacts and possible management responses. Austral Ecol 27(5):546–564
12. Zhang M, Liu N, Harper R, Li Q, Liu K, Wei X, Ning D, Hou Y, Liu S (2017) A global review on hydrological responses to forest change across multiple spatial scales: importance of scale, climate, forest type and hydrological regime. J Hydrol 546:44–59

13. Farooqi TJA, Abbas H, Hussain S (2020) The hydrological influence of forest harvesting intensity on streams: a global synthesis with implications for policy. Appl Ecol Environ Res 18(4):4987–5009
14. Kanianska R, Kizeková M, Nováček J, Zeman M (2014) Land-use and land-cover changes in rural areas during different political systems: A case study of Slovakia from 1782 to 2006. Land Use Policy 36:554–566
15. Bičík I, Kupková L, Jeleček L, Kabrda J, Štych P, Janoušek Z, Winklerová J (2015) Land use changes in Czechia 1845–2010. In: Land use changes in the Czech Republic 1845–2010. Springer, Cham, pp 95–170
16. Štěrba O, Měkotová J, Bednář V, Šarapatka B, Rychnovská M, Kubíček F, Řehořek V (2008) Říční krajina a její ekosystémy. Univerzita Palackého v Olomouci, Olomouc. 391 p. (in Czech)
17. Mitsch WJ, Gosselink JG (2000) The value of wetlands: importance of scale and landscape setting. Ecol Econ 35:25–33
18. Nienhuis NH (2008) Environmental history of the Rhine–Meuse Delta: an ecological story on evolving human-environmental relations coping with climate change and sea-level rise. Springer, Berlin
19. Verhoeven JTA, Setten TL (2010) Agricultural use of wetlands: opportunities and limitations. Ann Bot 105(1):155–163
20. Grinchenko OS, Sviridova TV, Kontroshchikov VV (2020) Long-term dynamics of ecosystems in the north of Moscow region (rationale for creation of the "crane country" nature park). Ecosyst Ecol Dynam 4(1):138–169
21. Pithart D, Dostál T, Langhammer J, Janský B (2012) Význam retence vody v říčních nivách. Daphne ČR – Institut aplikované ekologie. 141 p. (in Czech)
22. Starke P, Göbel P, Coldewey WG (2010) Urban evaporation rates for water-permeable pavements. Water Sci Technol 62(5):1161–1169
23. Wang L, Qu JJ (2007) NMDI: a normalized multi-band drought index for monitoring soil and vegetation moisture with satellite remote sensing. Geophys Res Lett 34(L20405)
24. Hale R, Scoggins M, Smucker NJ, Suchy AK (2016) Effects of climate on the expression of the Urban stream syndrome. Freshwater Sci 35(1):421–428
25. Vietz GJ, Walsh CJ, Fletcher TD (2016) Urban hydrogeomorphology and the Urban stream syndrome: treating the symptoms and causes of geomorphic change. Prog Phys Geogr Earth Environ 40(3):480–492
26. Murphy B, Wre D, Asce M (2020) Urban stream assessment procedure: a framework for assessing stream health in the Urban environment. Watershed management conference, pp 99–108
27. Hartmann P, Zink A, Fleige H, Horn R (2012) Effect of compaction, tillage and climate change on soil water balance of Arable Luvisols in Northwest Germany. Soil Tillage Res 124:211–218
28. European Commission (2013) Mapping and assessment of ecosystems and their services an analytical framework for ecosystems assessments under action 5 of EU biodiversity stategy to 2020: discussion paper-final April 2013. Publication Office, Luxembourg. https://doi.org/10.2779/12398
29. Hartmann T, Slavíková L, McCarthy S (2019) Nature-based flood risk management on private land: disciplinary perspectives on a multidisciplinary challenge. Springer, Cham
30. Wilkinson ME (2019) Commentary: Mr. Pitek's land from a perspective of managing hydrological extremes: challenges in upscaling and transferring knowledge. In: Hartmann T, Slavíková L, McCarthy S (eds) Nature-based flood risk management on private land disciplinary perspectives on a multidisciplinary challenge. Springer, Cham, pp 69–76
31. Shine C, de Klemm C (1999) Wetlands, water and the law. Using law to advance wetland conservation and wise use. IUCN environmental policy and law paper 38. 332 p
32. Hájek M, Vrabcová P (2020) Consideration of forest ecosystems services in environmental management accounting. Wood Res 65(1):135–148

33. RISWC (2020) SOWAC GIS geoportal – soil in maps. Research Institute for Soil and Water Conservation [online] Accessed 9 Apr 2020. https://geoportal.vumop.cz/index.php?projekt=vodni
34. CHMI (2020) Monthly and annual data – average air temperature and precipitation totals. Czech Hydrometeorological Institute [online] Accessed 9 Aug 2020. http://portal.chmi.cz/historicka-data/pocasi/mesicni-data/mesicni-prehledy-pozorovani
35. T.G.M. WRI (2006) Characteristics of streams and river basins in the Czech Republic. T.G.M. Water Research Institute (online). Accessed 8 July 2020 https://www.dibavod.cz/index.php?id=24
36. Rajnoch D (2015) Návrh PPO na zvolené části toku (Design of flood protection). Diploma thesis. Faculty of Civil Engineering, Institute of Water Structures, Brno University of Technology. 61 p. (in Czech)
37. NCA CR (2014) Habitat mapping layer [electronic database]. Version 2014. Prague. Nature Conservation Agency of the Czech Republic, Global Change Research Institute CAS
38. Pechanec V, Cudlín P, Machar I, Brus J, Kilianová H (2020) Modelling of the water retention capacity of the landscape. In: Zelenakova M, Fialová J, Negm A (eds) Assessment and protection of water resources in the Czech Republic. Springer water. Springer, Cham
39. US SCS (1956) National Engineering Handbook, section 5: hydraulics. United States. Soil Conservation Service
40. Janeček M, Kovář P (2010) Aktuálnost "Metody čísel odtokových křivek – CN" k určování přímého odtoku z malého povodí. Vodní hospodářství 7:187–190. (in Czech)
41. US SCS (1972) 'Hydrology, National Engineering Handbook, supplement A, section 4, chapter 10, Soil Conservation Service, U.S.D.A., Washington
42. Zhao Y, Wei F, Yang H, Jiang Y (2011) Discussion on using antecedent precipitation index to supplement relative soil moisture data series. Procedia Environ Sci 10:1489–1495
43. Janeček M (2012) Ochrana zemědělské půdy před erozí. Česká zemědělská univerzita, Praha. 113 p. (in Czech)
44. Pasák V (1983) Protection of agricultural soils against erosion. Research Institute of Melioration and Soil Protection, Prague. (in Czech)
45. Sun D, Yang H, Guan D, Yang M, Wu J, Yuan F, Jin C, Wang A, Zhang Y (2018) The effects of land use change on soil infiltration capacity in China: a meta-analysis. Sci Total Environ 626:1394–1401
46. Mishra SK, Sahu RK, Eldho TI, Jain MK (2006) An improved IaS relation incorporating antecedent moisture in SCSCN methodology. Water Resour Manage 20:643–660
47. ESA (2019) Copernicus – overview [online]. Accessed 9 Aug 2019. http://www.esa.int/Our_Activities/Observing_the_Earth/Copernicus/Overview4
48. Antognelli S (2019) NDVI and NDMI Vegetation Indices: Instruction for use [online]. Accessed 10 Aug 2019. https://www.agricolus.com/en/indici-vegetazione-ndvindmi-istruzioni-luso/
49. Seják J, Cudlín P, Pokorný J, Zapletal M, Petříček V, Guth J, Chuman T, Romportl D, Skořepová I, Vacek V, Vyskot I, Černý K, Hesslerová P, Burešová R, Prokopová M, Plch R, Engstová B, Stará L (2010) Hodnocení funkcí a služeb ekosystémů České republiky. Univerzita JE Purkyně v Ústí nad Labem, Fakulta životního prostředí 198 p. (in Czech)
50. European Environment Agency (2015) Water retention potential of Europe's forests. https://doi.org/10.2800/790618
51. El Kateb H, Zhang H, Zhang P, Mosandl R (2013) Soil erosion and surface runoff on different vegetation covers and slope gradients: a field experiment in southern Shaanxi Province, China. Catena 105:1–10

52. Marchi L, Borga M, Preciso E, Gaume E (2010) Characterisation of selected extreme flash floods in Europe and implications for flood risk management. J Hydrol 394:118–133
53. van Dijk AIJM, Bruijnzeel LA (2001) Modelling rainfall interception by vegetation of variable density using an adapted analytical model. Part 1. Model description. J Hydrol 247:230–238
54. Kermavnar J, Vilhar U (2017) Canopy precipitation interception in urban forests in relation to stand structure. Urban Ecosyst 20:1373–1387
55. Černý T, Dohnal M, Tesař M (2014) Význam intercepce v hydrologickém cyklu povodí pramenných oblastí. Stavební obzor 5–6:110–114

Using Landscape Connectivity to Identify Suitable Locations for Nature-Based Solutions to Reduce Flood Risk

Zahra Kalantari, Samaneh Seifollahi-Aghmiuni, Hanna Nordin von Platen, Matilda Gustafsson, Omid Rahmati, and Carla Sofia Santos Ferreira

Contents

Z. Kalantari (✉)
Department of Physical Geography and Bolin Centre for Climate Research, Stockholm University, Stockholm, Sweden

Navarino Environmental Observatory, Costa Navarino, Navarino Dunes, Messinia, Greece

KTH, School of Architecture and the Built Environment (ABE), Sustainable Development, Environmental Science and Engineering, Sustainability Assessment and Management, Stockholm, Sweden
e-mail: zahra.kalantari@natgeo.su.se; zahrak@kth.se

S. Seifollahi-Aghmiuni
Department of Physical Geography and Bolin Centre for Climate Research, Stockholm University, Stockholm, Sweden

Navarino Environmental Observatory, Costa Navarino, Navarino Dunes, Messinia, Greece
e-mail: samaneh.seifollahi@natgeo.su.se

H. N. von Platen and M. Gustafsson
KTH, School of Architecture and the Built Environment (ABE), Sustainable Development, Environmental Science and Engineering, Sustainability Assessment and Management, Stockholm, Sweden

Carla S. S. Ferreira, Zahra Kalantari, Thomas Hartmann, and Paulo Pereira (eds.), *Nature-Based Solutions for Flood Mitigation: Environmental and Socio-Economic Aspects*, Hdb Env Chem (2022) 107: 339–354, DOI 10.1007/698_2021_771,

Published online: 6 May 2021

Abstract Climate change and population growth are exacerbating environmental challenges and natural hazards. Extreme events such as flooding due to heavy precipitation are occurring more frequently and becoming more severe throughout the world. Future projections indicate increasing risks and serious threats to human societies, particularly those living in urban areas. To mitigate associated socio-economic challenges and support sustainable urban development, nature-based solutions (NBS) are being introduced in mitigation and adaptation strategies. However, NBS need to be carefully located and suitably designed to achieve their full potential in terms of flood mitigation. Landscape connectivity, addressing various landscape processes and components, can potentially be utilized for land management and biodiversity conservation. For instance, hydrological connectivity within a catchment, as part of the landscape connectivity, describes the main water flow pathways and areas where runoff tends to accumulate, and can thus be useful in identifying the best locations for NBS from a hydrological viewpoint. In this chapter, we describe use of landscape connectivity to identify suitable locations for NBS to reduce flood risk in urban areas, and exemplify the process for two distinct urban catchments, located in Sweden and Portugal. The results showed good usefulness of the method and revealed the importance of integrating connectivity mapping into future NBS planning practices and decision support systems.

Keywords Connectivity, Flood risk mitigation, Nature-based solutions, Urban areas, Wetlands

1 Introduction

Connectivity is defined as the degree to which a geomorphic system, such as a hydrological catchment, facilitates transfer and movement of water and sediments through coupling relationships among its components [1, 2]. Landscape connectivity is originally defined as "the degree to which the landscape facilitates or impedes movement among resource patches [3]." According to this definition, landscape connectivity combines a description of the physical structure of the landscape with

O. Rahmati
Soil Conservation and Watershed Management Research Department, Kurdistan Agricultural and Natural Resources Research and Education Center, AREEO, Sanandaj, Iran

C. S. S. Ferreira
Department of Physical Geography and Bolin Centre for Climate Research, Stockholm University, Stockholm, Sweden

Navarino Environmental Observatory, Costa Navarino, Navarino Dunes, Messinia, Greece

Research Centre of Natural Resources, Environment and Society (CERNAS), Polytechnic Institute of Coimbra, Coimbra Agriculture School, Bencanta, Coimbra, Portugal human community's response to that structure. It includes the type, amounts, and arrangement of habitat and land-use on the landscape and population dynamics [4]. This chapter focuses on structural aspects of landscape connectivity, representing the potential for a specific sediment particle or water molecule to move through the compartments of a landscape system at different temporal scales [5, 6]. The connectivity at multiple temporal scales depends on the interactions between catchment components, structural characteristics of catchment surfaces (e.g., topography, roughness), and processes driven by fluxes of water and sediment (e.g., erosion and sedimentation) [7, 8]. Thus, connectivity is not static, but rather varies over time and space due to the interactions between external forcing (mainly precipitation and temperature), landscape properties (i.e., structural connectivity), and the magnitude of water and sediment fluxes (i.e., functional connectivity). These ultimately determine the evolution of landforms and the changes in landscape properties caused by erosional and sedimentation processes [8]. Current ongoing climate and human-induced changes can significantly modify connectivity in catchments, due to their impact on these processes and interactions.

A clear understanding of landscape connectivity can support efficient and sustainable management of catchments with regard to land and water resources [9], which are globally undergoing significant change due to ongoing climate and land-use changes driven by population growth. Many parts of the transport infrastructure and built environment in urban areas are vulnerable to weather extremes because of the associated risk for flooding, landslides, and erosion [10]. Global warming will have a direct effect on future precipitation patterns, as a warmer climate will increase evapotranspiration and atmospheric moisture load, which in turn will increase the frequency of intense precipitation events. This will increase the probability of natural hazards such as flooding [11, 12]. In urban areas, maintenance costs of transport infrastructures, which are characterized by long lifetimes and high investment costs, due to weather stresses account for 30–50% of their installation costs (EC, 2013). In addition, land-use changes such as soil sealing and expansion of impervious land surfaces in urban areas increase the vulnerability to pluvial flooding [13, 14].

Conventional engineering solutions to problems caused by climate and human-induced changes in an urbanized catchment may not always be sustainable and cost-effective [15, 16]. Instead, the European Commission (EC) advocates strategies inspired and supported by nature, the so-called nature-based solutions (NBS) [17]. Natural and constructed wetlands, ponds, canals, and ditches provide various ecosystem services (e.g., flood regulation, water quality improvement, increased biodiversity) and are commonly used as NBS in urban areas. Wetlands in particular decrease the connectivity within an urbanized catchment and thus reduce the flood risk, through mobilizing water and sediment movement and storage capacity. The greatest advantage of NBS wetlands is their multi-functionality. Apart from providing flood regulatory services in an urban environment by e.g., mimicking the function of natural wetlands, they provide several co-benefits that can strengthen urban resilience and mitigate loss of biodiversity, global warming effects, and threats to human health. These co-benefits are essential for sustainable urban development, which in a future with increased urbanization will become even more important as a key factor in creating livable and attractive cities.

Developing sustainable cities and communities, as identified in the United Nations Sustainable Development Goals and stressed in Agenda 2030, requires more resilient flood control management to cope with increased urbanization and hydro-climatic changes, for which implementation of NBS is suggested as an efficient solution. Large-scale wetlands are internationally acknowledged as NBS measures for flood risk reduction in urban areas [18]. Knowledge about wetland effectiveness and co-benefits provided by large-scale wetlands has increased over the years, enabling possible mainstreaming in policy and planning practice. In fast-growing urban areas, where new housing areas are needed to meet the increased demand, space-efficient solutions for flood regulation services and contamination control are required. If these are implemented correctly, they have great potential to meet the principal goals formulated for NBS by [17] which are *Enhancing sustainable urbanization, restoring degraded ecosystems, developing climate change adaptation and mitigation, and improving risk management and resilience*.

Considering the dominant impact of global warming in increased frequency of flood events, rapid urbanization, and human developments, and associated implications for natural soil processes within urbanized catchments, integrating consideration of structural landscape connectivity into mitigation and adaptation strategies, recently proposed worldwide, can significantly support better design and implementation of these strategies in urban areas. It can thus enhance urban resilience to climate challenges under future developments. This chapter investigates the application of landscape connectivity aspects on flow processes to identify suitable locations for NBS (e.g., wetlands) for flood risk mitigation in urban catchments.

2 Connectivity and Dis-connectivity

Connectivity has recently received widespread scientific attention in the field of controlling runoff and soil erosion, where prediction of surface runoff patterns and sediment transport are important in urban risk mitigation [19, 20]. During recent decades, various conceptual frameworks have been established to describe landscape connectivity in terms of hydrological and sediment transport processes [21]. For example, [1] define connectivity in hydrology and geomorphology under three categories: *landscape connectivity*, addressing physical coupling of landforms within a catchment; *hydrological connectivity*, referring to the passage of water across a landscape, which generates catchment runoff responses; and *sedimentological connectivity*, relating to physical transfer of sediment and pollutants through a landscape. These three categories of connectivity highlight the degree of links and connections among inter-system components of a hydrological catchment.

Evaluations of landscape connectivity generally concentrate on the *structural connectivity* (network structure) and the *functional connectivity* (dynamic processes) of a catchment. Studies of structural connectivity aim to evaluate catchment properties (e.g., topography, surface roughness, soil properties, vegetation types and patterns, drainage network) that control the processes of water and sediment transfer and to analyze the spatial patterns and processes that influence flow paths in the catchment landscape. On the other hand, studies of functional connectivity aim to assess the way in which processes operate dynamically within the catchment (e.g., runoff and sediment delivery) and evolve over time at a specific spatio-temporal scale [22]. The structural and functional types of connectivity are often examined separately, but they interact actively with each other and should be studied in combination [23].

The nature and continuity of links between catchment compartments are controlled by different sets of processes at different positions in the catchment, so that fluxes may be connected (coupled) or disconnected (decoupled) over different time frames. Some catchments may have strong links and a high degree of coupling in their processes (i.e., high connectivity), while others may not. Catchment dis-connectivity is defined as the features and processes that disrupt flows, and thus decrease water and sediment transport in a catchment [24] . Various landforms can create dis-connectivity in a catchment and are generally divided into three categories: buffers, barriers, and blankets [6]. Buffers are landforms (e.g., dams, woody debris, and sediment slugs) that prevent water and sediment flow from entering channel networks, such as rivers and streams in the catchment (e.g., topographical properties such as slope and drainage area). Barriers are landforms that can disrupt water and sediment flows moving along the channel network when they reach it. Blankets are features that smother other landforms, drape channel surfaces, and affect the accessibility of water and sediment to entrainment (e.g., floodplains). The strength of coupling between landscape compartments is dictated by the type and distribution of the landforms creating dis-connectivity in the catchment. Dis-connectivity is higher in catchments with more complex morphology [6]. The increasing understanding of landscape connectivity has resulted in different methods and procedures to assess catchment hydrological processes, with particular focus on development of indices to describe landscape connectivity. For instance, sediment connectivity index (IC index) describes how well different parts of the catchment are linked together in terms of sediment transport (driven by runoff) and connectivity [25]. The index can be derived from high-resolution digital elevation models (DEMs) as the only data source. Use of IC index can reveal possible links between structural connectivity characteristics such as hillslope, hydrological network, and features acting as storage areas [26]. It has been applied in the field of sediment transport, but also in a broader context for flood risk assessment [27, 28].

3 Nature-Based Solutions (NBS) for Flood Risk Mitigation in Urban Areas

Nature-based solutions are inspired by nature [17]. They were initially used to address agricultural problems, including food security and water resource management [29]. NBS have been developed and applied in different contexts over the past

20 years, but are relatively new in urban planning practices [28]. The driving force for applying NBS in urban planning has been recognition of the concept by major international organizations, e.g., the EC, International Union for Conservation of Nature and Natural Resources (IUCN), and World Bank [30].

In a hydrological catchment, NBS can be used to decrease connectivity by disconnecting water and sediment fluxes across the landscape [31]. This category of NBS, referred to as landscape solutions (including wetlands, forests, ponds, and grassed waterways), is multifunctional, providing flood regulation services in urban and non-urban contexts. It also provides other co-benefits, such as improved water quality due to the natural features of landscape solutions and their ability to retain nutrients and contaminants, increase biodiversity, and supply recreational services [32]. In an urban context, the limited amount of available space requires a balance between green (natural) and gray (built) infrastructures. Constructed wetlands composed of ponds and small-scale wetlands linked together by canals/ditches, called "hybrid infrastructures" [33], have proven to be effective solutions for environmental and societal challenges in these areas [34]. In a non-urban context, constructing large-scale wetlands can support floodplain restoration and make the landscape more resilient to climate change challenges. Such large-scale actions will also positively influence the whole catchment, through flood risk reduction [17].

For successful implementation of NBS in urbanized catchments, as disconnecting features for flood risk reduction [21], it is important to analyze the catchment using a systems approach, where mapping of connectivity is central [35]. It is also essential to adapt planned solutions to the specific local site conditions, including local climate, ecosystems, and management strategies, in order to avoid generalized solutions, achieve the full potential of NBS, and minimize the costs [34]. Connectivity has been suggested as a useful characteristic and indicator in identifying suitable natural locations for implementing NBS such as wetlands in catchments with urban and non-urban contexts, in order to mitigate flood risk [34, 36].

The placement and size of wetlands are two important factors in achieving their full potential in terms of flood risk mitigation and nutrient control [37]. Due to climate change and large-scale land- and water-use changes, the importance of suitable layout design for wetlands to address hydrological and sediment fluxes and linkages in a catchment has been highlighted [38, 39]. Using a landscape connectivity metric, the main sediment and hydrological links in the landscape can be identified, the major flow paths through the catchment landscape can be detected, and the most suitable location for implementing large-scale mitigation strategies (e.g., wetland construction) can be determined, in order to achieve large-scale flood control and nutrient removal by these landscape features [1, 39].

To highlight the potential for integrating landscape connectivity as a local characteristic with land management strategies, in the following sections we provide examples of use of IC index to identify suitable locations for wetland construction in two urban catchments, Bällstaån catchment in the Stockholm region, Sweden, and Ribeira dos Covões catchment around the city of Coimbra, Portugal. For these two areas, the spatial pattern of connectivity across the catchments was determined using IC index, in order to highlight the importance and assess the potential of considering landscape connectivity in flood risk mitigation strategies, such as implementing NBS. These two urban areas were selected due to their flood-prone characteristics and distinct urbanization levels, driven by the high urban development in Bällstaån and rapidly growing urbanization in Ribeira dos Covões, despite different climate conditions and environmental challenges. Investigating the application of connectivity aspects in these two distinct urban catchments provided a good opportunity to understand the suitability of the IC index methodology for identification of optimal locations for NBS in urban environments.

4 Application of Connectivity Definition for NBS Implementation in Urban Areas

4.1 Study Areas

Figure 1 shows geographical location and land-use cover for the Bällstaån catchment in Stockholm County, Sweden (Fig. 1a, b) and the Ribeira dos Covões catchment in Coimbra, Portugal (Fig. 1c, d). The Bällstaån catchment is located almost in the center of Stockholm County, to the north-west of Stockholm city. The Ribeira dos Covões catchment is located to the south-west of Coimbra city, in the Center region. The main stream in the Bällstaån catchment has a total length of 10.5 km, starting in Viksjö, Järfälla, and flowing through Tensta and Sundbyberg before reaching Lake Mälaren, the source of drinking water for Stockholm city. The area of this catchment is approximately 39 km^2. The Ribeira dos Covões catchment occupies an area of approximately 6 km^2 and contains a stream network composed of one perennial main stream and several ephemeral tributaries, all draining to the Mondego river floodplain in the north of the catchment. With regard to land-use, almost 75% of the Swedish catchment is already covered by built environment, mostly housing and commercial areas, but also some industries close to the stream [40] (Fig. 1b). The Portuguese catchment has experienced significant urban expansion, with urban area increasing from 6% to 40% between 1958 and 2012 (Fig. 1d). During the same period, agricultural area declined from 48% to 4%, and woodland area decreased from 46% to 40%, with the former natural forests of oak and mixed woodland being replaced by commercial eucalyptus and pine plantations [41].

In September 2015, parts of the Swedish catchment were flooded due to a heavy rainfall event of 80 mm per hour. A number of dwellings in the towns of Bromsten and Spånga were seriously damaged, and schools in Järfälla, near the stream, had to close due to flooded cellars [42, 43]. The Portuguese catchment has also suffered from floods in recent years, as a consequence of short but intense precipitation events. Regardless of infrastructure and management of stormwater runoff, extensive areas of impervious paved surfaces in these two catchments lead to high rates of overland flow during heavy precipitation events [13]. This increases the risk of flooding and contributes to mobilization of pollutants to water bodies, thus

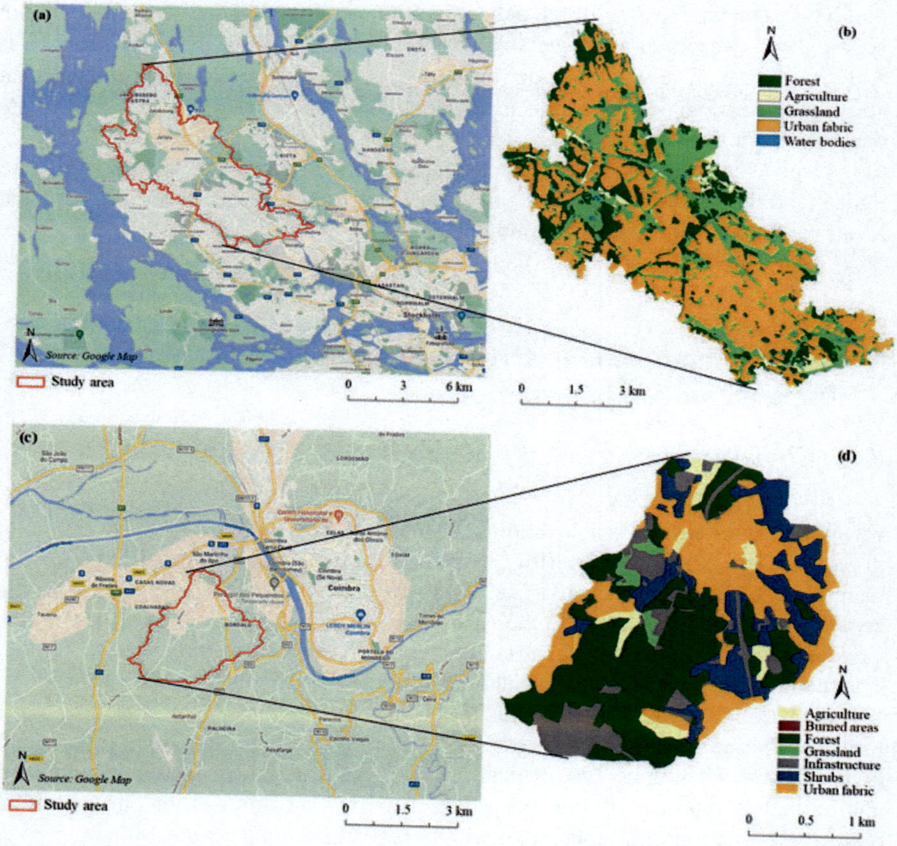

Fig. 1 Location of (**a**) Bällstaån catchment in relation to Stockholm city center, Sweden, and (**c**) Ribeira dos Covões catchment in relation to Coimbra city center, Portugal. Land-use map for (**b**) Bällstaån catchment, based on 2016 survey data and (**d**) Ribeira dos Covões catchment, based on 2012 survey data

threatening natural ecosystems and human health. This is particularly relevant in the Swedish catchment, since Lake Mälaren supplies the drinking water for Stockholm city [40].

4.2 *Connectivity Mapping*

The IC index developed by [25] can characterize structural properties of a catchment and support analysis for improved landscape management. Connectivity can be integrated with urban flood risk mitigation practices by using a landscape connectivity map, which can be developed based on IC calculation for various parts of the catchment. The IC index is calculated using Eq. (1), following the approach of [44], by combining land-use and topography characteristics:

$$IC = \log_{10}\left(\frac{D_{up}}{D_{dn}}\right) \tag{1}$$

where D_{up} and D_{dn} are the upslope and downslope characteristics, respectively, in the catchment (Fig. 2).

In order to use Eq. (1), a reference element needs to be selected in the catchment, which can be, e.g., the main stream or the catchment outlet. The degree to which one part of a catchment is able to connect to another part depends on land-use and hydrological conditions at spots across that specific part. Slope gradient of the catchment part is also relevant, as any water and sediment transfer will soon come to a halt on a part with zero slope [44].

The value of IC index varies within the range [−∞,+∞], with high values indicating higher connectivity. The upslope characteristic (D_{up}) is defined as the potential of contributing sediment produced upslope to downward reference points, calculated as:

$$D_{up} = \bar{W} \cdot \bar{S} \cdot \sqrt{A} \tag{2}$$

where $\bar{W}$ is average weighting factor, $\bar{S}$ is mean slope gradient of the upslope contributing area (m/m), and A is the upslope contributing area (m^2).

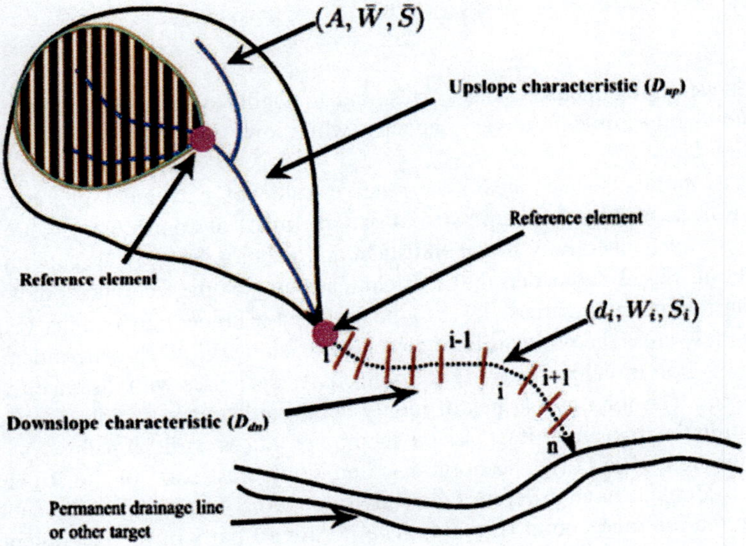

Fig. 2 Schematic illustration of variables and catchment components used for calculation of sediment connectivity index (IC index) (adapted from Crema & Cavalli, 2018)

The downslope component (D_{dn}) is defined by the travel (water or sediment) flow path length to reach the nearest reference point, calculated as:

$$D_{dn} = \sum_i \frac{d_i}{W_i S_i} \tag{3}$$

where i is the segment number through the travel path, d is length of each segment in the travel flow path according to the steepest downslope direction, W is weighting factor of segment i, and S is slope gradient of the pathway in segment i (Fig. 2).

In order to identify a weighting factor based on catchment characteristics, [45] suggest using a local measurement of topographical surface roughness, the so-called roughness index (RI). RI is derived from the standard deviation of the residual topography within a few meters scale. The residual topography is calculated as the difference between the original DEM and a refined version derived by averaging DEM values over an accurate moving cell window size. This is necessary in order to avoid the effect caused by large-scale topography, i.e., slope gradient. This refinement allows DEM to be used as the only input when calculating IC index. Therefore, the weighing factor and RI in a catchment can be quantified based on Eqs. (4) and (5), respectively, where 2.5 m resolution is considered for RI:

$$W_{\text{Cavalli}} = 1 - \left(\frac{RI}{RI_{MAX}}\right) \tag{4}$$

$$RI = \sqrt{\frac{\sum_{i=1}^{25} (x_i - x_m)^2}{25}} \tag{5}$$

High weighting factor values correspond to highly connected areas, i.e., high water flow and sediment delivery capacity, while low values correspond to weakly connected areas.

The IC index is calculated as a relative value of catchment characteristics, resulting in a catchment-specific spatial pattern (map) of connectivity. The spatial distribution of connectivity in the Bällstaån and Ribeira dos Covões catchments is presented in Fig. 3. Considering the catchment outlet as the reference point, in the Bällstaån catchment the areas closer to the outlet have higher connectivity (Fig. 3a), since in these areas the probability of discharge reaching the catchment outlet is high because of short travel pathways. The connectivity decreases with distance from the outlet point. The long red stripes of highly connected streams in low connectivity spots within the map are artifacts, as the method produces artificial pathways in very low slope areas (Fig. 3a). The north-east and south-west parts of the Ribeira dos Covões catchment have different connectivity patterns when taking the catchment outlet as the reference point (Fig. 3c). The north-east parts of the catchment have higher connectivity than the south-west parts. Areas with low connectivity adjacent to areas with high connectivity can be detected mostly in central parts of the

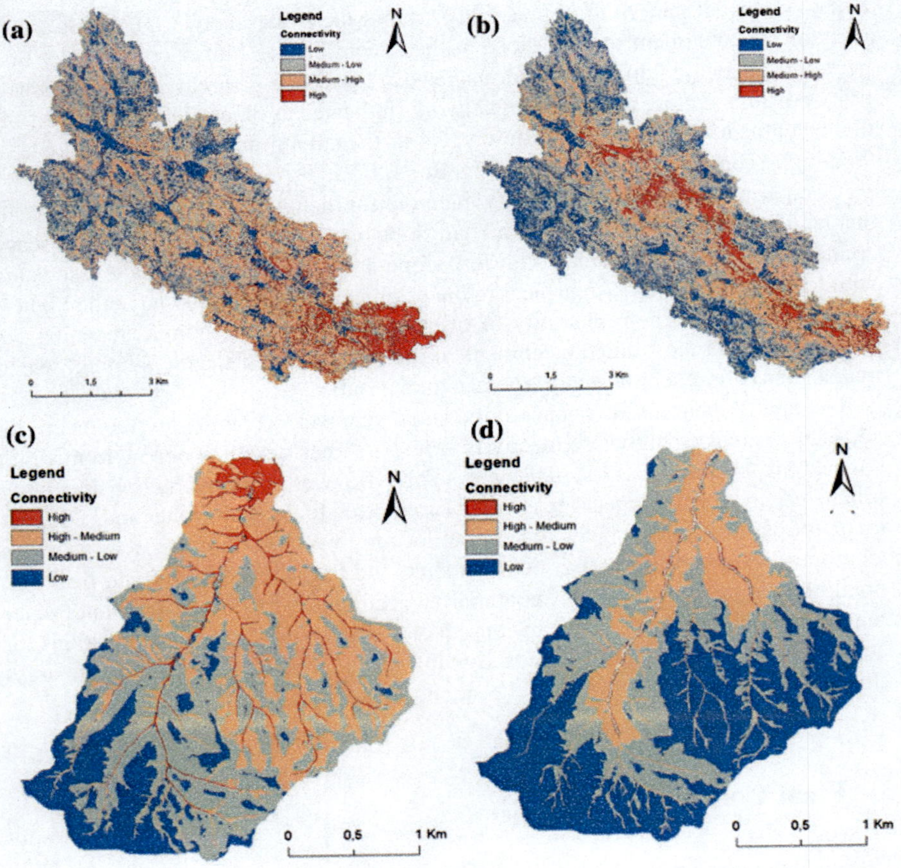

Fig. 3 Connectivity mapping based on IC index for (**a**, **b**) the Bällstaån catchment in Sweden and (**c**, **d**) the Ribeira dos Covões catchment in Portugal, when taking (**a**, **c**) the outlet as the reference point and (**b**, **d**) the main stream network as the reference point. Blue and red areas represent low and high connectivity, respectively

catchment. Some areas close to the western stream network with low elevations are also evident as areas with low connectivity. Furthermore, areas of low connectivity located on high elevations are evident both east and west of the perennial stream (Fig. 3c).

Considering the main stream network in the Bällstaån catchment as the reference point (Fig. 3b), connectivity is associated with the probability of discharge at any catchment point reaching the stream, which is higher at areas closer to the stream network and lower for areas located further away from the stream network. Some areas with low connectivity in the Bällstaån catchment are also located close to the stream (Fig. 3b). Long red stripes of highly connected streams in low connectivity areas can be seen in Fig. 3b. The mapping for Ribeira dos Covões catchment reveals a distinct spatial pattern of connectivity around the stream network (Fig. 3d), with parts of the catchment located close to the stream having higher connectivity than parts further away. Some minor anomalies can be seen, e.g., areas with low connectivity positioned close to the stream network (indicated as blue patches in Fig. 3d). In this catchment, major overland flow paths to the stream network are detectable as "red/high" connectivity tributaries (Fig. 3c, d).

Comparing results of connectivity mapping for the two urban catchments reveals that landscape topography (elevation) and slope play a key role in defining landscape connectivity. In general, areas with low slope and/or high elevation are associated with lower connectivity. Areas near to the catchment outlet usually have the highest connectivity, since the probability of overland flow from these areas reaching the outlet is high. In the studied catchments these areas are also steeply sloping, which makes them highly connected to the catchment outlet.

The spatial patterns of connectivity mapping reveal areas of low connectivity adjacent to areas of higher connectivity (Fig. 3). Areas with low or low to medium connectivity are generally suitable areas for NBS wetland construction in highly urbanized catchments. These areas are not connected to the main water and sediment transport pathways in such catchments, so locating wetlands in these areas provides temporary water storage and a sediment sink, and thus prevents overland flow and flooding. These areas have the potential to accumulate and delay overland water runoff and sediment flows, and can consequently be considered suitable locations for implementing NBS such as wetlands. Suitable areas for wetlands are limited to areas close to the stream network, since the stream is the main reason for flooding.

5 Final Considerations

This chapter shows the value and potential of using landscape connectivity with its structural focus in a catchment as a characteristic for determining the optimal location of NBS, so that their full potential can be exploited. In general, areas located in or adjacent to the stream network within a catchment are associated with higher flood risks. These areas usually have low slope and/or elevation, and low connectivity, allowing accumulation of water and sediment flows. Locations around the stream network (usually with low connectivity, but adjacent to areas with high connectivity) are often identified as suitable areas for implementing NBS such as wetlands. This highlights the relevance of using a metric of connectivity for flood risk mitigation practices in urbanized catchments. Although areas around the stream networks remain vulnerable to flooding, they have great potential for implementation of NBS wetlands and mitigation of downslope floods. In particular, when the stream network is low-order, with low discharge originating mainly from overland flow and stormwater runoff, NBS can be an appropriate strategy to regulate, treat, and protect catchment areas against flooding and contaminant transport.

Catchment-scale analysis of applying a connectivity index provided a conceptual understanding of the interactions within urban catchments. A great advantage with developing and using connectivity mapping is the possibility of identifying flow patterns and detecting potential catchment storage and discharge areas. This information can be used for a better understanding and analysis of water and sediment movement within a catchment. The use of IC index for connectivity mapping, reviewed in this chapter, has also been assessed in catchments of the Eastern Alps and resulted in realistic general connectivity patterns in agreement with field observations [25]. Also, it has been evaluated for identifying flood risks at the road-stream intersections, crucial locations where water and debris can accumulate and cause failures of the existing infrastructures, for two urban areas in southwest of Sweden. The results highlighted enhancements of accuracy and performance for statistic prediction models integrating IC index as a physical catchment characteristic. However, it only addresses structural connectivity in a catchment, while assessments of functional connectivity, composed of catchment-specific dynamic processes and conditions such as land-use, are also required in order to select suitable locations for NBS implementation.

The greatest challenge to successful implementation of NBS wetlands is reluctance to mainstream this solution in urban planning practice. To overcome this challenge, the total ecosystem effects generated by NBS in the whole catchment can be evaluated and compared with the effects before implementation. Implementation can also be improved by documenting strategies and designs for different local conditions. This can increase confidence in using NBS instead of conventional methods.

Landscape connectivity is an emergent property of species-landscape interactions, which results from the interaction between a behavioral process (movement) and the physical structure of the landscape. Broadly, two types of landscape connectivity as structural and functional can be considered. Structural connectivity just focuses on physical relationships and interactions, ignoring behavioral responses to landscape structure and features. Functional connectivity addresses dynamic processes on the landscape. Connectivity mapping using the IC index, used in this chapter is more related to structural connectivity which depends on the complexity of the landscape. For instance, in highly urbanized areas, the complexity of water and sediment transport will increase due to man-made infrastructures such as stormwater systems and roads, which cannot be considered in connectivity mapping based on IC index. This might constrain the usefulness of integrating connectivity definition into urban planning practices.

Acknowledgments This study arose from two research projects: 1) a project funded by Stockholm Region, KTH dnr. A-2019-0247, SLL dnr. LS 2018-0736; and 2) a project supported by the Portuguese Foundation for Science and Technology, through post-doctoral grant SFRH/BPD/120093/2016 to the last author. The authors would like to acknowledge networking support by the COST Action CA16209: Natural Flood Retention on Private Land (LAND4FLOOD).

References

1. Bracken LJ, Croke J (2007) The concept of hydrological connectivity and its contribution to understanding runoff dominated geomorphic systems. Hydrol Process 21(13):1749–1763
2. Heckmann T, Cavalli M, Cerdan O, Foerster S, Javaux M, Lode E, Smetanová A, Vericat D, Brardinoni F (2018) Indices of sediment connectivity: opportunities, challenges and limitations. Earth Sci Rev 187:77–108
3. Taylor PD, Fahrig L, Henein K, Merriam G (1993) Connectivity is a vital element of landscape structure. Oikos 68(3):571–572
4. Taylor PD, Fahrig L, With KA (2006) Landscape connectivity: a return to the basis. Connectivity conservation, pp 29–43
5. Hooke J (2003) Coarse sediment connectivity in river channel systems: a conceptual framework and methodology. Geomorphology 56(1–2):79–94
6. Fryirs K (2013) (Dis) Connectivity in catchment sediment cascades: a fresh look at the sediment delivery problem. Earth Surf Process Landf 38(1):30–46
7. Bracken LJ, Turnbull L, Wainwright J, Bogaart P (2015) Sediment connectivity: a framework for understanding sediment transfer at multiple scales. Earth Surf Process Landf 40(2):177–188
8. Llena M, Vericat D, Cavalli M, Crema S, Smith MW (2019) The effects of land use and topographic changes on sediment connectivity in mountain catchments. Sci Total Environ 660:899–912
9. Hălbac-Cotoară-Zamfir R, Keesstra S, Kalantari Z The impact of political, socio-economic and cultural factors on implementing environment friendly techniques for sustainable land management and climate change mitigation in Romania (2019). Sci Total Environ 654:418–429. https://doi.org/10.1016/j.scitotenv.2018.11.160
10. Michielsen A, Kalantari Z, Lyon SW, Liljegren E (2016) Predicting and communicating flood risk of transport infrastructure based on watershed characteristics. J Environ Manag 182:505–518. https://doi.org/10.1016/j.jenvman.2016.07.051
11. Rahmati O, Yousefi S, Kalantari Z, Uuemaa E, Teimurian T, Keesstra S, Pham TD, Tien Bui D (2019) Multi-hazard exposure mapping using machine learning techniques: a case study from Iran. Remote Sens 11:1943. https://doi.org/10.3390/rs11161943
12. European Environment Agency (EEA) (2013) Europe must adapt to stay ahead of a changing climate. https://www.eea.europa.eu/media/newsreleases/europe-must-adapt-to-stay. Accessed 17 Jul 2020
13. Ferreira CSS, Walsh RPD, Nunes JPC, Steenhuis TS, Nunes M, de Lima JLMP, Coelho COA, Ferreira AJD (2016) Impact of urban development on streamflow regime of a Portuguese peri-urban Mediterranean catchment. J Soils Sediments 16:2580–2593
14. Ahlmer AK, Cavalli M, Hansson K, Koutsouris AJ, Crema S, Kalantari Z (2018) Soil moisture remote-sensing applications for identification of flood-prone areas along transport infrastructure. Environ Earth Sci 77:533. https://doi.org/10.1007/s12665-018-7704-z
15. Kabisch N, Korn H, Stadler J, Bonn A (2017) Nature-based solutions to climate change adaptation in urban areas—linkages between science, policy and practice. In: Kabisch N, Korn H, Stadler J, Bonn A (eds) Nature-based solutions to climate change adaptation in urban areas. Theory and practice of urban sustainability transitions. Springer, Cham
16. Page J, Mörtberg U, Destouni G, Ferreira CSS, Näsström H, Kalantari Z (2020) Open-source planning support system for sustainable regional planning: a case study of Stockholm County, Sweden. Manuscript submitted for publication in Environment and Planning B: Urban Analytics and City Science. https://doi.org/10.1177/2399808320919769
17. European Commission (EC) (2015) Towards an EU research and innovation policy Agenda for nature-based solutions & re-naturing cities: final report of the Horizon 2020 Expert Group on 'Nature-based Solutions and Re-naturing Cities'. Directorate-General for Research and Innovation, Climate Action, Environment, Resource Efficiency and Raw Materials, Luxembourg (Publications Office of the European Union)

18. Kalantari Z, Ferreira CSS, Deal B, Destouni G (2019) Nature-based solutions for meeting environmental and socio-economic challenges in land management and development. Land Degradat Dev:1–4. https://doi.org/10.1002/ldr.3264
19. Liu Z, Xiu C, Ye C (2020) Improving urban resilience through green infrastructure: an integrated approach for connectivity conservation in the central city of Shenyang, China. Special issue on new models, new technologies, new data and applications of urban complexity from spatio-temporal perspectives. J Complex. 1653493. https://doi.org/10.1155/2020/1653493
20. Huck A, Monstadt J, Driessen P (2020) Building urban and infrastructure resilience through connectivity: an institutional perspective on disaster risk management in Christchurch, New Zealand. Cities, 98, 102573. https://doi.org/10.1016/j.cities.2019.102573
21. Kalantari Z, Ferreira CSS, Koutsouris AJ, Ahmer A-K, Cerdà A, Destouni G (2019) Assessing flood probability for transportation infrastructure based on catchment characteristics, sediment connectivity and remotely sensed soil moisture. Sci Total Environ 661:393–406. https://doi.org/10.1016/j.scitotenv.2019.01.009
22. Lopez-Vicente M, Ben-Salem N (2019) Computing structural and functional flow and sediment connectivity with a new aggregated index: a case study in a large Mediterranean catchment. Sci Total Environ 651(Part 1):179–191
23. Strogatz SH (2001) Exploring complex networks. Nature 410(6825):268
24. Fryirs K, Brierley GJ, Preston NJ, Kasai M (2007) Buffers, barriers and blankets: the (dis)-connectivity of catchment-scale sediment cascades. Catena 70:49–67
25. Cavalli M, Trevisani S, Comiti F, Marchi L (2013) Geomorphometric assessment of spatial sediment connectivity in small alpine catchments. Geomorphology 188:31–41
26. Cavalli M, Crema S, Marchi L (2014) Guidelines on the sediment connectivity ArcGis toolbox and stand-alone application. Technical report, Sediment management in Alpine basins (CNR-IRPI Padova - PP4)
27. Kalantari Z, Cavalli M, Cantone C, Crema S, Destouni G (2017) Flood probability quantification for road infrastructure: data-driven spatial-statistical approach and case study applications. Sci Total Environ 581:386–398
28. Kalantari Z, Ferreira CSS, Page J, Goldenberg R, Olsson J, Destouni G (2019) Meeting sustainable development challenges in growing cities: coupled social-ecological systems modeling of land use and water changes. J Environ Manag 245:471–480. https://doi.org/10.1016/j.jenvman.2019.05.086
29. Blesh JM, Barrett GW (2006) Farmers' attitudes regarding agrolandscape ecology: a regional comparison. J Sustain Agric 28(3):121–143
30. Mittermeier RA, Totten M, Pennypacker LL, Boltz F, Mittermeier CG, Midgley G, Rodriguez CM, Prickett G, Gascon C, Seligmann PA, Langrand O (2008) Climate for life. Conservation International, Washington
31. Kalantari Z, Ferreira CSS, Walsh RPD, Ferreira AJD, Destouni G (2017) Urbanization development under climate change: hydrological responses in a peri-urban Mediterranean catchment. Land Degrad Dev 28(7):2207–2221
32. The Economics of Ecosystems & Biodiversity (TEEB) (2011) TEEB manual for cities: ecosystem services in urban management. http://www.teebweb.org/publication/teeb-manual-for-cities-ecosystem-services-in-urban-management/. Accessed 20 Jul 2020
33. O'Hogain S, McCarton L (2018) Hybrid infrastructure: local, regional and global potential of nature based solutions. In: A technology portfolio of nature based solutions. Springer, Cham. https://doi.org/10.1007/978-3-319-73281-7_6
34. WWAP (2018) The United Nations world water development report 2018: nature-based solutions for water. UNESCO, Paris
35. Machovina B, Feeley KJ (2017) Restoring low-input high-diversity grasslands as a potential global resource for biofuels. Sci Total Environ 609:205–214
36. Ghajarnia N, Destouni G, Thorslund J et al (2020) Data for wetlandscapes and their changes around the world. Earth Syst Sci Data 12:1083–1100. https://doi.org/10.5194/essd-12-1083-2020

37. Swedish Environmental Protection Agency (Naturvårdsverket) (2009) Rätt våtmark på rätt plats. Rapport 5926. Naturvårdsverket förlag. https://www.naturvardsverket.se/Documents/publikationer/978-91-620-5926-2.pdf. Accessed 20 Jul 2020
38. Czuba JA, Hansen AT, Foufoula-Georgiou E, Finlay JC (2018) Contextualizing wetlands within a river network to assess nitrate removal and inform watershed management. Water Resour Res 54(2):1312–1337
39. Hansen AT, Dolph CL, Foufoula-Georgiou E, Finlay JC (2018) Contribution of wetlands to nitrate removal at the watershed scale. Nat Geosci 11(2):127–132
40. Stråe D, van der Nat D, af Petersens E (2014) Bällstaåns avrinningsområde, planeringsunderlag - PM. WRS & Miljöförvaltningen Stockholm stad. http://miljobarometern.stockholm.se/content/docs/vp/bal/Ballstaan_planeringsunderlag_2014.pdf. Accessed 20 Jul 2020
41. Ferreira CSS, Walsh RPD, Steenhuis TS, Ferreira AJD (2018) Effect of Peri-urban development and lithology on streamflow in a Mediterranean catchment. Land Degrad Dev 29:1141–1153
42. Järfälla Tidning (2015) Skolorna håller stängt efter översvämningar i Järfälla: "En meter vatten i källaren". https://www.stockholmdirekt.se/nyheter/skolorna-haller-stangt-efter-oversvamningar-i-jarfalla-en-meter-vatten-i-kallaren/aRKoig!@pViTyAW4NXWhq4Pruj6iw/. Accessed 28 May 2018
43. Mitt i Stockholm (2017) "Nu skiter jag i ån". https://mitti.se/nyheter/nu-skiter-jag-i-an/. Accessed 28 May 2018
44. Borselli L, Cassi P, Torri D (2008) Prolegomena to sediment and flow connectivity in the landscape: a GIS and field numerical assessment. Catena 75(3):268–277
45. Cavalli M, Tarolli P, Marchi L, Dalla Fontana G (2008) The effectiveness of airborne LiDAR data in the recognition of channel-bed morphology. Catena 73(3):249–260

From Sponge Cities to Sponge Landscapes with Nature-Based Solutions: A Multidimensional Approach to Map Suitable Rural Areas for Flood Mitigation and Landscaping

Filippo Carlo Pavesi and Michele Pezzagno

Contents

Abstract Floods are threatening communities. The benefits deriving from the creation of hydraulic risk mitigation systems, based also on green infrastructures, are well known and the examples of such measures are spread worldwide. A breakthrough in urban planning is represented by the *Sponge City* policies, but rural areas could become a network of nature-based water management systems, too. Therefore, the chapter proposes a multidimensional approach to map rural areas of particular interest for flood retention and landscaping. The method maps soil hydrological behaviour together with ecological and landscape peculiarities at regional scale. The output is the *Sponge Landscape Suitability Map*, which classifies

F. C. Pavesi (✉) and M. Pezzagno
DICATAM, Università degli Studi di Brescia, Brescia, Italy
e-mail: filippo.pavesi@unibs.it; michele.pezzagno@unibs.it

Carla S. S. Ferreira, Zahra Kalantari, Thomas Hartmann, and Paulo Pereira (eds.),
Nature-Based Solutions for Flood Mitigation: Environmental and Socio-Economic Aspects,
Hdb Env Chem (2022) 107: 355–376, DOI 10.1007/698_2021_767,

Published online: 6 May 2021

rural lands basing on their suitability for the implementation of nature-based measures. Afterwards, the chapter proposes a method to select the best Nature-Based Solutions to mitigate the hydrological risk basing on the local context. Combining the *Sponge Landscape Suitability* mapping with the selection of nature-based measures represents an innovative approach to spatial planning, contributing to achieve multiple benefits within hydraulic risk management and eco-landscape planning with a holistic perspective. The entire approach was applied to the Lombardy region (Italy) case study, with an in-depth analysis of Brescia province.

Keywords Hydraulic risk management, Nature-based solutions, Sponge landscape, Suitability Map

1 Introduction

Worldwide data on natural disasters highlights that cities and rural areas are increasingly exposed to the risk of loss or damage. In particular, analysis of Emergency Events Database (EM-DAT) data, from the beginning of the twentieth century until October 2019, shows two important aspects: firstly, from the second half of the twentieth century there has been an increase of flood and storm events; secondly, floods are the most widespread natural disaster and (along with storms) the most important natural hazard in terms of socio-economic damage. In particular, from 1998 to 2017, floods and storms represented, respectively, the 43.4% and 28.2% of all recorded disasters, affected 2.0 billion and 726 million people, and caused economic losses of US$ 656 and 1,330 billion [1].

To prevent communities from cultural, social, human, ecosystem, and economic losses, city plans should allow urban development with the foresight to ensure a reduction in exposure and vulnerability, thus preventing the creation of new disaster risk. However natural disasters continue to afflict communities exposed to risk [2]. The transition from *Disaster Management* to *Disaster Risk Governance*, urged by the Sendai Framework for Disaster Risk Reduction 2015–2030 [3], and integrated into the Sustainable Development Agenda [4], pushes researchers and decision-makers to identify new modalities to effectively protect communities and strengthen their resilience. *Multi-Layer Safety* (MLS) is a good practice since it integrates protection measures and emergency response into spatial planning [5–7]. To face the hydrological-hydraulic risk management challenge, accelerated by climate change [8] and land use change [9], it is urgent to integrate risk culture at all levels of urban and spatial planning, promoting a holistic approach for land management and conservation. In this framework, it is also important to remember the limits of the traditional (grey) infrastructures to mitigate floods [10–16].

At international level, the benefits deriving from the creation of an hydraulic risk mitigation system, based also but not exclusively on green infrastructures, are well known [17]. The EU Biodiversity Strategy for 2030 has recently reconfirmed the potential of Green Infrastructure (GI) design, stressing that, if "the biodiversity crisis and the climate crisis are intrinsically linked", the solutions to the problems are linked, too. In fact, "nature is a vital ally in the fight against climate change" and solutions based on this awareness, for instance Nature-Based Solutions (NBS), "will be essential for emission reduction and climate adaptation" [18]. Green Infrastructures (GI) can achieve multiple benefits, contributing to mitigate the impacts of natural disasters and, simultaneously, providing habitats for biodiversity and other ecosystem services. The role of Nature-Based Solutions (NBS) in protecting urban and rural areas from flooding has been clarified by hydraulic engineering and ecology studies. However, due to the enormous fragmentation of rules, responsibilities, and competencies in planning, their use is not adequately promoted and addressed by spatial planning strategies. An integrated land use policy is the only way to promote sustainable urbanization and increase cities resilience. The EU Biodiversity Strategy for 2030 supports the transition to adaptive territories, focusing on: the realization of a coherent network of protected areas; bringing nature back to agricultural land; addressing land take and restoring soil ecosystems; greening urban and peri-urban areas; reducing pollution; and restoring freshwater ecosystems.

A breakthrough in urban planning is represented by the *Sponge City* policies introduced in China [19]. Such kind of cities are planned and designed to address problems related to stormwater management and control, enhancing hydraulic risk resilience [20]. In sponge cities, innovative technical solutions, not necessarily only nature-based, are developed to make use and create a network of *blue* and *green* urban spaces in order to retain water on-site during flooding phenomena. Hence, the Sponge city concept integrate ideas from eco-hydrology, environmental and societal well-being, and climate change adaptation within the urban land use planning process [21].

Since hydrological and hydraulic dynamics generally involve huge areas not only urban, but also peri-urban and rural, flooding risk management solutions should also be pushed in peri-urban and rural areas, making evidence, at global scale, of the fundamental role played by "Landcare" policies [22], which are characterized by more attention to land take and land degradation phenomena [23], and by natural resources management. The *Sponge* concept can be applied also at larger spatial scale [24], combining the increasing need of land "to be like a sponge" with the opportunity to design high quality and more resilient landscapes, called *Sponge Landscapes* [25]. Thus, peri-urban and rural areas could become a network of nature-based water management systems that can support urbanized areas resilience, but since some areas are more suitable than others for retaining water, it is of paramount importance to identify them in order to design effective Nature-Based Solutions.

Therefore, the present chapter seeks to answer two main questions: (1) how to choose the peri-urban and rural areas that could better work like a sponge? (2) how to develop these areas in order to mitigate hydraulic and hydrologic risk and hazard?

Attempting to answer the first question, this chapter applies the *Spongescaping* Approach. The *Spongescaping* Approach is a method that classifies rural areas basing on their hydrological and landscape peculiarities (e.g. cultural, ecological, and landscape aspects). The main output of this approach is the *Sponge Landscape Suitability Map,* which merges soil permeability, soil drainage efficiency, and landscape peculiarity data to identify areas that can be more efficient in mitigating the hydrogeological risk and that have, simultaneously, important landscape peculiarities [25].

Once having classified areas suitable to become *Sponge Landscapes* (with the *Sponge Landscape Suitability Map*), the chapter seeks to answer the second question identifying the best Nature-Based Solutions to mitigate hydrological risk and to enhance ecosystems living conditions, with particular attention to the local peculiarities of each area. The so-called Natural Water Retention Measures were selected, since besides aiming "to protect and manage water resources and address water-related challenges by restoring or maintaining ecosystems as well as natural features and characteristics of water bodies using natural means and processes" [26], they provide other ecosystem services such as increasing biodiversity, improving the aesthetic of places, offering recreational and leisure opportunities, improving water quality and groundwater recharge, and reducing agricultural sediment loss [27–29]. Since the European Natural Water Retention Measures platform classifies each measure with qualitative values related to its capacity to provide every single ecosystem service, the chapter proposes a method to convert the qualitative values into numerical ones in order to calculate the multifunctional potential of each Natural Water Retention Measure.

Hence, the chapter proposes a top-down approach that is the result of the combination of two methodologies. Firstly, applying the *Spongescaping* Approach, it detects peri-urban and urban areas with hydrological and landscapes qualities which make them more suitable to reduce hydrological risk; secondly, after having converted the qualitative Natural Water Retention Measures classification (defined by the European Natural Water Retention Measures platform) into a quantitative one, it selects the Natural Water Retention Measures that can better fit and develop an area, basing on the local peculiarities of the area of interest (e.g. agriculture landscape, forest landscape).

The approach was tested and applied to the case study of the *Lombardy region (Italy),* with an in-depth analysis of the *Brescia Province.*

2 Material and Methods

2.1 Study Area

Lombardy is a region (NUT 2) of northern Italy extending for 23.863 km^2 [30] and featured by mountainous areas (40.5% of the regional area), a vast plain area (47.1%), and hilly zones (12.4%) [31]. In detail, from north to south, the region is characterized by six main morphological bands: the massive mountain range of the Alps; the sub-mountainous bend of Prealps, with the glacial lakes system; the hill belt; the Po river alluvial plain, divided into high plain and low plain and crossed by

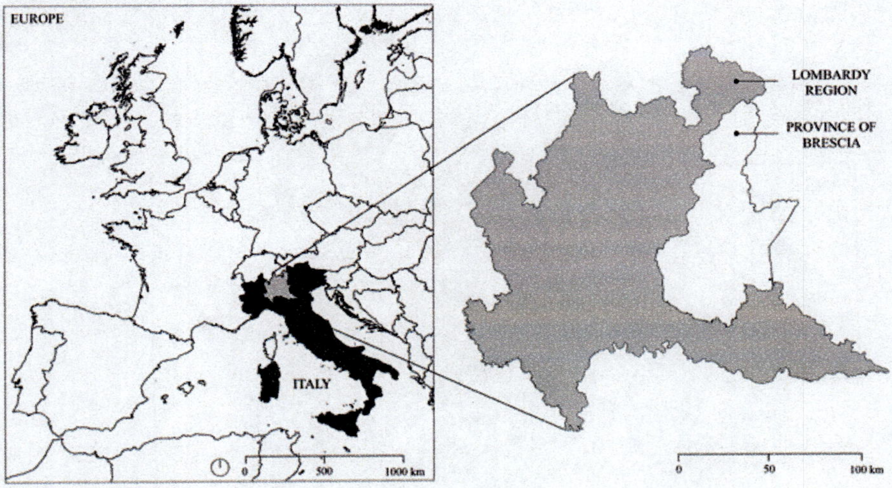

Fig. 1 Localization of Italy, the Lombardy region and the province of Brescia

numerous river valleys flowing into the Po river; and the Apennine mountainous system [31] (Figs. 1 and 2).

Lombardy counts the largest population of any Italian region (9.70 million inhabitants) [30] and has a population density of 406 inhab/km^2 [30], highly located between the sub-mountainous and the Po plain bands, in which urban settlements generated a dense urbanized net that has fragmented the agricultural landscape of the plain. This area hosts almost eight million of inhabitants [31], making it one of the densest in Europe. Most widespread in the fertile Po plain, agricultural land represents the 48% of the regional area [32], while mountainous bends are characterized by vast wooded areas. More than one-third of the regional area is at risk of floods [33].

Brescia is one of the major cities of Lombardy and the chief city of the *Province of Brescia*. The province is located in the north-eastern side of the region and represents the most extended province of Lombardy (4,785.62 km^2) [30] and the second province per number of inhabitants (1,238,044 inhabitants, 258 inhab/km^2) [30] (Figs. 1 and 2). It stretches from the Alps to the Po plain, featuring mountainous landscapes in the north, highly urbanized area nested in an agricultural landscape in the middle, and rural areas in the south. Almost 30% of this area is at risk of floods [33].

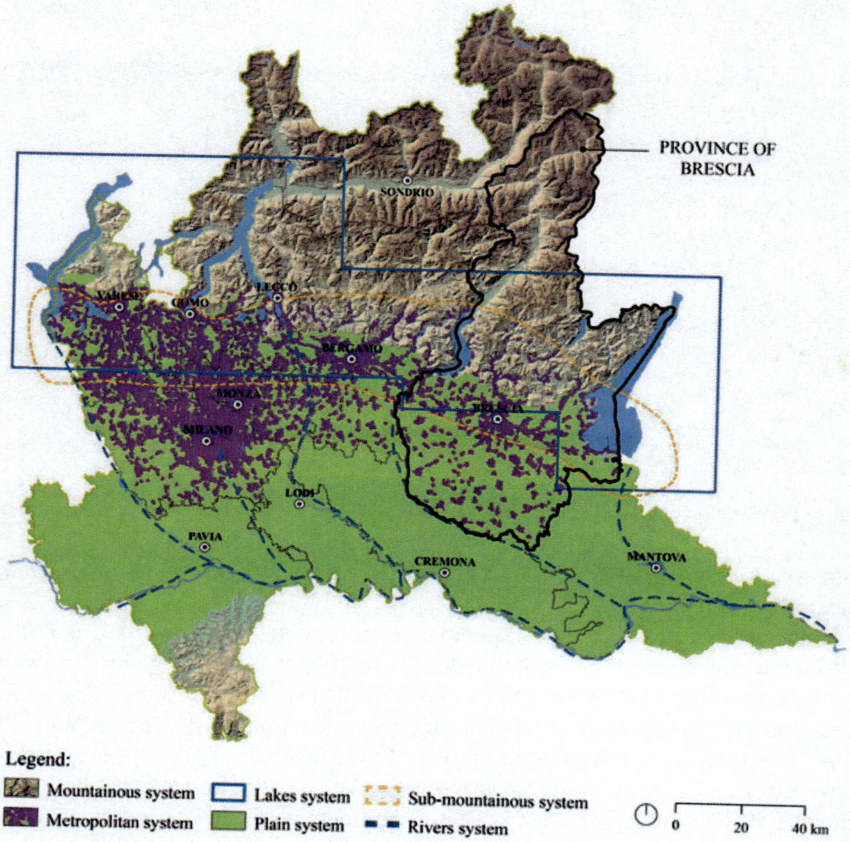

Fig. 2 Landscape system of Lombardy region and province of Brescia. (adapted from Regione Lombardia, PTR, Atlante di Lombardia, 2011)

2.2 *The* Spongescaping *Approach and the Construction of the* Sponge Landscape Suitability Map *(SLSM)*

The *Spongescaping Approach* is a process that allows to realize a map, the *Sponge Landscape Suitability Map*, in which peri-urban and rural lands are classified basing on their landscape and hydrological peculiarities.

The constructive process of the *Sponge Landscape Suitability Map* requires three main steps:

1. The realization of the *Sponge Map* (SM): which considers soil hydrological characteristics;
2. The realization of the *Eco-Landscape Map* (ELM): which considers land ecological and landscape values;

3. The realization of the *Sponge Landscape Suitability Map* (SLSM): which shows areas with higher soil hydrological characteristics and ecological and cultural values, thus suitable to be developed with nature-based measures.

To apply this method, a geospatial approach is fundamental (GIS mapping) being aware, as mentioned in a recent review on methods for landscape mapping [34], that "early examples of characterization of large regions or continents [35, 36] were coarse in typology and rather inaccurate, partly due to a lack of systematic digital information with a high-spatial accuracy and computer-supported data processing" [37]. The procedure is designed for geo-datasets realized at regional or sub-regional scale (1:50,000, 1:25,000, 1:10,000 scale), relevant for spatial and landscape planning.

2.2.1 *Sponge Map* (SM)

The *Sponge Map* (SM) is designed to respond to "the need for the land to be a sponge" and is realized processing, with a GIS software, soil permeability and land use related soil drainage efficiency data.

Soil permeability data could be constructed with field measurement, but this process is complex, expensive, and laborious [38]. Moreover, a field data collection can be carried out only for small areas; therefore, it is not the appropriate method to realize analysis at regional or sub-regional scale. At regional scale, soil datasets generally present estimated values of soil permeability [39], which are characterized by less accuracy but wider coverage, whilst others relate hydrological data to land use (e.g. Soil Survey Geographic (SSURGO) database – LUCAS database) [40].

In order to obtain soil permeability values, this study selected Hydrologic Soil Groups data, since they also consider the depth of the aquifer.

Soil drainage efficiency is related to land use. Land Use (LU) maps usually classify land according to Corine Land Cover (CLC) classification 2018. Corine Land Cover uses a 3-level hierarchical classification system at EU scale, and up to 5-level at the regional scale to provide more details. To describe soil drainage efficiency a *Land Use Drainage efficiency index* (LUDei) was defined, applying to each land use a drainage coefficient derived by regional experimentation [41]. Index values refer to the drainage capacity of the different land uses and vary in a range of 0–100%, from non-draining surfaces (sealed soil) to fully draining surfaces. For natural and agriculture land uses generally was assigned a 100% of drainage surface, for anthropogenic land uses index was calculated on the basis of average values of draining surface of the elements.

To obtain a *Sponge Map*, permeability and drainage values can be easily processed converting vectorial data in raster data (pixels, spatial resolution 10 m). The use of a raster calculator allows to calculate values in case of overlapping raster coverages. The *Sponge Map* is directly based on *Sponge values* (Sv), which are obtained by multiplying *Soil Permeability values* (SPv) by *Land Use Drainage efficiency index* (Fig. 3, Eq. 1).

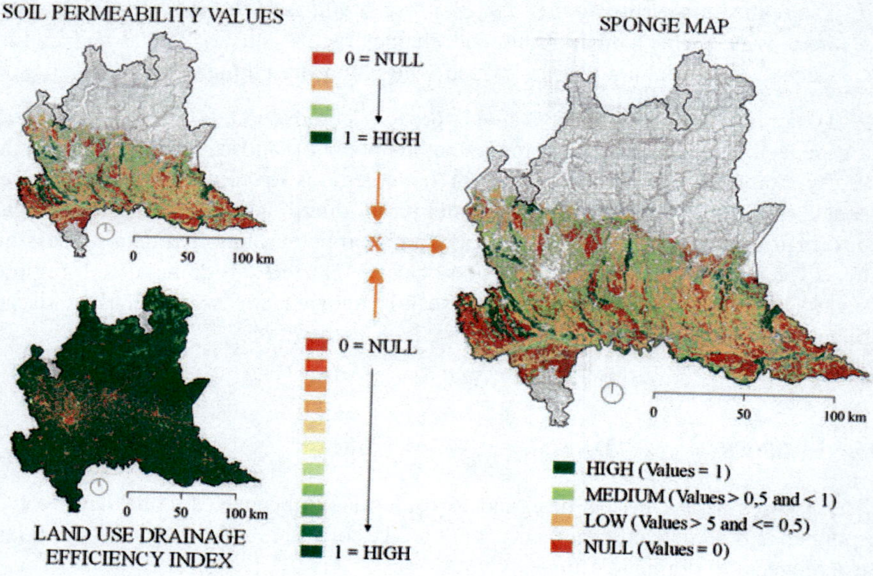

Fig. 3 The constructive process of the *Sponge Landscape Suitability Map* of the Lombardy region

Therefore, the *Sponge value* (Sv) can be defined as:

$$Sv = SPv \times LUDei \tag{1}$$

Sv*:* Sponge value. SPv: Soil Permeability value. LUDei: Land Use Drainage efficiency index.

If one of the two factors is zero, the sponge effect is null. For instance, this happens with sealed soils or poorly permeable soils.

In detail, to realize the *Sponge Map* (SM) of the Lombardy region more steps were necessary due to segregation of information and differences of maps graphic scales. In fact, in Lombardy there are two different soil permeability maps for the mountain environment and for the plain one, therefore two different *Sponge Maps* were realized.

- Plain *Soil Permeability values* (SPv):

  The shapefile of the pedological map of the Lombardy region classifies 538 Soil Units at 1:50,000 (the shapefile is freely downloadable from the regional geo-portal [42]). Soils units consist of one or more soils types and have no data on permeability. Permeability is available in another regional dataset (not available on the regional geo-portal) that identifies 636 types of soil described by permeability values, runoff index, hydrologic group, and depth of the aquifer. Since some Soil Units include more than one type of soil of the Hydrologic Soil Groups classification, to join data it was necessary to consider the Hydrologic Soil

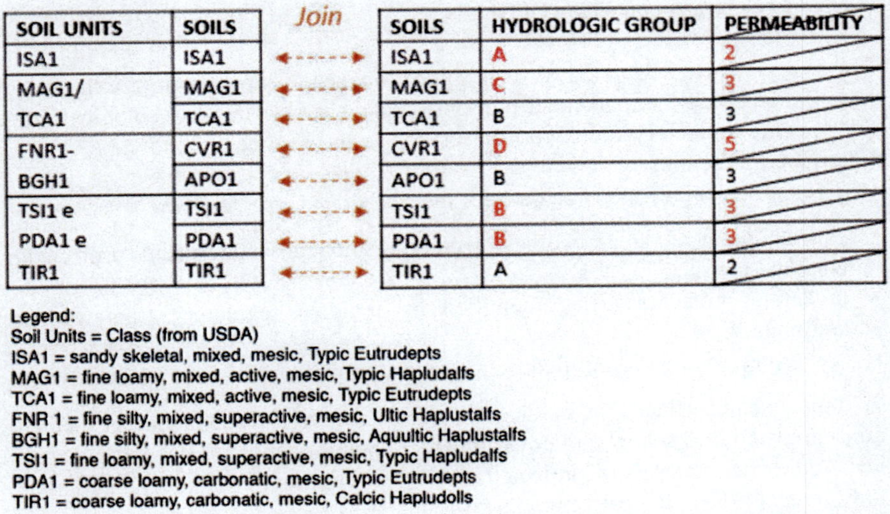

Fig. 4 Method of assigning of Hydrologic Group value to Soil Units by joining data in a GIS process

Groups values of each type of soil in the Soil Units. The method proposed assigns to Soil Units the worst value, in favor of safety (Fig. 4).

- Mountain *Soil Permeability values* (SPv):

  There is a shapefile of the hydrological map of the Lombardy region (freely downloadable from the regional geo-portal [43]) that classified land units at the graphic scale of 1:10.000. Soil units have different values (high, medium, or low) for the soil surface and the substratum. Thus, in order to link a unique soil permeability value to a specific soil unit it is necessary to combine surface and substratum values: firstly, is necessary to reclassify qualitative values in numerical values (high = 1, medium = 0.5, and low = 0); afterwards, to obtain the unique value for the soil unit, the surface and substratum values must be multiplied. The process produces a second vector map regarding soil permeability of mountain areas.

Regarding the *Land Use Drainage efficiency index* (LUDei), it derives from Lombardy Landscape Plan's official documentation (Strategic Environmental Assessment, Environmental Report, Annex G [44]). Therefore, for the entire Lombardy region it is possible to create a map based on land use drainage efficiency index. This index is related to the land use classes of the Lombardy region Land Use Map (DUSAF). DUSAF shapefile, freely downloadable from the regional geo-portal [45], classifies land according to Corine Land Cover Classification 2018 (up to the fifth level). In this case, it was not necessary to reclassify the values and this process produces a third vector map for the entire Lombardy Region.

2.2.2 *Eco-Landscape Map* (ELM)

The *Eco-Landscape Map* (ELM) is designed to take the "opportunity for the land to be a new landscape" and includes portions of land characterized by ecological values and specific regulatory framework in relation to the European Landscape Convention (ELC) principles.

The fundamental geospatial data for the *Eco-Landscape Map* are:

- strongly protected areas: ecological, cultural, and landscape heritage protected by National laws;
- moderate protected areas: identified by regional regulatory framework on landscaping;
- low protected areas: identified by regulatory framework on ecological networks.

It must be underlined that, in the geo-database, ecological and landscape elements are segregated into spatial and environmental planning, and cultural and geographical studies (i.e. datasets of protected areas [46], cultural heritage [47], agriculture [48], forest [49], ecological networks [50]), therefore to use these datasets at regional scale, their aggregation is necessary. This study aggregated the datasets according to the Level of Protection of the considered elements.

Therefore, the *Eco-Landscape Map* is the result of the geospatial polygon aggregation, realized with a GIS software, of protected areas' maps and identifies the whole areas with ecological and landscape peculiarities.

2.2.3 *Sponge Landscape Suitability Map* (SLSM)

Intersecting with a GIS software the *Sponge Map* and the *Eco-Landscape Map*, the *Sponge Landscape Suitability Map* (SLSM) is obtained (Fig. 5, Table 1). The *Sponge Landscape Suitability Map* (SLSM) shows the permeability and drainage characteristics of the protected areas, hence the suitability of a protected area to be developed in order to enhance landscape flood resilience.

Basing on *Sponge values*, the maps classified areas into four classes of suitability:

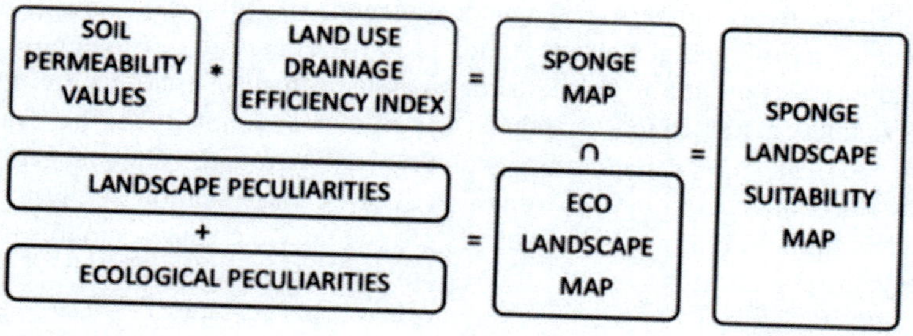

Fig. 5 Methodological scheme for the realization of the *Sponge Landscape Suitability Map*

Table 1 Maps necessary for the realization of the *Sponge Landscape Suitability Map* (Typical reference scale: 1:10,000/1:50,000; Source: regional or sub-regional geoportal)

Variable	Cartography	Type of data
Permeability	Pedological map	Vector and raster
Drainage efficiency	Land use map	Vector and raster
Landscape peculiarities	Landscape elements map	Vector
Ecological peculiarities	Ecological network map	Vector

- *high*: with *Sponge values* equal to 1 means high land suitability for sponge land design with Nature-Based Solutions;
- *medium*: with *Sponge values* ranging from 0.5 to 1 means medium land suitability for sponge land design with Nature-Based Solutions;
- *low*: with *Sponge values* ranging from 0 to 0.5, low high land suitability for sponge land design with Nature-Based Solutions;
- *null*: with *Sponge values* equal to 0 means that the selected area is not suitable for sponge land design with Nature-Based Solutions;

2.3 The Selection of Local-Based Nature-Based Solutions

One of the most significant ecosystem services for the *Sponge Landscape* research is the *flood risk reduction ecosystem service* that is strongly related with the benefits deriving through the mechanism of water retention, such as slowing, storing, and reducing runoff superficial runoff. For this reason, the Nature-Based Solutions selected by the present study are the Natural Water Retention Measures, which, besides providing a vast fan of ecosystem services, are all designed to retain water basing on natural processes.

The European Natural Water Retention Measures platform classifies measures per sector (agriculture, forest, hydro morphology, urban) and per benefits with an in-depth analysis of the wide range of ecosystem services provided by different measures [51, 52]. Moreover, the platform classifies each measure by a qualitative value related to its capacity to provide every single ecosystem service.

Since the most suitable areas to become *Spongescapes,* detected with the *Sponge Landscape Suitability Maps,* are featured by woods and intensive crops, this study has selected just the Natural Water Retention Measures that can be implemented in agriculture and forest landscapes. Therefore, twenty-seven Natural Water Retention Measures were selected (thirteen for the agriculture sector and fourteen for the forest one). Furthermore, since the present study is focused on flood risk mitigation, just the following ecosystem services were selected to be analysed:

- flood risk reduction (the most significant for the *Sponge Landscape* research),
- biodiversity preservation,
- climate change adaptation and mitigation,

- erosion/sediment control, and,
- filtration of pollutants.

Subsequently, the qualitative values provided by the European Natural Water Retention Measures platform were transformed into numerical values to calculate the multifunctional potential of the best European Natural Water Retention Measures to reduce the hydrological risk. The platform publishes matrix tables combining measures (rows) and ecosystem services (field). The capacity of every single measure to provide each ecosystem service is described by coloured cell: green cell = high capacity; light green cell = medium capacity; yellow cell = low capacity; red cell = irrelevant. The conversion process consists in replacing coloured cells with numbers: green cells = 3; light green cells = 2; yellow cell = 1; red cells = 0.

3 Results

3.1 *The* Spongescaping *Approach and the* Sponge Landscape Suitability Map *(SLSM)*

3.1.1 Regional Level

At regional level, the present study was limited to the realization of the *Sponge Maps* (Fig. 6). The maps reveal that the plain area is generally characterized by medium and low *Sponge values* (Sv): medium permeability and drainage efficiency soils are interspersed with low and null ones, and just some narrow spits of land are marked

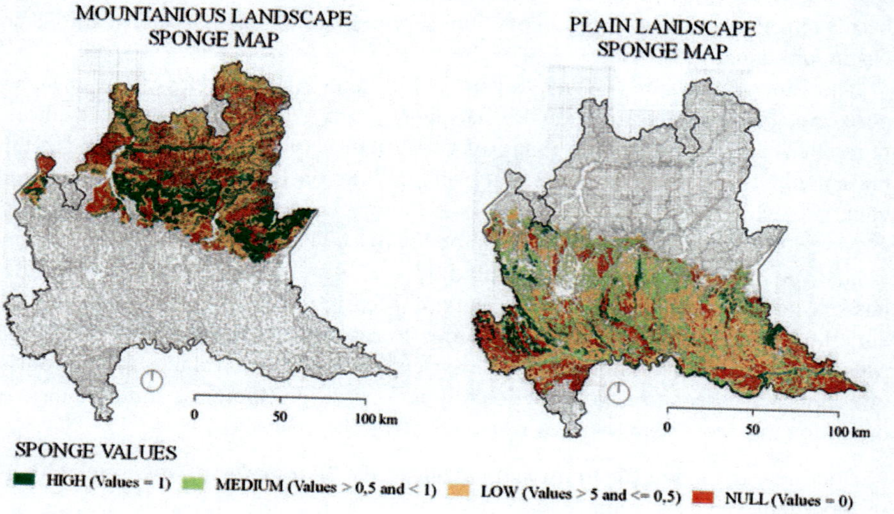

Fig. 6 Mountainous and plain landscape *Sponge Maps* of Lombardy region

by high values. In addition, note that as latitude decreases the sponge effect decreases, too. This is related to geological and pedological conditions of the land.

The alpine landscape features vast spatial extents with medium and null *Sponge values* (Sv), nevertheless it is also characterized by wide areas with high *Sponge values* (Sv). Moreover, as latitude increases, there is a tendency to decrease of the sponge effect.

Therefore, Lombardy region seems to feature a median band with higher *Sponge values,* which include the alpine landscape as well as the plain one.

3.1.2 Province Level

The resulting *Sponge Maps* reveal that the plain area is generally characterized by low *Sponge values* (Sv): medium permeability and drainage efficiency soils are interspersed with low and null ones, and just some narrow spits of land are marked by high values (Fig. 7).

The alpine landscape (Fig. 8), by contrast, features vast areas with high *Sponge values* (Sv) juxtaposed to as many with medium and null permeability and drainage efficiency ones.

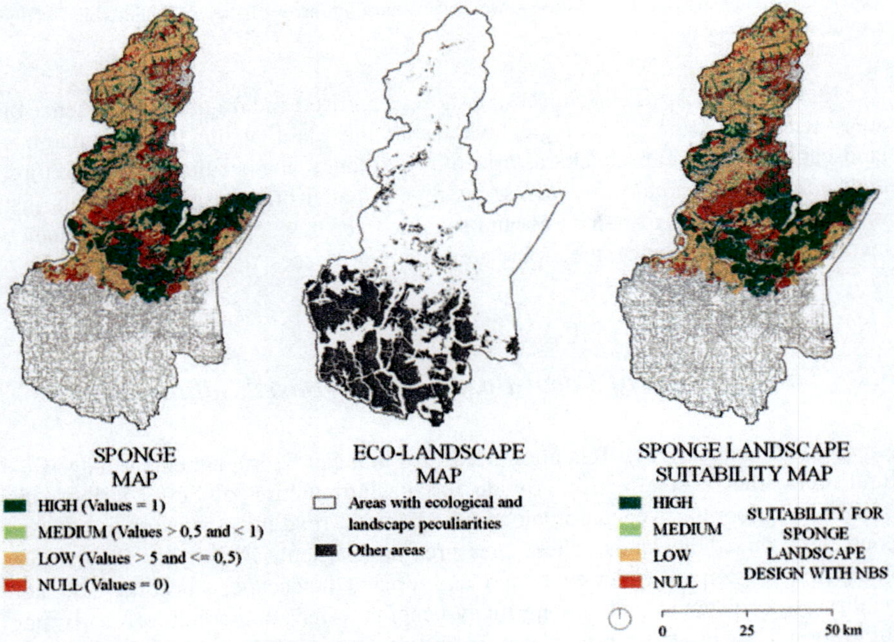

Fig. 7 The realization of the *Sponge Landscape Suitability Map* of the alpine landscape of the province of Brescia

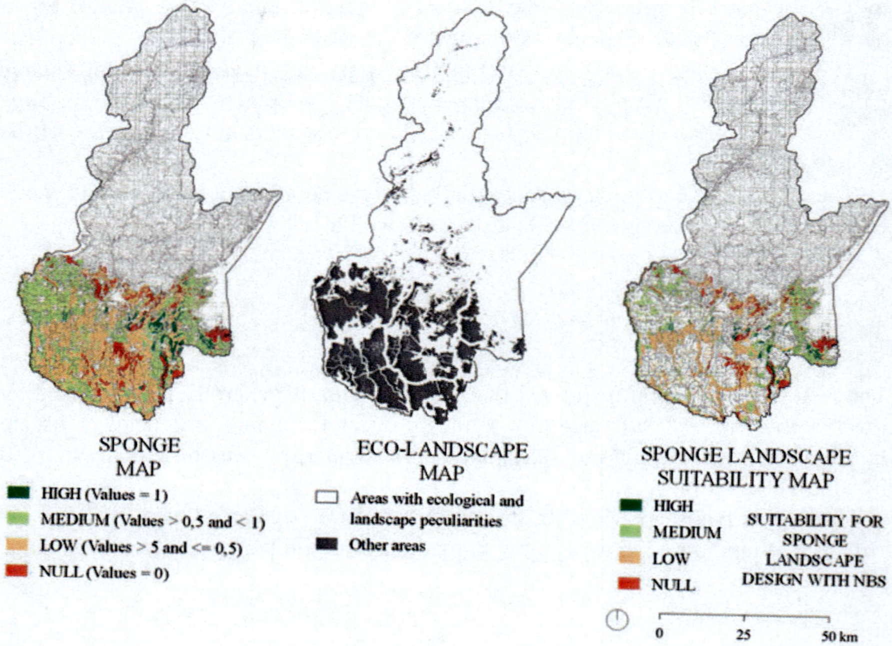

Fig. 8 The realization of the *Sponge Landscape Suitability Map* of the plain landscape of the province of Brescia

Moreover, the *Eco-Landscape Maps* shows a reduced and fragmented presence of areas with landscape or ecological values in the plain, while the mountainous landscape is almost entirely characterized by eco-landscape peculiarities. Therefore, the *Sponge Landscape Suitability Maps* reveal that, while the plain presents just small patches and narrow strips of land suitable to be developed in order to enhance landscape flood resilience, the alpine area has huge spaces.

3.2 The Selection of Local-Based Nature-Based Solutions

Among the Natural Water Retention Measures that can be implemented in agricultural areas, the most effective provide for the introduction of buffer strips with natural and blooming vegetation along field borders, road infrastructure and watercourses or on arable lands. These measures are particularly suitable for *Sponge Landscape* design in agricultural and anthropized landscape, since they do not require extensive areas and, among many benefits, offer good conditions for water infiltration, slow down superficial runoff, and decrease soil erosion. Also, meadows and pastures can enhance water infiltration and crops along the contour lines or arranged on terraces can reduce soil erosion and surface runoff (Table 2). Results for

Table 2 Multifunctional potential of Natural Water Retention Measures for the agriculture sector. Source: Natural Water Retention Measures platform data reprocessed

Capacity of NWRM for agriculture to provide ecosystem services (high values for flood risk reduction and multifunctional values are in bold)	Flood risk reduction	Biodiversity preservation	Climate change adaptation and mitigation	Erosion/ sediment control	Filtration of pollutants	Multi-functional value (sum of previous values)
Buffer strips and hedges	**3**	1	2	3	3	**12**
Green cover	**3**	1	2	3	3	**12**
Meadows and pastures	**3**	0	2	3	2	10
Early sowing	**3**	0	2	3	2	10
Intercropping	2	2	1	2	3	10
Strip cropping along contours	2	0	0	3	2	7
Traditional terracing	2	0	0	3	2	7
Controlled traffic farming	2	0	0	2	2	6
Reduced stocking density	2	0	0	2	2	6
Mulching	2	0	0	1	0	3
Crop rotation	1	1	0	1	2	5
No-till agriculture	0	2	2	3	2	9
Low-till agriculture	0	0	1	0	0	1

Natural Water Retention Measures are ordered (high to low) firstly by flood risk reduction ecosystem service values, secondly by multifunctional values. About flood risk reduction and improving other ecosystem services, most effectiveness Natural Water Retention Measures are those positioned in the upper rows

Capacity of Natural Water Retention Measures to provide ecosystem service: 3 = High; 2 = Medium; 1 = Low; 0 = Irrelevant. Capacity of NWRM to provide multiple ecosystem services: from 12 to 15 = High; from 8 to 11 = Medium; from 4 to 7 = Low; from 0 to 3 = Irrelevant

the forest areas show that the realization of water retention ponds can reduce the peak flow and highlight the importance of the afforestation. In fact, afforestation of the catchments increases infiltration and the ability of soil to store water while trees canopies diminish the impact of raindrops on barren surfaces, which reduces runoff and soil erosion. Trees along the watercourses also represent a solution to conserve watercourse quality and reduce banks erosion (Table 3).

Table 3 Multifunctional potential of Natural Water Retention Measures for the forest sector. Source: Natural Water Retention Measures platform data reprocessed

Capacity of NWRM for forest to provide ecosystem services (high values for flood risk reduction and multifunctional values are in bold)	Flood risk reduction	Biodiversity preservation	Climate change adaptation and mitigation	Erosion / sediment control	Filtration of pollutants	Multifunctional value (sum of previous values)
Peak flow control structures in managed forests	**3**	2	0	3	3	11
Afforestation of reservoir catchments	2	3	3	3	3	**14**
Targeted planting for "catching" precipitation	2	2	3	3	3	**13**
Continuous cover forestry	2	3	3	2	2	**12**
Sediment capture ponds	2	3	1	1	3	10
Forest riparian buffers	2	3	0	1	2	8
Coarse woody debris	2	3	0	1	0	6
Land use conversion	1	3	3	3	3	**13**
Maintenance of forest cover in headwater areas	1	2	3	3	3	**12**
Trees in urban areas	1	3	3	1	3	11
Urban forest parks	1	3	3	1	2	10
Overland flow areas in peatland forests	1	2	1	3	3	10
"Water sensitive" driving	1	3	0	3	2	9
Appropriate design of roads and stream crossings	1	3	0	3	1	8

Natural Water Retention Measures are ordered (high to low) firstly by flood risk reduction ecosystem service values, secondly by multifunctional values. About flood risk reduction and improving other ecosystem services, most effectiveness Natural Water Retention Measures are those positioned in the upper rows Capacity of Natural Water Retention Measures to provide ecosystem service: 3 = High; 2 = Medium; 1 = Low; 0 = Irrelevant. Capacity of NWRM to provide multiple ecosystem services: from 12 to 15 = High; from 8 to 11 = Medium; from 4 to 7 = Low; from 0 to 3 = Irrelevant

4 Discussion

Regarding the *Spongescaping* Approach and the *Sponge Landscape Suitability Map*, the proposed method has allowed to classify and select the most suitable rural areas to be developed with nature-based solutions for water management, therefore suitable areas to create a *Spongescape,* basing on soil hydrological properties and land ecological and landscape values.

The *Sponge Landscape Suitability Map* represents a novelty in hydraulic risk management field. The Flood Directive imposes to map flood hazard and risks, but this new map moves further beyond the demands of the Directive detecting, at spatial scale, suitable areas to mitigate hydraulic risk and, at the same time, to re-design the landscape. The *Spongescaping* Approach can improve the *Multi-Layer Safety* Approach and achieve the objective of reaching a sustainable spatial layout.

The case study demonstrates that the methodology is easily replicable at regional as well as at provincial scale, making evidence of the importance of spatial data at regional scale in order to achieve innovative results for spatial and landscape planning. Nevertheless, it has highlighted some limitations, one of which is the lack of a unique soil permeability geo-data set. In fact, Lombardy region has a soil permeability geo-data set concerning the mountainous environment and a separated one concerning the plain. Moreover, these geo-data are characterized by different scales and values. The analysis of the *Sponge Landscape Suitability Maps* reveals that, besides both the alpine and plain landscapes present suitable areas, the alpine one offers major opportunities to develop *Sponge Landscapes*. These areas are generally valleys or prealpine calcareous reliefs, characterized by extensive woods. Due to their spatial dimensions, vast *Spongescapes* could be realized. On the other hand, the plain offers less fragmented suitable spaces characterized by intensive crops. Hence, the plain landscape needs not extensive but targeted measures. Consequently, to reduce hydrological risk, to improve environmental conditions and, thus, to identify the most suitable nature-based solution to realize, it is of paramount importance to take into consideration consolidated land uses as well as the spatial dimensions of the areas suitable to become *Spongescapes.*

The *Spongescaping* Approach is different from the *Sponge City* one, as it considers rural and peri-urban areas and exclusively provides Nature-Based Solutions. This novel approach can contribute in the same way to the mitigation of hydraulic risk, but it can contribute more to restore the ecosystems and qualify the landscape.

Regarding the selection of local-based Nature-Based Solutions, Natural Water Retention Measures contribute to re-design the historical and traditional landscape and, for the case study of the Lombardy region, contributes to the re-design of the traditional agricultural landscape of the Po river valley. Results show that afforestation of areas in the plains contributes to reduce hydraulic risk and, simultaneously, to diversify the monoculture landscape and to create new livable habitats for the fauna, and to increase biodiversity. In general, the results put in evidence how today's agricultural practices, deriving directly from traditional ones, could improve territorial resilience. In the forest sector, the most effective Natural Water Retention Measures are those that provide for engineered ponds to reduce the runoff, afforestation of previously bare or heavily eroded areas in reservoir catchment, and tree-covered areas alongside streams and other water bodies. These measures are also suitable for *Sponge Landscape* design, as they offer good conditions for water infiltration and evapotranspiration, as well as for the improvement of the water quality. Also, these Natural Water Retention Measures contribute in re-designing the wooded landscape, which mitigates the hydraulic risk. In the mountainous environment, measures aim at accumulating water and avoiding runoff. For instance, realizing ponds upstream offers benefits deriving from the mechanism of water retention and good conditions for fauna (i.e. endemic species), or watering points for grazing animals. The proposed method has the advantages of allowing a preliminary assessment of the best Natural Water Retention Measures to design a *Sponge Landscape*, but has also some limits. The method is theoretical and is not directly applied to a case study. The effectiveness of the measures must be verified both on the basis of the local context and considering the effects of climate change [53]. Some measures may be more efficient than reported in the project, such as the capacity of meadows and pastures (agriculture Natural Water Retention Measures) to provide biodiversity preservation or the capacity of land use conversion (forest Natural Water Retention Measures) to provide flood risk reduction. Too efficient Natural Water Retention Measures could retain too much water upstream, releasing just few water downstream. Hence, it is necessary to evaluate and balance the benefits locally, with the contribution of specialists.

Both the realization approach of the *Sponge Landscape Suitability Map* and the selection method of Natural Water Retention Measures represent an improvement in spatial and landscape planning, but it is thanks to the combination of the two methodologies that it is possible to reach the most benefits. The *Sponge Landscape Suitability Map* identifies areas suitable to design measures that can mitigate hydraulic risk, but it is only through the rational selection of the Natural Water Retention Measures that the most efficient solutions can be identified, both the hydraulic and the eco-landscape ones. Therefore, the combination of the *Spongescaping* Approach with the selection of Natural Water Retention Measures can represent a novel approach to strengthen both spatial and landscape planning.

5 Conclusions

Floods are threatening communities. Several urban planning policies (e.g. *Sponge cities*) are mainstreaming flood risk management measures as ordinary planning tools, with a holistic perspective that includes risk management, environmental and social well-being, ecosystem services and biodiversity conservation, while spatial policies (e.g. Italian ones) are still providing a sectorial planning that deal separately with aspects like hydrogeological risk, eco-landscape quality, landscape protection, town planning, and city development.

The present study demonstrates that working in parallel with hydrological and eco-landscape aspects at regional and provincial scale is possible. Rural areas suitable to enhance floods resilience and eco-landscape quality can be detected and, therefore, a net of interconnected nature-based solutions can be planned. If the sponge function becomes one of the structural aspects of the landscape, together with the ecological, aesthetics and cultural ones, and land use, a *Sponge Landscape* can be realized. This system of values must result in a holistic planning approach: landscape planning may be characterized by a cross-sectional orientation, integrating different disciplines (risk management, biodiversity, ecosystem services) into a unique knowledge system. In this holistic perspective, combining the Sponge Landscape Suitability mapping with the selection of Natural Water Retention Measures represents an innovative approach to spatial planning that can contribute to achieve multiple benefits within hydraulic risk management and eco-landscape planning.

Moreover, the research shows that an effective *Sponge Landscape* requires nature-based measures realized basing on local peculiarities. Planning a *Sponge Landscape* (*Spongescaping*) in a mountainous area is different than planning it in a plain, therefore different Natural Water Retention Measures are required. Type of land use and spatial dimension of the suitable areas need to be considered. In fact, since alpine suitable areas are wider and characterized by high fragmentation of private property, to realize a *Spongescape* extensive landscape governance processes that involve the private sector are necessary. Measures must be implemented at the local scale and policymakers must be involved in the planning process: local governance actions are fundamental to share a unique *Spongescaping vision*.

Therefore, to be effective, landscape planning must be based on a collaborative strategy among plans at different scales (i.e. regional/local planning policies must deal with spatial planning, landscape, civil protection, landslides, and floods). The output of such an approach is a rich, varied, multifunctional, and resilient landscape [54]. In order to achieve a sustainable development vision, spatial and landscape planning have to manage water and soil as complementary elements of urban and rural areas. The *Sponge Landscape* concept could help to achieve some objectives foreseen in the new Common Agricultural Policy 2021–2027 referred to climate change action, environmental care, and to preserve landscapes and biodiversity.

References

1. UNISDR, CRED (2017) Economic losses, poverty and disasters 1998–2017
2. Field CB, Barros V, Stocker TF et al (2012) Managing the risks of extreme events and disasters to advance climate change adaptation: special report of the intergovernmental panel on climate change. Cambridge University Press
3. UNISDR (2015) Chart of the Sendai framework for disaster risk reduction. Unisdr
4. UNISDR (2015) Disaster risk reduction and resilience in the 2030 agenda for sustainable development

5. Gersonius B, Veerbeek W, Subhan A, et al (2011) Toward a more flood resilient urban environment: the Dutch multi-level safety approach to flood risk management. In: Otto-Zimmermann K (ed) Resilient cities: cities and adaptation to climate change. Proceedings of the geodesign the multi-layered water safety global forum 2010. Springer, Netherlands, pp 273–282
6. Sophronides P, Steenbruggen J, Scholten HJ, Giaoutzi M (2016) Geodesign the multilayered water safety. Res Urban Ser. https://doi.org/10.7480/rius.4.825
7. Rijke J, Smith JV, Gersonius B et al (2014) Operationalising resilience to drought: multi-layered safety for flooding applied to droughts. J Hydrol. https://doi.org/10.1016/j.jhydrol.2014.09.031
8. UN Environment (2019) Global environment outlook – GEO-6: healthy planet, healthy people. Cambridge University Press, Cambridge
9. Du J, Cheng L, Zhang Q et al (2019) Different flooding behaviors due to varied urbanization levels within River Basin: a case study from the Xiang River basin, China. Int J Disaster Risk Sci 10:89–102. https://doi.org/10.1007/s13753-018-0195-4
10. Scholten T, Hartmann T, Spit T (2019) The spatial component of integrative water resources management: differentiating integration of land and water governance. Int J Water Resour Dev. https://doi.org/10.1080/07900627.2019.1566055
11. Zoppi C (2020) Ecosystem services, green infrastructure and spatial planning. Sustain
12. Pinto LV, Ferreira CSS, Pereira P et al (2020) Integration of ecosystem services and green and blue infrastructures concepts in the land use planning process: the Coimbra case study. PRO 30:90. https://doi.org/10.3390/proceedings2019030090
13. Piro P, Turco M, Palermo SA, et al (2019) A comprehensive approach to stormwater management problems in the next generation drainage networks. In: Internet of things
14. Menoni S (2018) Integrated knowledge in climate change adaptation and risk mitigation to support planning for reconstruction. In: Smart, resilient and transition cities
15. La Greca P, La Rosa D, Martinico F, Privitera R (2011) Agricultural and green infrastructures: the role of non-urbanised areas for eco-sustainable planning in a metropolitan region. Environ Pollut. https://doi.org/10.1016/j.envpol.2010.11.017
16. Murgante B, Scardaccione G, Las Casas G (2009) Building ontologies for disaster management: seismic risk domain. CRC Press, Taylor & Francis, London
17. European Environment Agency (2017) Green infrastructure and flood management: promoting cost-efficient flood risk reduction via green infrastructure solutions
18. European Comission (2020) Bringing nature back into our lives: an EU biodiversity strategy for 2030
19. Zevenbergen C, Fu D, Pathirana A (2018) Sponge cities: emerging approaches, challenges and opportunities
20. Hora KER, Sales MM (2019) For more sponge cities. In: de Oliveira FL, Mell I (eds) Planning cities with nature. Theories, strategies and methods. Springer, Berlin, pp 251–263
21. Chan FKS, Griffiths JA, Higgitt D et al (2018) "Sponge City" in China—a breakthrough of planning and flood risk management in the urban context. Land Use Policy 76:772–778. https://doi.org/10.1016/j.landusepol.2018.03.005
22. Wilson GA (2004) The Australian Landcare movement: towards 'post-productivist' rural governance? J Rural Stud 20:461–484. https://doi.org/10.1016/j.jrurstud.2004.03.002
23. Ronchi S, Salata S, Arcidiacono A et al (2019) Policy instruments for soil protection among the EU member states: a comparative analysis. Land Use Policy. https://doi.org/10.1016/j.landusepol.2019.01.017
24. Pavesi FC (2020) SPONGE LAND(SCAPE) Prime indicazioni per la pianificazione d'area vasta. Sperimentazioni attraverso il caso di studio della Regione Lombardia. Tesi di dottorato di ricerca in Ingegneria Civile, Ambientale, della Cooperazione Internazionale e di Matematica, Università degli Studi di Brescia
25. Pavesi FC, Barontini S, Pezzagno M (2020) "Sponge land (scape)": an interdisciplinary approach for the transition to resilient communities. In: EGU general assembly 2020

26. Office International de l'Eau Natural Water Retention Measure. http://nwrm.eu/concept/3857. Accessed 9 Oct 2020
27. European Union (2014) EU policy document on natural water retention measures by the drafting team of the WFD CIS Working Group Programme of Measures (WG PoM)
28. Collentine D, Futter MN (2018) Realising the potential of natural water retention measures in catchment flood management: trade-offs and matching interests. J Flood Risk Manag
29. Hartmann T, Slavikova L (2018) How private land matters in flood risk management
30. ISTAT (2019) Classificazioni statistiche e dimensione dei comuni al 01/07/2020
31. Regione Lombardia (2010) Piano Territoriale Regionale. Paesaggistico, Piano
32. AA. VV (2020) Il sistema agro-alimentare della Lombardia. Rapporto 2019, 1st ed
33. Trigila A, Iadanza C, Bussettini M, Lastoria B (2018) Dissesto idrogeologico in Italia: pericolosità e indicatori di rischio - Edizione 2018
34. Simensen T, Halvorsen R, Erikstad L (2018) Methods for landscape characterisation and mapping: a systematic review. Land Use Policy. https://doi.org/10.1016/j.landusepol.2018.04.022
35. Meeus JHA (1995) Pan-European landscapes. Landsc Urban Plan. https://doi.org/10.1016/0169-2046(94)01036-8
36. Lioubimtseva E, Defourny P (1999) GIS-based landscape classification and mapping of European Russia. Landsc Urban Plan. https://doi.org/10.1016/S0169-2046(99)00008-0
37. Mücher CA, Klijn JA, Wascher DM, Schaminée JHJ (2010) A new European landscape classification (LANMAP): a transparent, flexible and user-oriented methodology to distinguish landscapes. Ecol Indic. https://doi.org/10.1016/j.ecolind.2009.03.018
38. Vienken T, Dietrich P (2011) Field evaluation of methods for determining hydraulic conductivity from grain size data. J Hydrol. https://doi.org/10.1016/j.jhydrol.2011.01.022
39. Singh VK, Kumar D, Kashyap PS et al (2020) Modelling of soil permeability using different data driven algorithms based on physical properties of soil. J Hydrol. https://doi.org/10.1016/j.jhydrol.2019.124223
40. da Anjinho PS, Barbosa MAGA, Costa CW, Mauad FF (2021) Environmental fragility analysis in reservoir drainage basin land use planning: a Brazilian basin case study. Land Use Policy 100:104946. https://doi.org/10.1016/j.landusepol.2020.104946
41. Regione Lombardia (2014) Documento preliminare della variante finalizzata alla revisione del piano Territoriale Regionale comprensivo del Piano Paesaggistico regionale
42. Regione Lombardia (2013) Basi informative dei suoli. http://www.geoportale.regione.lombardia.it/metadati?p_p_id=PublishedMetadata_WAR_geoportalemetadataportlet&p_p_lifecycle=0&p_p_state=maximized&p_p_mode=view&_PublishedMetadata_WAR_geoportalemetadataportlet_view=editPublishedMetadata&_PublishedMetadata_WAR
43. Regione Lombardia (1987) Base informativa della cartografia Geoambientale - Carta idrologica con indicazioni della permeabilità. http://www.geoportale.regione.lombardia.it/metadati?p_p_id=PublishedMetadata_WAR_geoportalemetadataportlet&p_p_lifecycle=0&p_p_state=maximized&p_p_mode=view&_PublishedMetadata_WAR_geoportalemetadataportlet_view=editPublishedMetadata&_PublishedMetadata_WAR
44. Regione Lombardia (2017) Variante al Piano Paesaggistico Regionale. https://www.sivas.servizirl.it/sivas/#/login/schedaProcedimento?idProcedimento=1&idPiano=93300
45. Regione Lombardia (2018) Uso e copertura del suolo 2018 (DUSAF 6.0). http://www.geoportale.regione.lombardia.it/metadati?p_p_id=PublishedMetadata_WAR_geoportalemetadataportlet&p_p_lifecycle=0&p_p_state=maximized&p_p_mode=view&_PublishedMetadata_WAR_geoportalemetadataportlet_view=editPublishedMetadata&_PublishedMetadata_WAR
46. Regione Lombardia (2020) Aree protette. http://www.geoportale.regione.lombardia.it/metadati?p_p_id=PublishedMetadata_WAR_geoportalemetadataportlet&p_p_lifecycle=0&p_p_state=maximized&p_p_mode=view&_PublishedMetadata_WAR_geoportalemetadataportlet_view=editPublishedMetadata&_PublishedMetadata_WAR

47. Regione Lombardia (2017) Vincoli paesaggistici. http://www.geoportale.regione.lombardia.it/metadati?p_p_id=PublishedMetadata_WAR_geoportalemetadataportlet&p_p_lifecycle=0&p_p_state=maximized&p_p_mode=view&_PublishedMetadata_WAR_geoportalemetadataportlet_view=editPublishedMetadata&_PublishedMetadata_WAR
48. Regione Lombardia (2019) Valore agricolo suoli 2018. http://www.geoportale.regione.lombardia.it/metadati?p_p_id=PublishedMetadata_WAR_geoportalemetadataportlet&p_p_lifecycle=0&p_p_state=maximized&p_p_mode=view&_PublishedMetadata_WAR_geoportalemetadataportlet_view=editPublishedMetadata&_PublishedMetadata_WAR
49. Regione Lombardia (2020) Piani di Indirizzo Forestale. informazioni/Enti-e-Operatori/agricoltura/boschi-e-foreste/piani-indirizzo-forestale/piani-indirizzo-forestale
50. Regione Lombardia (2011) Rete Ecologica Regionale (RER). http://www.geoportale.regione.lombardia.it/metadati?p_p_id=PublishedMetadata_WAR_geoportalemetadataportlet&p_p_lifecycle=0&p_p_state=maximized&p_p_mode=view&_PublishedMetadata_WAR_geoportalemetadataportlet_view=editPublishedMetadata&_PublishedMetadata_WAR
51. Office International de l'Eau (2015) European NWRM platform
52. European Commission (2015) Selecting, designing and implementing Natural Water Retention Measures in Europe
53. Mooney H, Larigauderie A, Cesario M et al (2009) Biodiversity, climate change, and ecosystem services. Curr Opin Environ Sustain
54. von Haaren C, Galler C, Ott S (2008) Landscape planning. The basis of sustainable landscape development. Leipzig

Part III
Socio-Economic Aspects of NBS

A Cooperative Game for Upstream–Downstream River Flooding Risk Prevention in Four European River Basins

Weronika Warachowska, Xana Alvarez, Nejc Bezak, María Gómez-Rúa, Andrea Janeiro-Otero, Piotr Matczak, Juan Vidal-Puga, and Vesna Zupanc

W. Warachowska (✉)
Faculty of Geographical and Geological Sciences, Adam Mickiewicz University in Poznań, Poznań, Poland
e-mail: warachowska@amu.edu.pl

X. Alvarez
Engineering Department of Natural Resources and Environment, School of Forestry Engineering, University of Vigo, Vigo, Spain
e-mail: xaalvarez@uvigo.es

N. Bezak
Faculty of Civil and Geodetic Engineering, University of Ljubljana, Ljubljana, Slovenia
e-mail: nejc.bezak@fgg.uni-lj.si

M. Gómez-Rúa
Faculty of Economics and Business, University of Vigo, Vigo, Spain
e-mail: mariarua@uvigo.es

A. Janeiro-Otero
Faculty of Environment and Natural Resources, University of Freiburg, Freiburg im Breisgau, Germany

P. Matczak
Faculty of Sociology, Adam Mickiewicz University in Poznań, Poznań, Poland
e-mail: matczak@amu.edu.pl

J. Vidal-Puga
Facultade de Ciencias Sociais e da Comunicación, Universidade de Vigo, Vigo, Spain
e-mail: vidalpuga@uvigo.es

V. Zupanc
Biotechnical Faculty, University of Ljubljana, Ljubljana, Slovenia
e-mail: vesna.zupanc@bf.uni-lj.si

Carla S. S. Ferreira, Zahra Kalantari, Thomas Hartmann, and Paulo Pereira (eds.), *Nature-Based Solutions for Flood Mitigation: Environmental and Socio-Economic Aspects*, Hdb Env Chem (2022) 107: 379–398, DOI 10.1007/698_2021_766,

Published online: 6 May 2021

Contents

Abstract This chapter tests whether a group of landowners living in the upstream part of a river basin could change land use to increase retention and thus decrease flood risk of the other group living in downstream parts of the river basin.

A cooperative game theory model combined with cost–benefit analysis is applied in four river basin settings in Europe: Stille Oder (Germany), Pysznica (Poland), Vipava (Slovenia), and Lea (Spain). These settings demonstrate various characteristics differentiating among catchments in terms of their size and land use, such as agricultural, forestry, and urbanisation.

Analysis reveals that in two of four river basins – Stille Oder (Germany) and Pysznica (Poland) – it is possible to find a mutually beneficial compromise between landowners to change land use (afforestation), which is economically reasonable for both of them, leading to flood risk reduction.

The cost–benefit analysis was applied to estimate the possible total benefit of afforestation that was an input data to the game theory model. The model applied in this chapter offers insights for flood risk reduction relying on nature-based solutions. It determines the benefits of cooperation that can be achieved by decision-making process participants separately and their coalition when cooperating. The sharing-rule function can help planners to distribute the total benefits from flood loss reduction among landowners fairly. Afforestation appears a feasible method for flood risk management.

The chapter also formulates further directions for game theory application in the field of environmental chemistry such as transportation of pollutants during flood events.

Keywords Cost-benefit analysis, Flood risk management, Game theory, Land use change

1 Introduction

Flood is the most significant hydrological hazard worldwide in terms of risks to life and property [1, 2]. Climate change and increasing exposure of people and assets increase the probability of future flood events that lead to a reduction in safety for local populations and higher costs for flood damage [3–5]. It is also expected that the damage caused by floods will increase in the coming decades, influencing infrastructure and the health and lives of the affected people [6, 7].

To cope with the growing flood risk, it is necessary to establish feasible flood protection measures. Recent developments suggest that nature-based solutions could play an important role instead of or in addition to traditional engineered approaches. Controlled flood retention is one strategy considered to have potential for flood risk management [8]. While nature-based solutions are claimed to provide effective solutions, they require land to reach the desired capacity, and land use planning needs to be involved in the development of solutions. Taking into account inconsistent approaches of spatial planners and water engineers appears a difficult task [9]. Furthermore, spatial planning involves multiple stakeholders (such as sectors, interest groups, and individuals) driven by diverse interests that need to be integrated [10]. How to align these interests is subject to much debate, from theoretical and practical points of view. Consequently, a defined public interest concept needs to be developed [11, 12] that can provide legitimate solutions [13].

For flood risk management, the river basin scale is fundamental. The relation between the downstream and upstream of a river basin needs to be considered in terms of expected damage reduction, because actions undertaken upstream influence the risk of flood downstream. In general, downstream areas benefit from upstream flood retention services [14], whereby downstream homeowners, commercial businesses, public institutions, and infrastructure operators benefit directly from the reduction in flood risk. Moreover, landowners of flood-protected land, both agricultural and (undeveloped) building land, benefit indirectly from upstream flood retention, as land located in flood-prone areas would be exposed to lower flood risk or even classified outside flood hazard zones; thus, it may become legally suitable for development. This is usually accompanied by a significant appreciation in property value [8].

Afforestation in the upstream part of a river basin is usually considered for increasing retention capacity. However, the introduction of upstream flood retention requires a change in land use that involves costs. Therefore, convincing upstream landowners to modify their land use becomes a crucial step in establishing protection measures. Property rights and fairness are key in such negotiations. While one agent is expected to act for the good of others, the question of strategic behaviour becomes imminent. A proper distribution of resources, welfare, rights, duties, and opportunities need to be considered within a comprehensive framework to solve common distributive problems [15]. In addition, change in land use is strictly regulated in many countries. For instance, in Galicia (Spain) forest policy allows agricultural land that has been explicitly abandoned for more than 10 years to be afforested [16]. At the same time, agricultural land abandonment is the largest land use change process in Europe. More specifically, in the study area of Galicia, it is estimated the abandonment rate will reach 44% by 2030 [17]. This suggests significant potential for afforestation as a flood risk management measure.

In this study, a game theory model is applied to four different European basins to simulate a decision-making process aimed at reducing the negative consequences of flood. The costs and benefits of actions/inactions are examined in terms of potential cooperation between both upstream and downstream agents. This approach is based on game theory, a mathematical tool that enables analysis to solve allocation problems where two or more agents have their own interests, both seeking to maximise benefits. Here, a cooperative 'game' is defined, where 'players' cooperate and a mutually beneficial compromise is possible.

2 Game Theory in Flood Issues

Game theory is an analytical tool that enables interactions of rational players pursuing their interests to be modelled. Hence, it is a suitable tool to be applied in research on water resource management such as conflicts on irrigation or transboundary water conflicts, as well as for flood risk management, which involves parties with a conflict of interest. For instance, communities occupying both riversides may compete in heightening their levees. In this game, increasing safety on one riverside decreases safety on the other [18]. The same situation can occur when considering the upstream–downstream distinction. Machac et al. [19] discuss scenarios for negotiations between upstream and downstream from a game theory perspective. The authors analysed how changes in conditions (such as a preference for upstream or downstream) influence the outcomes of the game.

Many types of games have been developed and can be applied to specific conflict situations related to flood risk management. Parrachino et al. [15], Zara et al. [20, 21] provide the basics and a review of some applications of cooperative game theory to issues of water resources. There is also a wide literature devoted to the study of allocation problems to solve issues related to transboundary rivers using cooperative game theory. Applications include water resource development [22], water allocation [23], pollution control costs [24], and flood cost sharing [25]. Non-cooperative game theory has also been applied to water management problems [26], water right conflicts [27], and efficient allocation of water [28]. Gómez-Rúa [29], van den Brink et al. [30], and Sun et al. [31] address the problem of sharing the cost of cleaning a polluted river, for example, using environmental taxes. Béal et al. [32] and Beard [33] provide surveys on the use of cooperative game theory to model water allocation problems.

There is an important distinction between cooperative and non-cooperative games. In non-cooperative games, players compete and make decisions independently, whereas in cooperative game players make decisions together [26]. Hui et al. [18] argue that cooperative games involve methods of optimisation that assume perfect cooperation between players. However, in many areas of natural resource management, non-cooperation is players' dominant strategy [34]. For example, the prisoner's dilemma and other non-cooperative types of games depict such situations leading to non-Pareto optimal results.

To make further progress in this field of research, a multi-model and multidisciplinary approach is recommended [35]. In particular, flood damage is attributed to increasing exposure due to high population growth and economic development in flood-prone areas [36–38]; therefore, awareness of floods and effectiveness of flood protection measures are also taken into account as factors that influence the decisions of private landowners in land management cases [39].

Álvarez et al. [40] apply game theory to study the problem of incentivising land owners to use their land in a way that reduces flood risk. Mitigating flood risk has numerous benefits; for instance, a reduction of the costs derived from flooding. Using the game theory framework, a wide literature related to transboundary rivers exists that studies different problems associated with the river. This literature developed in two directions. First, models were proposed that study how to share the costs of cleaning a river among the regions located along it. Second, other models have been proposed for studying how to share water resources among the different regions located along a river.

The first problem consists of two main approaches: some studies consider a river a segment divided into different regions and assume that the cost of cleaning each region is exogenously given (observable) [41–44]. These studies propose different allocation rules for distributing the cost of cleaning the river among the regions. The second approach is taken in Gengenbach et al. [45] and van der Laan and Moes [46], where the cost allocation method adopted is thought to affect the decision of each region about how much waste to discharge.

For other types of problems, the focus is on analysing water allocation and achieving fair distribution of welfare resulting from distributing river water among different regions. The first paper by Ambec and Sprumont analysed how water should be allocated among agents, proposing monetary transfers among them from the point of view of the game theory [47]. Several other papers followed considering this topic [48, 49].

3 Method

This study applies the cooperative game theory model combined with cost–benefit analysis. The main goal is to use a sharing-rule function to distribute the total benefit among agents. A sharing-rule function determines the way benefits of cooperation should be shared for a given situation modelled by a cooperative game. A cooperative game determines the benefits of cooperation that can be achieved by: (a) each agent separately, and (b) each coalition when cooperating. This enables us to determine the most stable and fair share given by the sharing-rule function. The sharing rule can be used to set compensations and incentives to achieve a fair allocation of costs and benefits, as proposed in the cooperative game theory model presented by Álvarez et al. [40]. The model establishes distribution rules that satisfy a core idea of stability, namely that no agent or group of agents can find themselves in a worse position than working separately. Álvarez et al. [40] propose three such rules. The first rule is the most favourable possible for the upstream agent. The second is the most favourable possible for the downstream agent. Finally, the third

Table 1 Area for each agent in selected basins (ha)

	Stille Oder (Germany)	Pysznica (Poland)	Vipava (Slovenia)	Lea (Spain)
Agent 1 (upstream)	5,729	3,517	38,605	13,219
Agent 2 (downstream)	4,229	3,026	19,317	2,011
Total	9,958	6,543	57,922	15,230

balances both approaches by taking a compromise solution between the previous two.

This study is explorative; therefore, the model is simplified with the following assumptions: (1) there are only two agents in the drainage basin; (2) two main land uses are considered (defined as 'forest' and 'other'). According to Bentley and Coomes [50], afforestation of lands previously degraded by agriculture helps to repair the soil so that it can retain more water and reduce the flow of the nearby river.

Agents in a so-called flood game (as defined in [40]) are spatial units located in different parts of the drainage basin. In this study, for each of four selected basins, two agents (players) were defined as decisions-making process participants. Agent represent the regions including all subbasins in each part of the drainage basins and was delineated as follows: (a) upstream agent, located in the upstream part of the river where the flood risk is low and flood protection measures and actions are to be undertaken (agent 1); (b) downstream agent, located in the downstream part of the river where the flood risk is high (agent 2). Agents have the freedom to change the use of their own land and the right to deny any change on their own land that they do not agree with. Although game theory enables to represent each landowner as a separate agent, the approach had to be simplified. Because this analysis is the first attempt to apply this model, the agents represent collectives of landowners located in each part of the drainage basin. Delimitation for each basin was considered separately. The main factor was delimitation of the flood extent based on flood hazard maps at both a European and global scale based on streamflow data from the European and Global Flood Awareness System (Flood Maps). The shape and size of the basin were also considered.

Areas for each agent for each selected basin are presented in Table 1.

To apply the game theory model, we need to consider the worth of upstream and downstream agents (w_1 and w_2, respectively) when acting individually, and the worth of both agents when they cooperate (w_{12}). For the former, we have $w_1 = \max(A; F)$, and for the later, $w_{12} = \max(A, F + B)$, where:

$A =$ How much agent 1 gets if it does not change the land use to forest.

$F =$ How much agent 1 gets if it does change the land use to forest.

$B =$ Benefit provided by the decrease in flood damage due to agent 1 changing the land use to forest.

We also normalise $w_2 = 0$, since it does not play a role in the share.

Given these values, a stable sharing rule should provide the following payoff allocation (x_1 and x_2):

- Agent 1: $x_1 = w_1 + (w_{12} - w_1 - w_2)\ d$
- Agent 2: $x_2 = w_2 + (w_{12} - w_1 - w_2)\ (1 - d)$

where d is a value between 0 and 1. For $d = 1$, we obtain the most favourable deal for agent 1. For $d = 0$, we obtain the most favourable deal for agent 2. For $d = 0.5$, we obtain a compromise deal.

4 Study Areas

Four European river basins were subjected to analysis: the Stille Oder river basin (Germany), the Pysznica river basin (Poland), the Vipava river basin (Slovenia), and the Lea river basin (Spain) (see Fig. 1). The criteria for the selection of case studies

Fig. 1 Location of the studied basins: (**a**) Stille Oder, (**b**) Pysznica, (**c**) Vipava, (**d**) Lea

were as follows: an area where a significant flood risk exists and where a potential upstream–downstream conflict could be present was selected, the idea was to capture different climate zones and different hydro-meteorological conditions within the Europe and finally the selected catchment need to have the required data available.

The Stille Oder river, also known as Mucker, is a former branch of the Oder River. It is located in the north-east part of the federal state of Brandenburg in Germany, as part of the Oderbruch, a former delta of the Oder river. Today, the Oder's main channel is restrained to the eastern edge of the depression, and the remnants of the former branches bear designations like the Stille Oder. Approximately 86% of the basin area consists of non-irrigated arable land, with another 11% of pasture, 3% of discontinuous urban fabric, and the remaining consisting of small percentages of agricultural land with significant areas of natural vegetation, broad-leaved and coniferous forests, inland marshes, and water courses [20, 21]. The Oderbruch suffered from heavy flooding in 1785, 1838, 1947, 1981/82, 1997, and 2010, the most recent event reaching a water level above 7 m due to rainfall of up to 200 l/m^3 [51].

The Pysznica River basin is a right tributary of the Parsęta River located in the north-west of Poland. Dominant types of land use in the catchment include agricultural areas (74%), which is mainly non-irrigated arable land, pastures, complex cultivation patterns, and land principally occupied by agriculture, with significant areas of natural vegetation. Complementary types of land use are broad-leaved, coniferous and mixed forests (24%) and discontinuous urban fabric (2%) [20, 21]. According to Polish maps of flood risk and flood danger [52], flood risk on the Pysznica river catchment is low; however, it is assumed it will increase significantly over the next 10 years.

The Vipava River catchment (upstream of the Miren discharge gauging station) is located approximately 1 km before the border with Italy and 2.5 km before the confluence with the Soča River. The annual maximum discharge at the location of the Miren station can be as much as 400 m^3/s, while minimum annual flows can be less than 1 m^3/s [53]. Thus, the difference between minimum and maximum flows is quite large, which is a consequence of rainfall generation mechanisms in the area where extreme rainfall events are relatively frequent. Forest covers approximately 65% of the Vipava River catchment and agricultural areas around 32%, while urban areas represent approximately 3% of the total area [20, 21]). Since the climate is Mediterranean, the agriculture is well-developed in the area and at specific locations supported by irrigation systems.

The Lea River basin, located in Galicia (North-western Spain), is a tributary of the Miño River in the upper part of the basin. The river catchment is associated with complex cultivation patterns (38%), forests and semi natural areas (59%), land principally occupied by agriculture (1.5%), and artificial surfaces (1.5%) [20, 21]. According to Spanish maps of flood risk [52], it exhibits a medium risk of flooding for the lower basin and a very low risk for the upper river basin. Therefore, it would not change its level of risk for the next 10 years.

5 Costs and Benefits

Costs related to flood damage can be all assigned to agent 2, since according to the flood maps, the risk of flooding only occurs in the downstream part of the river basin. Agent 1's strategy for initial land use is defined as the initial state where none of the costs or benefits appear. The flood risk has not been reduced, no costs are incurred, and the payoffs are normalised to zero. Payoffs for agent 1's forest strategy constitute the difference between flood damage before and after land use change. The total benefit derived from land use change has been assigned to agent 1, as all activities related to change of land use are undertaken only in the upstream part of the basin.

A cost–benefit analysis has been conducted for a time period of 100 years, which means the most important aspects (described in the following) can be captured. Notice that money in the present is worth more than the same amount in the future because of both inflation and earnings from alternative investments that could be made during the 100-year period. For example, any investor would prefer to get 100 € today than 100 € next year. We expect, however, that there is an amount (e.g. 105 €), so that an average investor would be indifferent between obtaining 100 € today and 105 € next year. In that case, we say that that the money has a yearly discount rate of 5%. In the economic literature, a standard way to compare cash flows in different periods of time is by the net present value (NPV), which represents the value inflows in present currency.

Concerning the discount rate for the analysis, the '*Guide to Cost Benefit Analysis of Investment Projects*' proposed a 5.5% discount rate for cohesion countries and 3.5% for other countries for the 2007–2013 period. However, taking a 100-year time horizon, the discount rate applied was 3.5% for all four basins. Similar values were adopted in other studies [54]. All costs and benefits were assigned to three main groups: (a) expected flood damage related to initial land use; (b) expected costs and benefits related to initial land use; and (c) expected costs and benefits related to land use change. These groups are presented in the following description.

(a) Expected flood damage for initial land use, including all the costs related to potential damage caused by flood both before and after land use change. Calculations (before land use changes) were conducted on the basis of global flood depth-damage functions developed by Huizinga et al. [55]. The damage curves depict fractional damage as a function of water depth as well as the relevant maximum damage values for specific assets and land use classes. Damage curves and maximum damage values were adjusted for local circumstances for each of the four analysed basins. Flood extent was attributed following the flood maps. Equation (1) displays the formula for calculating expected damage for initial land use.

$$T1 = A * D * M \tag{1}$$

where: $T1$ = total damage [€], A = area covered by specific impact category (Residential, Commercial, Industrial, Agriculture, Infrastructure) [ha],

D = damage function (adjustment for specific flood depth), M = max damage (according to EU flood depth-damage functions) [€/ha].

Flood damage after land use change was calculated, based on the assumption of Salazar et al. [56] regarding the influence of afforestation on peak discharge reduction. Specific peak discharges for each basin were compared to the function defined by Salazar et al. [56] on the basis of case study analysis in different European hydro-climatological regions. Then, the flood damage after land use change was estimated, assuming the total damage would decrease by the same percentage as the peak discharge. Although this method does not allow for costs to be precisely specified, it is still possible to estimate the general tendency for how afforestation influences flood risk and flood damage.

(b) Expected costs and benefits related to initial land use.

The total benefit of initial land use was calculated using the NPV, and by including (1) potential benefit from harvested crops (Eq. 2) defined as a generalised benefit from agricultural land, (2) costs of land cultivation (Eq. 3), and (3) subsidies for agricultural activities. Equation (4) presents the formula for calculating the total benefit from initial land use. All three equations consider the discount rate for 100 years period, including the first year for which the initial costs, benefits, and subsidies were defined.

Costs and benefits were considered only for part of the area that is meant to be afforested, located upstream.

$$T2 = A * P * \frac{1 - (1/(1+d))^{101}}{1 - (1/(1+d))} \tag{2}$$

where: $T2$ = total benefit from harvests [€], A = area cover by agriculture [ha], P = price in 2020 [€/ha], d = discount rate (3.5%)

$$T3 = A * C * \frac{1 - (1/(1+d))^{101}}{1 - (1/(1+d))} \tag{3}$$

where: $T3$ = total cost of cultivation [€], A = area cover by agriculture [ha], C = cost of land cultivation in 2020 [€/ha], d = discount rate (3.5%).

$$\mathrm{T} = \mathrm{T2} - \mathrm{T3} + \mathrm{S} \tag{4}$$

where: T = total benefit from initial land use [€], $T2$ = total benefit from harvests [€], $T3$ = total cost of cultivation [€], S = agricultural subsidies [€].

(c) Costs and benefits related to land use change.

The main assumption of the study is that the change of land use upstream would reduce flood risk and limit flood damage downstream. To assess the positive potential influence of land use change, the costs and the benefits were analysed, including (1) cost of land use change and land cultivation after change, including one-off investment costs and land cultivation for the whole period of analysis (Eq. 5); (2) subsidies for afforestation; and (3) benefits from harvesting (Eqs. 6 and 7). Equation (8) displays the formula for the total benefit from afforested land. Equations (6), (7), and (8) use the NPV that was also used in Eqs. (2) and (3). However, here the benefit flow total value of harvest is null during the first years and increases steadily until reaching the optimal flow in 10 years' time.

In just the same way as the case of initial land use, costs and benefits were considered only for the area that is meant to be afforested. The subsidies were not included for the Slovene case because afforestation is not governmentally supported. For Germany, Poland, and Spain, national and regional government support is provided, which includes one-off support for afforestation either care or maintenance bonus for a 5–20 year period.

$$T4 = A * I + A * M * \frac{1 - (1/(1+d))^{101}}{1 - (1/(1+d))} \tag{5}$$

where: $T4$ = total cost of land use change and management [€], A = area meant to change land use [ha], I = investment costs (once for the whole area) [€/ha], M = management costs [€/ha], d = discount rate (3.5%).

$$T5 = \text{NPV}\ (A * W * P, d) \tag{6}$$

$$T6 = \text{NPV}\ (A * P, d) \tag{7}$$

where NPV(X,d) is the function used to compute the NPV depending on the value of the 100-year cash flow (X) and the discount rate (d = 3.5%), and where: $T5$ = benefit from harvests for Poland and Germany [€], $T6$ = benefit from harvest for Slovenia and Spain [€], A = area covered by forest [ha], W = amount of wood [m^3/ha], P = price of wood [€/m^3] or [€/ha].

$$\text{TF} = T5 - T4 + S \text{ or } \text{TF} = T6 - T4 + S \tag{8}$$

where: TF = total benefit of land use change [€], $T4$ = total cost of land use change and management [€], $T5$ = benefit from harvests for Poland and Germany [€], $T6$ = benefit from harvests for Slovenia and Spain [€], S = subsidies [€].

6 Results

Due to defined flood risk in each of the analysed regions, the damage caused by flood events were estimated as input data for the game theory model. The differences in total damage values for each basin are related to the land use structure, the area of flooded land, and the depth of flood. The results of calculations for flood damage before land use change (Eq. 1) are presented in Table 2.

The highest costs in the Vipava River basin (Slovenia) reflect the significant area covered by residential and industrial buildings, for which the highest maximum damage values were defined. The Stille Oder basin costs were mostly derived from agriculture as this comprises almost 95% of land use. Relatively small damages counted for the Lea river basin (Spain) are related to the area least endangered by flood risk of all the analysed basins. The least damage quantified for the Pysznica River basin (Poland) are the results of the small water depth and the high percentage of land covered by agricultural areas.

The results of cost–benefit analysis are the payoffs of the game for each agent in case of two undertaken strategies. The results of game theory model application and payoffs for each agent in the four analysed basins are presented in Table 3.

Differences between the benefits from initial land use for each basin are related to the area covered by agricultural land, the type of crop, and possible subsidies for agricultural activities. The difference in the total benefit between Germany and the other countries is mainly related to the fact that almost 95% of the basin area is covered by non-irrigated arable land. However, this result is only achieved with the

Table 2 Flood damage related to initial land use

	Damage (million €)			
Damage class	Stille Oder (Germany)	Pysznica (Poland)	Vipava (Slovenia)	Lea (Spain)
Residential buildings	0.091	0.143	50.315	0.244
Industrial buildings	–	–	26.916	1.242
Agriculture	10.719	0.323	0.246	0.071
Infrastructure	0.010	0.048	1.708	0.064
Total	10.820	0.514	79.185	1.621

Table 3 The share of costs and benefits (million €)

	Stille Oder (Germany)	Pysznica (Poland)	Vipava (Slovenia)	Lea (Spain)
The benefit for agent 1 if does not change the land use *(A)*	51.522	5.303	0.506	1.620
The benefit for agent 1 if does change the land use *(F)*	30.955	12.667	−5.209	9.944
Benefit from flood damage decrease if agent 1 change the land use *(B)*	1.407	0.093	1.822	0.073
The worth of agent 1 without cooperation (w_1)	51.522	12.667	0.506	9.944
The worth of both agents when they cooperate (w_{12})	51.522	12.760	0.506	10.017
Payoff allocation (x_1)	51.522	12.667 + 0.093d	0.506	9.944 + 0.073d

Table 4 Benefit from land use change (million €)

	Stille Oder (Germany)	Pysznica (Poland)	Vipava (Slovenia)	Lea (Spain)
Cost of land use change and land cultivation	3.714	5.605	9.744	1.908
Subsidies	10.017	4.362	–	2.868
Benefits from harvesting	24.652	13.910	4.536	8.984
Total benefit	30.955	12.667	−5.209	9.944

Table 5 Benefit transfer and possible land use change (million €)

	Stille Oder (Germany)	Pysznica (Poland)	Vipava (Slovenia)	Lea (Spain)
Land use change	No	Yes	No	Yes
Transfer more favourable to agent 1	–	0.093	–	0.073
Transfer more favourable to agent 2	–	No transfer	–	No transfer
Compromise deal	–	0.0465	–	0.0365

help of state payments (decoupled farm payment, compensation payments, and subsidies).

Differences in the benefits of land use change are influenced by relatively high subsidies for afforestation in Germany and a lack of them in Slovenia, which is depicted in Table 4 (results of the application of Eqs. 2 and 3).

Possible benefit distribution was analysed to assess the possibility of land use change. Table 5 presents the information about possible land use change and the transfer that agent 2 should make to agent 1 for compensation for land use change.

According to the information presented in Table 5, it was assumed that land use change in the Stille Oder (Germany) and Vipava (Slovenia) river basins is not a probable scenario. For both upstream and downstream players, land use change is unfavourable and no benefit is obtained. Therefore, the analysis of possible benefit transfer was performed only for the Pysznica (Poland) and Lea (Spain) river basins. In both cases, the benefit transfer direction from downstream to upstream agent is presented and the compromise deal constitutes a transfer of 0.0465 and 0.0365 million € (for Poland and Spain, respectively), which would gratify both players in the basin. Note that the compromise deal constitutes the half of the transfer more favourable to agent 2 (from agent 1). It is also the half of the benefit from flood damage decrease if agent 1 changes the land use. According to above the compromise deal is directly related to avoided damages caused by flood that appears downstream when upstream agent decides to change the land use.

7 Discussion

The main aim of this study was to investigate the potential cooperation between decision-making process participants to distribute the total costs and benefits related to land use change that leads to flood risk reduction. The findings of the analysis depict that in two of four analysed river basins it is possible to find a mutually beneficial compromise among landowners for flood risk reduction if land use change (afforestation) is economically reasonable for both agents.

This study offers a methodological contribution to establishing and applying distribution rules for sharing the benefits and the costs related to flood risk reduction and land use change.

The chapter presents the results of the application of a game theory model on four European basins, offering the first empirical approach to the theoretical model. Therefore, analysis was based on a number of assumptions. First, flood damage (both before and after land use change) was estimated rather than precisely modelled. Although the applied method does not allow direct specification of the costs, it is possible to estimate the general tendency for flood damage change. This simplification may influence the final result of the analysis; therefore, it is recommended for future analysis to apply combined hydrological and hydraulic models to accurately define the losses caused by flood events. Second, this work applies to only one scenario and a 100-year time horizon. Multiple scenarios, assuming different time spans, different land uses, or a different course of afforestation could enrich the analysis.

Third, this analysis relies on two players. As a future line of research, the number of players included in the game theory model could reflect the number of landowners in the basins, as this would imply a more complex model and consequently more precise results. However, this would require detailed land ownership structure analysis and adaptation of the model to account for local conditions. It should also be underlined that subsidies play a crucial role in the structure of costs and benefits, and local or national governments should be considered as a separate agent.

Fourth, cost–benefit analysis (CBA) could be extended by ecosystem co-benefits or regulating services like water quality improvement provided by reforestations. Game theory has significant potential in the field of water quality changes and transportation of pollutants during flood events. For example, Alcalde-Unzu et al. [41] use the clean-up cost vector to estimate the transfer rate of the waste in a polluted river. They use estimation to share of cost of cleaning the river. On the other side, Wei and Luo [57] focuses on how to reach a balance between the sustainable development of local economy and the effective protection of water resources from an ecological perspective for the local government, and how to maximise the profit of the local firm in an ecological compensation system. Besides a reduction in the risk of flooding, afforestation entails several other benefits, such as improving the landscape and the environment, and providing a source of income for forest owners. The payment for environmental services (PES) can be considered a method to incorporate services provided by the environment into calculations of costs and benefits. Moreover, PES could encourage forest owners to maintain or implant forests by compensating them at equivalent or better rates than other activities that would otherwise provoke deforestation [58]. Thus, owners who are located in strategic areas for flood risk reduction (such as upstream) may consider reforestation as a viable alternative for land use. PES can be estimated through game theory and can be considered a way to assess and plan an efficient forest policy.

A possible negative aspect is that if reforestations are carried out without planning it is possible that the flow of a river is reduced (even disappearing) in regions where there are water shortages. Therefore, it is important to consider the impact on regional water availability. Bentley and Coomes [50] point out that afforestation of lands previously degraded by agriculture helps to repair the soil, enabling it to retain more water and reduce the flow of the nearby river. If this action were carried out in natural grasslands where the soil is in good condition, the flow of the river would be considerably reduced.

Within the planning and management of these reforestations, and taking into account the criteria of improving the water quality, one strategy commonly advanced to achieve this goal is the management of riparian vegetation [11]. Several studies have documented that riparian forest can strongly influence the chemical content of adjacent streams [59, 60], particularly through the removal of nutrients in runoffs from agricultural uplands [61]. Therefore, vegetation restoration and management in riparian areas is widely recommended and promoted, especially in agricultural areas [62], but also in those areas of medium-high risk of flooding. In the four basins in this study, a restoration of the riparian vegetation could be planned (both in forest areas and in those lands for agricultural use) and framed within the proposed reforestation. Accordingly, the ecosystem services of riparian vegetation can be helped through the improvement of chemical water quality in streams, while reducing the risk of flooding in these areas. Conversely, the managed change of agricultural land to forest cover proposed in this study is recommended to address the issue of high nitrate in groundwater, ensuring good quality groundwater in the long term [63]. This would lead to savings in the treatment of drinking water, since Lopez et al. [64] have found a positive and significant effect of local forest cover on water treatment cost savings. Although this study does not focus on the specific benefits that changes in land use can generate in water quality, it implies that such effects can be highlighted.

8 Conclusion

The chapter presents an application of the game theory concept to four catchments located in parts of Europe with diverse climate characteristics. The investigation revealed that in two of four cases (Poland and Spain) mutually beneficial compromise between landowners to change land use (afforestation) could be detected, while Germany and Slovenia would not benefit from such a change, due to the considerable influence of subsidies.

Presented results reflect the possible direction for further actions in compensation for establishing new flood protection measures, however the undertaken scope of analysis, based on several assumptions such us limited number of agents and simplified flood risk assessment could influence the results. We recommend further, investigations using a larger number of agents and more detailed analysis (e.g. more detailed definition of the flood risk before and after afforestation or investigation of other measures) in order to enhance the knowledge about the upstream–downstream relationship in the flood risk management.

Acknowledgments The study was conducted as part of a Short-Term Scientific Mission (STSM) within the LAND4FLOOD: Natural Flood Retention on Private Land (CA16209) Cost Action. The project is based on work supported also by: 'GEO+' – high quality interdisciplinary doctoral studies programme conducted at the Faculty of Geographical and Geological Sciences of the Adam Mickiewicz University in Poznań (no. POWR.03.02.00-00-I039/16); project P2-0180 founded by Slovenian Research Agency; ARRS research programme P4-0085 Agroecosystems; project no. ECO2017-82241-R founded by Spanish Ministerio de Economía, Industria y Competitividad; project no. ED431B 2019/34 founded by Xunta de Galicia and project no. R815 131H 64502 founded by Conselleira de Educación, Universidade e Formación Profesional, Xunta de Galicia, España.

References

1. Downton MW, Pielke RA (2005) How accurate are disaster loss data? The case of US flood damage. Nat Hazards 35:211–228
2. Hirabayashi Y, Mahendran R, Koirala S, Konoshima L, Yamazaki D, Watanabe S, Kanae S (2013) Global flood risk under climate change. Nat Clim Chang 3(9):816–821
3. Jania J, Zwoliński Z (2011) Ekstremalne zdarzenia meteorologiczne, hydrologiczne i geomorfologiczne w Polsce (Extreme meteorological, hydrological and geomorphological events in Poland). Landform Anal 15:51–64
4. Kundzewicz ZW, Kanae S, Seneviratne SI, Handmer J, Nicholls N, Peduzzi P, Sherstyukov B (2014) Flood risk and climate change: global and regional perspectives. Hydrol Sci J 59(1):1–28
5. Milly PCD, Wetherald RT, Dunne KA, Delworth TL (2002) Increasing risk of great floods in a changing climate. Nature 415(6871):514–517
6. Cook HF (2017) The protection and conservation of water resources. Wiley, Hoboken
7. Jongman B, Winsemius HC, Aerts JC, de Perez EC, van Aalst MK, Kron W, Ward PJ (2015) Declining vulnerability to river floods and the global benefits of adaptation. Proc Natl Acad Sci 112(18):E2271–E2280
8. Löschner L, Nordbeck R, Schindelegger A, Seher W (2019) Compensating flood retention on private land in Austria: towards polycentric governance in flood risk management? Landsc Archit Front 7(3):32–45
9. Hartmann T, Driessen P (2013) The flood risk management plan: towards spatial water governance. J Flood Risk Manag 10(2):145–154
10. Forester J (2004) Reflections on trying to teach planning theory. Plann Theory Pract 5 (2):242–251
11. Alexander ER (2002) The public interest in planning: from legitimation to substantive plan evaluation. Plan Theory 1(3):226–249
12. Moroni S (2004) Towards a reconstruction of the public interest criterion. Plan Theory 3 (2):151–171

13. Hegger DLT, Driessen PPJ, Dieperink C, Wiering M, Raadgever GTT, van Rijswick HFMW (2014) Assessing stability and dynamics in flood risk governance: an empirically illustrated research approach. Water Resour Manag 28(12):4127–4142
14. Seher W, Löschner L (2016) Balancing upstream-downstream interests in flood risk management: experiences from a catchment-based approach in Austria. J Flood Risk Manag. https://doi.org/10.1111/jfr3.12266
15. Parrachino I, Zara S, Patrone F (2006) Cooperative game theory and its application to natural, environmental and water resource issues : 2. Application to natural and environmental resources. Policy research working paper no. 4073, wps4073. World Bank, Washington
16. Xunta de Galicia. Ley 7/2012, de 28 de junio, de montes de Galicia. https://www.xunta.gal/dog/Publicados/2012/20120723/AnuncioC3B0-050712-0001_es.html. Accessed Aug 2020
17. Perpiña Castillo C, Coll Aliaga E, Lavalle C, Martínez Llario JC (2020) An assessment and spatial modelling of agricultural land abandonment in Spain (2015–2030). Sustainability 12 (2):560
18. Hui R, Lund JR, Madani K (2015) Game theory and risk-based leveed river system planning with noncooperation. Water Resour Res. https://doi.org/10.1002/2015WR017707
19. Machac J, Hartmann T, Jilkova J (2017) Negotiating land for flood risk management: upstream-downstream in the light of economic game theory. J Flood Risk Manag 11(1):66–75
20. Zara S, Dinar A, Patrone F (2006) Cooperative game theory and its application to natural, environmental, and water resource issues: 2. Application to natural and environmental resources. World Bank
21. European Environment Agency (2020) CORINE Land Cover (CLC) 2018, Kopenhagen K, Denmark. CRC/TR32 Database (TR32DB). https://land.copernicus.eu/pan-european/corine-land-cover/clc2018
22. Suzuki M, Nakayama M (1976) The cost assignment of the cooperative water resource development: a game theoretical approach. Manag Sci 22:1081–1086
23. Wang LZ, Fang L, Hipel KW (2003) Water resources allocation: a cooperative game theoretic approach. J Environ Inf 2(2):11–22
24. Shi G, Wang J, Zhang B, Zhang Z, Zhang Y (2016) Pollution control costs of a transboundary river basin: empirical tests of the fairness and stability of cost allocation mechanisms using game theory. Environ Manag 177:145–152
25. Abraham A, Ramachandran P (2020) A solution for the flood cost sharing problem. Econ Lett 189:109030
26. Madani K (2010) Game theory and water resources. J Hydrol 381(3):225–238
27. Zanjanian H, Abdolabadi H, Niksokhan MH, Sarang A (2018) Influential third party on water right conflict: a game theory approach to achieve the desired equilibrium (case study: Ilam dam, Iran). Environ Manag 214:283–294
28. Gudmundsson J, Hougaard JL, Ko CY (2019) Decentralized mechanisms for river sharing. J Environ Econ Manag 94:67–81
29. Dosskey MG, Vidon P, Gurwick NP, Allan CJ, Duval TP, Lowrance R (2010) The role of riparian vegetation in protecting and improving chemical water quality in streams. J Am Water Resour Assoc 46(2):261–277
30. van den Brink R, He S, Huang J-P (2018) Polluted river problems and games with a permission structure. Games Econ Behav 108:182–205
31. Sun P, Hou D, Sun H (2019) Responsibility and sharing the cost of cleaning a polluted river. Math Methods Oper Res 89(1):143–156
32. Béal S, Ghintran A, Rémila E, Solal P (2013) The river sharing problem: a survey. Int Game Theor Rev 15(03):1–19
33. Beard R (2011) The river sharing problem: a review of the technical literature for policy economists. University Library of Munich
34. Carraro C, Marchiori C, Sgobbi A (2005) Applications of negotiation theory to water issues, World Bank policy res. Working pap. 3641, World Bank, Washington

35. Teng J, Jakeman AJ, Vaze J, Croke BFW, Dutta D, Kim S (2017) Flood inundation modelling: a review of methods, recent advances and uncertainty analysis. Environ Model Softw 90:201–216
36. Bouwer LM (2011) Have disaster losses increased due to anthropogenic climate change? B Am Metereol Soc 92
37. Neumayer E, Barthel F (2011) Normalizing economic loss from natural disasters: a global analysis. Global Environ Change 21(1):13–24
38. Visser H, Petersen AC, Ligtvoet W (2014) On the relation between weather-related disaster impacts, vulnerability and climate change. Clim Chang 125:461–477
39. Di Baldassarre G, Viglione A, Carr G, Kuil L, Salinas JL, Blöschl G (2013) Socio-hydrology: conceptualising human-flood interactions. Hydrol Earth Syst Sci 17:3295–3303
40. Álvarez X, Gómez-Rúa M, Vidal-Puga J (2019) River flooding risk prevention: a cooperative game theory approach. J Environ Manag 248:109284
41. Alcalde-Unzu J, Gómez-Rúa M, Molis E (2015) Sharing the costs of cleaning a river: the upstream responsibility rule. Games Econ Behav 90:134–150
42. Gómez-Rúa M (2013) Sharing a polluted river through environmental taxes. SERIEs 4:137–153
43. Ni D, Wang Y (2007) Sharing a polluted river. Games Econ Behav 60:176–186
44. van den Brink R, van der Laan G (2008) Comment on "Sharing a polluted river". Mimeo
45. Gengenbach MF, Weikard HP, Ansink E (2010) Cleaning a river: an analysis of voluntary joint action. Nat Resour Model 23:565–589
46. van der Laan G, Moes N (2012) Transboundary externalities and property rights: an international river pollution model. Tinbergen discussion paper 12/006-1. Tinbergen Institute and VU University, Amsterdam
47. Ambec S, Sprumont Y (2002) Sharing a river. J Econ Theory 107:453–462
48. Ambec S, Ehlers L (2008) Sharing a river among satiable agents. Games Econ Behav 64:35–50
49. Wang Y (2011) Trading water along a river. Math Soc Sci 61:124–130
50. Bentley L, Coomes DA (2020) Partial river flow recovery with forest age is rare in the decades following establishment. Glob Chang Biol 26(3):1458–1473
51. Deutsches Gewässerkundliches Jahrbuch Elbegebiet, Teil III 2014. (PDF)
52. Xunta de Galicia, Vicepresidencia e Consellería de Presidencia, Administracións Públicas e Xustiza, Dirección Xeral de Emerxencias e Interior (2016) Plan Especial de Protección Civil ante el Riesgo de Inundaciones de Galicia. https://ficheiros-web.xunta.gal/cpapx/plans-de-emerxencia/inungal_cas.pdf. Accessed 21 Aug 2020
53. ARSO (2020) Slovenian Environment Agency. https://www.arso.gov.si/
54. Dittrich R, Ball T, Wreford A, Moran D, Spray CJ (2018) A cost-benefit analysis of afforestation as a climate change adaptation measure to reduce flood risk. Journal of Flood Risk Management, e12482.
55. Huizinga J, De Moel H, Szewczyk W (2017) Global flood depth-damage functions: methodology and the database with guidelines. Joint Research Centre, Sevilla
56. Salazar S, Frances F, Komma J, Blume T, Francke T, Bronstert A, Bloschl G (2012) A comparative analysis of the effectiveness of flood management measures based on the concept of "retaining water in the landscape" in different European hydro-climatic regions. Nat Hazards Earth Syst Sci 12:3287–3306
57. Wei C, Lou CC (2020) A differential game design of watershed pollution management under ecological compensation criterion. J Clean Produc 274
58. Couto Pereira SN (2010) Payment for environmental services in the Amazon forest: how can conservation and development be reconciled? J Environ Dev 19(2):171–190
59. Riis T, Kelly-Quinn M, Aguiar FC, Manolaki P, Bruno D, Bejarano MD, Portela AP (2020) Global overview of ecosystem services provided by riparian vegetation. Bioscience 70 (6):501–514
60. Vidon P, Allan C, Burns D, Duval TP, Gurwick N, Inamdar S, Sebestyen S (2010) Hot spots and hot moments in riparian zones: potential for improved water quality management. J Am Water Resour Assoc 46(2):278–298

61. Dosskey MG (2001) Toward quantifying water pollution abatement in response to installing buffers on crop land. Environ Manag 28(5):577–598
62. NRC (National Research Council) (2002) Riparian areas: functions and strategies for management. National Academies Press, Washington
63. Zhang H, Hiscock KM (2011) Modelling the effect of forest cover in mitigating nitrate contamination of groundwater: a case study of the Sherwood sandstone aquifer in the East Midlands, UK. J Hydrol 399(3–4):212–225
64. López-Moreno JI, Beguería S, García-Ruiz JM (2006) Trends in high flows in the central Spanish Pyrenees: response to climatic factors or to land-use change? Hydrol Sci J 51 (6):1039–1050

Win–Win for Everyone? Reflecting on Nature-Based Solutions for Flood Risk Management from an Environmental Justice Perspective

Maria Kaufmann, Sally Priest, Paul Hudson, Lukas Löschner, Pavel Raška, Arthur Schindelegger, Lenka Slavíková, Ružica Stričević, and Tim Vleesenbeek

M. Kaufmann (✉) and T. Vleesenbeek
Institute for Management Research, Radboud University, Nijmegen, The Netherlands
e-mail: m.kaufmann@fm.ru.nl; tim.vleesenbeek@student.ru.nl

S. Priest
Flood Hazard Research Centre, Middlesex University, London, UK
e-mail: S.Priest@mdx.ac.uk

P. Hudson
Institute of Environmental Sciences and Geography, University of Potsdam, Potsdam-Golm, Germany
e-mail: phudson@uni-potsdam.de

L. Löschner
University of Natural Resources and Life Sciences – BOKU, Institute of Spatial Planning, Environmental Planning and Land Rearrangement – IRUB, Vienna, Austria
e-mail: lukas.loeschner@boku.ac.at

P. Raška
Department of Geography, Faculty of Science, J. E. Purkyně University, Ústí nad Labem, Czechia
e-mail: pavel.raska@ujep.cz

A. Schindelegger
TU Wien, Institute of Spatial Planning, Land Policy and Land Management, Vienna, Austria
e-mail: arthur.schindelegger@tuwien.ac.at

L. Slavíková
Lenka Slavíková, Institute for Economic and Environmental Policy, Faculty of Social and Economic Studies, J. E. Purkyně University, Ústí nad Labem, Czechia
e-mail: lenka.slavikova@ujep.cz

Carla S. S. Ferreira, Zahra Kalantari, Thomas Hartmann, and Paulo Pereira (eds.),
Nature-Based Solutions for Flood Mitigation: Environmental and Socio-Economic Aspects,
Hdb Env Chem (2022) 107: 399–424, DOI 10.1007/698_2021_759,

Published online: 11 April 2021

Contents

Abstract Nature-based solutions (NbS) are often framed positively in terms of win–win options or no-regret measures. However, are NbS equally beneficial for everyone? Are burdens and benefits of NbS really equally distributed and projects embraced by everyone? Is the process leading to the implementation of NbS always fair and inclusive? This chapter provides a broad overview of different environmental justice issues, critically reflecting on NbS through recognition justice, procedural justice, and distributive justice. Whereas the current critical literature focuses particularly on urban NbS, this chapter focuses on the wider translocal consequences of NbS projects. The theoretical reflections are illustrated with case studies of NbS from various countries: the recognition of marginalised women in Vietnam in mangrove restoration projects, the challenges when introducing procedural justice in implementing NbS in Serbia, the legal injustices locals are faced in the Czech Republic when they want to implement NbS, the trade-off between public collective and individual economic interest when implementing a sand nourishment project in the Netherlands, and the development of a beneficiary-pays based upstream–downstream compensation scheme in Austria.

Keywords Distributive justice, Environmental justice, Nature-based solutions, Procedural justice

1 Nature-Based Solutions and Environmental Justice: An Emerging Topic

Nature-based solutions (NbS), i.e. measures that 'are inspired by, supported by, or copied from nature' [1], are highly endorsed by policymakers on various political levels [2, 3]. The European Commission promotes NbS as a way to improve all three dimensions of sustainability, as NbS can: '[...] simultaneously meet environmental, social and economic objectives'. [1]. In general, NbS are often framed positively in

R. Stričević
Ružica Stričević, Faculty of Agriculture, University of Belgrade, Belgrade, Serbia
e-mail: sruzica@agrif.bg.ac.rs

terms of 'solutions' or 'win–win' options. By framing these measures as solutions, we tend to ignore their potential adverse impacts on certain communities or consider them to be apolitical and neglect the power dynamics at play during the decision-making processes. Randrup et al. [4] confirm that many NbS projects start from an expert-driven problem definition adopting an apolitical and technocratic top-down approach, whereas Kotsila et al. [5] stress that '"umbrella concepts" like NbS are inherently political, since they represent ideological commitments, inform institutional practices, and produce certain imaginaries of nature and its functions'. (ibid. p. 15). However, as observed with other flood management decisions (e.g. [6–8]), NbS have the potential to cause procedural injustices and/or negative outcomes which may lead to a redistribution of social and environmental inequalities.

A critical literature on NbS is slowly emerging [5, 9, 10]. These scholars criticise that NbS promote a utilitarian approach with neoliberal values (such as a focus on quantifiable benefits, profit, quick economic returns, and growth), which, however, ignores the inherent socio-environmental inequalities and injustices and related adverse societal consequences [5]. For instance, studies have shown that NbS can enhance the attractiveness of areas, which can result in increases in land and property prices and subsequently in rents or values. In turn, this can lead to the displacement of residents who cannot afford these costs anymore into areas of lower residential quality, which eventually reinforces community segregation [9, 11]. These dynamics are described under the terms: ecogentrification [12], ecological [13], or environmental gentrification [14].

This critical literature tends to focus on NbS in urban areas. However, due to their nature, the implementation of NbS has the potential to create both positive and negative effects at a wider spatial scale. This contribution adds to the existing literature on NbS by exploring those solutions implemented in rural areas and the translocal justice issues arising in these dimensions, for example, between urban and rural, upstream and downstream (Case 5) or coastal areas and their hinterland (Case 4), or within rural areas (e.g. different land-uses) (Cases 2 and 5). We focus on NbS interventions in the context of flood risk management. The aim of this chapter is explicitly to provide a broad overview of different justice issues in relation to NbS: recognition, procedural, and distributive justice. Herein, we reflect on the existing theory on environmental justice in the context of NbS, which are illustrated with practical case studies on NbS. Case 1 illustrates the importance of recognising the role of marginalised groups, namely women, when implementing NbS. Cases 2 and 3 zoom in on procedural justice issues, namely the difficulty of implementing public participation in Serbia (Case 2) and the institutionalised injustices that aggravate landowner's possibilities to implement NbS on their private land (Case 3). Finally, Cases 4 and 5 focus on distributive justice outcomes: Case 4 illustrates the negative consequences of coastal NbS for local mussel industry, and Case 5 describes how upstream retention basins are co-financed by the downstream communities that benefit from this measure.

2 Three Types of Environmental Justice

To reflect on NbS from a critical perspective, we are inspired by the literature on environmental justice, which distinguishes three foci concepts: Justice of recognition, procedural and distributive justice [15].

2.1 *Justice of Recognition*

Justice of recognition focuses on (mis)recognition of or (dis)respect for particular groups (see overview below) [15]. Recognition justice stresses that social differences exist and are attached to both privilege and oppression. A lack of recognition of group differences, with a focus on homogeneity, is considered to be part of the reason for unjust procedures and unjust distribution of burdens and benefits [16]. Stakeholders that are not even participating cannot voice their concerns, hence they are unlikely to be considered in the final implementation. A general principle is that 'all those affected by a decision should be involved to some degree in making that decision' [17]. Of course, even though stakeholders have been approached, they may choose not to participate. One may think of several reasons for this non-participation but research in the context of NbS is missing:: genuine disinterest; personal problems (e.g. health); lack of understanding about the potential implications of a project for them; inappropriate communication channels (e.g. digital media for elderly); or structural injustices that prevent people from participating (e.g. because they have to work five jobs due to the socio-economic system).

Research on recognition often focuses on the exclusion or inclusion of particularly minorities or vulnerable groups. In the context of NbS this might include:

- Gender, particularly women, and their specific position are often excluded (Case 1 describes a case study on NbS where the position of women is recognised);
- Indigenous groups, whose traditional water rights are being constrained by NbS or whose spiritual conception of nature is being ignored and in the worst case violated [18];
- The inclusion or exclusion of deprived households who have to relocate due to increasing house prices [11];
- The elderly as particular vulnerable groups, but also as groups with a potentially strong relationship to the area [11];
- The youth as the future generation that have to carry the burden/benefits of NbS;
- Certain economic sectors (e.g. farmers, fishers, etc.) whose livelihoods are influenced by NbS measures (see Cases 1 and 4);
- Ethnical minorities as research has shown that different ethnical groups use natural areas differently [19] and would consequently have different demands on the design of NbS.

Case 1 emphasises the importance of recognising the vulnerable position of women when restoring mangrove forests in Vietnam. It illustrates how women are normally misrecognised, and how the ResilNam project consciously recognised the marginalised position of women, so that they were more actively included in the implementation process, which led to an alleviation of injustices.

Case 1 Recognition Justice: The Recognition of Marginalised Women in Vietnam

This case study illustrates how NbS can be an avenue for community-based adaptation [23, 24], which can potentially address injustices if the concerns of marginalised groups (e.g. women) are focused upon. The ResilNam project [25] centred on improving flooding resilience via mangrove restoration in two coastal communes along the Tam Gang lagoon (Thua Thien Hue Province, Vietnam). The ecosystem services provided by the lagoon support about 300,000 individuals. Mangroves increase resilience by offering flood protection, improving lagoon ecosystems, and potential eco-tourism. In this project, mangrove restoration was combined with the active engagement of residents, Women's Union, local leaders, and researchers to understand, enhance, and use local knowledge while mitigating gender-influenced injustices and increase resilience. Involvement of local residents defined the core ecosystem services and impacts to consider while building the capacity of the Women's Union to engage in disaster risk management (DRM) and training local women in developing mangrove based eco-tourism. By recognising the marginalised position of women, DRM decision-making embraced a more holistic and comprehensive contextual understanding of the problems of women and possible solutions. The ResilNam project can illustrate this with two examples.

Example 1: There is preparedness training for rapid flood recovery leading to (compounding) injustices. This is because the training participants have tended to be male due to participant selection (e.g. physical labour expectations) or as (domestic) responsibilities prohibit participation. This has led to differences in preparedness confidence and capacities between men and women. Several community-wide meetings saw men being more proactive and confident due to previous engagement, thereby linking recognition injustices with flood preparedness. The ResilNam project addressed this by boosting the recognition and prominence of the Women's Union in DRM (and women overall). This was through extensive stakeholder engagement and building bridges boosting the Women's Union's DRM role directly (e.g. an agreement to work with DRM) and indirectly by generating a women-focused avenue for, previously underexploited, DRM networking. This addressed the problem of who participates in, and is recognised, as part of DRM. A core element of the successful engagement was to organise meetings in a way that maximises the likelihood of women taking part. To promote women's active participation in capacity-building programmes, activities were scheduled according to their schedules (e.g. overall free days, half-day workshops), involved active social components (e.g. karaoke) to reduce the perceived burden of participation, and were limited only to women for a more inviting atmosphere. Moreover, the Women's

Union stated that the key condition of improving women's role in DRM is gender equality. Therefore, communication efforts also aimed to increase awareness of including both men and women in participation. Overcoming these injustices as part of the ResilNam initiative created an environment for producing independently organised DRM actions outside of the project [25].

Example 2: DRM tends towards a utilitarian framework. Decisions are made when an action's positive impacts outweigh its negative impacts to improve social welfare. Impacts are often considered to be equal across stakeholders, an often-incorrect assumption as disaster impacts are often gendered [26–29]. Therefore, traditional approaches to DRM tend towards producing unjust outcomes due to these differential impacts or conflicting DRM objectives. For example, the traditional Vietnamese focus on flood infrastructure magnifies negative social and environmental impacts (such as the loss of ecosystems) for marginalised groups [30, 31]. Maintaining the status quo as post-flood reconstruction focuses on infrastructural but not environmental damage. Rather any environmental damage is left to the lagoon's natural processes. This recovery strategy increases (long-term) impacts on those who are more reliant on the lagoon's ecosystem and ecosystem service. Local women, as discovered through community-engagement, tend to be more heavily impacted by flooding for this reason. Therefore, improving lagoon ecosystems begins to address this imbalance. Ecosystem services may not have a high DRM priority due to prioritising on tangible outcomes.

In addition to improving the ecosystem services upon which women are more reliant, the distributive imbalances need to be recognised and recorded to demonstrate these impacts and progress in achieving just outcomes. One approach was to link welfare and rates of flood recovery across genders with flood experiences and the role of ecosystem services [32]. These results show that women suffered larger welfare impacts from flooding combined with slower recovery rates and demonstrated a positive link between welfare and ES quality. The second was a discrete choice experiment valuing the respondents' willingness-to-pay towards mangrove maintenance. This increased policymaker tangibility of the value of these ecosystem services. Women's willingness-to-pay for the mangroves was (overall) significantly higher [25]. These approaches indicated that women benefit more from this particular NbS creating more inclusive DRM strategies. Therefore, considering impacts as gender-equal ignores the potential to empower women through ES that benefit them relatively more, especially if tied to tangible alternative livelihood sources (e.g. eco-tourism).

The type of NbS also led to different social conflicts due to gender differences in fishing techniques. The communities revealed that in their perception women tended to use traditional methods, while men tended to rely on more extractive/destructive approaches. This generates conflict as the traditional methods benefit more readily from the project's long-term ES improvements, while the extractive techniques negatively interact with the mangroves over the short term. This is because the area given over to the mangroves will no longer be suitable for their fishing techniques, limiting short-term fishing potential. Developing alternative ES-linked incomes sources (e.g. eco-tourism) create incentives that could mitigate these conflicts through tangible immediate economic value. Furthermore, the benefits from eco-tourism are more valued by women, further addressing distributive injustices, especially when combined with the training events to boost their capacity to exploit future opportunities. This case highlights the influence of recognition justice on procedural and distributive justice outcomes of NbS.

2.2 *Procedural Justice*

Procedural justice focuses on the fairness of decision-making [15]. In the following, we will discuss three main issues of procedural justice: information, participation, and access to legal processes (for an overview, see [17]).

The availability of appropriate, sufficient, and accurate *information* for all participants is often considered crucial for procedural justice as it helps to ensure transparency, effective participation, and informed consent [17]. Of course, as Simcock [17] argues, terms like 'appropriate', 'sufficient', and 'accurate' information are contested. Issues that need to be considered hereby are, for example, whether information are provided in an understandable language that is also appropriate for lay people; or whether information events are organised at times that consider people's working hours or other periods (e.g. holiday period) and at locations that are easily accessible for the stakeholders with different economic capabilities, e.g. accessible by public transport.

Participation relates to the extent that different participants' opinions, suggestions, and concerns are considered in the decision-making process [17]. A person or collective can exercise different degrees of influence in a decision-making process (see also [33]). Simcock [17] broadly distinguishes 'listen as a spectator', i.e. a stakeholder receives information about a decision, but has no influence over that decision; 'consultative influence', i.e. a stakeholder is able to give their opinion on an issue but the final decision is ultimately made by others; and 'direct authority', i.e. a stakeholder is able to formally shape the outcome of the decision-making process – either by taking the decision individually or by sharing power with others in a democratic process (such as voting). These different degrees of influence are similar to Rowe and Frewer's [34] forms of public participation. They distinguish a communication mode, which is a one-way flow of information from public authorities to stakeholders; a consultation mode, where stakeholder provides feedback to the plans of public authorities, and finally a co-productive mode, where goals and outcomes are jointly formulated. Public participation, however, is also influenced by historical and cultural developments, some countries have a long tradition with public participation with experienced public authorities and emancipated citizens, whereas other countries have less experience. Case 2 describes the challenges related to introducing public participation in Serbian water management. The case illustrates that public participation is influenced by cultural norms and tradition and that countries where such a tradition is missing, procedural justice is difficult to achieve.

Research has demonstrated the advantages of stakeholder participation. Short et al. [35] describe the implementation of NbS in the Stroud Valleys, UK. They concluded that involvement of local stakeholders through regular and directed interactions (also in form of field visits) resulted in collective learning and enabled the project to better deal with complex issues 'through a shared repertoire of resources and practices' (ibid. p. 244). Local citizens formed four flood action groups that had been actively involved in the implementation of NbS. For example, they were involved in the recruitment of the Project Officer and citizens had a representative in the strategic group overseeing the planning and implementation. The Project Officer approached each landowner or land manager one by one to plan the details of the NbS. The landowners could also choose (if feasible) the contractor [35].

An important issue underlying procedural justice are power imbalances between stakeholders and the government authorities, but also between stakeholders. A prominent problem is 'elite capture' [36] that means only interested (often highly educated) stakeholders participate (see Case 2). Additionally, it is important to consider whether the same weight is given to participants' opinions or whether certain stakeholders are privileged [37, 38]. In the context of NbS the balance of power over decision-making may be even more complex than other flood risk management situations. The implementation of NbS will often involve the use of private land (e.g. for flood storage) and, as such, require the engagement and negotiation with (often) multiple landowners. This challenging process of negotiating the acquisition or temporary use of private land will likely exacerbate power imbalances, with some able to exert greater power over decisions than others. The legal framework plays hereby an important role as it creates certain rights and obligations, which can alleviate or create injustices. Case 3 illustrates how private landowners in the Czech Republic want to implement NbS on their private land and the justice issues arising from these approaches. On the one hand, it describes how landowners are faced with increasing costs and high administrative burdens when due to bureaucratic regulations and rules. On the other hand, it also illustrates how these bottom-up, self-governance approaches can exclude stakeholders that are affected by these private measures from the decision-making processes.

Access to legal process describes the availability of independent grievance mechanisms that are accessible, fair, transparent, and effective. In other words, the ability of stakeholders to appeal certain plans [37, 38]. That implies also adequate and fair negotiations for compensation arrangements to restore livelihoods or compensate for particular burdens or services. The cases presented in Case 4 and 5 touch upon grievance mechanisms and compensation regulations in the Netherlands and Austria, respectively. In both cases, NbS conflicted with economic interests of individuals but eventually the courts decided in favour of the collective public interest.

Whether recognition and procedural justice are distinct is contested. Whereas Young [20] and Schlosberg [21] see recognition justice as a separate feature of justice that cannot be solely assumed and the lack of recognition as a reason for injustice, Miller [22] considers recognition an integral part of procedural justice. Either way, recognition is closely linked to procedural justice (see also [17]).

Case 2 The Challenges in Introducing Procedural Justice in the Context of NbS in Serbia

Serbia is threatened by floods from large international rivers such as the Danube. From 1960 to 2012 there were 73 floods in Serbia, of different intensity and spatial impact [39], followed by 3 more catastrophic floods in 2014, 2017, and 2020. Therefore, the hydrological status of rivers is continuously monitored, and special attention is paid to flood defences. Experiences gained during the floods of the Danube in 1965 influenced the strengthening and upgrading of grey infrastructure (e.g. dams, dikes) on all great rivers. The defence infrastructure mitigated numerous floods; however, in 2014, the hydrological situation was such that neither the retention basins nor the embankments could retain the complete flood wave to mitigate adverse effects.

After the flood in June 2014, the Government of the Republic of Serbia launched the *Program for the Reconstruction of Damaged Water Facilities and Elimination of the Consequences of Floods*. Within this programme, the consequences of the flood were surveyed. Additionally, the Republic of Serbia became an official candidate for the membership of the European Union; thus, it had to transpose EU-legislation into domestic law, including the principles of the EU-Water Framework Directive (WFD) and the EU Floods Directive. The EU-legislation forced Serbia to prepare flood risk maps [39] and broaden its flood risk management measures, considering now, among others, also nature-based solutions (NbS). The legislation also required Serbia to adjust its flood risk governance by introducing public participation processes; a requirement which had largely been missing [40]. This case explores Serbia's efforts to deliver effective procedural justice for NbS.

As the Kolubara catchment area was most affected from floods in 2014. It was decided to prepare a study to analyse flood risk and management options in the catchment [41]. This study aimed to develop the concept of integrated flood protection, including structural measures (e.g. technical measures, such as upgrading and reconstruction of flood defence facilities) and non-structural measures (e.g. anti-erosion watershed management, NbS for water retention). It also was one of the first efforts in Serbia to introduce public participation in the construction process and flood risk management process.

Public participation was undertaken in cooperation with representatives of local governments and the appointed commissioners for public relations. Meetings were organised with stakeholders that were potentially affected by floods: holders of social functions, businessmen, representatives of political parties, nongovernmental organisations, citizens, and socially vulnerable population (unemployed and elderly). These stakeholders have been contacted by the appointed commissioners for the public relations who were trained to distribute the questionnaire to the targeted public. The method of data collection was adapted to the real circumstances in the field, even though workshops with local governments and stakeholders were organised, the questionnaire was considered most appropriate to collect and analyse data. The questionnaire was analysed by experts using a multi-criteria evaluation.

A number of observations can be made with regard to procedural justice. The participating respondents were mostly highly educated respondents and representatives of local self-governments, which hints at elite capture. These respondents were not familiar with the flood risk maps and the maps of retention areas generated by '*the Study for the Improvement of Water Protection in the Kolubara Basin*' [41]. This illustrates the challenges of communicating often complex concepts with lay people, and, as such, communication needs careful preparation by public authorities. Slightly less than half of the respondents agreed that uninhabited parts of the territory as future retention areas could be flooded (probably they were not farmers). This illustrates that public participation processes can potentially create new injustices for individuals not involved but still potentially affected by measures, like in this case the allocation of burdens between rural and urban areas. The respondents believe that their personal influence and power in solving problems in the city is very weak. They see politicians and powerful businessmen as the most influential in decision-making. This might also explain why public participation conducted by the local communities had a weak response. Neither local government nor members of the public were used to these kinds of public participation processes and therefore did not understand its benefits or relevance.

The introduction of public participation and activities has been envisaged by the Serbian Waters Law [40] since 2010. Previously, the public had only been engaged in flood risk management in relation to post-event compensation with limited planning or pre-implementation consultation or participation. The novelty of this engagement is, in part, a barrier to its success and for the Serbian authorities to discharge responsibilities for ensuring procedural justice. Local communities, entities, or stakeholders were not sufficiently aware of the possibilities of influencing the planning and implementation of flood defences [41]. Additionally, NbS are new measures to be applied in the Kolubara river basins and as such, experience of these measures by both the public and professionals is limited. Therefore, the public are even less likely to appreciate any benefits or challenges caused by their use. It may be concluded that there is no culture of public participation in water management, which is reflected in a low awareness of citizens and limited expectation to be consulted, as well as limited experience of local governments (and a lack of qualified staff) with engaging public participation and experience with processing the findings.

Although efforts to ensure procedural justice within this case studied have been met with limited success, as public participation becomes more common and a mainstream for flood risk management, the public may become more and more interested and emancipated in water projects. Internet technology is increasingly influencing the population to be informed in time and react to plans and events that are made in their environment. A good example are the volunteers that protested against the construction of mini-hydropower plants on small rivers [42, 43], and their voice has been adopted. The government recognises the importance of public participation and is taking steps to improve this process, including organisation of events in the water sector. One such meeting where stakeholders had the opportunity to express their opinions, views, and concerns was held in the National Assembly on the topic of 'Public listening – State of water in Serbia' [44]. However, there is still a lack of experience in the joint work of state bodies and the public in general public interest and engagement, requiring skills and capacities to be improved before public participation in Serbia is able to deliver effective procedural justice.

Case 3 Legal Injustices Faced by Private Landowners When Implementing NbS in the Czech Republic
The last three decades have been marked by severe riverine and flash floods in Czechia [45, 46]. Although the current political discourse has turned to alarming droughts, the very recent flash floods in eastern and northern Czechia in June 2020 [47] reminded us that floods and droughts must be considered a hydrological continuum, and that these extreme events should be managed by improving the overall water retention capacity of the Czech landscape. Among the measures favoured by the research community and promoted by governmental authorities and NGOs are NbS in the form of natural water retention measures, which consist of small pools and other water bodies dispersed in upper catchments. Despite uncertainties about the upscale effects of these measures on flood wave attenuation [48], they are generally believed to positively affect water recharge and mitigate extremes by increased water retention, alongside other functions, such as biodiversity conservation. For this reason, the establishment and restoration of pools and other small water bodies in Czechia is supported by environmental strategies and financed by EU funding and state budgets under the umbrella programme '*Establishment and Restoration of Pools, Wetlands and Peatbogs'* [49].

Yet, the recent experience gathered by multiple case studies in the Czech countryside reveal key institutionally embedded barriers in planning and decision-making which can add to the ongoing debate in flood risk governance. The studies were conducted by researchers from J. E. Purkyně university in the last 5 years and were based on semi-structured interviews with various actors initiating establishment of pools, on in situ observations, and on analysis of recent policy documents and legislation. The revealed barriers record unequal distribution of implementation burdens among stakeholders and identify the institutional conflicts resulting from mismatched procedural arrangements and value incommensurability. In this respect, these barriers serve as narratives of injustice related to legally anchored imbalance of capability to co-produce and co-decide about the NbS. The issues of unequal distribution of implementation burdens will be shown referring to the availability of land (privately or state-owned), funding (from EU, municipality or private), and the intentions/benefits (common environmental goods or individual benefits). The injustices emerge at or between these three issues.

The first study reports the effort of a private farmer (see also [50]), who decided to establish a number of pools and wetlands on his private pastures and meadows (*land*) in a legally protected landscape area and using his own financial resources (*funding*) in order to improve water recharge and landscape water retention capacity (*intention*). In respect of procedural justice, the scheme has resulted in a rather limited community participation as it was initiated and supervised by the farmer. Although the measures could have potential broader environmental effects outside of his land, it was the farmer who decided who to invite to participate and how to address the results of eventual negotiations. This indicates that bottom-up initiatives are not necessarily more inclusive as they may either leave out locals or limit their participation. While this approach allowed the farmer to effectively avoid complicated funding procedures while still contributing to overall environmental quality of the land, he was also burdened with additional procedural costs to designate administratively the pools and wetlands as ecologically significant elements. Without undertaking this, these areas would have been excluded (as payment is allocated per area of cultivated land and only designated ecologically significant features) and he would have received proportionally lower funding based on the Single Area Payment Scheme of the Common Agricultural Policy. Taking into account Wilkinson's [51] note that a larger volume of diffused water bodies must be established should they have profound effect on water retention and attenuation of peak discharges, the case study indicates that such costs and burdens may disincentivise minor landowners with lower financial resources and social capital acting as initiators of NbS.

The second study documents a collaborative effort of a governmental environmental agency – National park Czech Switzerland (*intention*) with private landowners (*land*) to establish pools and wetlands upstream. Here, the efforts recognised a variety of stakeholders. The intention was primarily motivated by conservation efforts and was realised when an agreement to apply for funding was reached with a landowner (in other cases the intention failed due to landowner's request for costly buy-outs of the land). The major reported barrier concerns procedural arrangements infringing upon the landowner, who is obliged to provide all documentation for funding. Depending on a type and size of the water body, this documentation may include confirmations issued by Building Authorities, Water Catchment Authorities, Forest Agencies, and other subjects. If applying for funding, the landowner is obliged to follow the procedure regarding the design of the pools (e.g. [52]), which itself refers to more than 10 legal acts or legally binding standards. Also, after completion, the landowner is responsible to comply with restricted uses of the pools that may affect the distribution of economic, social as well as aesthetic values of their own land (the pools cannot be used as biotope swimming pools, or as water source for cattle, for example). Such restrictions are considered by some landowners as disproportionate burdens compared to their willingness to implement NbS for sake of common environmental good. For these administrative burdens, some of the intended projects that originally found a common support from environmental agencies and private landowners failed to be completed.

The last case study traces an emerging plan of a private landowner, acting also in charge as a municipal representative, to restore and establish pools on his own land (*land*). His motivation is rather complex – to increase landscape diversity, water retention capacity, and also provide a sustainable water resource for gamekeeping (*intention/benefit*). According to in situ observations, the proposed measures will improve the environmental status of the land if realised. While *funding* is not a key issue (may either be private or public), the case indicates a disproportionate capability of locals to conduct such plans depending on their social capacity and power. The results of this particular project cannot be anticipated, but generally the formal status seems to favour this initiator over the other locals in terms of co-decisions about location of pools and therefore their environmental effects on a particular piece of land (thus also the distribution of benefits from such NbS). In addition, the initiator who is in a formal position could be in advantage to distribute the time, and eventually financial costs, for related administrative procedures between his own private resources and the municipal budget. This implies that NbS are more feasible for elites endowed with formal power. Despite overall environmental benefits of the proposed measures, such situations may imply conflict of interests.

Summing up the presented cases, they indicate that reaching the public goods provided by NbS requires accordance among sound intentions, available land, and funding. The eventual discrepancies among these issues may induce a lack of governmental support for grassroots efforts or raise procedural barriers among the public sector organisations and authorities. These barriers are legally anchored and seem to result from complex, yet fragmented environmental and planning policy and legislation (cf. [53]) causing disproportionate administrative costs to initiators of the NbS and to landowners.

Conversely, increasing agreement among the inputs, which may facilitate establishment of pools, may have other potential adverse impacts. First, they may cause recognition and procedural injustices by leaving out some of those, who might have benefited from participation on establishment or restoration of pools. Second, the facilitating effect of accordance among the inputs advantages those who have sufficient financial resources, have social capital (ties with experts and officers, formal positions, etc.), and intend to perform the NbS on their own land (primarily for their environmental benefit).

2.3 *Distributive Justice*

Distributive justice describes the allocation of burdens and benefits [15] of particular NbS projects. The main benefits of NbS are, ideally, reduced flood risk, but also an increase in recreational space with the associated amenity values [35], as well as increased habitat and biodiversity (ibid). Carnelli [54] argues that NbS projects can have substantial social and cultural benefits as well and, when the community is actively involved, they can become owners of the NbS project and foster synergies between various interests (e.g. flood management and biodiversity conservation) [35]. Additionally, NbS projects can have economic benefits. Short et al. [35] stress that by using local materials and local companies to implement NbS, the economic benefits feed back into the local community and the environmental impacts are reduced. This approach, according to Short et al., also strengthens the local knowledge capacity to work on NbS in the future.

However, NbS projects can also cause substantial burdens, such as: relocation of houses and displacement of inhabitants, which can have negative consequences on people's identity and social networks, loss of land (reserved for retention) or reduced property and land value, loss of access to livelihoods (e.g. reduced crop yields, other natural resources), or increasing flood risk for certain groups (e.g. farmers whose land is used for retention). Next to the burdens/benefits related to the outcome of an NbS project, there could also be burdens related to the process of its implementation (e.g. temporal disruption and air pollution for citizens caused by lorries that transport material), these burdens are often overlooked. The distributional effects can take place on different levels: between rural and urban areas, between upstream and downstream, within rural areas, etc. Case 4 focuses particularly on the negative impacts of NbS on individual coastal livelihoods for the protection of the hinterland. It illustrates how mussel farmers might be financially burdened by a sand nourishment project, which might negatively impact their mussel grounds. It also emphasises the difficulties to arrange compensation for the negative consequences of NbS as the causality of effect and damage might be difficult to prove.

The distribution of burdens and benefits is influenced by the underlying dominating rationale. In the political philosophy literature, a number of ideal typical rationales can be distinguished:

- Elitist/libertarian justice focuses on the principle of 'maximum liberty'. It is based on the idea that people are entitled to what they have achieved individually due to their merit or rank and that the government should not intervene [55, 56]. It often is associated with the beneficiary pay's principle (ibid.). The principle can be applied on an individual level, but also on the collective level within the implementation of NbS (see Case 5 as an example). It can be exemplified by a downstream community (who would benefit from the risk reduction) paying, or at least contributing towards, the costs of implementation of NbS in the upstream environment.
- Utilitarian justice is based on the principle of 'maximising utility', that is, redistributing collective resources to achieve the maximum societal benefits [56, 57]. The focus here is on preventing the most damages; therefore, NbS should be implemented on the basis of pure benefit–cost analysis, with lower value land being used to protect higher value land/assets. An example might be poor quality agricultural land being used for flood storage to reduce damages to an adjacent heavily populated urban area.
- Rawlsian 'maximin rule' states that: resources should be distributed so that they favour the most vulnerable, i.e. this principle focuses on absolute vulnerabilities and neglects that people can be vulnerable to different degrees [56, 58]. NbS implementation should be used to reduce the risk to the most vulnerable (although there would be the need to clarify who the most vulnerable are). This could focus on those areas with the lowest income or social capital (i.e. those less able to help themselves).
- The egalitarian principle builds on the notion of equal opportunity for every citizen in terms of distributional outcomes. It implies a public responsibility to

provide a certain level of safety or well-being [56, 59]. NbS should be used to even out flood risk and to ensure that where possible all have a flood risk below a certain level.

Case 5 is an example of a beneficiary-pays approach. It focuses on the upstream burdens caused by NbS in form of land reserved for flooding and how this is compensated by the downstream community. It describes how an innovate corporative was set up to organise that the downstream beneficiaries are compensating the farmers carrying the burdens upstream.

Case 4 Distributional Trade-Offs Between Public-Collective and Private-Economic Interests When Implementing NbS in the Netherlands
The Roggenplaat sand nourishment project is NbS implemented in the western part of the Netherlands. It reflects the distributional trade-offs between public-collective and private-economic interests in the context of NbS.

The Roggenplaat is a sandbank of high ecological value (e.g. stop over for migratory birds, and resting spot for seals) in the Oosterschelde. The island contributes to the safety of the coast as it breaks the waves and reduces the impact on nearby dikes [60, 61]. Sand nourishment (a total of 1.3 million m^3) was applied on seven different parts on the 1,460 ha big island [62]. The sand nourishment project has been mainly financed by European and national taxes [62], but it was co-funded by crowdfunding (300 donors raising 13,500 euro), which highlights the public support for the project [63]. Nevertheless, despite the wide support, the project is not endorsed by the 89 mussel growers in the Oosterschelde, represented by the *PO Mosselcultuur* (in Dutch: *Producentenorganisatie van de Nederlandse Mosselcultuur*), due to potential negative impacts on their economic activities. The companies are renting plots from the government and use those to grow mussels for consumption. Together with import and processing facilities, the average annual turnover is 200–250 million euros, making it an important economic sector for the province of Zeeland. Moreover, next to the financial importance of the sector, it has cultural significance as a 'characteristic' sector for Zeeland, contributing to a positive image [63].

The sand nourishment project could potentially negatively affect the quality of the mussel growing plots in two ways. First, the nourishment could lead to the relocation of the sand on the plots. If sand coverage on the plots is too high, the mussels would suffocate and die. This could be disastrous for plot owners due to the impact on annual turnover [63]. Second, the current coming from the Roggenplaat brings important nutrients for the mussels on the plots, enhancing their quality. The nourishment of the Roggenplaat could lead to a change in currents and therefore negatively affect the growth of the mussels. This would result in a reduced mussel volume, quality, and value [63, 64].

This situation is normally covered by a compensation regulation (in Dutch: *nadeelcompensatie*), i.e. citizens and companies who encounter temporary or permanent disadvantages from state-implemented measures have the right for reimbursement of damage (e.g. loss in turnover) [61]. This compensation regulation would also be applicable for the NbS project. However, the mussel growers argued that this compensation arrangement was not appropriate to compensate for losses. In a regular situation, the causality between the state-implemented measures and the negative effects needs to be visible in order to determine if the measure taken is really the cause of the negative effect. In the case of the Roggenplaat, this causal link can be difficult to prove due to the dynamic character of the natural system, that is, the link between the economic damage and the sand nourishment. They were concerned that the organisation implementing the project would contest any compensation and argue that the damage or the reduced mussel yield could also be caused by other issues. Hence, the mussel growers called for a new, customised arrangement that would ensure their compensation. They filed an objection for the issuing of these permits, not with the intention of preventing the project, but to ensure that they would be adequately compensated. They argued that the initiator of the project is also morally responsible for any negative consequences and should financially compensate them, considering that many mussel farmers are small family businesses and that economic loss could be disastrous for these families and their livelihood. Despite these concerns, however, the Ministry of Infrastructure and Water Management and the Province of Zeeland rejected their objections. The mussel sector did unsuccessfully appeal against this at the Council of State, the highest court in the Netherlands. The Council of State argued that the chances of damage were very small and that if damage should occur the current regulations would offer enough room to implement measures or ask for compensation [62, 63, 65].

Finally, a part solution was found and the Ministry agreed to provide exchange plots, so that in case the current plots are damaged by implementing the sand nourishment project, the mussels can be relocated. However, the mussel growers never saw this as a real possibility as the quality of these exchange plots is not as good as the nutrient-rich current plots. Hence, the usefulness of these plots remains uncertain: Can the potentially negative impact be identified in time to move the mussels?, How to prove the negative impact stems from the sand nourishment project?, Is the quality of the new plots really appropriate?, Are sufficient plots available for all mussel farmers? [63]. This case highlights that the increased uncertainty related to NbS projects, which stems from the dynamic and unpredictable effects of NbS measures, can cause new challenges for compensation arrangements. The case illustrates the distributional trade-offs: collective interests are put above individual economic livelihoods as well as illustrative of legal processes that procedural justice in the Netherlands allows. Particularly the temporal scale is interesting as the effects might only be visible in the future, also raising the issue of inter-generational equity.

Case 5 The 'Beneficiary-pays' Approach for Upstream–Downstream Flood Protection in Austria

Given its location along the Alpine ridge and the topographic confinements for settlement and infrastructure development, Austria is highly prone and exposed to a range of gravitational and hydro-meteorological hazards. Flooding constitutes by far the most frequent and damaging type of hazards, ranging from small scale fluvial and pluvial flood events connected mainly to extensive local precipitation to large-scale events along the country's major rivers, first and foremost at river Danube. The necessity to mitigate flood damage and provide flood-protected areas for land development led to the emergence of a complex flood risk management system. Relying traditionally on grey infrastructure [66], over the years and in response to a sequence of disastrous flood events in the late 1990s and early 2000s, protection schemes have become more versatile, integrating spatial planning and NbS in a more anticipatory flood risk management [67]. Today, flood risk management in Austria is characterised by the aim to prevent flood damage (rather than responding to disasters) and, where possible, a stronger emphasis on controlled flood storage on open land to complement linear flood defence measures.

Accordingly, the combination of structural and non-structural risk reduction measures based on established evaluation methods (a utilitarian-based cost–benefit analysis) has been a recent focus of state-led flood prevention. The costs of flood prevention/protection measures (initial investments and maintenance costs) are typically split among the different levels of government, with the possibility to also involve local beneficiaries. This distribution is regulated in a separate federal legal Act [68], which allows project-specific amendments and adaptations in financing actual schemes. Normally, costs are shared among the federal government, the provinces, and the municipalities, while in some cases private beneficiaries, including infrastructure operators (e.g. rail, road), can also be committed to contribute to the funding. Further, the Federal Water Act [69] importantly provides the possibility to organise private contributions (i.e. from the land and property owners) to finance flood protection via water cooperatives. Depending on the specific project design and the statutes of the cooperative, its members can be committed to cover a share of the investment and the maintenance costs of the flood protection scheme. Once such a cooperative is established, it is possible to declare mandatory membership for beneficiaries, but voluntary membership normally prevails.

Although the possibility to establish water cooperatives exists in the Water Act since the 1950s, it only quite recently has been applied for the purpose of flood risk management, specifically to balance upstream/downstream interests between the providers and the beneficiaries of flood protection measures [70–72]. One prominent case where a flood protection scheme has been realised via a cooperative, using elements of the 'beneficiary-pays' principle is in the municipality of Altemarkt im Pongau in the Austrian state of Salzburg. The alpine municipality – a popular (winter) tourism destination with approx. 4,500 inhabitants located in the upper reach of the river Enns – was repeatedly affected by smaller flood events but has been spared from some of the major floods which affected other parts of the country during the past decades. However, a comprehensive flood assessment revealed an extensive flood exposure even to more frequent events (1-in-30 years), which implied prohibiting development for large parts of the central village located in the valley basin.

With the need to instal flood protection measures, a discussion and planning process started leading to negotiations on project implementation and financing. Based on the results of flood hazard analysis, local action started with awareness-raising measures and a town meeting, recognising the need to include public participation as part of procedural justice efforts. The outcome was an agreement to develop a flood risk reduction project to protect the municipality's settlement area (including more than 1,200 residents and a total of 350 buildings) against a 100-year design flood. In addition to linear measures (levee, river widening) the project features a large retention basin in an agricultural grassland area to store flood water upstream and reduce damage in the downstream settlement area. Given its past experiences with funding flood protection measures along a smaller tributary of the Enns river, the municipality decided to cover only part of the costs for the flood protection scheme along the river Enns transferring the remaining share of the costs to the beneficiaries of the flood protection measures.

In terms of risk communication, the municipality physically pegged and marked the inundation lines of the 30-/100-year flood and continuously informed the public on the progress in planning a comprehensive protection scheme. The actual technical planning was conducted by the responsible state authority and local communication carried out together with representatives of the municipality (esp. mayor, municipal council, head of office) [73]. With regarding to realising the flood protection scheme, the first challenge was to acquire the land needed for a river widening and a controlled retention basin. The negotiation with the agricultural landowners in the planned retention basin turned out to be very difficult as they had to accept controlled and potentially more damaging flooding on their land (higher flood depths, longer duration of flood storage). Some of the affected farmers were, therefore, initially unwilling to provide any land for flood protection. Their resistance was overcome after long negotiations with a generous offer that guarantees farmers annual payments for the next 100 years (irrespective of any flooding) in addition to flood damage compensation in an actual event. Additionally, a farm had to be relocated and the buildings and land compensated [74]. Overall, these negotiations increased the project costs significantly. The second challenge was the implementation of the 'beneficiary-pays' principle by establishing a water cooperative with more than 1.200 members in 2018. The individual contributions of local beneficiaries were calculated according to the existing real estate value and exposure [73]. Furthermore, the initial investments and maintenance costs for the implemented measures are also covered by the annual contributions.

Establishing the cooperative with beneficiary contributions was a time-consuming act and was not based on necessity, but rather the local political conviction that beneficiaries should contribute to the risk reduction they experience from the planned measures. Due to the large number of beneficiaries, it took more than 2 years to formally establish the cooperative. While the majority of beneficiaries voluntarily joined the cooperative, several designated members of the cooperative resisted the payments and formed a citizen initiative with the aim to reduce the contributions of beneficiaries. Given the provisions of the Federal Water Act and the overwhelming support for the cooperative, the opposing beneficiaries were legally obligated to join and contribute their share to the flood protection scheme.

The case study presents an interesting flood risk governance approach that seeks to actively involve the beneficiaries of a protection scheme also relying on NbS to achieve distributive justice. Nevertheless, the cooperative only includes homeowners located within the calculated 100-years flood sparing future developers and investors in the former flood area from contributing (which may lead to future distributional injustices) and also excluding other benefits (recreational value, etc.). Additionally, the integration in the scheme was exclusively based on the calculated risk reduction. Overall, a distributive effect among upstream and downstream landowners was achieved based on a long and exhausting participation process that, in the end, only accounts for a small share of the initial project costs. Nevertheless, a feeling of shared ownership and responsibility was established among the affected inhabitants.

3 Concluding Remarks

The reflections and cases presented in this chapter illustrate that even though NbS are often framed as apolitical, do-no-harm 'solutions', justice issues play a role in all stages of NbS from planning over financing to implementation. NbS are not significantly different from other climate adaptation or flood risk management strategies in terms of justice [75–78], but similar justice issues arise and need to be considered. Hence, the cases presented reinforce the need for paying increased attention to issues of justice in the context of NbS to ensure fairness in processes and outcomes. NbS have the potential to benefit, as well as disadvantage, stakeholders at various levels and temporal and geographical scales.

Table 1 summarises the different issues of justice (distributive, procedural, and recognition justice) specifically relevant for NbS that were identified in the cases. Whereas Case 1 focused much more on recognition justice, Cases 2 and 3 zoom more in on procedural justice issues and Cases 4 and 5, in turn, concentrate on issues of distributive justice. Of course, to some degree all justice dimensions come back in all of the three cases.

The various dimensions of distributive justice are not mutually exclusive, but can overlap, meaning multiple types of distributive injustice need considering for each NbS measure implemented. Two issues of distributive justice seem particularly relevant in the context of NbS. First, the impact of NbS implemented in rural (and maybe also in certain urban) locations needs to be considered at a much broader spatial scale as they may lead to translocal distribution of burdens and benefits. All stakeholders that have an interest or who might be (directly or indirectly) affected should ideally be included in the decision-making process. Second, the dynamic and (partly) unpredictable nature of NbS (i.e. including unforeseen impacts or ineffectiveness) demands the longer-term consideration (and monitoring) of distributive outcomes as burdens and benefits might only emerge some years after implementation. Additionally, many NbS are novel and therefore untested over both the longer term and in different biophysical and social contexts; hence, the breadth of evidence of the impacts and efficacy is lacking. This reinforces the need for monitoring not only their effectiveness but also the distributive outcomes over the longer term.

Table 1 Overview of environmental justice issues in the context of NbS

Type	Example(s)
Recognition justice	
Recognition	• Excluding certain groups in public participation (e.g. less highly educated, Case 2) or the economic concerns of certain groups (mussel growers, Case 4) • Romanticising and preference of nature-based practices or lifestyles in contrast to more-invasive (but potentially widely spread) practices (Case 1)
Procedural (in)justice	
Information	• NbS dynamics and impacts often uncertain and difficult to provide reliable information (e.g. the causality between NbS and impact is more difficult to demonstrate) (Case 4) • The (positive and negative) impacts may be less tangible and therefore difficult to articulate (Case 1)
Participation	• Offers possibilities for co-creation and co-financing due to synergies of NbS projects (eco-tourism) (Case 1; Case 5) • Danger of elite capture due to complexity of NbS projects (Case 2) • Potential exclusion of people that cannot financially participate in beneficiary-pays approaches (Case 5) • Bottom-up initiatives of private landowners can exclude other potentially affected stakeholders (Case 3) • Newness of measure and complex dynamics of some NbS hampers recruitment of the public (Case 2)
Access to legal processes	• More difficult to prove causality (e.g. link between NbS and negative impact) in the context of some NbS, which may limit opportunities for accessing justice by legal processes (Case 4) • Courts may give preference to collective interests above individual interests (e.g. economic interests (Case 1, Case 4)) and land-use rights (Case 5) • Existing regulations might disproportionally burden landowners who want to implement NbS on their land (Case 3)
Distributive (in)justices	
Spatial geographical (in)justice, i.e. the benefits and burdens vary *spatially*	• Rural-urban differences (Case 2) • Coast and hinterland (Case 4) • Upstream–downstream (Case 5)
Social (in)justices, i.e. the benefits and burdens vary *socially and economically*	• Some groups favoured/marginalised over others [e.g. gender, Case 1; level of education,

(continued)

Table 1 (continued)

Type	Example(s)
	Case 2; economic livelihoods (e.g. fishers, Case 1; mussel farmers, Case 4; or farmers, Case 5); access to (administrative) power or influence, Case 3]
(Private) individual vs. collective (in)justices, i.e. distribution of burdens between those benefiting from measures and those burdened	Collective above individual • Compulsory purchase of private land for good of the community (farmers Case 5) • Potential negative consequences of NbS on certain individual's livelihoods (fishers, Case 1; mussel farmers, Case 4) • Potential negative consequences of NbS on private land for the collective (Case 3) • National taxpayers funding NBS with only a few benefitting (Case 4 – despite some crowdfunding)
Temporal (in)justices, i.e. distribution of burdens and benefits varies in time	• Future generations benefitting from ecosystem services once nature-based solution established (long-term) (Case 1) or having to carry the burden of negative effects (Case 4) • Future generations: lost opportunities because land is used for NbS and not available for farming • Degree of potential disruption to activities – temporarily during implementation or permanent or temporary impacts of NbS (e.g. temporary flooding of land or permanent land acquisition) – linked to this is whether 'compensation' is applied proactively or retrospectively (Case 4) • Uncertain long-term effects of NbS (Case 4)

In the context of procedural justice, four issues seem particularly relevant for NbS. First, it may be more difficult to prove the causality between a negative impact and the implementation or ineffectiveness of NbS in comparison to grey infrastructure (see Case 4). Therefore, opportunities for accessing justice by legal processes may be limited in some cases. Second, NbS seem to have the possibility to provide synergies with other interests (e.g. tourism, recreation, or biodiversity), maybe more so than traditional technical flood defence infrastructure. Hence, there are – theoretically – more opportunities for co-creation or co-financing. But these need to be actively sought out and the pitfalls of public participation, e.g. elite capture, avoided. Additionally, these indirect benefits need to be fully included in cost–benefit analyses. Third, the communication of reliable information to stakeholders can be hampered by uncertainty of the effectiveness and intangibility often associated to NbS. Fourth, landowners may choose to implement NbS on their private land using private resources, but these bottom-up initiatives might still have positive or negative consequences for other stakeholders. Representative engagement of all stakeholders may be lacking, which suggests that often-praised bottom-up initiatives are not necessarily inclusive. This is of course particularly relevant, if the measures on private land are financed wholly or partly by public money. Here, the individual could potentially place burdens on the collective, whereas other examples show that the individual has to carry the burden for the collective (e.g. mussel farmer case – Case 4, or farmers in Austria/Czech Republic – Cases 3 and 5). Hence, the relation between collective and individual is not straightforward in the context of NbS.

With regard to recognition justice, NbS also seem to face similar problems to other flood risk management approaches. However, what might be particularly relevant for NbS could be a potential tendency to romanticise nature-based practices and livelihoods. As such, these may neglect the socio-economic contexts that potentially force (the majority of) people to adapt more-invasive practices (e.g. in the context of fishing) and their worsening their socio-economic position even more.

The cases show that some injustices are recognised, for example, the position of women in the Vietnam case or the burdens carried by upstream communities in the Austrian case; however, there are certainly injustices that remain hidden as they are not yet discussed either in the literature or in practice. That emphasises the need of actors responsible for the planning and implementation of NbS to be sensitive to issues of justice and to map the various (potential) injustices emerging in their NbS project in a participatory way. Even though injustices cannot always be completely avoided, being aware of them and open to them, helps to identify potential solutions or mitigations (e.g. beneficiary-pays compensation in the Austrian case or gender-targeted participation as shown by the Vietnam case) to alleviate existing injustices.

References

1. EU (2015) Towards an EU research and innovation policy agenda for nature-based solutions & re-naturing cities. Brussels
2. IPCC (2018) Summary for policymakers. In: global warming of 1.5°C. Intergovernmental panel on climate change. https://doi.org/http://www.ipcc.ch/publications_and_data/ar4/wg2/en/spm.html
3. IPBES (2018) Summary for policymakers of the regional assessment report on biodiversity and ecosystem services for Asia and the Pacific of the intergovernmental science-policy platform on biodiversity and ecosystem services
4. Randrup TB, Buijs A, Konijnendijk CC, Wild T (2020) Moving beyond the nature-based solutions discourse: introducing nature-based thinking. Urban Ecosyst. https://doi.org/10.1007/s11252-020-00964-w
5. Kotsila P, Anguelovski I, Baró F et al (2020) Nature-based solutions as discursive tools and contested practices in urban nature's neoliberalisation processes. Environ Plan E Nat Space:251484862090143. https://doi.org/10.1177/2514848620901437
6. Kaufmann M, Priest SJ, Leroy P (2018) The undebated issue of justice: silent discourses in Dutch flood risk management. Reg Environ Chang 18:325–337. https://doi.org/10.1007/s10113-016-1086-0
7. Penning-Rowsell EC, Pardoe J (2012) Who benefits and who loses from flood risk reduction? Environ Plan C Gov Policy 30:448–466. https://doi.org/10.1068/c10208

8. Johnson C, Penning-Rowsell EC, Parker D (2007) Natural and imposed injustices: the challenges in implementing 'fair' flood risk management in England. Geogr J 173:374–390. https://doi.org/10.1111/j.1475-4959.2007.00256.x
9. Kabisch N, Frantzeskaki N, Pauleit S et al (2016) Nature-based solutions to climate change mitigation and adaptation in urban areas and their rural surroundings. Ecol Soc 21. https://doi.org/10.5751/ES-08373-210239
10. Sekulova F, Anguelovski I (2017) The governance and politics of nature-based solutions. Work Pap Naturvation. https://doi.org/RBM EXP. No. 980177
11. Haase D, Kabisch S, Haase A et al (2017) Greening cities – to be socially inclusive? About the alleged paradox of society and ecology in cities. Habitat Int. https://doi.org/10.1016/j.habitatint.2017.04.005
12. Irvine KN, Warber SL, Devine-Wright P, Gaston KJ (2013) Understanding urban green space as a health resource: a qualitative comparison of visit motivation and derived effects among park users in Sheffield, UK. Int J Environ Res Public Health. https://doi.org/10.3390/ijerph10010417
13. Dooling S (2009) Ecological gentrification: a research agenda exploring justice in the city. Int J Urban Reg Res 33(3):621–639
14. Checker M (2011) Wiped out by the "Greenwave": environmental gentrification and the paradoxical politics of urban sustainability. City Soc. https://doi.org/10.1111/j.1548-744X.2011.01063.x
15. Walker G (2012) Environmental justice: concepts, evidence and politics. Routledge, New York
16. Young IM (2000) Inclusion and democracy. Oxford University Press, Oxford
17. Simcock N (2016) Procedural justice and the implementation of community wind energy projects: a case study from South Yorkshire, UK. Land Use Policy 59:467–477. https://doi.org/10.1016/j.landusepol.2016.08.034
18. Gunn AS, Mccallig C, Ethics S, Autumn N (1997) Environmental values and environmental law in New Zealand. Ethics Environ 2:103–120
19. Byrne J, Wolch J (2009) Nature, race, and parks: past research and future directions for geographic research. Prog Hum Geogr. https://doi.org/10.1177/0309132509103156
20. Young IM (1990) Justice and the politics of difference. Princton University Press, Princton
21. Schlosberg D (2001) Three dimensions of environmental and ecological justice. In: European Consortium for Political Research Annual Joint Sessions, Grenoble, France, 6–11 April 2001. Workshop: the nation-state and the ecological crisis: sovereignty, Economy and Ecology
22. Miller D (2003) A response. In: Bell DA, De-Shalit A (eds) Forms of justice: critical perspectives on David Miller's political philosophy. Rowman and Littlefield, Lanham
23. Reid H (2016) Ecosystem- and community-based adaptation: learning from community-based natural resource management. Clim Dev. https://doi.org/10.1080/17565529.2015.1034233
24. Reid H (2009) Community-based adaptation to climate change. Particip Learn Action 60:11–33
25. DKKV (2019) Strong roots, strong women. Women and ecosystem-based adaptation to flood risk in Central Vietnam. Bonn
26. Neumayer E, Plümper T (2007) The gendered nature of natural disasters: the impact of catastrophic events on the gender gap in life expectancy, 1981-2002. Ann Assoc Am Geogr. https://doi.org/10.1111/j.1467-8306.2007.00563.x
27. CSRD (2015) Gender needs and roles in building climate resilience in the city of hue, Vietnam Asian cities climate. London, UK
28. Gaillard JC, Sanz K, Balgos BC et al (2017) Beyond men and women: a critical perspective on gender and disaster. Disasters. https://doi.org/10.1111/disa.12209
29. Cutter SL (2017) The forgotten casualties redux: women, children, and disaster risk. Glob Environ Chang. https://doi.org/10.1016/j.gloenvcha.2016.12.010
30. Renaud FG, Sudmeier-Rieux K, Estrella M (2013) The role of ecosystems in disaster risk reduction. United Nations University Press, Tokyo
31. Stone R (2016) Dam-building threatens Mekong fisheries. Science 354(6316):1084–1085

32. Hudson P, Pham M, Bubeck P (2019) An evaluation and monetary assessment of the impact of flooding on subjective well-being across genders in Vietnam. Clim Dev. https://doi.org/10.1080/17565529.2019.1579698
33. Arnstein SR (1969) A ladder of citizen participation. J Am Plann Assoc. https://doi.org/10.1080/01944366908977225
34. Reed MS, Vella S, Challies E, de Vente J, Frewer L, Hohenwallner-Ries D, van Delden H (2018) A theory of participation: what makes stakeholder and public engagement in environmental management work? Restor Ecol 26:7–17. https://doi.org/10.1111/rec.12541
35. Short C, Clarke L, Carnelli F et al (2019) Capturing the multiple benefits associated with nature-based solutions: lessons from a natural flood management project in the Cotswolds, UK. L Degrad Dev 30:241–252. https://doi.org/10.1002/ldr.3205
36. Ribot JC (2006) Choose democracy: environmentalists' socio-political responsibility. Glob Environ Chang
37. Johnson C, Tunstall S, Priest S et al (2008) Social justice in the context of flood and coastal Erosion risk management: a review of policy and practice. Defra, London
38. Green C (2007) Mapping the field: the landscapes of governance. Report for the SWITCH Project
39. Serbian Government (2020) Preliminary flood risk assessment for the Republic of Serbia. http://www.rdvode.gov.rs/doc/dokumenta/6.2.1 Znacajna poplavna podrucja za teritoriju Republike Srbije.pdf
40. Serbian WFD (2020) Interested in the future of Serbian Water Resource Management? http://wfd-serbia.eu/2020/03/03/interested-in-the-future-of-serbian-water-resource-management
41. Babić Mladenovic M et al (2016) Study of flood protection improvement in the Kolubara river catchment area. Preliminary report. https://studijakolubara.srbijavode.rs/izvestaji_o_rezultatima_studije/Друга-фаза/preliminarni_izvestaj/
42. Todorovic I (2020) Environmentalist groups unite to protest small hydropower, pollution in Serbia. Balk. Green Energy News
43. NGO Defence River Stara Planina (2020) River Stara Planina Mountain. https://novastaraplanina.com/en/
44. Serbian Parlament (2019) Održano javno slušanje na temu "Stanje voda u Srbiji". http://www.parlament.gov.rs/Održano_javno_slušanje_na_temu_Stanje voda_u_Srbiji.37315.941.html
45. Šercl P, Stehlík J (2003) The August 2002 flood in the Czech Republic. Geophys Res Abstr 5:404
46. Blöschl G, Kiss A, Viglione A et al (2020) Current European flood-rich period exceptional compared with past 500 years. Nature. https://doi.org/10.1038/s41586-020-2478-3
47. Floodlist (2020) Czech Republic – Deadly Flash Floods in East. http://floodlist.com/europe/czech-republic-flash-floods-olomouc-june-2020
48. Lane SN (2017) Natural flood management. Wiley Interdiscip Rev Water. https://doi.org/10.1002/wat2.1211
49. AOPK ČR (2020) Tvorba a obnova tůní, mokřadů a rašelinišť [Establishment and restoration of pools, wetlands and peatbogs]. http://www.dotace.nature.cz
50. Slavíková L, Raška P (2019) This is my land! Privately funded natural water retention measures in the Czech Republic. In: Hartmann T, Slavíková L, McCarthy S (eds) Nature-based flood risk management on private land. Springer
51. Wilkinson ME (2019) Commentary: Mr. Pitek's land from a perspective of managing hydrological extremes: challenges in upscaling and transferring knowledge. In: Hartmann T, Slavíková L, McCarthy S (eds) Nature-based flood risk management on private land. Springer
52. AOPK ČR (2014) Standardy péče o přírodu a krajinu – Vytváření a obnova tůní [Standards for Nature and Landscape Management – Creation and restoration of pools]. https://standardy.nature.cz/res/archive/155/020271.pdf?seek=1394520652
53. Aubrechtová T, Semančíková E, Raška P (2020) Formulation matters! The failure of integrating landscape fragmentation policy. Sustain. https://doi.org/10.3390/SU12103962

54. Carnelli F (2018) Slowing down the flood, naturally. The integration of local knowledges into flood risk governance: insights from south West England and North Italy. University of Milan-Bicocca
55. Nozick R (1974) Anarchy, state, utopia. Basic Books, New York
56. Davy B (1997) Essential injustice : when legal institutions cannot resolve environmental and land use disputes. Springer, New York
57. Mill JS (2010) Utilitarianism, liberty and representative government. Wildside Press, Milton Keynes
58. Rawls J (1973) A theory of justice. Harvard University Press, Cambridge
59. Sen A (2010) The idea of justice. Penguin, London
60. Rijkswaterstaat (2016) Projectplan Waterwet: Projectplan voor "Zandsuppletie Roggenplaat." The Hague
61. Rijkswaterstaat (2019) Schadevergoeding in de vorm van nadeelcompensatie en planschade. https://www.rijkswaterstaat.nl/over-ons/contact/schade-en-compensatie/nadeelcompensatie.aspx
62. Wesseling M (2019) Zand moet bedreigde Roggenplaat redden. Trouw
63. Vleesenbeek T (2020) Building with nature on the Roggenplaat. A policy arrangement for the sand nourishment project on the Roggenplaat. Radboud University
64. Rijkswaterstaat (2016) Risico beoordeling van de Roggenplaat suppletie. Rijkswaterstaat, The Hague
65. Modde M (2018) Mosselkwekers vechten suppletie Roggenplaat aan bij Raad van State. PZC
66. Rauter M, Schindelegger A, Fuchs S, Thaler T (2019) Deconstructing the legal framework for flood protection in Austria: individual and state responsibilities from a planning perspective. Water Int. https://doi.org/10.1080/02508060.2019.1627641
67. Nordbeck R, Steurer R, Löschner L (2019) The future orientation of Austria's flood policies: from flood control to anticipatory flood risk management. J Environ Plan Manag. https://doi.org/10.1080/09640568.2018.1515731
68. WBFG (1985) Wasserbautenförderungsgesetz [Federal Hydraulic Engineering Development Act]
69. WRG (1959) Wasserrechtsgesetz [Federal Water Act]
70. Nordbeck R, Löschner L, Scherhaufer P et al (2018) Hochwasserschutzverbände als Instrument der interkommunalen Kooperation im Hochwasserrisikomanagement. Österreichische Wasser- und Abfallwirtschaft
71. Seher W, Löschner L (2018) Balancing upstream–downstream interests in flood risk management: experiences from a catchment-based approach in Austria. J Flood Risk Manag 11:56–65
72. Thaler T (2014) Developing partnership approaches for flood risk management: implementation of inter-local co-operations in Austria. Water Int. https://doi.org/10.1080/02508060.2014.992720
73. Löschner L, Nordbeck R, Schindelegger A, Seher W (2019) Compensating flood retention on private land in Austria: towards polycentric governance in flood risk management. Landsc Archit Front 7:32–45. https://doi.org/10.15302/j-laf-1-020004
74. Schindelegger A (2019) Absiedlung als Planungsinstrument: Planerische Aspekte zu Siedlungsrückzug als Naturgefahrenprävention. TU Wien
75. Eriksen S, Aldunce P, Bahinipati CS et al (2011) When not every response to climate change is a good one: identifying principles for sustainable adaptation. Clim Dev 3:7–20. https://doi.org/10.3763/cdev.2010.0060
76. Paavola J, Adger WN (2006) Fair adaptation to climate change. Ecol Econ 56:594–609. https://doi.org/10.1016/j.ecolecon.2005.03.015
77. Adger WN (2006) Fairness in adaptation to climate change. MIT Press, Cambridge
78. Anguelovski I, Shi L, Chu E et al (2016) Equity impacts of urban land use planning for climate adaptation: critical perspectives from the global north and south. J Plan Educ Res 36:333–348. https://doi.org/10.1177/0739456X16645166

What Nature-Based Flood Protection Solutions Are Best Perceived by People? Lessons from Field Research in Czechia

Jan Macháč, Jan Brabec, Marek Hekrle, and Alena Vacková

Contents

J. Macháč (✉), J. Brabec, M. Hekrle, and A. Vacková
University of J. E. Purkyně in Ústí nad Labem, Faculty of Social and Economic Studies, Ústí nad Labem, Czechia
e-mail: jan.machac@ujep.cz; brabec@ireas.cz; Marek.Hekrle@ujep.cz; vackova@e-academia.eu

Carla S. S. Ferreira, Zahra Kalantari, Thomas Hartmann, and Paulo Pereira (eds.), *Nature-Based Solutions for Flood Mitigation: Environmental and Socio-Economic Aspects*, Hdb Env Chem (2022) 107: 425–446, DOI 10.1007/698_2021_763,

Published online: 6 May 2021

Abstract Nature-based solutions (NBS) can be used as alternatives to technical measures for mitigating the landscape impact of increasingly frequent and severe floods. This chapter aims to fill the existing knowledge gap regarding public preferences and awareness of the implementation of NBS and types of cultivated crops (TCC) that may be used to mitigate pluvial floods by exploring preferences of the residents in the Oleška river basin in Czechia. An image-assisted questionnaire survey and a discrete choice experiment were used, with focus also on residents' willingness to pay. Data were collected through face-to-face surveys ($n = 256$) with residents of the Oleška river basin.

Among the 9 examined NBS (linear and point) and 3 TCC, afforestation and grassing along the stream were best perceived by residents from the purely aesthetical point of view and, together with dry polders, they were considered the most relevant solution for flood mitigation. Based on the discrete choice experiment, linear and point NBS are preferred over TCC measures. However, implementation of any studied NBS was perceived better than a situation without any NBS. Local residents' preferences should be reflected as one of the criteria in planning and selecting the available efficient measures that may be implemented in response to climate change.

Graphical Abstract

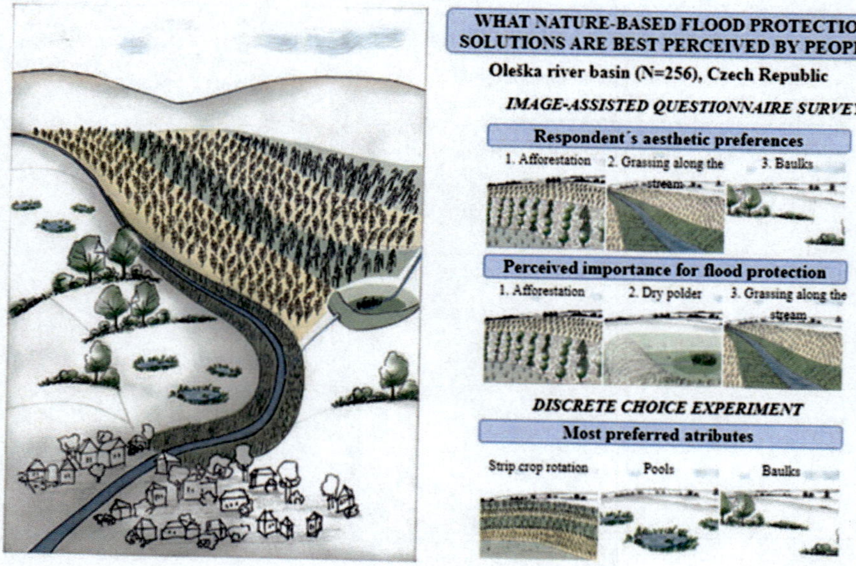

Keywords Aesthetic preferences, Choice experiment, Floods, Nature-based solutions, Public perception, Willingness to pay

1 Introduction

With respect to the projected increasing impact of climate change, especially rising frequency and severity of floods [1, 2], it is reasonable to expect that various measures including nature-based solutions (NBS) and change of type of cultivated crops (TCC) on agricultural land will play an important role in flood protection in the future. NBS are increasingly recognized in river basins as a viable alternative to technical flood management solutions or as their complements on agricultural land [3–5].

To ensure that NBS will be well-perceived by different stakeholders, their preferences need to be integrated into NBS planning and implementation. Although previous research has found that public support plays a crucial role in flood management planning, there are still many unknowns regarding the public preferences towards NBS [4–7]. Consequently, the implementation of NBS is often dependent on decision makers and less on consultation with other affected stakeholders [8].

The importance of public preferences is thus often overlooked and not fully reflected in flood management planning. Understanding people's perceptions of various NBS could be a crucial baseline for decision makers to identify the best possible flood mitigation plan. Ensuring inclusion of public preferences makes these flood mitigation processes more acceptable to public and hence better supported [1, 3, 7]. Moreover, evaluation of effectiveness of flood risk management strategies is strongly influenced by public perceptions and attitudes [5, 9].

However, flood risk protection and level of provision of additional ecosystem services differ among various NBS. Due to the need for adaptation to climate change, the emphasis is usually on regulatory and provisioning ecosystem services such as flood risk mitigation, water retention, erosion protection, water purification, and soil fertility rather than on cultural ecosystem services [10]. The importance of aesthetic value as a part of cultural ecosystem services of NBS is consistently deemed less important on agricultural land and not often fully reflected in flood mitigation planning, even though unfavourable aesthetics can be considered the most commonly voiced complaint about NBS projects [4]. In addition, flood mitigation planning seems often to be driven by factors such as project costs, pressing need for action, technical capacity, previous experience, or availability of funds [3, 4]. What may be a clear objective for decision makers may not necessarily be perceived as aesthetically positive, and therefore might be less accepted by the public [11]. Aesthetically pleasing solutions will ensure public support, which is important for implementation of consistent and sustainable flood protection [12, 13].

Based on literature review, there also seems to be a gap in empirical studies that investigate public preferences across more than just a few alternatives of NBS [14, 15]. The main focus of the existing studies is either on general functions of NBS and benefits provided [16–18], or the studies do not examine in detail which of the commonly considered NBS are best perceived by people [4].

Therefore, the aim of this chapter is to analyse public preferences regarding various NBS measures commonly considered in Czechia for flood mitigation.

Besides, implementation of these measures is technically possible in the studied river basin. Additional focus was on what agricultural landscape functions are perceived as the most important. For easier implementation and public acceptance of flood-related NBS, it is important to know whether people associate floods with agricultural functions. If not, it shows a higher need to communicate their more frequent use to the public.

2 Study Area

The Oleška river is a left-side tributary of the Jizera river in the northern part of Czechia (Fig. 1). The length of the river is 36.4 km. The river basin, which is oriented in the SE-NW direction, has 221.5 km of streams and covers about 171 km^2 with an elevation ranging from 541 to 315 m above sea level. It is the most important tributary of the Jizera. This area was chosen as a case study within the Czech-German cross-border project STRIMA II: Saxon-Czech Flood Risk Management II, and was selected as an appropriate basin to show a multidisciplinary approach to identifying, analysing, and selecting suitable flood mitigation solutions that are, at the same time, technically feasible and effective, and correspond to local residents' preferences. This chapter presents the project results related to the residents' preferences.

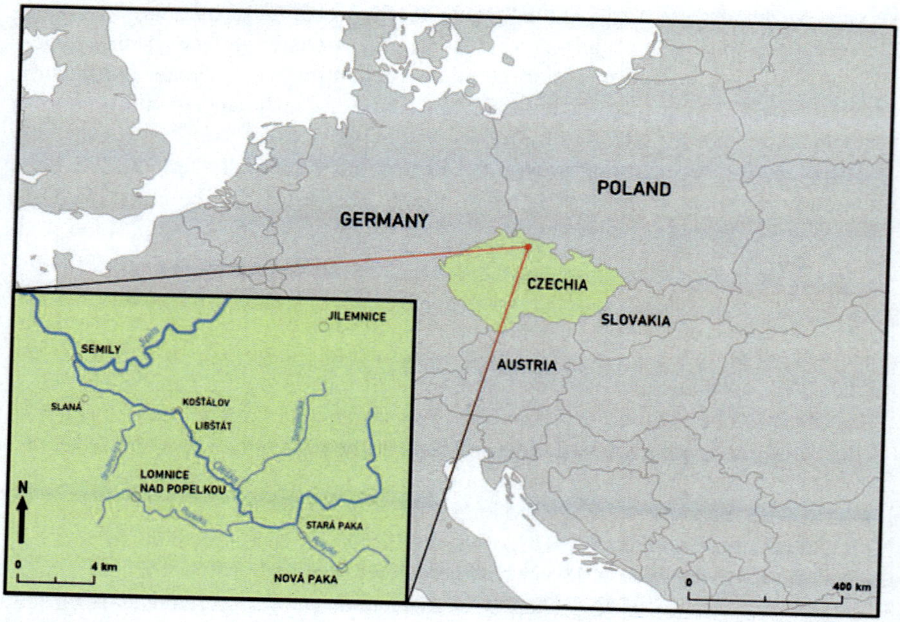

Fig. 1 Location of the Oleška river basin study area in Czechia

Based on Liberecký Kraj [19], the Oleška flows through the upland area of the Krkonoše foothills with rural development and discontinuous forestation, especially on the steeper slopes above the river. The area belongs to the Liberec region on the border of the Semily and Jičín districts around the towns of Lomnice nad Popelkou and Nová Paka. It features watercourses up to the fourth order, so the river network can be considered developed. This is also confirmed by the great dynamics and rapid course of erosion processes in the whole catchment, which has an impact on sediment transport to the river network. In total, there are 136 water reservoirs in the area.

Significant tributaries of the Oleška itself are the Tampelačka, which drains approximately 28.2 km^2 of the sub-basin, and the Popelka, which drains 27.7 km^2 of the sub-basin as a left-hand tributary. The slope curve of Oleška and its main tributaries can be considered balanced, without major fractures and gradients [20].

Regarding land use, the Oleška river is surrounded mostly by agricultural land and partly by urban areas, including the towns of Nová Paka, Semily, Lomnice nad Popelkou, Stará Paka, comprising 9094, 8367, 5541, and 2081 residents, respectively, and several small villages with up to 1,000 residents. There are in total 16 municipalities located directly in the hydrological basin with a total of 26 thousand residents. In the wider interest area, there are another 11 municipalities with additional 17 thousand residents [21] which are not located in the hydrology river basin but are in a close proximity and the residents often visit or own the land in the river basin. The study area is characterized by a relatively low population density (4.1 residents/ha in the river basin and 4.5 residents/ha for areas of interest, also including Semily and other border municipalities). The population is very unevenly distributed and concentrated especially along rivers. The age structure of the population corresponds to the age structure for Czechia as a whole [21]. The demographic ageing is present mostly in villages and not as significantly in larger towns [21].

In terms of land use, based on Czech Technical University in Prague [22], the catchment comprises mainly five categories: forest and scrubs (34%), permanent grassland (29%), arable land (almost 26%), built-up land (5%), and gardens (5%) (Figs. 2 and 3).

The shape of the river is mostly natural in the upper reaches, whereas the lower reaches were partly regulated in the last century, including regulation and straightening in the urban areas [19, 20]. Based on Liberecký Kraj [23], historically the biggest fluvial flooding events recorded in the Oleška river basin were in the 1940s. As one of the major tributaries of the Jizera river, the Oleška contributes to the fluvial floods in the lower reaches of this river. Water from the snow that melts in the Krkonoše mountains can raise the water level in spring, but summer storms are also an important source of floods [23]. A rise in the water level can cause damage to residential homes and infrastructures. In the last two decades, this occurred in 2000, 2013, and 2017 [24]. In terms of fluvial floods, structures in the immediate vicinity of the stream in places with a low unregulated bank are endangered by floods, especially in the municipalities of Stará Paka, Libštát, Košťálov, and Slaná [19, 25]. According to the flood protection plan of the town of Semily [26], the whole Oleška valley is flooded one to three times a year, including flooding of

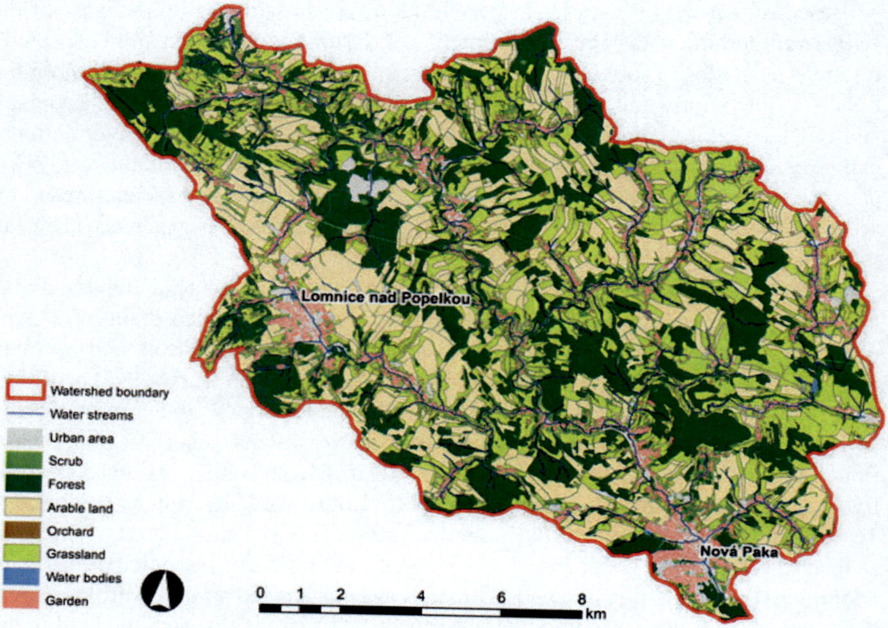

Fig. 2 Spatial distribution of land use categories in the Oleška river basin [22]

houses in Košťálov. The reason is insufficient riverbed capacity. Based on the document that characterizes floods in the area [27], the most common flood damage in the Oleška river basin is caused by pluvial floods, driven by shorter periods with heavy rainfall that cause flash floods and soil erosion. Therefore, soil loss and sediment transport to the river represent a major issue [22, 28].

Current flood protection is mainly associated with the current water reservoirs, strengthening of the banks in the municipalities located along the river, and land consolidation. According to field research, agricultural measures arising from zoning plans of flood-affected municipalities include numerous measures which should complement the existing ones (e.g. grassing along the stream, grassing of fields with higher slope, and contour tillage implemented on part of the land) in the form of a flood protection system for the whole Oleška river basin [29]. This system of measures covers not only landscape revitalization, but also design of new water retention measures (polders) and increasing of capacity of culverts and bridges. As stated in the zoning plans [28, 29], the current agriculture and forest management affects flood hazard. A range of measures and agricultural technologies are mentioned in the plans as suitable for implementation. The following measures are specifically mentioned in relation to soil erosion and flood protection: grassing, contour tillage, composition and types of cultivated crops, tree species composition, swales, and grassing along the streams [28, 29]. Due to the hilly terrain, attention should be paid to growing crops on sloping land according to the zoning plans.

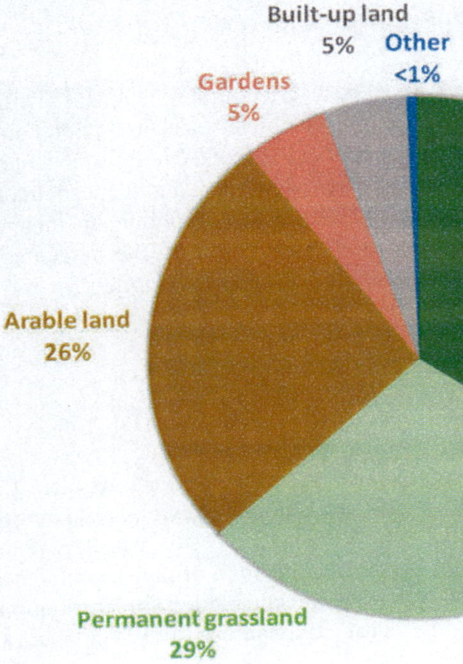

Fig. 3 Land use in the Oleška river basin

Given the current land use (Fig. 2), the implementation of technical measures located on the stream and measures on agricultural land must be considered. With regard to pluvial floods and to reducing both flood damage and erosion processes [22], the emphasis of this chapter is on NBS and TCC on agricultural land.

3 Methodology

Local residents' preferences regarding NBS and TCC used for flood mitigation were investigated using anonymized questionnaire-based interviews. Apart from questions covering information on the respondents' sociodemographic characteristics, the following topics were studied: respondents' opinions on the most important agricultural functions, evaluation of nature-based flood protection solutions from the aesthetic point of view using an image-assisted questionnaire survey, analysis of respondents' preferences using a choice experiment, and eliciting respondents' previous experience with floods.

3.1 Most Important Agricultural Landscape Functions

The respondents were shown nine preselected agricultural functions: growing crops; living space for animals; living space for plants; climate protection; flood protection; recreational and leisure time; landscape cultivation; contribution to local economy; and crops for energy use. These functions were described, and the respondents were asked to select the three functions that are the most important in their opinion. Evaluation of this part of the questionnaire was based on the percentage of an agricultural function being selected among the three most important.

3.2 Design of Image-Assisted Questionnaire Survey

Based on literature review, on recommendations included in zoning plans [28, 29], geographical conditions of the study area and authors' previous research on flood protection and agricultural measures [17, 30, 31], 12 types of measures (9 different NBS and 3 TCC) were selected for further identification of public preferences (see Table 1). The emphasis was put on NBS that are suitable for implementation in the studied river basin and are commonly considered and used in Czechia and Central Europe in general [22].

To support the assessment of public preferences regarding various NBS and TCC, an image-assisted questionnaire survey [32–34] was developed. At the beginning of the interview, the different types of measures and their functions were introduced to randomly selected respondents. Considering a possible bias due to disparities in the quality of drawn images, illustrated images of NBS and TCC were created in cooperation with a graphic designer (see Fig. 4). The main focus was on the use of the same colours, the same backgrounds, and elimination of factors that can bias the perception.

Trained interviewers then asked the respondents to rate every NBS and TCC from 1 (lowest value) to 10 (highest value) according to how much they would personally enjoy the solution in landscape from the aesthetic (visually pleasing) point of view. This part of the survey was therefore focused exclusively on aesthetic perception of various NBS/TCC. Respondents were then asked to pick three measures from the

Table 1 List of possible measures (NBS and TCC) investigated in this study to mitigate floods

Nature-based solutions (NBS)		Types of cultivated crops (TCC)
Grassing	Afforestation	Grain field
Contour tillage	Baulks	Maize field
Pools	Wetland	Strip crop rotation
Swale	Dry polders	
Grassing along the stream		

Fig. 4 Example of illustrated images used in the image-assisted questionnaires: Dry polders (left) and strip crop rotation (right)

same list of measures that they thought are the most important solutions for flood protection. Simple statistics were used to evaluate this part of the questionnaire. An average score was computed for each NBS and TCC in the aesthetic part and the percentage of a flood mitigation measure being selected among the three most important was used to evaluate the importance.

3.3 Design of Discrete Choice Experiment

To learn more about residents' preferences regarding NBS and TCC, an image-assisted discrete choice experiment was additionally used in the interviews. The method allowed us to discover what type of NBS and TCC residents enjoy in their neighbourhood and prefer over others. This analytical tool is based on consumer behaviour theory [35] and random utility theory [36]. That implies that people view each good not as a whole but as described by several attributes that characterize the good and these attributes are also the source of utility. The approach allows consumers to compare different types of NBS and decide which NBS they prefer over the others.

Therefore, respondents were asked to make choices between two options describing a hypothetical landscape using the same set of attributes but differing in the levels of the attributes. In these experiments, respondents express their preference by choosing such a combination of measures that brings them the highest possible utility [37], and their choices can be used in econometric models to find out more about general preferences of the public.

Respondents were first introduced to the design of the choice experiment and were provided with a thorough description of the selected attributes and their levels. The selected attributes were (1) TCC (i.e. maize field, grain field, and strip crop rotation); (2) a point NBS (i.e. none, dry polders, and pool); (3) a linear NBS (i.e. none, baulk, and grassing along the stream); and (4) an annual financial contribution (Fig. 5). Based on consultation with project partners from STRIMA II, the measures were selected from a catalogue of NBS flood protection measures

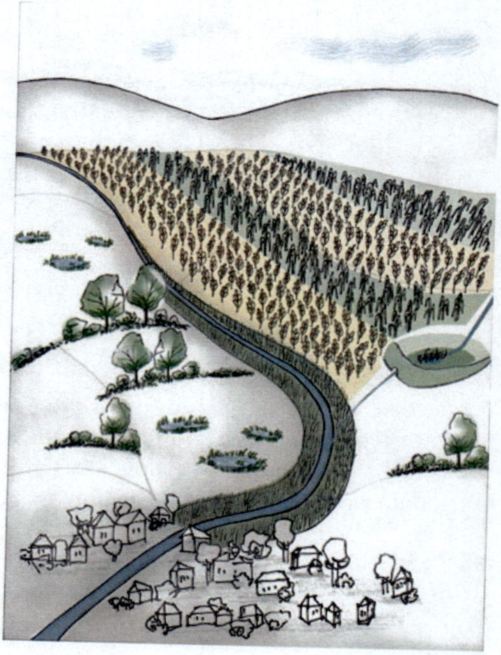

Fig. 5 Landscape with possible attributes used in the discrete choice experiment – measures include strip crop rotation, dry polders, pools, baulks, and grassing along the stream

[31] as appropriate measures for flood mitigation, technically feasible, and corresponding to local conditions of the Oleška river basin.

Money raised from the financial contributions would be used by a regional authority to invest in flood-mitigating NBS or TCC. The contribution was explained to the respondents as a new form of a local tax, similar to other local taxes that are paid directly to municipalities such as property tax and unlike most of the taxes that are paid into the national budget. The financial contribution form was chosen because the money raised should be used at the local level for covering flood mitigation measures directly in the region. Respondents were instructed to act as if the payment was required and had a real impact on their personal/family budgets. The inclusion of the payment vehicle allows us to express respondents' willingness to pay (WTP) for moving from one level of a specific attribute to another.

Using an orthogonal design, 18 combinations of various attribute levels were generated and paired into 9 choice cards such as the one shown in Fig. 6, which gave the respondents a choice between options described as Landscape A and Landscape B. Each respondent was asked to choose one of the options from each of the 9 choice cards or was allowed to opt out and choose neither if he/she felt that the combinations of attributes on the presented card do not provide enough value for the financial contribution or are simply not appealing enough. To ensure consistency of the results, all respondents were shown the same cards in the same order and the answers were recorded by trained interviewers using tablets. After proper recoding, the answers were used as inputs for a conditional logit model [37]. This model

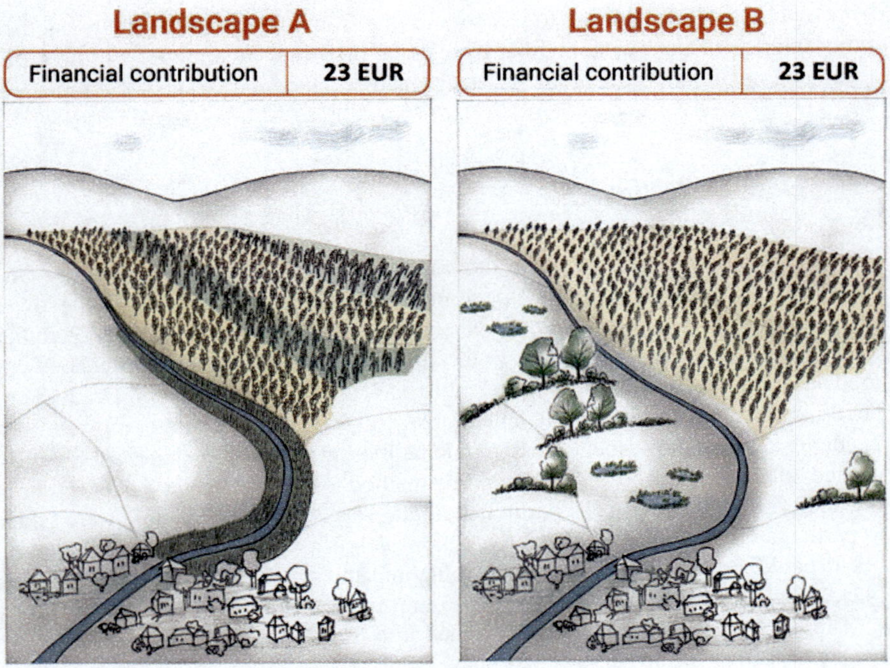

Fig. 6 Example of one set of cards used in the discrete choice experiment

determines the probability of a choice of a given alternative from the pool of all available alternatives based on the attribute levels used to describe such alternative [38] using Eq. (1):

$$\boldsymbol{\pi}_{ij} = \frac{\exp\left(x'_{ij}\boldsymbol{\beta}\right)}{\sum_{j} \exp\left(x'_{ij}\boldsymbol{\beta}\right)} \tag{1}$$

where $x'_{ij}\boldsymbol{\beta}$ represents the observable part of respondent's utility, which together with unobservable random element form the total utility. The outputs of the logit model allow to determine a marginal rate of substation between any of the attributes involved in the design that keeps the respondent's utility level constant. That implies that one can learn about the trade-off between attributes and consequently about the importance the individual attributes play in the decision-making process. For an overview of various models see [39].

Latent Gold software was used to estimate the model which determines the relationships between the individual levels of each attribute. These levels are ranked based on respondents' preferences. The respondents' WTP was calculated by dividing the coefficient of an attribute by the cost coefficient. This allows to determine how much people are willing to pay for moving from one attribute level to another, other things constant, meaning that the change predicted by a model will be of the estimated magnitude if there are no other changes happening at the same time.

3.4 Data Collection

The interviews were performed in three municipalities within the case study area along the Oleška river (one in the upper reaches of the Oleška river, one in the lower reaches, and one in the middle), where 126 interviews were made. Another 130 respondents were interviewed in the Jizera river basin past its confluence with the Oleška in three more towns (Železný Brod, Turnov, and Mnichovo Hradiště) to cover preferences in a wider catchment area. Respondents were randomly selected residents of the above-mentioned municipalities. The questionnaire survey was carried out using the face-to-face interview method with trained interviewers recording answers using tablets. The participation rate was 29%, as 877 people were asked to participate, but only 256 agreed to do so, which is a rather low outcome [40, 41], but it could be considered a standard outcome in comparison with other studies [42]. Interviews lasted between 10 and 15 min in most cases.

The data collection process was divided into three phases. During the development and testing of the survey (first phase, June 2019), the clarity of the questions, pictures, and choice experiment was verified and adjusted as necessary, both outside and inside the case study area. Respondents ($n = 30$) from this early survey stage are not included in the number of respondents presented above. In the second phase of data collection (July 2019), it turned out that the financial contribution had previously been set too low and the respondents ($n = 15$) did not view the attribute as binding and did not consider it at all. Therefore, the levels of the attribute were adjusted while the other attribute levels or parts of the survey were not modified. For that reason, the choice experiment is evaluated only on responses from the adjusted questionnaire (third phase, August 2019), which accounts for 241 responses. The results on agricultural landscape functions, and perceived aesthetics and importance of NBS are based on the complete set consisting of 256 responses (second and third phases).

4 Results

4.1 Sample Characteristics of Respondents

Women accounted for 62% of the respondents. The average age was 42, with 39 being the median value. Approximately 42% of the respondents were full-time employees, 7% were self-employed, 17% were pensioners, and 14% were on parental leave. The remaining 20% were part-time employees, students, working students, working pensioners, unemployed, and homemakers (listed in the order of frequency).

Approximately 29% of the respondents had a university degree, 66% had graduated from secondary school, and 5% had only primary education. Roughly 28% of the respondents had a monthly income lower than the minimum wage in Czechia in 2019 (520 EUR), 40% of the respondents had an income between the minimum and median wage for Czechia in 2019 (approximately 1,100 EUR). Income higher than the median wage was reported by 18% of the respondents and 14% refused to state their income. The low proportion of respondents with incomes higher than the median is probably due to the fact that the survey was conducted in smaller municipalities far from larger cities, where incomes are generally lower.

Almost 54% of the respondents stated that they had personal experience with river floods and 44% had encountered mud washed off adjacent fields.

4.2 *Agricultural Landscape Functions*

Only 23% of the respondents ranked flood protection among the three most important agricultural landscape functions, which means it was chosen significantly less frequently than the four other agricultural functions listed in Table 2. Almost three quarters of the respondents associated the basic function of agricultural landscape with growing crops, and selected this function as one of the three most important functions. Living space for animals is important for more than 50% of the respondents. Living space for plants and climate protection almost tied at third place (38% and 37% of the respondents selected these functions). The other functions were perceived as even less important than flood protection (<20%). The complete ranking can be seen in Table 2.

Table 2 Importance of agricultural landscape functions identified by respondents in questionnaires

Functions	Perceived importance selected as top three [%]
Growing crops	73.4
Living space for animals	56.6
Living space for plants	38.3
Climate protection	37.5
Flood protection	23.1
Recreation and leisure	20.0
Landscape cultivation	19.9
Contribution to local economy	16.8
Crops for energy use	14.1

Table 3 Evaluation of investigated NBS aesthetics and importance for flood protection, based on questionnaire survey

Nature-based solution	Average aesthetics score (1: lowest–10: highest)	Importance for flood mitigation – selected as top three [%] (rank)	
Afforestation	8.3	52.7	(1)
Grassing along the stream	7.9	34.8	(3)
Baulks	7.8	25.4	(6)
Pools	7.6	18.0	(9)
Dry polders	7.3	40.6	(2)
Wetland	7.3	23.8	(8)
Grassing	7.2	26.2	(5)
Swale	7.1	16.8	(10)
Contour tillage	6.9	30.9	(4)
Strip crop rotation	6.6	25.0	(7)
Grain field	5.6	6.6	(11)
Maize field	4.6	2.3	(12)

4.3 Perceived Aesthetics and Importance of Measures

As can be seen in Table 3, afforestation received the highest average score in the aesthetic evaluation. Many measures got scores between 7 and 8, including grassing along the stream, baulks, pools, dry polders, wetland, grassing, and swales. Less popular were contour tillage and strip crop rotation, which scored between 6 and 7. The other TCC (i.e. fields of grain or maize) appeared to be the least favourite with an average score around 5. These results imply that respondents prefer aesthetics of solutions that are not directly related to agricultural techniques and types of growing crops.

When asked which of the measures respondents consider to be the most important solutions for flood protection, the top three measures selected were afforestation (selected by 53% of the respondents), dry polders (41%), and grassing along the stream (35%). Pools and wetlands were ranked at the bottom despite being quite popular in the previous question. Similarly, contour tillage did not receive a high aesthetic score but was considered quite important for flood mitigation. These results suggest that the respondents do not perceive aesthetics and importance for flood mitigation as equal, but rather distinguish between them. Afforestation seems to be the only measure viewed both aesthetically appealing and important for flood mitigation; this possibly also applies to grassing along the stream and dry polders.

4.4 Preferences and Willingness to Pay for NBS and TCC

A conditional logit model was estimated based on the data collected from the choice experiment. Only a sample of 203 respondents was used as not all information were

Table 4 Coefficients of the conditional logit model for individual levels of attributes investigated in the choice experiment ($n = 185$)

Attribute	Attribute level	Coefficient (standard error)	WTP [EUR] (standard error)
TCC	Strip crop rotation	0.9151*** (0.2025)	50.7*** (11.7)
	Grain field	−0.1899*** (0.0699)	−10.5* (5.7)
	Maize field	−0.7252*** (0.1925)	−40.2*** (7.5)
Point NBS	No point measure	−0.477*** (0.0883)	−26.4** (12.1)
	Pool	0.3018*** (0.0901)	16.7* (8.7)
	Dry polder	0.1752*** (0.046)	9.7** (4.6)
Linear NBS	No linear measure	−1.1122*** (0.203)	−61.6*** (16.6)
	Baulk	0.8945*** (0.2026)	49.6*** (11.3)
	Grassing along the stream	0.2177*** (0.051)	12.1** (6.1)
Financial contribution		−0.0007** (0.0003)	
Optout		−5.0517*** (0.5387)	−280*** (96)
R^2	0.1204		

Significance levels: *** 0.01; ** 0.5; * 0.1

available for thirty-eight of the respondents (mainly socio-demographics). Approximately additional 9% of the respondents were then filtered out because their choices were recognized by the model as rather random and not based on the levels of presented attributes. Also, respondents quite often did not choose any of the two offered options and used the possibility to opt out and choose neither. The results of the choice experiment (coefficient and WTP) for the remaining 91% of the respondents ($n = 185$) are shown in Table 4.

All the coefficients of the conditional logit model are statistically significant at the 5% confidence level and reflect the respondents' true preferences. The payment coefficient is significant and negative as expected since paying more money is associated with negative utility. The remaining coefficients express respondents' relative opinions on various NBS. When it comes to TCC, it is quite obvious that the respondents liked strip crop rotation the best. The coefficients for grain field and maize field are negative with maize being disliked even more than grain. The situation regarding point measures is not as clear with only a small difference between coefficients for pools and dry polders, but both options are clearly superior to having no measure implemented. In a similar fashion, no measure is the least preferred among the linear NBS. However, in this case baulks were preferred over grassing along the stream.

Comparing attributes according to the importance attributed to them by the respondents in the decision-making process (Fig. 7), it was found that people mostly cared about situations with no linear measures, strip crop rotation, baulks, maize fields, and a financial contribution. Strip crop rotation and baulks had a positive effect on the respondents' decisions, meaning they were more likely to pick a landscape with these attributes. On the other hand, no linear measures, maize fields,

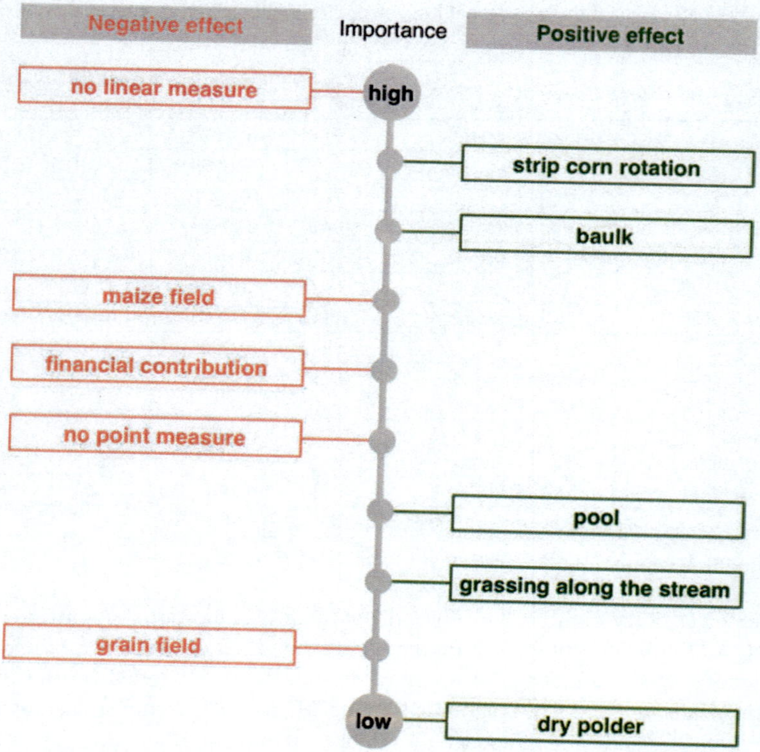

Fig. 7 Ranking of importance of attributes and their effects in decision-making process

and the financial contribution had a negative effect on the decisions as it was rather important for the respondents to choose a landscape without these attributes (or with a lower price). Compared to attributes stated above, no point measures, pools, grain fields, grassing along the stream, and dry polders were less important in the decision-making process. The described importance is shown in Fig. 7.

The findings are similar to the respondents' preferences based on the image-assisted questionnaire survey, except strip crop rotation. The right-hand part of Table 4 also shows that respondents are willing to contribute financially to have the preferred measures implemented. It is possible to determine the WTP based on the difference between the WTP estimates for the individual related attribute levels. Differences between the attributes are shown in Fig. 8. Each column in Fig. 8 compares differences in WTP for selected attributes. For example, it can be seen from column 2 (point NBS) that respondents are willing to pay 7 EUR each year to have a pool (most preferred attribute level) over a dry polder, 36 EUR each year to have a dry polder over no point measure (least preferred point NBS), and 43 EUR to have a pool over no point measure.

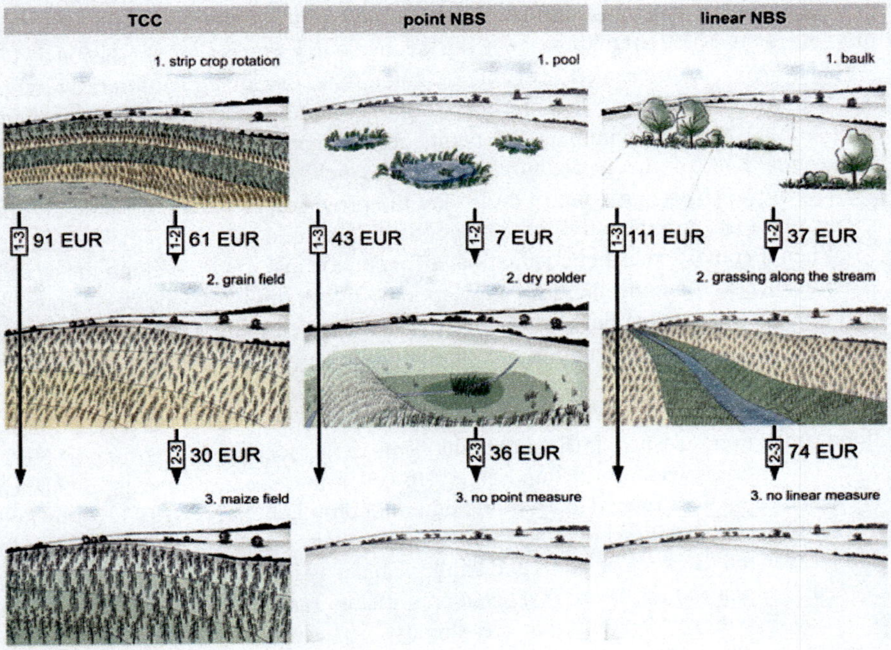

Fig. 8 Differences in respondents' annual willingness to pay for TCC and two types of NBS attributes (columns) and associated attribute levels (lines)

5 Discussion

Public acceptance is very important for effective implementation of NBS in a flood mitigation strategy. That is closely related to two aspects: aesthetic evaluation and perceived importance when it comes to actual flood mitigation. From 12 various commonly implemented NBS, the participants selected afforestation, dry polders, and grassing along the stream as the three most important and effective solutions for flood mitigation (Table 3). Especially afforestation and dry polders are widely used in Czechia, which indicates that the views of the public agree with the Czech implementation strategy. The respondents' ranking of NBS measures is based on awareness of flood mitigation effectiveness, which may not correspond to the actual effectiveness of these solutions. As some studies show, people tend to underestimate effectiveness of NBS and trust them to a lower extent than to traditional protection Schemes [43, 44]. However, it turns out that in a direct comparison, NBS are still preferred over grey infrastructures [45]. The real effectiveness of each individual NBS for flood mitigation is influenced by many local conditions and factors such as soil type, slope, crop type, measure extent, position in the river basin, and distance from a water body [22, 30, 46]. The perceived importance of flood protection measures depends also on respondents' education, personal experience with a specific solution, and media influence (scope and frequency of information about the implemented NBS in the mass media such as newspapers or television) [47–49]. However, as presented in Table 2, other functions of agricultural landscape are identified by residents as more important than flood protection, namely growing crops, living space for animals and plants, and climate protection. Thus, although other functions of the agricultural landscape such as flood protection are also perceived positively, growing of crops and the provision of habitat for animals and plants dominate significantly in the respondents' choices.

When it comes to aesthetic valuation, afforestation and grassing along the stream were also ranked among the three best options. Interestingly, dry polders, grassing and especially contour tillage were not perceived so well in the aesthetic ranking as opposed to the perceived importance ranking. However, it looks like any NBS is better than none as the absence of NBS lowered the probability of selecting such an option in the choice experiment (see Fig. 8). As in other studies of perception of flood protection measures [50], our results confirm that local residents support NBS, e.g., for capturing water in the landscape. However, there seems to be an imbalance between preferences related to TCC and other flood mitigation NBS. TCC (grain and maize, strip crop rotation) together with contour tillage seem to be less preferred in comparison to NBS such as pools and baulks. This may imply that respondents in the case study area value diversity of elements in the agricultural landscape. People value adding those elements that are already in use in Czechia. These elements complement the not-so-popular TCC (crop diversity), which corresponds to the results of a study that assessed landscape preferences in northeastern Germany (except livestock, which is not considered in this study) [51]. The landscape preferences were the lowest for livestock, followed by crop diversity and linear NBS. Point NBS were preferred the most [51]. Unfortunately, methods used in this case study (choice experiment) do not allow us to perform a cross-comparison between attributes and therefore to conclude whether point NBS are also more preferred than linear NBS or TCC. The preferences for individual measures in both attributes differ based on aesthetic perception and perceived importance. It may only be concluded that diversity and the addition of NBS elements are preferred over TCC.

It should be noted that the survey of people's preferences reflected all potential solutions without analysing any implementation barriers in the case study area. As stated by Spegel [52], residents' preferences may not reflect technical and other feasibility of a solution and therefore their preferences may be unreasonable/difficult to implement. In practice, it is often better to define a smaller number of solutions that are feasible and enforceable with regard to landowners and farmers, which may not be the most preferable solution by residents. If the measures are to be implemented, it is also necessary to know whether there is at least some overlap in the aesthetic preferences related to specific NBS with the actual effectiveness and other ecological and hydrological impacts of these NBS [11].

Although residents' preferences and their aesthetic evaluation should be included in NBS planning for flood mitigation, preferences of other stakeholders such as landowners, farmers, or local governments should be considered as well. What is perceived the best by residents may not necessarily be best perceived by farmers, for whom measure implementation may reduce the revenues and increase maintenance costs. As stated by Nesshöver et al. [7], the final output of the planning process (implemented solutions) should be the result of a broad set of societal trade-offs and compromises. As stated by Janssen et al. [6], implementation of NBS is also connected to making trade-offs between other functions and levels of benefits (private vs. societal). The involvement of residents and other stakeholders in planning is important not only in the form of examining their preferences, but also in the design of solutions and supporting a decision which reflects the diversity of the community [4]. However, the implementation of NBS and TCC in the Oleška river basin cannot be based solely on the preferences presented in this study, and should also reflect their potential efficiency for mitigating floods and other environmental and social aspects and trade-offs (e.g. hydrological impact, preferences of other stakeholders).

6 Conclusion

This chapter analysed the perceived importance and aesthetic value of 12 different types of NBS and TCC in the context of their flood protection function. A multi-method approach was introduced to explore residents' preferences related to NBS and TCC, specifically aesthetic evaluation and perception of effectiveness of individual NBS regarding flood protection. Using a choice experiment, we combined these aspects to determine the residents' willingness to pay for some of the individual NBS and TCC.

The study revealed that aesthetic preferences and importance for flood mitigation match for the most popular NBS. Afforestation, dry polders, grassing along the stream and possibly baulks were viewed as both aesthetically appealing and effective solutions. The opposite may be said about TCC (maize field, grain field), which were unpopular from the aesthetic point of view and also perceived as less effective than NBS in mitigating floods.

The results of the choice experiment confirm the above-mentioned findings. With the exception of afforestation (not studied), the same exact measures were found most popular when respondents were asked to choose between landscapes described by hypothetical combinations of NBS and TCC. It is also clear that people are willing to pay to have additional NBS implemented in the landscape. WTP is positive and (mostly) significant when comparing an individual NBS with a situation without the measure or with a TCC measure.

It is clear from the analysis that there is a good basis for acceptance of NBS among the residents in the Oleška river basin. Although flood protection is not viewed as one of the crucial agricultural functions, and growing crops or living space for animals and plants were viewed as more important, the preferences show people's positive attitude towards NBS, which should ensure their easier implementation in future.

Acknowledgements We would like to thank the Cooperation Programme Free State of Saxony – Czech Republic 2014-2020 for financing the project "STRIMA II: Saxon-Czech Flood Risk Management II" (grant number 100282105), and also all the respondents in this survey as well as the interviewers. This chapter was written as part of COST-Action LAND4FLOOD (Natural Flood Retention on Private Land).

References

1. Kabisch N, Frantzeskaki N, Pauleit S et al (2016) Nature-based solutions to climate change mitigation and adaptation in urban areas: perspectives on indicators, knowledge gaps, barriers, and opportunities for action. Ecol Soc 21:39. https://doi.org/10.5751/ES-08373-210239
2. Vallecillo S, Kakoulaki G, La Notte A et al (2020) Accounting for changes in flood control delivered by ecosystems at the EU level. Ecosyst Serv 44:101142. https://doi.org/10.1016/j.ecoser.2020.101142
3. Brillinger M, Dehnhardt A, Schwarze R, Albert C (2020) Exploring the uptake of nature-based measures in flood risk management: evidence from German federal states. Environ Sci Pol 110:14–23. https://doi.org/10.1016/j.envsci.2020.05.008
4. Loos JR, Rogers SH (2016) Understanding stakeholder preferences for flood adaptation alternatives with natural capital implications. Ecol Soc 21:32. https://doi.org/10.5751/ES-08680-210332
5. Santoro S, Pluchinotta I, Pagano A et al (2019) Assessing stakeholders' risk perception to promote nature based solutions as flood protection strategies: the case of the Glinščica river (Slovenia). Sci Total Environ 655:188–201. https://doi.org/10.1016/j.scitotenv.2018.11.116
6. Janssen S, Vreugdenhil H, Hermans L, Slinger J (2020) On the nature based flood defence dilemma and its resolution: a game theory based analysis. Sci Total Environ 705:135359. https://doi.org/10.1016/j.scitotenv.2019.135359
7. Nesshöver C, Assmuth T, Irvine KN et al (2017) The science, policy and practice of nature-based solutions: an interdisciplinary perspective. Sci Total Environ 579:1215–1227. https://doi.org/10.1016/j.scitotenv.2016.11.106
8. Gutman J (2019) Commentary: Urban Wetlands Restoration as NBS for Flood Risk. Nat-Based Flood Risk Manag Priv Land Discip Perspect Multidiscip Chall 127. https://doi.org/10.1007/978-3-030-23842-1_13
9. Venkataramanan V, Lopez D, McCuskey DJ et al (2020) Knowledge, attitudes, intentions, and behavior related to green infrastructure for flood management: a systematic literature review. Sci Total Environ 720:137606. https://doi.org/10.1016/j.scitotenv.2020.137606
10. Qiu J (2019) Effects of landscape pattern on pollination, pest control, water quality, flood regulation, and cultural ecosystem services: a literature review and future research prospects. Curr Landsc Ecol Rep 4:113–124. https://doi.org/10.1007/s40823-019-00045-5
11. Junker B, Buchecker M (2008) Aesthetic preferences versus ecological objectives in river restorations. Landsc Urban Plan 85:141–154. https://doi.org/10.1016/j.landurbplan.2007.11.002
12. Frantzeskaki N (2019) Seven lessons for planning nature-based solutions in cities. Environ Sci Pol 93:101–111. https://doi.org/10.1016/j.envsci.2018.12.033
13. Andersson E, Borgström S, McPhearson T (2017) Double insurance in dealing with extremes: ecological and social factors for making nature-based solutions last. In: Kabisch N, Korn H, Stadler J, Bonn A (eds) Nature-based solutions to climate change adaptation in urban areas: linkages between science, policy and practice. Springer, Cham, pp 51–64. https://doi.org/10.1007/978-3-319-56091-5_4
14. Arfaoui N, Gnonlonfin A (2020) Supporting NBS restoration measures: a test of VBN theory in the Brague catchment. Econ Bull 40:1272–1280

15. Ryffel AN, Rid W, Grêt-Regamey A (2014) Land use trade-offs for flood protection: a choice experiment with visualizations. Ecosyst Serv 10:111–123. https://doi.org/10.1016/j.ecoser.2014.09.008
16. Huang Y, Tian Z, Ke Q et al (2020) Nature-based solutions for urban pluvial flood risk management. WIREs Water 7:e1421. https://doi.org/10.1002/wat2.1421
17. Macháč J, Trantinová M, Zaňková L (2020) Externalities in agriculture: how to include their monetary value in decision-making? Int J Environ Sci Technol. https://doi.org/10.1007/s13762-020-02752-7
18. McVittie A, Cole L, Wreford A et al (2018) Ecosystem-based solutions for disaster risk reduction: lessons from European applications of ecosystem-based adaptation measures. Int J Disaster Risk Reduct 32:42–54. https://doi.org/10.1016/j.ijdrr.2017.12.014
19. Liberecký kraj (2020) Charakteristika vodních toků (Characteristics of watercourses). https://povodnovyportal.kraj-lbc.cz/charakteristiky-vodnich-toku. Accessed 20 Aug 2020
20. Šebesta D (2016) Geomorfologické poměry povodí Olešky (Geomorphological conditions of river basin of the Oleška river. https://www.vcm.cz/documents/1043/02-sebesta_prace_a_studie_23_2016.pdf. Accessed 10 Aug 2020
21. Czech Statistical Office (2020) Počet obyvatel v obcích – k 1.1.2020 (The number of residents in municipalities – as of January 1st 2020). https://www.czso.cz/csu/czso/pocet-obyvatel-v-obcich-k-112019. Accessed 23 Aug 2020
22. Czech Technical University in Prague (2020) Studie erozně-odtokových poměrů v povodí "Oleška". (Study of erosion-runoff conditions in the Oleška river basin). http://storm.fsv.cvut.cz/projekty/strima-ii/?lang=en. Accessed 10 Aug 2020
23. Liberecký kraj (2020) Rozsah ohrožení (Extent of a threat). https://povodnovyportal.kraj-lbc.cz/rozsah-ohrozeni. Accessed 23 Oct 2020
24. Povodňový portál (2020) Historické zkušenosti s povodněmi (Historical experience with floods). https://www.vop-povodnovyportal.cz/povodnovy-plan/semily-548/zkusenosti-s-povodnemi. Accessed 23 Oct 2020
25. Povodňový portál (2020) Stará Paka. https://www.povodnovyportal.cz/povodnovy-plan/stara-paka-495/. Accessed 5 Nov 2020
26. Vrabcová (2017) Protipovodňový plán města Semily (Flood protection plan of the town of Semily) . https://www.semily.cz/assets/File.ashx?id_org=14724&id_dokumenty=6831. Accessed 5 Nov 2020
27. STRIMA II (2020) Charakteristika povodní (Flood charasteristics). https://www.strima.sachsen.de/download/Charakteristika_povodni_final.pdf. Accessed 23 Oct 2020
28. Vojtěch M (2015) Územní plán Stará Paka (Zoning plan of Stará Paka). https://www.starapaka.cz/soubory/prilohy/2019/2019-05-15_OPATRENI.pdf. Accessed 5 Nov 2020
29. Koutová A (2012) Územní plán Košťálov (Zoning plan of Košťálov). https://www.semily.cz/customers/semily/ftp/File/uzemni_planovani/uzemni_plan_obce/kostalov/07_oduvodneni_UP_Kostalov.pdf. Accessed 5 Nov 2020
30. Bauer M, Dostal T, Krasa J et al (2019) Risk to residents, infrastructure, and water bodies from flash floods and sediment transport. Environ Monit Assess 191:85. https://doi.org/10.1007/s10661-019-7216-7
31. Czech Technical University in Prague (2019) Catalog of nature-based flood protection measures (Katalog přírodě blízkých protipovodňových opatření). Czech Technical University in Prague, Prague. http://storm.fsv.cvut.cz/data/files/STRIMAII/katalogPBPO.pdf. Accessed 6 Apr 2020. Accessed 5 November 2020
32. Daniels B, Zaunbrecher BS, Paas B et al (2018) Assessment of urban green space structures and their quality from a multidimensional perspective. Sci Total Environ 615:1364–1378. https://doi.org/10.1016/j.scitotenv.2017.09.167
33. Casado-Arzuaga I, Madariaga I, Onaindia M (2013) Perception, demand and user contribution to ecosystem services in the Bilbao metropolitan greenbelt. J Environ Manag 129:33–43. https://doi.org/10.1016/j.jenvman.2013.05.059

34. Lindemann-Matthies P, Briegel R, Schüpbach B, Junge X (2010) Aesthetic preference for a Swiss alpine landscape: the impact of different agricultural land-use with different biodiversity. Landsc Urban Plan 98:99–109. https://doi.org/10.1016/j.landurbplan.2010.07.015
35. Lancaster KJ (1966) A new approach to consumer theory. J Polit Econ 74:132–157. https://doi.org/10.1086/259131
36. McFadden D (1973) Conditional logit analysis of qualitative choice behavior. In: Zarembka P (ed) Frontiers in econometrics. Academic Press, New York, pp 105–142
37. Louviere JJ, Hensher DA, Swait JD (2000) Stated choice methods: analysis and applications. Cambridge University Press, Cambridge
38. Hauber AB, González JM, Groothuis-Oudshoorn CGM et al (2016) Statistical methods for the analysis of discrete choice experiments: a report of the ISPOR conjoint analysis good research practices task force. Value Health 19:300–315. https://doi.org/10.1016/j.jval.2016.04.004
39. Vojáček O, Pecáková I (2010) Comparison of discrete choice models for economic environmental research. Prague Econ Pap 19:35–53. https://doi.org/10.18267/j.pep.363
40. Arnberger A, Eder R (2015) Are urban visitors' general preferences for green-spaces similar to their preferences when seeking stress relief? Urban For Urban Green 14:872–882. https://doi.org/10.1016/j.ufug.2015.07.005
41. Macháč J, Hekrle M, Meyer P et al (2020) Cultural ecosystem services and public preferences: how to integrate them effectively into Smart City planning? In: 2020 smart city symposium prague (SCSP). pp 1–6
42. Derkzen ML, van Teeffelen AJA, Verburg PH (2017) Green infrastructure for urban climate adaptation: how do residents' views on climate impacts and green infrastructure shape adaptation preferences? Landsc Urban Plan 157:106–130. https://doi.org/10.1016/j.landurbplan.2016.05.027
43. Chou R-J (2016) Achieving successful river restoration in dense urban areas: lessons from Taiwan. Sustainability 8:1159. https://doi.org/10.3390/su8111159
44. Martinez-Juarez P, Chiabai A, Suárez C, Quiroga S (2019) Insights on urban and periurban adaptation strategies based on stakeholders' perceptions on hard and soft responses to climate change. Sustainability 11:647. https://doi.org/10.3390/su11030647
45. Wong-Parodi G, Klima K (2017) Preparing for local adaptation: a study of community understanding and support. Clim Chang 145:413–429. https://doi.org/10.1007/s10584-017-2088-8
46. Keesstra S, Nunes J, Novara A et al (2018) The superior effect of nature based solutions in land management for enhancing ecosystem services. Sci Total Environ 610–611:997–1009. https://doi.org/10.1016/j.scitotenv.2017.08.077
47. Duan J, Wang Y, Fan C et al (2018) Perception of urban environmental risks and the effects of urban green infrastructures (UGIs) on human Well-being in four public green spaces of Guangzhou, China. Environ Manag 62:500–517. https://doi.org/10.1007/s00267-018-1068-8
48. Bubeck P, Botzen WJW, Aerts JCJH (2012) A review of risk perceptions and other factors that influence flood mitigation behavior. Risk Anal 32:1481–1495. https://doi.org/10.1111/j.1539-6924.2011.01783.x
49. O'Donnell EC, Lamond JE, Thorne CR (2017) Recognising barriers to implementation of blue-green infrastructure: a Newcastle case study. Urban Water J 14:964–971. https://doi.org/10.1080/1573062X.2017.1279190
50. Vávra J, Lapka M, Cudlínová E, Dvořáková-Líšková Z (2017) Local perception of floods in the Czech Republic and recent changes in state flood management strategies. J Flood Risk Manag 10:238–252. https://doi.org/10.1111/jfr3.12156
51. Häfner K, Zasada I, van Zanten BT et al (2018) Assessing landscape preferences: a visual choice experiment in the agricultural region of Märkische Schweiz, Germany. Landsc Res 43:846–861. https://doi.org/10.1080/01426397.2017.1386289
52. Spegel E (2017) Valuing the reduction of floods: public officials' versus citizens' preferences. Clim Risk Manag 18:1–14. https://doi.org/10.1016/j.crm.2017.08.003

Legal Implications of Natural Floods Management: Lithuania Case Study

Katažyna Bogdzevič, Marius Kalinauskas, Miguel Inacio, Eduardo Gomes, and Paulo Pereira

Contents

Abstract Floods are a reoccurring global phenomenon. Practical implementation of flood risk management relies heavily on grey infrastructure with a lesser focus on nature-based solutions (NBS). With the rising awareness about ecosystem services, natural flood management measures are getting more attention from scientists and policymakers. Lithuanian authorities in the national flood risk management plan provide four NBS for flood management: afforestation, wetlands restorations, agri-environmental measures, and water retention in urban areas (e.g., ponds). However, implementation of the NBS requires more land than grey infrastructure. In the case of Lithuania, in some instances, the NBS require a change of the land use or impacts upon the private land. Therefore, it can be influenced by the legal regulations related to land use planning and protection of private property rights. The problems may occur in case of the insufficient, incomplete, incoherent, or contradictory legal

K. Bogdzevič
Institute of International and European Union Law, Mykolas Romeris University, Vilnius, Lithuania

M. Kalinauskas, M. Inacio, E. Gomes, and P. Pereira (✉)
Environmental Management Laboratory, Mykolas Romeris University, Vilnius, Lithuania
e-mail: pereiraub@gmail.com

Carla S. S. Ferreira, Zahra Kalantari, Thomas Hartmann, and Paulo Pereira (eds.), *Nature-Based Solutions for Flood Mitigation: Environmental and Socio-Economic Aspects*, Hdb Env Chem (2022) 107: 447–467, DOI 10.1007/698_2021_768,

Published online: 6 May 2021

framework. The analysis revealed that the Law on the Land, the Law on the Special Land Use Conditions, and the Law on the Territorial Planning are insufficiently coordinated, which may influence implementation of the NBS on the private property. The afforestation is foreseen only for the public land. However, those lands are mainly used for the agriculture, and in this case possibilities to change the land use are restricted. A similar problem may occur if the area for the afforestation lays within the protected area. The Law on the Protected Areas allows land use changes only in exceptional circumstances; the flood mitigation is not one of them. The incoherence or incompleteness of law, which can hamper implementation of NBS, can occur not only in Lithuania. Therefore, the study can be a starting point for further investigations and solutions in this regard.

Keywords Afforestation, Agri-environmental measures, Land use planning, Legal regulations, Natural flood management

1 Introduction

Floods are natural phenomena and cannot be entirely prevented. At the same time, they significantly impact societies, the environment, and the economy. There is considerable evidence that the number of extreme events will increase in the future [1]. Different measures can be taken in order to reduce flood risk and its adverse consequences. Flood risk management measures include primarily grey infrastructure (GI) such as pipes or dikes [2]. However, lately, natural flood management (NFM) has received greater attention [3–5]. The NFM can be defined as "the alteration, restoration, or use of landscape features to reduce flood risk" [6]. NFM is based on measures and techniques that work with catchment-wide "natural hydrological and morphological processes, features, and characteristics to manage the sources and pathways of flood waters" [7]. NFM uses natural features such as woodlands, ponds, wetlands, and river meanders [8], as well as saltmarshes [9] or dunes [10]. Therefore, in the literature, it is considered a type of NBS [11]. The use of nature to manage the flood risk is considered as more sustainable [3] and beneficial both for the environment [8, 12] and for human well-being [13]. However, implementation of the NFM can be hindered due to the legal limitations [14], for instance, related to the land use changes. NFM measures include, for instance, re-meandering river channels, planting forests [14], and the creation of wetland habitats [15]. Implementing these measures can lead to land use changes and sometimes interfere with private land. Therefore, the legal framework related to NFM is essential for ensuring its effectiveness and efficiency [16]. The variety of measures and techniques used in NFM can require the application of different laws, for instance, laws regulating spatial planning, forestry, protected areas, agriculture, green area management, transportation, or insurance and private property protection. In order to be successful, the NBS for flood mitigation should consider local socioecological systems [17]. NFM also requires cooperation from landowners and communities [18].

Floods are a recurrent problem in Lithuania [19]. However, prior to implementing the European Union Floods Directive [20], Lithuanian authorities did not undertake any steps towards the planning of flood risk management [21]. The first flood risk management plan (FRMP) [22] has been introduced during the third stage of implementation of the FD. This plan is a supplementary document for the 2017–2023 Water Development Program (WDP) [23] and the Action Plan for its implementation [24]. The FRMP describes flood mitigation measures that Lithuanian authorities plan to implement. However, so far, it remains unclear how the implementation of these measures should operate within the existing laws. This chapter aims to analyze the legal implications of NFM using an example of Lithuania. The authors strive to identify legal gaps and to propose their solutions. The national legal framework is analyzed in the context of its completeness and coherence. The analysis is performed using comparative, functional, and formal-dogmatic legal methods. First, the legal framework has been identified. The analysis of the laws encompassed both the interpretation of the content and the meaning of the particular provisions, their place within the legal framework, analysis of their possible application and relation to each other. This allowed to assess whether the provisions are coherent. Moreover, provisions of particular laws are analyzed regarding the overall aim of the flood management related laws and policies. This aim can be defined as flood protection and management. The doctrinal (formal) analysis of the law is supported by the practical implications of its implementation. This allowed to assess if the legal framework is sufficient and complete. Finally, to some extent, Lithuanian national regulations are compared to the foreign regulations. This analytical framework, applied for the Lithuanian case study, can be potentially used for the further analysis of foreign national laws or even international laws. This would allow to conduct broader comparative analysis in the future.

2 Study Area

Lithuania is located in the northeast part of Europe (Fig. 1) and is bordered in the south by Belarus, Russia, and Poland and in the north by Latvia. It has a total area of 65,300 km^2. Administratively, it is divided into 10 counties, 60 municipalities, and 557 elderships. In 2019 had an approximate population of 2.791.903 inhabitants that are mainly concentrated in the cities of Vilnius (556.767), Kaunas (287.191), and Klaipeda (148.511). The population density is estimated to be 43 inhabitants per km^2. The resident population in Lithuania shows a decreasing trend as a consequence of migration ([25], https://osp.stat.gov.lt/). In 2018, according to Corine Land Cover, the largest land use in Lithuania was Cropland (51.86%), followed by Woodland and Forest (34.75%), Grassland (7.02%), Urban (3.40%), Rivers and Lakes (1.99%), Wetlands (0.87%), and Sparsely Vegetated Areas (0.10%). The most important river basins in Lithuania are Nemunas (46,746 km^2), Venta

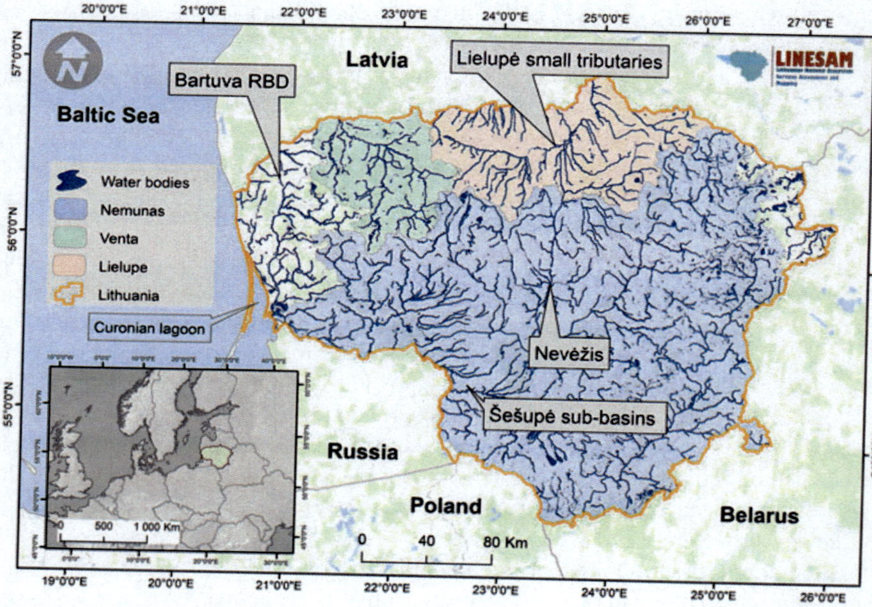

Fig. 1 Study area and the largest basins in Lithuania

(5,179.07 km^2), and Lielupė (8,852.16 km^2) [23] (Fig. 1). Floods in Lithuania are frequent after spring snowmelt and can produce some material losses. Between 1922 and 2010, there is a decreasing trend in the frequency and magnitude of spring floods. Also, as consequence of climate change the flood peak time is changing from March to February [21].

3 Legal Framework in Lithuania

NFM can require changes in the land use, which can affect both state-owned and private land. Therefore, the legal framework includes legal acts related to flood risk management, adopted mainly due to the implementation of the FD in Lithuania, land use planning and related laws, and laws related to private property. Legal acts related to flood risk management are used to identify which NBS Lithuanian authorities plan to implement, and to what extent. The other legal acts are used to analyze the legal challenges or legal constraints that can hamper the implementation of the particular NBS. The analysis includes potential incompleteness and insufficiency of laws as well as possible contradictions and incoherence of the legal framework.

The most important legal acts are presented in Table 1. They have been selected using the following criteria: the law or the legal act makes a direct reference to flood risk management, the legal act includes provisions on land use planning, the legal acts that directly or indirectly affect land use, and the legal acts related to the

Table 1 Legal framework

	Name of the law or the legal act	Scope of application
Legal acts related to flood risk management	Resolution of the Government of the Republic of Lithuania on Approval of 2017–2023 Water Development Program [23]	The water development program sets the goals, objectives, and desired results of the Lithuanian water sector until 2023. These goals and objectives are supposed to be in line with other related policies based on the country's traditions, European Union (EU) legal norms, international conventions, resolutions, agreements, and programs
	Decree of the minister of environment of the Republic of Lithuania and the minister of agriculture of the Republic of Lithuania regarding approval of the action plan for the implementation of the water development program for 2017–2023 [24]	The action plan for the water development program provides a list of measures and timeline of their implementation. Part of the measures applies to the flood risk management
Land use planning and related laws	Law on the Land [26]	The law regulates the relations related to land ownership, land use, as well as land management and administration of the land in Lithuania, its exclusive economic zone and the continental shelf in the Baltic Sea
	Law on the Forests [27]	The law regulates the restoration, protection, and use of forests and creates legal preconditions of management of forests, irrespectively of the ownership
	Law on the Territorial Planning [28]	The law regulates the land use planning in Lithuania, the continental shelf, and the territories of the exclusive economic zone in the Baltic Sea and establishes the rights and obligations of the persons participating in this process
	Law on the Protected Areas [29]	The Law regulates the system of protected areas and associated public relations, the legal bases for the determination and establishment of protected areas, change of boundaries, change of status, protection, management, and control, regulates activities in these areas, and establishes areas of international importance, including the European Ecological Network "Natura 2000" areas, as well as the development and regulation of the natural framework

(continued)

Table 1 (continued)

	Name of the law or the legal act	Scope of application
	Law on the Special Land Use Conditions [30]	The law establishes special land use conditions, specifies the territories in which these conditions are applicable, regulates the establishment of these territories, and establishes the rights and obligations of persons participating in this process
Laws related to private property	Constitution of the Republic of Lithuania [31]	The constitution is the main act in Lithuanian legal framework, it ensures protection of private property
	Civil Code of the Republic of Lithuania [32]	The civil code is the basis for the interpretation of the nature of property law and includes the norms regulating property relations

protection of private property. Therefore, only the legal acts that impact the implementation of the NBS have been presented.

Flood risk management in Lithuania is planned at the national level, and the main institution responsible therein is the Ministry of Environment [33]. Lithuania has prepared one FRMP [22] for all four river basin districts (RBD), identified according to the WFD: the Dauguva RBD, Lielupė RBD, Nemunas RBD, and Venta RBD. The authorities identified 54 river sections in which flood events with significant adverse consequences may occur [22]. All the Baltic Sea coastal areas and the Curonian Lagoon are identified as a potential territory for flood events. This territory is a part of the Nemunas RBD (Fig. 1).

It is worth mentioning that the legal status of the FRMP is not precisely clear. The authorities claim that it is an integral part of the WDP and the Action Plan [22]. However, none of these documents mentions the FRMP [21]. Moreover, differently than WDP and its Action Plan, the FRMP was approved neither by the decree of the minister nor the government's resolution. Therefore, it is important that at least partly the measures provided within the FRMP are incorporated into the WDP and its Action Plan. The FRMP provides general guidelines regarding flood risk management. The other two documents mentioned above give more details regarding the practical implementation of those measures – the territories where the measures will be implemented, and a timeline of implementation.

The FRMP differentiates five types of measures for the flood risk management: preventive measures, engineering flood protection measures (e.g., dikes), non-structural flood protection measures to reduce the existing flood risk by natural run-off management in the river basin, preparedness measures, and recovery measures [22]. The FRMP provides four non-structural flood protection measures: afforestation, restoration of wetlands, agri-environmental measures (e.g., meadow management), and water retention in urban areas (e.g., ponds), which are considered as NBS [34–37]. However, the analysis will focus on three of them, namely:

afforestation, agri-environmental measures, and water retention in urban areas, leaving the restoration of the wetland out of analysis due to the absence of plans by the Lithuanian authorities to implement this measure in the future.

4 Natural Flood Management Measures in Lithuanian Flood Risk Management Plan

NBS for flood management are, in fact, multifunctional and provide several ecosystem services, for instance, biodiversity, drought control, or recreational opportunities [38]. Forest cover is an effective measure to protect from flooding [34]. However, its effectiveness depends on several factors, such as previous land use [39], species selection, and planting type [40]. Moreover, planting trees' costs are immediate, and the effects are uncertain, and its benefits may occur in the indefinite future [41]. Nevertheless, the afforestation's use as a flood management measure is getting more attention in European policy [42, 43]. Afforestation is an NFM measure that is used in other countries. For instance, countries such as Bulgaria and Romania have also introduced afforestation in their flood risk management plans, prepared pursuant to Article 7 of the EU Floods Directive [20, 44].

Therefore, it is not surprising that the Lithuanian FRMP, which is partly implemented through the Action Plan, identifies afforestation as one of the measures to protect against flooding. However, a closer look at the Action Plan reveals that this measure is limited to the state-owned land in the Bartuva RBD, Lielupė small tributaries, Nevėžis, and Šešupė sub-basins (Fig. 1). In this context, it is a great weakness of the Action Plan that it requires to implement measures without conducting a cost–benefit analysis. It requires to identify state free plots in order to plant trees. Nevertheless, the fact that there are state free plots to implement afforestation plans, this does not mean that they are located in areas where flood retention is important. Whereas afforestation as a flood protection measure can be effective only if performed in the proper place. The areas where floods can be retained more effectively can be located in private land.

Another important NBS included in FRMP is the restoration of wetlands, which can play a significant role in flood and water quality management [35]. Wetlands are particularly helpful in storing the water during the snowmelt and high rainfalls [36]. The Lithuanian authorities recognize the value of the wetlands in floods management. However, the extent of the application of this measure is very limited. It is only focused on protected areas and only applied to the extent that has already been implemented. This means that although the authorities understand the significance of restoring wetlands, they do not plan to extend this measure's implementation in the nearest future.

Other NBS are related to agriculture. References to the agri-environmental solutions usually mean sustainable and environmentally friendly agriculture practices [45, 46]. Agri-environmental measures include organic farming, reduction of fertilizers and pesticides, crop rotation, providing biodiversity benefits [46]. Notably, more extensive and sustainable agriculture practices can reduce the risk of flooding and pollution [36]. In the case of Lithuania, the agri-environmental measures, which are supposed to contribute to flood risk management, are provided within the Rules on implementation of the 2014–2020 Rural Development Program's measure "Agri-environment and climate" [47]. The rules mentioned above provide subsidies for the following activities: management of natural and semi-natural meadows, management of specific meadows, extensive management of wetlands, conservation of endangered bird reed canary grass, strips of woody plants or fields on arable land, protection of water bodies against pollution and soil erosion in the arable land, maintenance of drainage ditches slopes [47]. These activities are also included in the programs of the measures for all four RBDs, which are important to maintain the coherence between the legal acts.

Finally, it is important to underline some NBS for the flood management applicable to cities. Due to the increasing density of buildings, climate change, land use changes, flooding in urban areas is becoming a challenging issue [48]. The significant increase of flooding in urban areas is also assigned to the decrease of green areas within the cities and their disconnection [49]. This can be tackled through planning the open green spaces in the cities [48]. It is also beneficial to increase water storage ponds, ditches, or bunds [37]. These spaces have the potential to absorb and retain water, which is important in terms of flood protection and mitigation.

Lithuanian authorities plan to apply afforestation mainly in state-owned land. Agro-environmental measures and urban water retention measures can be applied also on private land [22]. Unfortunately, the FRMP does not envisage any partnership or cooperation between public authorities and private landowners, which can result in difficulties or even impossibility to apply NBS on private land. This, in turn, can significantly limit the extent of NBS implementation for flood management because it requires more land than the traditional grey infrastructure [11]. Flooding in urban areas is rather common in Lithuania, with several urban areas developed in flood-prone locations (pluvial and fluvial). In Fig. 2, it can be observed that the major cities (Kaunas,[1] Klaipeda,[2] and Vilnius[3]) were developed in places naturally vulnerable to pluvial floods (water accumulation areas) and impermeabilized water streams. Since soil sealing increases flood risk [50], without a proper sewage network, it is very likely that these areas will significantly be affected by waterlogging problems. Kaunas and Silute[4] areas are historically affected by fluvial

[1] http://tarpukaris.autc.lt/en/search/image-archive/37/flood-in-kaunas-y-1926-jonavos-str.

[2] https://en.delfi.lt/culture/sleet-inundates-klaipeda-city-south-turns-into-a-lake.d?id=76330011.

[3] https://en.delfi.lt/culture/tuesdays-downpour-flooded-cellars-interrupted-traffic-power-supplies-photos.d?id=75198976.

[4] https://www.15min.lt/en/article/in-lithuania/flooded-silute-district-declares-state-of-emergency-525-293339?fbclid=IwAR3tuPravxuuzL9arwHOI_WGox3mE364_1Wu64IE4qYqy2ot_XMK40xJEr0.

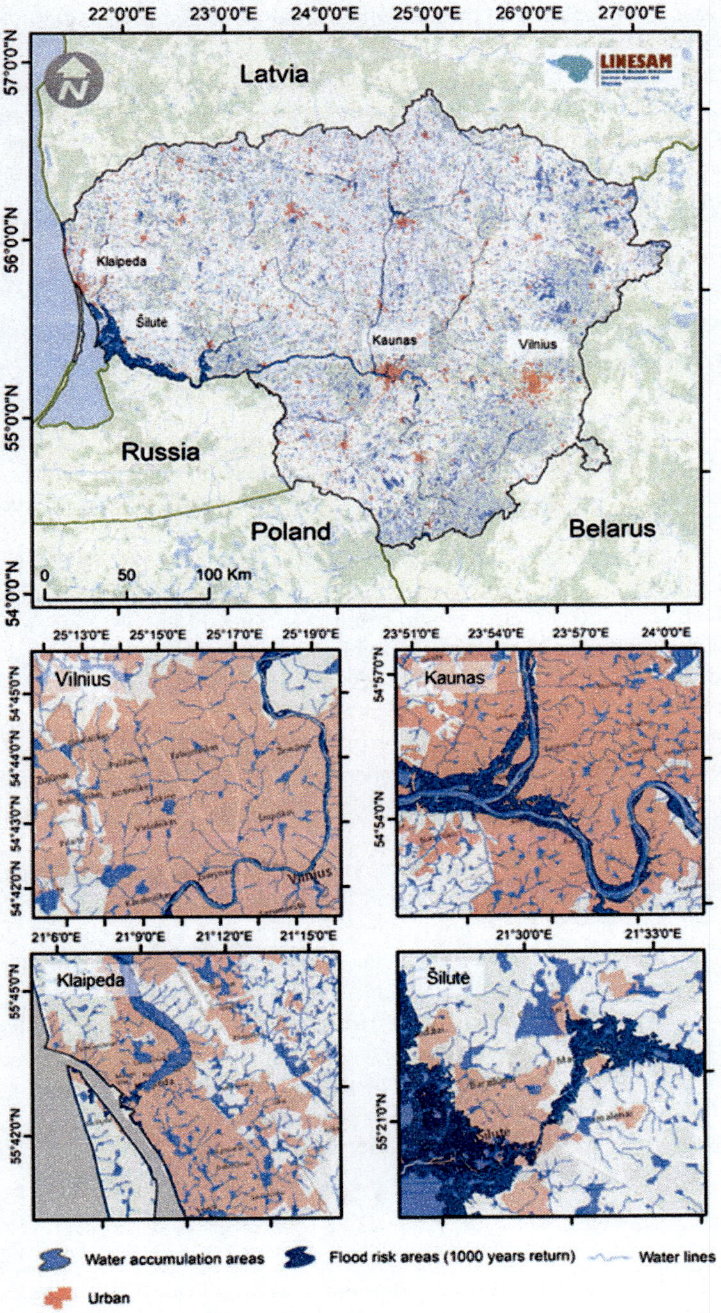

Fig. 2 Build up areas in 2018 (Corine land cover; https://land.copernicus.eu/pan-european/corine-land-cover) located in flood risk (return period of 1,000 years) and water accumulation areas. Water accumulation areas were assessed using the topographic wetness index (TWI) calculated in QGIS

floods, and an important part of these urbaņ areas is located in the 1,000-year return period flood.

In Lithuanian urban areas, floods typically cause property damages, road disruptions, energy supply shortages, and sewage system collapse [51]. Therefore, it is not surprising that national authorities are working on the solutions for managing floods in cities, including NBS oriented measures. The FRMP requires to promote water retention infrastructures in collection basins (e.g., lakes, ponds).

5 Incompleteness and Insufficiency of Law as a Challenge for the Implementation of the NBS

The significance of law in flood management cannot be overestimated neither on international nor national levels [52]. The implementation of any NFM measure may be hampered due to the existence of incomplete, insufficient, contradicting, or incoherent legal acts.

As it was already pointed out, the main weakness of the FRMP, WDP, and its Action Plan are very limited provisions related to the implementation of the NBS on private land. This incompleteness will result in serious difficulties in the possibility of interfering with private property and oblige the owner to implement the NBS in his or her land. Against this background, if the implementation of the NBS is not provided by the law or does not make it possible to refer to the other documents (plans, strategies), which do provide for implementation of the NBS, there will be lack of the legal ground to require implementing NBS on the private property, which is constitutionally protected. However, it is not the only legal loophole that can jeopardize possible NBS implementation.

Constitutional protection of private property, in general, can raise difficulties in the regulation of flood plain development [53]. However, constitutional protection does not protect unrestrictedly private property from any interference of the State therein. For instance, in Germany, the interference can be justified by the public interest that prevails over the private interest [53]. In Poland, private property can be restricted if the public interest requires. However, the Constitution of Poland [54] provides an exhaustive list of the public interests that prevail over the protection of private property [55]. Whereas in Lithuania, the constitutional protection of the private property means that the restriction posed on the use of the private property has not only to be justified by the public interest, but they also shall be clearly provided by the law [56].

Fig. 2 (continued) using 30 m resolution digital elevation model (https://land.copernicus.eu/imagery-in-situ/eu-dem/eu-dem-v1.1). Areas with a TWI index >10 were considered as water accumulation areas

The Law on the Land [26] requires the owners to use the land according to the legally established land use, to follow the provisions of the Law on the Special Land Use Conditions [30], and the provisions of the Law on the Territorial Planning [28]. The Law on the Special Land Use conditions [30] requires considering its provisions during the land use planning process, which is regulated by the Law on the Territorial Planning. The Law on the Special Land Use Conditions [30] for the first time in the Lithuanian legislative history allows establishing special land use conditions for the flood risk territories, which makes imposing restrictions regarding land use easier. Particularly, due to the fact that Article 7 of the Law on the Special Land Use Conditions [30] allows establishing special land use conditions on particular territory without consent of the landowner. Although the legal provisions neither of the Law on the Special Land Use Condition nor the Law on the Territorial Planning [28] do not directly imply a possibility to impose the implementation of the NBS on the land, the use of which is restricted due to the flood risk, the experience of the other countries shows that in such circumstances is easier to encourage the owners to implement NBS [57].

An example from England and Scotland shows that the policymaker can encourage the landowners to change land use by offering them financial incentives [58]. The financial incentives may take the form of tax reduction, subsidies, tradable permits, and other similar measures [59]. However, the Lithuanian authorities provide for such measures to a very limited extent. In fact, only the Rules on implementation of the 2014–2020 Rural Development Program's measure "Agri-environment and climate" introduce incentives for implementing its particular activities, which were presented in the sect. 3 of the given Chapter. The beneficiaries of these incentives are private and legal persons working in agriculture [47]. Therefore, the success of the activities mentioned above will depend on the active (or inactive) participation of private persons in their implementation. Notably, the implementation of this measure is not adjusted to the needs of flood management. The remuneration is being paid for the implementation, for instance, meadows management irrespectively whether it will have any real consequences for the flood management or not. Therefore, the effectiveness of this measure in flood management can be minimal. Its application will primarily depend on accomplishing of the formal requirements to receive the remuneration by the private or legal person. The Rules on implementation of the 2014–2020 Rural Development Program's measure "Agri-environment and climate" [47] do not provide any special conditions or reliefs for the landowners or other users of the land, whose land is more valuable in terms of flood management. To this end situation in Lithuania is similar to other places, for instance, Scotland and England [58]. However, worth noticing that the EU encourages an increased integration of water policy objectives and other European policies such as the Common Agricultural Policy. This integration is particularly important since the CAP reform is a major driver of land use change [60].

The further example allows shifting the focus of the analysis to afforestation and continuing identification of the potential loopholes and insufficiencies in law. The FRMP provides that in Bartuva RBD, Lielupė small tributaries sub-basin, Nevėžis, and Šešupė sub-basins the forest cover is insufficient; therefore, afforestation is

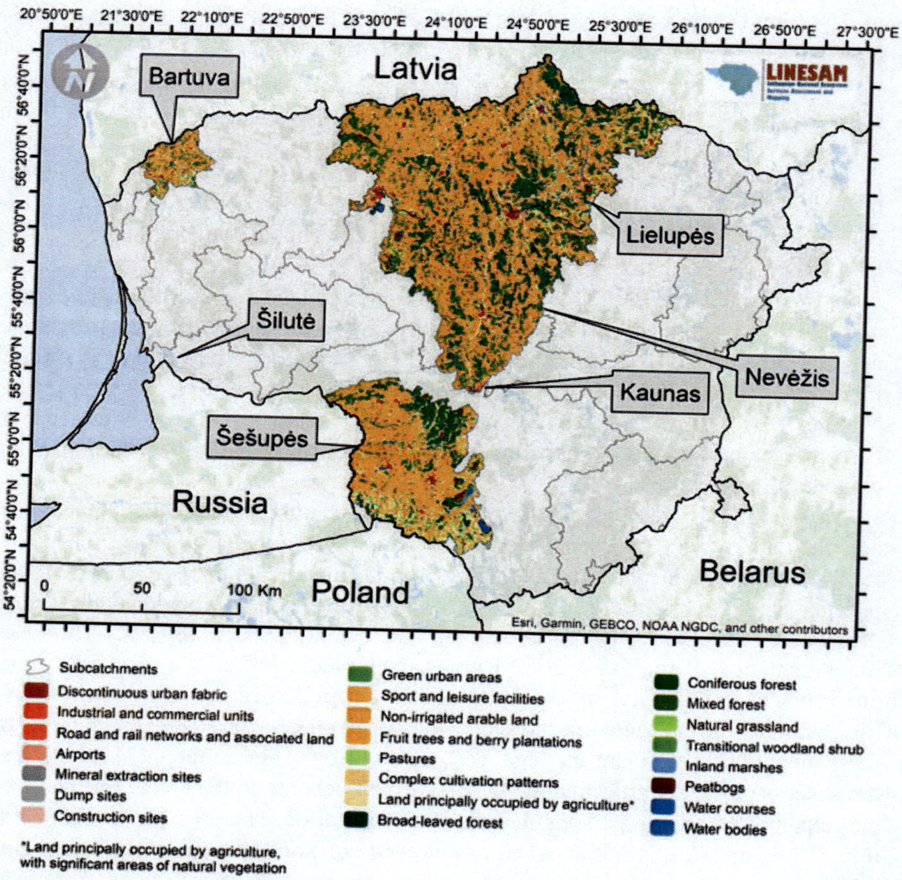

Fig. 3 Land use in Bartuva, Lielupės, Nevėžis, and Šešupės basins in 2018 (2018 Corine land cover data from https://land.copernicus.eu/pan-european/corine-land-cover)

required (Fig. 3). The afforestation as a flood protection measure is foreseen in the WDP, its Action Plan, and the FRMP, however, its implementation depends on further coordination between the legal provisions of the mentioned documents and other legal acts, particularly the Law on the Territorial Planning [28] and the Law on the Construction [61]. None of the two laws neither mentions floods management measures nor refers to the WDP, Action Plan, or even the Law on the Special Land Use Conditions [30]. The flood risk areas as well as territories that are supposed to be afforested have to be considered in the territorial planning process [62]. However, these provisions seem to be insufficient. First, requiring consideration of flood risk areas is not equal to requiring consideration of flood protection measures. The spatial document can include particular territory as flood risk area, which does not mean that the territory will be considered to be afforested. Moreover, presently afforestation is considered in different policies (e.g., WDP, The National Forestry Sector

Development Program 2012–2020 [63]), however, their relation to each other remains unclear. Therefore, it is very likely that afforestation planned as flood protection measure will not occur in the spatial planning documents. Therefore, it can be stated that existing legal framework implementing afforestation for flood management might occur insufficient for the effective implementation of the provided measure.

Implementation of the afforestation has two stages and engages multiple institutions. The first stage requires the State Forest Enterprise, which manages, uses, and disposes the state forests, to form plots, which are located within the territories provided in Fig. 3 and belong to the State Land Fund, which is responsible for the state land management, for the afforestation. The second stage includes planting trees. However, the Action Plan provides insufficient provisions regarding the process of planting trees. The Action Plan mentions planting trees in Bartuva RBD in its point no. 13.4, however, it misses giving further details about the process of planting trees in this RBD in its point no. 13.5. This can mean that until the year 2023, the afforestation in Bartuva RBD will not be implemented. Moreover, the incomplete land reform hampers the process of increasing the country's afforestation. It creates problems related to the transfer of vacant state land fund to state forest managers for forest cultivation (11,457 ha of such land were transferred to forest enterprises in 2001–2010, but only 2,163 ha in 2006–2010, and only 33 ha in 2011).

Finally, incompleteness might be observed considering the implementation of NFM in Lithuanian cities. In order to implement the requirements of the FRMP related to the installation of ponds and other water bodies for the collection of surface water, they must be included in the design and construction conditions. However, Lithuanian legislator did provide the requirements mentioned above neither in the Law on the Territorial Planning [28] nor in the Law on the Construction [61]. Therefore, the "hard law" instruments, which would provide a solid ground for implementing the measures mentioned above, are missing. In such a case, planning ponds in the cities can be hampered because developers will hardly use expensive city territory for other purposes than those strictly required by laws. To summarize, the problem of incompleteness and insufficiency of law occurs mainly regarding the possible implementation of NBS on private property. While regarding the implementation of the NBS on state-owned land, only some insufficiency in legal regulations can be observed.

6 Contradicting and Incoherent Laws as a Challenge for the Implementation of the NBS

In legal theory, coherence of law is strictly linked with its systematic nature, which means that different parts of the whole legal system harmonically link with each other [64] and do not contradict each other [65]. Contradiction and incoherence of law, unwelcome phenomena in general, can impede the implementation of the NBS.

The coherence of law is particularly important in flood risk management since incoherence can result in increased vulnerability to flood events [66]. Therefore, the aim of this Section is twofold: to analyze whether the provisions of particular legal acts, important for the implementation of the NBS, are not contradictory towards each other and if they are parts of a coherent system, particularly whether they contribute to the achievement of the main goal, which is flood risk management. It is worth noting, the main goal mentioned before derives from the legal acts, which are regulating or influencing flood risk management, that can be seen as a system. The analysis of the Lithuanian legal acts revealed several contradictions that can impede the implementation of the NBS.

Potential contradictions can occur if, for the implementation of the NBS, the land use change is required. This is particularly relevant for the afforestation. Due to the finite amount of land, a competition between land uses is inevitable [67]. In some instances, land use changes can be complicated. For example, some of the protected areas are located within the basin and sub-basins that are supposed to be afforested (Fig. 3). Therefore, a potential conversion of the existing land use to the forest can create a conflict between the protected area conservation [29] and planned natural flood management. According to the Lithuanian Law on Protected Areas [29], any activities in the protected area require to consider the provisions of the Law on Special Land Use Conditions [30], the Law on the Environmental Protection [68], the Law on Protected Species of Animals, Plants and Fungi [69], the Law on the Forests [27], the Law on the Territorial Planning [28], and other laws related to environmental protection and environmental impact assessment. The difficulties can arise since afforestation can change environmental conditions and affect, for example, protected habitats. In such a situation as presented above, to achieve coherence of law can be difficult owing to the contradiction of desired behavior (planting forest v. not changing land use). Therefore, the solution could be to compare the costs and benefits of both implementing afforestation and not changing the land use in each particular case.

Moreover, contradictions can occur considering the potential change of the agricultural land to forest. The land use of Bartuva RBD, Lielupė small tributaries sub-basin, Nevėžis, and Šešupė sub-basins, where the afforestation is planned, are shown in Fig. 3. Cropland use (non-irrigated arable land) is the most common in all the catchments. Forest areas have much less cover compared to agriculture. According to the Law on the Land [26], arable land with soil productivity higher than the national average, as well as land with drainage systems, must be used in such a way that its area is not reduced, except for ecologically depleted areas of the natural framework, and soil properties are not impaired. Therefore, although possible, it can be problematic to introduce the afforestation in a land with high productivity.

In the catchments considered for afforestation plans, most land use in a buffer area of 12 m (Table 2) is agricultural (Non-irrigated arable land). This means that these areas' capacity to retain floods is reduced, especially if they are managed with heavy machinery known to increase soil compaction and reduce the capacity to store water [70]. It is important to increase the areas of riparian forests to increase the

Table 2 Land use area in Bartuva, Lielupės, Nevėžis, and Šešupės in 12 m water courses buffer area. For cultivated water courses a buffer area of 12 m is recommended (https://climate-adapt.eea.europa.eu/metadata/adaptation-options/establishment-and-restoration-of-riparian-buffer-s). Data in %. Source: Data from Corine Land Cover 2018 (https://land.copernicus.eu/pan-european/corine-land-cover)

Land use/Catchments	Bartuva	Lielupės	Nevėžis	Šešupės
1.1.2. Discontinuous urban fabric	1.01	1.21	1.00	2.59
1.2.1. Industrial and commercial units	0.16	0.20	0.32	0.45
1.2.2. Road and rail networks and associated land	0.14	0.04	0.12	0.03
1.2.4. Airports	–	0.01	0.14	0.03
1.3.1. Mineral extraction sites	–	0.06	0.02	–
1.3.2. Dump sites	–	0.01	0.06	–
1.3.3. Construction sites	–	–	–	–
1.4.1. Green urban areas	0.14	0.09	0.09	0.01
1.4.2. Sport and leisure facilities	–	0.08	0.09	0.03
2.1.1. Non-irrigated arable land	31.70	41.58	41.43	48.44
2.2.2. Fruit trees and berry plantations	0.12	0.13	0.10	0.10
2.3.1. Pastures	8.61	4.47	4.05	6.64
2.4.2. Complex cultivation patterns	6.66	7.81	4.96	10.88
2.4.3. Land principally occupied by agriculture, with significant areas of natural vegetation	21.45	7.06	5.17	8.21
3.1.1. Broad-leaved forest	8.36	16.37	17.84	4.64
3.1.2. Coniferous forest	3.84	2.17	1.90	5.36
3.1.3. Mixed forest	11.40	11.90	13.58	7.32
3.2.1. Natural grassland	0.03	0.07	0.09	0.04
3.2.4. Transitional woodland shrub	6.18	6.08	8.16	3.67
4.1.1. Inland marshes	0.15	0.16	0.05	0.50
4.1.2. Peatbogs	0.19	0.50	0.83	1.07

capacity of flood retention capacity and reduce the amount of diffuse pollutants that reach water bodies. However, implementing these measures can be difficult to achieve since converting arable land to the forest can raise conflicts with local farmers because riparian areas are the most fertile for agriculture [71, 72]. The solution, in this case, could be rather the implementation of the agri-environmental measures, thereby maintaining fertile soil for agriculture and at the same time increasing it flood mitigation capability.

To summarize, achieving coherence of law can be hindered if the implementation of the NBS requires a change of land use. In such circumstances as presented above, although the particular provisions do not contradict each other, the achievement of the flood protection goal may conflict with other environmental goals, as, for instance, habitat protection.

7 Conclusions

Lithuanian authorities, in their flood risk management plan, provided mainly grey infrastructure measures for flood management. Nevertheless, there are also four NBS included, namely afforestation, agri-environmental measures, ponds for the water storage in cities, and restoration of wetlands. However, the last one will not be implemented in the nearest future, hence the analysis focused on the other three NBS. A further analysis revealed that the implementation of the NBS might be hindered owing to the drawbacks of the legal framework. First, there is an insufficient coordination between the Law on the Special Land Use Conditions, the Law on the Land, and the Law on the Territorial Planning. Particularly the last two should make a clear link to the Law on the Special Land Use Conditions. Such a clear link would allow more easily make restrictions related to the land use on private property, which in turn, as the foreign practice suggests, would allow for convincing easier the landowners to implement the NBS on their land. Moreover, those laws should make clear references to flood management documents: FRMP, WDP, and the Action Plan. In the most populated cities in Lithuania, measures provided within the FRMP can be insufficient and hence ineffective because they are not included in the Law on the Construction. In other words, the regulations requiring installing natural water storage places, such as ponds, are not strict enough to ensure their proper implementation, and the potential developer can be reluctant to use valuable urban land for water storage.

In the case of coherence of law, there are several doubts. There could occur a conflict between the objectives of the FRMP, the Law on the Land, and the Law on the Protected Areas. For instance, it is unclear whether the land identified as a free state-land is the best one for the afforestation. It seems that Lithuanian authorities decided to include some measures without proper evaluation of their future effectiveness. In the case of the afforestation, it is particularly important to plant trees in the areas that can give the best protection against the floods, and these are usually the areas close to the rivers. At this moment, the majority of the land bordering the rivers (namely Nevėžis, Lielupė, and Šešupė) are agriculture lands. The Law on the Land restricts the conversion of the agricultural land to other land use, hence the implementation of the afforestation in such land can be jeopardized. In some instances, the protected areas are located within the basin and sub-basins that are supposed to be afforested. Therefore, afforestation can also be incompatible with the Law of the Protected Areas [29] since it allows the activities in those areas as long as they do not infringe other environmental protection goals, e.g., protection of habitats. Finally, in the case of agri-environmental measures, the state chooses, for instance, to pay private and legal persons for proper meadows management. However, it is unclear if it is beneficial to pay for all the farmers or only for those whose land is important in terms of flood protection. Therefore, even if it may seem that Lithuanian authorities plan to implement several NBS for flood management at first glance, their effective implementation will require changes in the existing legal framework.

The analytical framework applied in this study allowed to assess whether the Lithuanian national law can hamper the implementation of NBS due to its potential incompleteness, incoherence, or insufficiency. Therefore, it can be potentially considered while assessing other national laws related to implementation of NBS in flood affected areas.

References

1. IPCC (2014) Climate change 2014: impacts, adaptation, and vulnerability; IPCC WGII AR5 chapter 23. IPCC. https://www.ipcc.ch/report/ar5/wg2/. Accessed 24 Aug 2020
2. Kundzewicz ZW (2002) Non-structural flood protection and sustainability. Water Int 27 (1):3–13. https://doi.org/10.1080/02508060208686972
3. Wingfield T, Macdonald N, Peters K, Spees J, Potter K (2019) Natural flood management: beyond the evidence debate. Area 51(4):743–751. https://doi.org/10.1111/area.12535
4. Holstead KL, Kenyon W, Rouillard JJ, Hopkins J, Galán-Díaz C (2017) Natural flood management from the farmer's perspective: criteria that affect uptake. J Flood Risk Manag 10 (2):205–218. https://doi.org/10.1111/jfr3.12129
5. Lane SN (2017) Natural flood management. WIREs Water 4(3):e1211. https://doi.org/10.1002/wat2.1211
6. Pescott O, Wentworth J (2011) Natural flood management. POST note 396. Parliamentary Office of Science and Technology. https://researchbriefings.files.parliament.uk/documents/POST-PN-396/POST-PN-396.pdf
7. Forbes H, Ball K, McLay F (2015) Natural flood management handbook. Scottish Environment Protection Agency (SEPA), Stirling, pp 6–130
8. Cook B, Forrester J, Bracken L, Spray C, Oughton E (2016) Competing paradigms of flood management in the Scottish/English borderlands. Disaster Prev Manag 25(3):314–328. https://doi.org/10.1108/DPM-01-2016-0010
9. Möller I, Kudella M, Rupprecht F, Spencer T, Paul M, van Wesenbeeck BK, Wolters G, Jensen K, Bouma TJ, Miranda-Lange M, Schimmels S (2014) Wave attenuation over coastal salt marshes under storm surge conditions. Nat Geosci 7:727–731. https://doi.org/10.1038/NGEO2251
10. Sigren JM, Figlus J, Armitage AR (2014) Coastal sand dunes and dune vegetation: restoration, erosion, and storm protection. Shore Beach 82:5–12
11. Hartmann T, Slavíková L, McCarthy S (2019) Nature-based solutions in flood risk management. In: Hartmann T, Slavíková L, McCarthy S (eds) Nature-based flood risk management on private land. Springer, Cham, pp 3–8. https://doi.org/10.1007/978-3-030-23842-1_1
12. Barber NJ, Quinn PF (2012) Mitigating diffuse water pollution from agriculture using soft-engineered runoff attenuation features. Area 44:454–462. https://doi.org/10.1111/j.1475-4762.2012.01118.x
13. de Bell S, Graham H, Jarvis S, White P (2017) The importance of nature in mediating social and psychological benefits associated with visits to freshwater blue space. Landsc Urban Plan 167:118–127. https://doi.org/10.1016/j.landurbplan.2017.06.003
14. Waylen KA, Holstead KL, Colley K, Hopkins J (2018) Challenges to enabling and implementing natural flood management in Scotland. J Flood Risk Manag 11(S2):S1078–S1089. https://doi.org/10.1111/jfr3.12301
15. Browder G, Ozment S, Rehberger Bescos I, Gartner T, Lange G-M (2019) Integrating green and gray: creating next generation infrastructure. World Bank and World Resources Institute, Washington
16. Sheehan J (2019) Commentary: swapping development rights to prevent flood plain development in flanders: a legal architecture perspective. In: Hartmann T, Slavíková L, McCarthy S

(eds) Nature-based flood risk management on private land. Springer, Cham, pp 99–103. https://doi.org/10.1007/978-3-030-23842-1_10
17. Ferreira CSS, Kalantari Z (2019) Commentary: the Blauzone Rheintal approach from a natural Hazard perspective – challenges to establish effective flood Defence management programs. In: Hartmann T, Slavíková L, McCarthy S (eds) Nature-based flood risk management on private land. Springer, Cham, pp 161–167. https://doi.org/10.1007/978-3-030-23842-1_17
18. Howgate OR, Kenyon W (2009) Community cooperation with natural flood management: a case study in the Scottish Borders. Area 41(3):329–340. https://doi.org/10.1111/j.1475-4762.2008.00869.x
19. Stonevicius E, Valiskevicius G (2018) Identification of significant flood areas in Lithuania. Water Res 45:27–33. https://doi.org/10.1134/S0097807817050116
20. Directive 2007/60/EC of the European Parliament and the Council of 23 October 2007 On the assessment and management of flood risks. Off J EC L 288, pp 27–34
21. Mikša K, Kalinauskas M, Inácio M, Pereira P (2021) Implementation of the European Union floods directive – requirements and national transposition and practical application: Lithuanian case-study. Land Use Policy 100. https://doi.org/10.1016/j.landusepol.2020.104924
22. FRMP (2017) Flood risk management plan for the Nemunas, Lielupė, Venta and Dauguva river basin districts. (Potvynių rizikos Nemuno, Lielupės, Ventos ir Dauguvos upių baseinų rajonuose valdymo planas), Vilnius
23. WDP. Resolution of the Government of the Republic of Lithuania of February 1, 2017 No. 88 On approval of 2017–2023 Water Development Program (Lietuvos Respublikos Vyriausybės nutarimas "Dėl vandenų srities plėtros 2017–2023 metų programos patvirtinimo"), Register of the legal acts (TAR), 2017-02-09, Nr. 2348
24. The Action Plan. Decree of the Minister of Environment of the Republic of Lithuania and the Minister of Agriculture of the Republic of Lithuania. No. D1–375/3D-312 of 5 May 2017 regarding Approval of the Action Plan for the Implementation of the Water Development Program for 2017–2023″ (LR Aplinkos ministro ir Lietuvos Respublikos žemės ūkio ministro įsakymas "Dėl vandenų srities plėtros 2017–2023 metų programos įgyvendinimo veiksmų plano patvirtinimo"), Register of the legal acts (TAR), 2017-05-08, No. 7777
25. Inacio M, Miksa K, Kalinauskas M, Pereira P (2020) Mapping wild seafood potential, supply, flow and demand in the Lithuania. Sci Total Environ 718:137356. https://doi.org/10.1016/j.scitotenv.2020.137356
26. The Republic of Lithuania Law on the Land (Lietuvos Respublikos Žemės įstatymas) No. I-446 of 26 April 1994, Official Journal (Valstybės žinios), 1994, No 34-620
27. The Republic of Lithuania Law on the Forests (Lietuvos Respublikos Miškų įstatymas) No. I-671 of 22 November 1994, Official journal (Valstybės žinios), 1994, No. 96-1872
28. The Republic of Lithuania Law on the Territorial Planning of 12 December 1995, No. I-1120 (Lietuvos Respublikos Teritorijų planavimo įstatymas), Official Journal (Valstybės žinios), 1995, No. 107–2391
29. The Republic of Lithuania Law on the Protected Areas of 9 November 1993, No. I-301 (Lietuvos Respublikos Saugomų teritorijų įstatymas), Official Journal (Valstybės žinios), 1993-11-24, No. 63–1188
30. The Republic of Lithuania Law on Special Land Use Conditions of 2018 No. XIII-2166 (Lietuvos Respublikos Specialių žemės naudojimo sąlygų įstatymas), Register of the legal acts (TAR), 2019-06-19, No. 9862
31. The Constitution of the Republic of Lithuania of 25 October 1992
32. Civil Code of the Republic of Lithuania of 18 July 2000, No. VIII-1864, Official Journal (Valstybės žinios), 2000-09-06, No. 74–2262
33. Flood Risk Assessment and Management Procedure (2009) Resolution of the government of the Republic of Lithuania no. 1558 of November 25, 2009. On Approval of the Description of the Flood Risk Assessment and Management Procedure (Lietuvos Respublikos Vyriausybės nutarimas "Dėl potvynių rizikos vertinimo ir valdymo tvarkos aprašo patvirtinimo"). Off J (Valstybės žinios) 1446376

34. Dadson SJ, Hall JW, Murgatroyd A, Acreman M, Bates P, Beven K, Heathwaite L, Holden J, Holman IP, Lane SN, O'Connell E, Penning-Rowsell E, Reynard N, Sear D, Thorne C, Wilby R (2017) A restatement of the natural science evidence concerning catchment-based 'natural' flood management in the UK. Proc R Soc A. https://doi.org/10.1098/rspa.2016.0706
35. Pattison-Williams JK, Pomeroy JW, Badiou P, Gabor S (2018) Wetlands, flood control and ecosystem services in the Smith creek drainage basin: a case study in Saskatchewan, Canada. Ecol Econ 147:36–47. https://doi.org/10.1016/j.ecolecon.2017.12.026
36. Howe J, White I (2003) Flooding, pollution and agriculture. Int J Environ Stud 60(1):19–27. https://doi.org/10.1080/00207230304746
37. Gunnel K, Mulligan M, Francis RA, Hole DG (2019) Evaluating natural infrastructure for flood management within the watersheds of selected global cities. Sci Total Environ 670:411–424. https://doi.org/10.1016/j.scitotenv.2019.03.212
38. Collentine D, Futter MN (2018) Realising the potential of natural water retention measures in catchment flood management: trade-offs and matching interests. J Flood Risk Manag 11 (1):76–84. https://doi.org/10.1111/jfr3.12269
39. Hümann M, Schüler G, Müller C, Schneider R, Johst M, Caspari T (2011) Identification of runoff processes. The impact of different forest types and soil properties on runoff formation and floods. J Hydrol 409(3):637–649. https://doi.org/10.1016/j.jhydrol.2011.08.067
40. Iacob O, Rowan JS, Brown I, Ellis C (2014) Evaluating wider benefits of natural flood management strategies: an ecosystem-based adaptation perspective. Nord Hydrol 45 (6):774–787. https://doi.org/10.2166/nh.2014.184
41. Dittrich R, Butler A, Ball T, Wreford A, Moran D (2019) Making real options analysis more accessible for climate change adaptation. An application to afforestation as a flood management measure in the Scottish Borders. J Environ Manage 245:338–347. https://doi.org/10.1016/j.jenvman.2019.05.077
42. European Commission (2009) White paper - adapting to climate change: towards a European framework for action, COM/2009/0147, European Commission, Brussels
43. European Commission (2013) Communication from the Commission to the European Parliament, the Council, the European Economic and Social Committee and the Committee of the Regions: An EU Strategy on adaptation to climate change. COM/2013/0216 final, Brussels
44. European Commission (2019) Report from the Commission to the European Parliament and the Council on the implementation of the Water Framework Directive (2000/60/EC) and the Floods Directive (2007/60/EC) Second River Basin Management Plans First Flood Risk Management Plans
45. European Commission (2003) Working document. The water framework directive (WFD) and tools within the common agricultural policy (CAP) to support its implementation, European Commission, Brussels
46. European Commission (2017) Science for Environment Policy (2017) Agri-environmental schemes: how to enhance the agriculture-environment relationship. Thematic Issue 57. Issue produced for the European Commission DG Environment by the Science Communication Unit, UWE, Bristol. Available at: http://ec.europa.eu/science-environment-policy
47. The Rules on implementation of the 2014–2020 Rural Development Program's measure "Agri-environment and climate" approved by the order of the Minister of Environment No. 3D/254 of 3 April 2015 (Lietuvos Kaimo plėtros 2014–2020 metų programos priemonės „Agrarinė aplinkosauga ir klimatas" įgyvendinimo taisyklės) TAR, 2015-04-07, Nr. 5319
48. Schuch G, Serrao-Neumann S, Morgan E, Low Choy D (2017) Water in the city: green open spaces, land use planning and flood management – an Australian case study. Land Use Policy 63:539–550. https://doi.org/10.1016/j.landusepol.2017.01.042
49. Donofrio J, Kuhn Y, McWalter K, Winsor M (2009) Water-sensitive urban design: an emerging model in sustainable design and comprehensive water-cycle management. Environ Pract 11 (03):179–189. https://doi.org/10.1017/S1466046609990263
50. Pereira P, Barcelo D, Panagos P (2020) Soil and water threats in a changing environment. Environ Res 186:109501. https://doi.org/10.1016/j.envres.2020.109501

51. Decree (2012) Decree of the Minister of Environment of the Republic of Lithuania No. D1–23 of 11 January 2012 regarding the approval of the report on preliminary flood risk assessment (LR Aplinkos ministro įsakymas dėl preliminaraus potvynių rizikos vertinimo ataskaitos patvirtinimo), Official Journal (Valstybės žinios), 2012, No. 9–348
52. Cirillo G, Albrecht J (2015) The importance of law in flood risk management. In: Brebbia CA (ed) River Basin management VIII. WITPRESS, pp 91–102. https://doi.org/10.2495/RM150091
53. Tarlock D, Albercht J (2018) Potential constitutional constrains on the regulation of floodplain development. J Flood Risk Manage 11:48–55. https://doi.org/10.1111/jfr3.12274
54. The Constitution of the Republic of Poland of 2 April 1997
55. Frankiewicz A (2009) Konstytucyjna regulacja własności w Rzeczypospolitej Polskiej. Studia Erasmiana Wratislaviensia 3:178–193
56. Ruling of the Constitutional Court of the Republic of Lithuania of 19 September 2002, No. 34/2000–28/01
57. Löschner L (2019) Commentary: a spatial planning perspective on privately funded natural water retention measures. In: Hartmann T, Slavíková L, McCarthy S (eds) Nature-based flood risk management on private land. Springer, Cham, pp 77–81. https://doi.org/10.1007/978-3-030-23842-1_1
58. Rouillard JJ, Ball T, Heal KV, Reeves AD (2015) Policy implementation of catchment-scale flood risk management: learning from Scotland and England. Environ Sci Pol 50:155–165. https://doi.org/10.1016/j.envsci.2015.02.009
59. Jack BK, Kousky C, Sims KRE (2008) Designing payments for ecosystem services: lessons from previous experience with incentive-based mechanisms. Proc Natl Acad Sci 105 (28):9465–9470. https://doi.org/10.1073/pnas.0705503104
60. Fish RD, Ioris AAR, Watson NM (2009) Integrating water and agricultural management: collaborative governance for a complex policy problem. Sci Total Environ 408 (23):5623–5630. https://doi.org/10.1016/j.scitotenv.2009.10.010
61. The Republic of Lithuania Law on the Construction (Lietuvos Respublikos statybos įstatymas) No. I-1240 of 19 March 1996, Official Journal (Valstybės žinios), 1996, No. 32-788
62. Decree of the Minister of Environment of the Republic of Lithuanian No D1–8 of 2 January 2014 on the approval of the Rules on Preparation of Integrated Spatial Planning Documents (LR Aplinkos ministro įsakymas dėl Kompleksinio teritorijų planavimo dokumentų rengimo taisyklių patvirtinimo) Register of the legal acts (TAR), 2014-01-06, No 25
63. Resolution of the Government of the Republic of Lithuania No 569 of 23 May 2012 on approval of the National Forestry Sector Development Program 2012–2020 (LR Vyriausybės nutarimas dėl nacionalinės miškų ūkio sektoriaus plėtros 2012–2020 metų programos patvirtinimo) Official Journal (Valstybės žinios), 2012-05-30, No 61–3058
64. Schiavello A (2001) On "coherence" and "law": an analysis of different models. Ratio Juris 14 (2):233–243. https://doi.org/10.1111/1467-9337.00179
65. Schiavello A (2011) Neil MacCormick's second thoughts on legal reasoning and legal theory. A defence of the original view. Ratio Juris 24(2):140–155. https://doi.org/10.1111/j.1467-9337.2011.00480.x
66. Metz F, Angst M, Fischer M (2020) Policy integration: do laws or actors integrate issues relevant to flood risk management in Switzerland? Glob Environ Change 61:101945. https://doi.org/10.1016/j.gloenvcha.2019.101945
67. Rounswell MDA, Reginster I, Araújo MB, Carter TR, Dendoncker N, Ewert F, House JI, Kankaanpää S, Leemans R, Metzger MJ, Schmit P, Smith P, Tuck G (2006) A coherent set of future land use change scenarios for Europe. Agric Ecosyst Environ 114(1):57–68. https://doi.org/10.1016/j.agee.2005.11.027
68. The Republic of Lithuania Law on the Environmental Protection (Lietuvos Respublikos Aplinkos apsaugos įstatymas) No. I-2223 of 21 January 1992, Official Journal (Lietuvos Aidas), 1992, Nr 20-0

69. The Republic of Lithuania Law on the Protected Species of Animals, Plants and Fungi (Lietuvos Respublikos saugomų gyvūnų, augalų ir grybų rūšių įstatymas) No. VIII-499 of 6 November 1997, Official Journal (Valstybės žinios), 1997-, No. 108-2727
70. Pereira P, Bogunovic I, Munoz-Rojas M, Brevik EC (2018) Soil ecosystem services, sustainability, valuation and management. Curr Opin Environ Sci Health 5:7–13
71. Bornschein A, Pohl R (2018) Land use influence on flood routing and retention from the viewpoint of hydromechanics. J Flood Risk Manag 11:6–14. https://doi.org/10.1111/jfr3.12289
72. Dixon SJ, Sear DA, Nislow KH (2019) A conceptual model of riparian forest restoration for natural flood management. Water Environ J 33:329–341. https://doi.org/10.1111/wej.12425

Sticks, Carrots, and Sermons for Implementing NBS on Private Property Land

Katažyna Bogdzevič and Marius Kalinauskas

Contents

Abstract Floods are natural phenomena that cannot be avoided. Lately, next to the traditional grey infrastructure, countries have started implementing nature-based solutions (NBS) for flood management. NBS need more land than traditional measures, and this land often belongs to private persons. This chapter aims to analyze what are the possibilities to implement NBS on private land. Usually, the governments can achieve their goals using policy instruments that are called "stick," "carrots," and "sermons." In terms of NBS "sticks" are expropriation and land-use restrictions, "carrots" – financial incentives, including payments for ecosystem services, and "sermons" – informational measures. Implementation of "sticks" from the legal perspective is the most complicated because it interferes with property rights protected internationally as human rights. The latter is not absolute, and the state can expropriate or restrict land-use provided the aim of this interference is

K. Bogdzevič (✉)
Institute of International and European Union Law, Mykolas Romeris University, Vilnius, Lithuania
e-mail: kbogdzevic@mrune.eu

M. Kalinauskas
Institute of Communication, Mykolas Romeris University, Vilnius, Lithuania
e-mail: m.kalinauskas@mruni.eu

Carla S. S. Ferreira, Zahra Kalantari, Thomas Hartmann, and Paulo Pereira (eds.), *Nature-Based Solutions for Flood Mitigation: Environmental and Socio-Economic Aspects*, Hdb Env Chem (2022) 107: 469–488, DOI 10.1007/698_2021_762,

Published online: 6 May 2021

justified, lawful, and proportionate. "Carrots" can be effective; however, they require long-term investments from the state. Whereas "sermons" are rather supplementary instruments supporting the implementation of "sticks" and "carrots."

Keywords Human rights, Nature-based solutions, Private land, Property rights

1 Introduction

Floods are natural phenomena with which humankind deals for centuries. Floods cause both people and economic losses, and it is projected that flood-related damages will increase in the future [1]. Among the reasons why flood-related damages continue to increase are intensive human activities in floodplains [2]. These activities include increased urbanization, reduction of forests and wetlands [2], and intense agriculture [3]. Therefore, reducing developments in floodplains can significantly reduce future flood damages [4]. Modern flood risk reduction strategies often include nature-based solutions (NBS) for flood mitigation. However, their implementation requires more space than traditional defense-oriented infrastructures [5].

It happens quite often that the land needed for the NBS is privately-owned [5]. Ownership rights are protected by the law, and interference into these rights, including floodplain development, can raise legal problems, particularly if the property rights are under constitutional or quasi-constitutional protection [6]. Implementation of NBS on state-owned land creates fewer difficulties than if private property is involved [7]. In the latter, the state can impose on landowners the restrictions through land-use planning, zoning [8], or expropriation [9], which would eventually lead to the implementation of NBS on privately-owned land. Implementation of NBS on private land can also be based on a voluntary basis. It is applied when the landowner decides to implement NBS on the private land [10]. In any case, an implementation of NBS on private property would need both legal justification for imposing on landowners the duties related to the land-use and strategies for encouraging the owners to implement NBS on their land.

This chapter aims to analyze different legal issues related to private property that can potentially hamper the implementation of NBS. Different policy instruments were considered. Those instruments are often classified as "carrots, sticks, and sermons" [11]. In this configuration, the "sticks" means – regulation, the "carrots" – economic means, and the "sermons" – information [11]. Figure 1 presents measures supporting the implementation of NBS on private land, dividing them into "sticks," "carrots," and "sermons." Further analysis follows this division.

First, attention will be paid to international, particularly human rights-related legal issues of protecting private property and possibilities for the authorities to interfere in private land using "sticks." For this purpose, both national and international legal provisions will be analyzed. Further, existing practices on how to deal with legal issues in order, on the one hand, to respect the rights of the landowners

STICKS	CARROTS	SERMONS
• Expropriation • Restrictions on land use • Pre-emption rights	• Compensations • Payments for ES (including subsidies) • Buying the land from landowners • Charges (e.g., tax reduction) • Land consolidation schemes	• Information • Knowledge sharing • Partnership for NBS

Fig. 1 Measures supporting the implementation of NBS on private land

and, on the other hand, to implement the NBS on private land will be analyzed. Therefore, it can be stated that the research will focus both on the detection of legal problems and their potential solutions. The analysis includes also other measures the so-called carrots and sermons, which are less coercive comparing to "sticks." This can be beneficial for the countries that are and will be implementing NBS for flood mitigation and the international authorities (e.g., the European Union) while creating new strategies and policies for flood protection.

The chapter presents possible obstacles that different states encounter and possible solutions. However, it does not claim to present an exhaustive list of neither. Moreover, greater attention has been paid to the protection of property as a human right since this issue in the context of the implementation of NBS seems to be overlooked by scholars.

2 Methodology

The analysis provided in this chapter is based on law, relevant case-law, policy documents, and literature. A search for the European Court of Human Rights' relevant case-law was conducted using the HUDOC database (https://hudoc.echr.coe.int/eng). Only the most important case-law was selected. Case-law of the other institutions and national laws was analyzed using the literature.

Legal provisions regarding private property have been qualitatively analyzed regarding their meaning, purpose (function), and possible application in the implementation of NBS. It is worth mentioning that the analysis encompasses both direct references to the particular legal acts as well as to their interpretation provided by different scholars and by relevant case-law. This analysis allowed to identify the possible scope of the protection of private property and more specific obstacles that can occur while implementing the NBS.

In order to present possible national solutions for the implementation of NBS on private land, several examples from the literature have been chosen. In order to identify relevant studies, keywords: nature-based solutions, private property, natural flood management were used. It is important to recognize that many studies addressing nature-based solutions for flood management mentioned private property regulation as potentially problematic but they did not elaborate on that issue [12, 13]. Therefore, only those studies were chosen for the analysis that provided at least minimal references to a possible solution regarding the implementation of NBS on private land. Minimal references mean that authors at least mention how to deal with the private property issues while implementing NBS even if it is not the main topic of their work. Solutions mentioned in the selected examples were divided according to different policy instruments that allow to reach the governmental goals (Fig. 1).

3 Application of Coercive Measures ("Sticks") for Implementation of NBS: In the Context of the Protection of Property Rights as Human Rights

"Sticks" are regulatory means, which can directly or indirectly hinder the implementation of NBS for flood management on private property. "Sticks" are considered to be measures that imply "a command-and-control strategy," and behaviors contradicting those measures can be unlawful [14]. It still needs to be answered how far the intervention to the private land can go using the coercive measures while implementing the NBS. This is an issue that can be discussed both from an international and national standpoint. An international level is important since it provides standards that the states have to comply with. For instance, in case of land-use and property rights, the land-use laws are national but the property rights are protected by international law, which usually prevails over the national law. National standards in this regard have to comply with international, and therefore, further, the bigger attention is paid to the latter.

Expropriation of private land for public goals can be mentioned first among the coercive regulatory measures [9]. In this case – first – does not mean that the authorities will apply it in the first place but from the perspective of the protection of private property it is the most coercive measure and the most interfering with the property rights, thus it is discussed first. Further examples will show, it is not the first measure that the state authorities will use. From the legal standpoint, expropriation is not an easy measure to apply. The expropriation can be de jure and de facto [6]. De jure expropriation shifts the property ownership and the land initially belonging to an individual person becomes an ownership of the state. In this situation, the state purchases the land from the owner for the public purpose [6]. The de facto (also called "indirect") expropriation can be considered if the state regulates the use of private property the way that it diminishes its value [15]. In any circumstances, the expropriation has to be considered the last resort measure [9]. Inclusion of expropriation into flood management or related plans can cause resistance within the society [16]. Even more than practical difficulties, an expropriation can raise legal concerns due to the protection of private property.

The regulatory measures also include obligations and constraints deriving from the land-use planning and zoning regulations, as well as pre-emption rights. They can also be considered "sticks" since they affect how the land can be used and, for instance, whether the owner will be able to get a profit from it or not. Therefore, further considerations are relevant not only in case of expropriation but also other regulatory constraints related to private land.

Provisions on the protection of private property can be found both in national laws, mainly in national constitutions [6] and international law. Article 17 of the Universal Declaration of Human Rights [17] stipulated that everyone has the right to own property and "*no one shall be arbitrarily deprived of his property.*" None of the further main global documents in the domain of human rights repeated this declaration [18]. Nevertheless, property rights are affected by international law, and particularly rules safeguarding human rights and investments [19]. On the regional level, Protocol No. 1 [20] to the European Convention of Human Rights and Fundamental Freedoms (ECHR) [21] safeguards the property right. Property is also protected under Article 21 of the American Convention on Human Rights [22]. Article 1 (Protection of property) of Protocol No. 1 states:

> "Every natural or legal person is entitled to the peaceful enjoyment of his possessions. No one shall be deprived of his possessions except in the public interest and subject to the conditions provided by law and by the general principles of international law. The preceding provisions shall not, however, in any way impair the right of a State to enforce such laws as it deems necessary to control the use of property in accordance with the general interest or to secure the payment of taxes or other contributions or penalties."

The European Court of Human Rights (ECHR) had already many occasions to elaborate on the meaning of the provisions provided above. The ECHR, in its case-law, interprets principles of the right to property according to the provisions mentioned above. The state can interfere in the exercise of the property right; however, the interference has to be lawful, proportionate, and justified by a public or general interest [23]. To consider interference as lawful, it has to have a legal basis, protecting the owner from arbitrariness [24]. Another important requirement is a legitimate general or public interest that needs to be protected [23]. Finally, it is essential to assess whether the restrictions imposed on the owner are proportionate to the protection of public interest that is pursued by the state [25]. It is worth mentioning that the ECHR first examines if the lawfulness and public interest occur and only an affirmative answer allows the Court to assess the proportionality [23]. The other question to be answered is: whether the implementation of the NBS can constitute such a legitimate general interest that would justify interference (expropriation, restrictions related to land-use) of the state authorities on the private land? Moreover, provided the answer is positive, whether protection of this interest is proportional to the interference on private land. To answer a question posed above a more in-depth look at the ECHR case-law is required. Particularly important are cases related to the environment and land-use planning.

For several occasions, the ECHR has been confronted with the conflicting values of the protection of the environment, on the one hand, and the right to property on the other [26, 27]. The ECHR has confirmed that although the ECHR does not provide general protection of the environment, it can be considered to be in the public interest [28]. On the contrary, housing development, for instance, cannot be considered strong public interest as environmental protection [29]. Therefore, it is worthy to assess whether the implementation of NBS can be considered as a public interest. The ECHR has explained that what is in a "public interest" has to be assessed by the national authorities since they know better about their society's needs [25]. According to the ECHR, the regional planning and environmental protection, where the society's general interest is at stake, give the state a wider margin of appreciation than when exclusively private interests are at stake [30]. Implementation of NBS for flood management encompasses both environmental protection elements [31] and protection of the society from floods [32]. NBS for flood management are important for the society's general interest, not exclusively private interest. It can be concluded that the implementation of NBS meets the criterion of a legitimate public interest. Following that the state could have a margin of appreciation to enforce the laws that will restrict the use of private land to implement the NBS. However, the latter is true provided the other criteria from Article 1 of Protocol No. 1 are met.

Another important criterion is the lawfulness. The existence of a legal basis in the national law is insufficient to meet the criterion mentioned earlier [23]. According to the ECHR, the law should be of a certain quality, requiring it to be "accessible to the persons concerned, precise and that the consequences of its application be foreseeable" as well as there should be "compatibility with the rule of law which includes freedom from arbitrariness" [33]. In order to meet the criterion of lawfulness, the interference has to conform with relevant legislation, for instance – land-use planning or designed to protect the environment [34]. Moreover, sometimes national laws can require the appropriate level of legislation if it interferes in a private property. For example, in Lithuania, any private property restrictions might be imposed only by law [35]. To ensure that the laws are known to the persons they relate to, it might be necessary to publish them in official journals or make them publicly available in other ways. The ECHR does not put strict requirements in this regard [23]. Foreseeability of law is no less important. In terms of implementation of NBS it can be achieved by making the information regarding, for instance, the flood risk areas publicly available. The latter can be easily achieved by publicly providing the information required due to the implementation of the EU Floods Directive (FD) [36], particularly – flood risk and flood hazard maps and flood risk management plans.

The last important criterion is proportionality; namely, there must be a "fair balance" between the public interest and the requirement of the protection of the individual's right to private property [23]. It is important to emphasize that a question of a fair balance is raised by the ECHR only if the earlier mentioned criteria of lawfulness and the public interest are met [37]. The value of this criterion cannot be overestimated since it is often decisive for evaluating whether Article 1 of Protocol No. 1 has been violated [23]. The proportionality test allows determining the extent to which the interference of the state restricted a person's exercise of property rights. Subsequently, these restrictions can be balanced against the importance of a legitimate public interest that justified the interference. The ECHR does not provide an exhaustive list of factors that might be considered for the purpose mentioned above. Usually, the ECHR assesses whether the authorities acted in good time and in "an appropriate and consistent manner" [23].

Some factors can be particularly important while conducting a proportionality test in case of restriction on private land resulted from the implementation of NBS. Among those factors is a possibility for a person whose property rights were affected by the state's interference to challenge the measure that was taken [27]. Another factor to be considered is whether the measures taken by the authorities were possibly the least intrusive [38]. Moreover, the compensation provided by the state is an important factor while assessing if the measures taken by the state were proportional. The ECHR usually considers that proportionality of applied measures is achieved if the compensation that has been paid for the owner of the property corresponds to its market value at the time of the expropriation [39]. The compensation shall cover the loss of the land itself and lost income if the owner of the land was pursuing business activities on it [40].

The ECHR acknowledges that the state has a broad margin of appreciation if it restricts private property use due to its land-use planning laws [41]. Moreover, restrictions imposed on landowners due to urban or regional planning are considered to be in general or public interest [41]. The state is still obliged to strike a fair balance between the protected public interest and the right to private property. However, the lack of compensation in such cases is insufficient to consider the measures as disproportionate [30]. The right to compensation will depend on the strength and significance of the public interest at stake [31]. It is important to emphasize that the ECHR case-law reveals that the Court, whilst analyzing the proportionality, always considers the individual cases' circumstances.

Worth mentioning that property rights protected by the Article 21 of the American Convention on Human rights [22] can also be restricted due to the "interest of society" if the restrictions are established by law and justly compensated [42].

The earlier analysis allows concluding that compliance with the criteria of lawfulness and public interest when considering the implementation of NBS for flood management can be achieved. At the same time, the criterion of proportionality requires more in-depth reflection. Is it possible to achieve a "fair balance" when interference to private land is required due to the implementation of the NBS? This question does not have an easy or unambiguous answer for several reasons. Unlike in the case of traditional engineered flood management structures, so far, there are no uniform guidelines providing standards neither for implementation nor for evaluation of the effectiveness of NBS [43]. This is why, it can be challenging to evaluate if the NBS, which require more land than traditional grey infrastructure, will provide similar or better result in terms of flood management [5]. In other words, it can be difficult to justify NBS using a positive cost–benefit ratio [5]. This circumstance can create an obstacle to restrict the use of private land for the purpose of implementation of NBS since it could be complicated to prove that NBS are the least intrusive measures that could be adopted for flood protection. As an additional justification, that the use of NBS for flood management and related restrictions posed on the private land can strive a "fair balance", can serve the environmental benefits provided by the NBS. It is claimed that other benefits of NBS should be considered in the cost–benefit analysis, which would enable “a more holistic comparison to traditional engineering approaches” [43].

The ECHR stated for several times that “in today’s society, the protection of the environment is an increasingly important consideration” [30]. Following that, if the implementation of NBS would encompass both the aim of the protection of the environment and flood protection to achieve a “fair balance” between property restrictions and the aim pursued by the state would be easier. Restrictions on the use of private land would be easier to justify if they were based on a state’s compliance with its international obligations, such as those under EU law. The burden of the state’s duty would be more significant since the implementation of the international obligation would serve the national and international community’s interests.

4 Measures for the Implementation of NBS for Flood Management on Private Land in the Practice of Different Countries

There is hardly one way for all states regarding the implementation of NBS on private land. This is particularly relevant in applying the coercive measures due to the differences in legal protection of private property in different countries and the percentage of the land belonging to the private persons thereof. For instance, in Scotland, around 50% of the land is privately-owned [44]. In such situation, the implementation of NBS can be subject to cooperation with hundreds of different landowners, which can be more complicated than implementing NBS on one owner’s land [45]. Despite the possible difficulties lately, the states are shifting slowly from the traditional flood protection measures using mainly grey infrastructure to nature-based solutions [46]. One of the issues that may occur is an impossibility to apply legal solutions authorizing the installation of the traditional flood protection measures to ensure natural flood management [13]. This can result from uncertainty regarding the efficiency of the nature-based flood protection measures [46]. Therefore, it seems even more valuable to analyze how different countries deal with implementing NBS on private land.

Analysis of the selected examples of implementation of NBS revealed that the measures used for NBS implementation on private land are similar in different countries (Table 1). The examples were selected from the literature and the most important criterion was whether the selected study addresses an issue of implementing NBS for flood management on private land. The majority of the examples are from Europe, some – from other parts of the world. They are not sought to be a systematic review of case studies in the selected area. The aim of the table below is to illustrate the theoretical considerations with practical examples.

Table 1 International experiences in implementing NBS on private land

Continent	Country	City/region	NBS or similar – land requiring – measures	If/how implemented on private land	Source
Africa	Mozambique	Quelimane and Nacala	Revegetation (particularly reforestation), soil bunds, rehabilitation of natural drains/streams. Goals of measures to alleviate flood and erosion risks; specifically, to attenuate peak flows and to control runoff volumes	Issues of private property not raised. However, it has been noted that engagement of the local community in the planning process is important—The land used for NBS – Owned mainly by the municipality	CES Consulting Engineers Salzgitter GmbH;Inros Lackner SE [51]
Asia	Bangladesh	Dhaka City	Measure suggested by the authors: Retention ponds as fisheries-recreational spots	As an economically feasible solution, purchasing the land by the state	Faisal et al. [52]
Europe	The Czech Republic	The western margin of the České Středohoří Mountains	Grasslands, orchards, small pools to support the retention function of the terrain	Measures are implemented on the private land by the initiative of the owner himself. In order to avoid complicated permitting procedures he created the pools of a size that did not require any permissions	Slavíková and Raška [10]
		Pilsen	Urban wetlands	Mainly implemented on publicly owned land. However, there are plans to implement on private land. For this purpose, the state is buying the land from private persons. However, the authorities are not willing to spend too much money on buying land. Therefore, the authors suggest that better communication and presentation of multiple co-benefits of NBS both to stakeholders and the authorities could help. As alternative measure, the authors suggest payments for ecosystem services	Macháč and Louda [53]

(continued)

Table 1 (continued)

Continent	Country	City/region	NBS or similar – land requiring – measures	If/how implemented on private land	Source
	Austria	Alpine region	Open spaces for flood retention and flood runoff	Implementation of the "Blue Zones" in local land-use plans by (re)zoning areas as open space. Landowners and other affected parties engaged in the planning process	Löschner et al. [48]
	Belgium	Flanders	Flood prone open area policy. An "open space area" – Can be designated for nature, agriculture, soft recreation, development either prohibited or limited	Government-based initiatives (land-use planning, zoning; reallocation of development rights) with financial compensation; land readjustment. To some extent planning on the flood prone areas became a responsibility of the Flemish government, which "which releases more local governments from making unpopular decisions in complex spatial implementation plan procedures." Drawback: High market prices and the financial burden for the state could be too significant	Crabbé and Coppens [49]
	Germany	North-Central Germany halfway between Hamburg and Berlin – the biosphere reserve Elbe-Brandenburg River landscape	Dike relocation. Turning agricultural land back into floodplain forest	Property issues have been solved through a land consolidation scheme. The land was purchased by the state. Tenants were compensated for the loss of agricultural land. Stakeholders were actively engaged in the planning process	Warner and Damm [54]

	The Netherlands	The eastern part of the Netherlands in the province of Gelderland (the project *Oekense Beek*)	Aim of the project: To increase the buffer capacity of the area; in other words – Increasing the sponge function of the stream.	Different instruments possible to use: (a) communicative instruments – Communication with landowners; (b) expropriation (to the very limited extent, mainly concerns the Natura2000 lands); (c) imposing on the landowners to change the designation of the land (very limited possibilities, concerns compensations); (d) financial instruments (buying or exchanging the land; compensations of lost profit, subsidies, or payment for particular services)	Kaufmann and Wiering [9]
North America	Canada	Great Toronto area	Recreational or green space areas created in the flood prone areas	Expropriation of 337 houses (providing compensations). Law was amended in order to allow the local authorities to acquire land through expropriation for flood management and recreational needs	Doberstein et al. [47]
	USA	California	Urban river reconfigurations (reconfigurations of waterways in order to allow waters to follow more natural paths)	Acquisition-based and regulatory land-use planning techniques were identified as possibly used to deal with implementation of the foresaw measures on the privately-owned land	Jacobson [50]

The most coercive measures – “sticks” – expropriation, land-use restriction through land-use planning, pre-emption rights were not popular among the countries. The expropriation has been used only once [47] and once suggested as a possible ultima ratio measure [9]. In contrast, land-use restrictions were used or suggested as possible to use several times [9, 48–50]. Worth noting that these measures often will require compensation.

Non-coercive measures, in other words, “carrots” and “sermons” are quite usual in dealing with landowners [9]. Usually, “carrots” mean financial incentives, including buying land from the landowners [50], compensations [9], subsidies, or payments for particular ecosystem services (PES) [55], charges, for instance, tax reduction [56]. However, although the compensations schemes are considered “carrots” in the literature [9, 11], it has to be kept in mind that, in many instances, they are closely connected to the expropriation and zoning [57] – with the typical coercive measures. This is the case, for instance, in Belgium [49].

From the authorities’ perspective, the aim of all those measures will be similar, namely to influence the behavior of the landowner to implement the NBS on their land, for instance, using their land for water storage [58]. From the perspective of the landowners, the measures have a different effect. In case of selling the land for the authorities (e.g., in case of the pre-emption rights), they cease to be the owners of the land. In case of other financial incentives, they stay with their land, however, the authorities expect from them particular behavior. Purchasing the land for the implementation of NBS could seem more expensive than other measures. At the same time, they are raising less legal questions since these measures directly transfer land ownership to the state [50]. Purchasing land as a measure for dealing with private land while implementing NBS for flood management was applied or suggested as feasible to apply in case studies (Table 1) from Bangladesh [52], the Czech Republic [53], Germany [54], and the Netherlands [9]. It was the most often occurring measure in case studies (Table 1) used by the authorities or suggested by the authors.

Aside from the purchasing land, other measures like land consolidation and transferable development rights (TDR) can be mentioned. Land consolidation is a regulatory measure which allows exchanging the land parcels [59]. This measure primarily served as a measure supporting farming structures, but with the time it changed into a multifunctional tool [60] that could be used for flood management purposes. In the latter case, the state is acquiring land in order to implement flood protection measures [8]. Possibility of land consolidation depends on the availability of the land [8]. Landowners can be interested in land consolidation since it allows negotiating with the state authorities regarding the compensations [8] and is less restrictive compared to expropriation [9].

TDR unlike land consolidation or land acquisition do not affect the ownership of the land since it is a market mechanism that allows transferring development rights from the area that is not supposed to be developed (“sending areas”) to the area that is designated for the development (“receiving areas”) [61]. TDR measures usually serve preserving environmental, agricultural, esthetic, or historical values [62], but this measure can also be used for flood management [49]. This measure can be attractive for the authorities because it does not require compensating the landowners [49]. In the case of TDR landowners from "sending areas" are compensated by the landowners from "receiving areas," who are acquiring development rights [63]. The shortcomings of the TDR include low motivation of the landowners in participating in TDR [63], which can be overcome by offering additional incentives [49], valuing of development rights [62], and transaction costs [63].

Subsidies and compensations, similarly to TDR, do not affect ownership rights. It is worth emphasizing that in this case, it is a pre-flood the compensation for the harm that landowner endures due to the implementation of flood risk management measures [57]. From the standpoint of the owner, the compensation and subsidies could seem unattractive. For instance, the value of compensation can depend on what burden or loss the landowners endured. It has to be considered that they also benefit from the actions of the state, aiming at flood protection [64]. The compensation can depend on whether the landowner suffered economic disadvantages, for instance, due to restrictions related to agriculture in the floodplain areas, or these were the restrictions related to settled areas as limitations of construction [65]. The actual loss of the landowner can be higher than the compensated one, for instance, if the agricultural land is changed to a forest, the landowner can lose the subsidies for agricultural activities [66]. Even from the perspective of the protection of property as a human right, the compensation shall be lawful and proportionate, which does not presume that 100% of loss needs to be compensated. Moreover, for the authorities, compensation schemes could also raise difficulties in allocating resources [13] since any financial incentives can become for the state too expensive [49], particularly because the authorities have to provide long-term financing mechanism in order to achieve success [67]. In the selected case studies (Table 1), compensation occurs mainly concerning the expropriation [47] and in case of imposing on the landowners to change the land-use [9].

The other incentive-based mechanisms can be placed under the umbrella of PES. These mechanisms can include subsidies, tradable permits, and market friction reductions [56]. PES are usually defined following the Wunder [68] as "(1) a voluntary transaction where (2) a well-defined environmental service (or a land use likely to secure that service) (3) is being 'bought' by a (minimum one) service buyer (4) from a (minimum one) service provider (5) if and only if the service provider secures service provision (conditionality)." PES are voluntarily used beyond the command-and-control scheme [68]. There are different types of measures, however in terms of flood protection, the most relevant is watershed protection-related PES, which according to Kumar et al. [69] "allows participants to pay upstream landowners for best practice land use which limits deforestation and land degradation to reduce risks, such as floods and soil erosion while maintaining aquifer quality." PES can be effective provided the landowners consider them credible [69]. Therefore, effectiveness will depend on the landowners' willingness to implement NBS on their land and the possibilities of the authorities to pay for the landowners for their behavior accordingly since, as it was already mentioned, any financial incentives imply long-term costs for the authorities. Noteworthy that PES, like other policies, can be considered cost-effective if they allow achieving the same results using other policies [56]. Despite possible obstacles, there are already examples of using PES for watershed protection [70]. In the selected case studies (Table 1) PES were not used; however, there were suggestions that they can be one of the measures supporting NBS implementation on private land [9, 53]. For instance, subsidies, typical economic policy instruments [11], can be considered particularly connected with agriculture as a part of the EU Common Agricultural Policy [44].

The effectiveness of any policy can be supported using "sermons" or, in other words, information. According to Vedung and van der Doelen [71], "information as a public policy instrument covers government-directed attempts at influencing people through the transfer of knowledge, communication of reasoned argument, and moral suasion in order to achieve a policy result." Increasing stakeholder awareness can be essential for the effectiveness of flood risk strategies [72], including the implementation of NBS, since the information helps to enhance their credibility. Moreover, information sharing and engaging the private sector in identifying risks as well as response measures and adaptation can help to mobilize more considerable investments in reducing vulnerability [73]. Finally, a partnership for NBS is gaining attention [74, 75]. Partnership assumes collaboration between policy-makers and stakeholders, which helps to create synergies between them by pooling together their knowledge, experience, capacities, and resources [75].

Partnership, engagement of stakeholders, and information and knowledge sharing are considered important in the selected case studies (Table 1). For instance, Macháč and Louda [53] observed that communication between decision-makers and stakeholders could be important for the implementation of NBS. Whereas Warner and Damm [54] noted that even if the landowner is self-motivated, he may need persuasion and formal agreements that would facilitate the involvement in pursuing NBS. Therefore, it seems that "sermons" are not the main but much supportive while implementing NBS on private land.

5 Concluding Remarks

Implementation of NBS in some instances can interfere with the property rights of landowners. The aim of the present study was to identify ways how NBS could be implemented on private land from a legal perspective. Possible ways to deal with the issue at stake include different policy instruments, which in the literature are often called "sticks," "carrots," and "sermons." They differ in terms of coerciveness. The most coercive are "sticks," which in the case of NBS for flood management can take the form of expropriation, land-use planning, or pre-emption rights. These are the instruments that are the most interfering with private land. They can infringe ownership as a human right (e.g., in European countries, parties to the ECHR). The ECHR requires that any state interference into property rights should be lawful, justified by a public or general interest, and proportionate. In the case of lawfulness, it is important where land-use restrictions are, if they are in, e.g., land-use planning laws; if they are precise, foreseeable and not arbitrary. This means that anybody can have access to that law, its application depends on objective criteria, and landowners can know in advance about the possible restrictions.

Public or general interest requires to prove that measures will serve the society or the community, not just a private person's interests. Environmental protection and protection from flooding without any doubts can be considered belonging to the public or general interest sphere. The most problematic can be the proportionality test since it is not always possible to justify that implementation of NBS considering cost–benefit analysis will be as effective as other measures requiring less land. Considering the case-law of the ECHR, implementation of "sticks" will require substantial justification from the authorities, particularly regarding proportionality of chosen measures. Concluding, it can be stated that using "sticks" to implement NBS on private land requires the state authorities to provide legal means that will not be easy to adopt. These would be the measures often limiting constitutional rights and can raise objections of the landowners since those measures will affect their property rights directly in a coercive manner. This is probably why in selected examples "sticks" are rarely used.

The "carrots" and "sermons" raise less legal difficulties since they are not coercive as "sticks." The landowners are either paid (in case of "carrots") or convinced (in case of "sermons") to implement NBS on private land. From the standpoint of the state, these measures have both advantages and disadvantages. Similarly, as in the case of sticks, they require finances, and for the long-term effect – long-term financing, which can raise difficulties in allocating resources. On the other hand, they allow the implementation of NBS having the consent, support, and motivation of the landowners for it. In selected examples, these measures were used or suggested to use multiple times. Therefore, bearing in mind potential difficulties in justifying "stick," the other two – "carrots" and "sermons" seem to be more oriented towards landowners.

So far, the implementation of NBS on private land is lacking sufficient attention from academia. The studies analyzing NBS for flood mitigation rarely address the issue of private land comprehensively. This study is a modest contribution to discussing how NBS for flood protection can be implemented on private land, particularly, what are the possibilities to justify "sticks" in order to comply with international human rights standards. Also the chapter gave an overview of other measures ("carrots" and "sermons") but it is still worth to continue and elaborate on the issue of implementation of NBS on private land in further studies.

References

1. IPCC (2014) Climate change 2014: impacts, adaptation, and vulnerability. IPCC WGII AR5 Chapter 23. IPCC. https://www.ipcc.ch/report/ar5/wg2/. Accessed 17 Nov 2020
2. Kundzewicz ZW, Kanae S, Seneviratne SI, Handmer J, Nicholls N, Peduzzi P, Mechler R, Bouwer LM, Anell N, Mach K, Muirood R, Brakenridge GR, Kron W, Benito G, Honda Y, Takahashi K, Sherstyukov B (2013) Flood risk and climate change: global and regional perspectives. Hydrol Sci J 59:1–28

3. Howe J, White I (2003) Flooding, pollution and agriculture. Int J Environ Stud 60(1):19–27. https://doi.org/10.1080/00207230304746
4. Johnson KA, Wing OEJ, Bates PD, Fargione J, Kroeger T, Larson WD, Sampson CC, Smith AM (2020) A benefit–cost analysis of floodplain land acquisition for US flood damage reduction. Nat Sustain 3:56–62. https://doi.org/10.1038/s41893-019-0437-5
5. Hartmann T, Slavíková L, McCarthy S (2019) Nature-based solutions in flood risk management. In: Hartmann T, Slavíková L, McCarthy S (eds) Nature-based flood risk management on private land. Springer, Cham, pp 3–8. https://doi.org/10.1007/978-3-030-23842-1_1
6. Tarlock D, Albercht J (2018) Potential constitutional constrains on the regulation of floodplain development. J Flood Risk Manage 11:48–55. https://doi.org/10.1111/jfr3.12274
7. Kapovic Solomun M (2019) Commentary: small retention in polish forests from a Forest management perspective—copying of existing could be right path. In: Hartmann T, Slavíková L, McCarthy S (eds) Nature-based flood risk management on private land. Springer, Cham, pp 45–51. https://doi.org/10.1007/978-3-030-23842-1_5
8. Löschner L (2019) Commentary: a spatial planning perspective on privately funded natural water retention measures. In: Hartmann T, Slavíková L, McCarthy S (eds) Nature-based flood risk management on private land. Springer, Cham, pp 77–81. https://doi.org/10.1007/978-3-030-23842-1_8
9. Kaufmann M, Wiering M (2019) Dilemmas of an integrated multi-use climate adaptation project in the Netherlands: the Oekense Beek. In: Hartmann T, Slavíková L, McCarthy S (eds) Nature-based flood risk management on private land. Springer, Cham, pp 193–207. https://doi.org/10.1007/978-3-030-23842-1_21
10. Slavíková L, Raška P (2019) This is my land! Privately funded natural water retention measures in the Czech Republic. In: Hartmann T, Slavíková L, McCarthy S (eds) Nature-based flood risk management on private land. Springer, Cham, pp 55–67. https://doi.org/10.1007/978-3-030-23842-1_6
11. Bemelmans-Videc ML, Rist RC, Vedung E (2010) Carrots, sticks and sermons: policy instruments and their evaluation. Comparative policy evaluation series, p 277
12. Kalantari Z, Ferreira S, Keesstra S, Destouni G (2018) Nature-based solutions for flood-drought risk mitigation in vulnerable urbanizing parts of East-Africa. Curr Opin Environ Sci Health 5:73–78. https://doi.org/10.1016/j.coesh.2018.06.003
13. Waylen KA, Holstead KL, Colley K, Hopkins J (2018) Challenges to enabling and implementing Natural Flood Management in Scotland. J Flood Risk Manag 11(S2):S1078–S1089. https://doi.org/10.1111/jfr3.12301
14. Jagers SC, Harring N, Matti S (2018) Environmental management from left to right – on ideology, policy-specific beliefs and proenvironmental policy support. J Environ Plan Manag 61(1):86–104. https://doi.org/10.1080/09640568.2017.1289902
15. Wagner J (1999) International investment, expropriation and environmental protection. Golden Gate Univ Law Rev 29(3):465–538
16. van Herk SJ, Rijke J, Zevenbergen C, Ashley R (2015) Understanding the transition to integrated flood risk management in the Netherlands. Environ Innov Soc Trans 15:84–100. https://doi.org/10.1016/j.eist.2013.11.001
17. The Universal Declaration of Human Rights adopted by the United Nations General Assembly in Paris on 10 December 1948
18. Katuoka S, Motuziene I (2020) Shareholders' rights in international law: (con)temporary reflections in the Diallo case. Entrepreneurship Sustain Issues 8(1):249–260. https://doi.org/10.9770/jesi.2020.8.1(17)
19. Sprankling JG (2014) The global right to property. Colum J Trans L 52(2):464. https://doi.org/10.1093/acprof:oso/9780199654543.003.0009
20. Protocol no. 1 to the European convention for the protection of human rights and fundamental freedoms of 20 March 1952, European Treaty Series – No 009
21. The European Convention of Human Rights and Fundamental Freedoms (1950)
22. Inter American Convention on Human Rights (1969) San José

23. ECHR (2020) Guide on article 1 of protocol no. 1 to the European convention on human rights. Protection of property. Council of Europe/European court of human rights
24. Mikša K, Kalinauskas M, Inácio M, Gomes E, Pereira P (2020) Ecosystem services and legal protection of private property. Problem or solution? Geogr Sustain 1(3):173–180. https://doi.org/10.1016/j.geosus.2020.08.003
25. ECHR (2016) Béláné Nagy v. Hungary [GC], no 53080/13, 13 December 2016
26. ECHR (2001) Chapman v. the United Kingdom [GC], no. 27238/95, ECHR 2001-I
27. ECHR (2018) GIEM SRL and others v. Italy (merits) [GC], nos 1828/06 and 2 others, 28 June 2018
28. ECHR (2003) Kyrtatos v. Greece, no. 41666/98, ECHR 2003-VI
29. ECHR (2019) Svitlana Ilchenko v. Ukraine, no 47166/09, ECHR 4 July 2019
30. ECHR (2010) Depalle v. France [GC], no. 34044/02, ECHR 2010
31. Moss T, Monstadt J (2008) Restoring floodplains in Europe: policy contexts and project experiences. IWA Publishing, London
32. European Commission (2015) Towards an EU Research and Innovation policy agenda for Nature-Based Solutions & Re-Naturing Cities. Final report of the Horizon 2020 Expert group on 'Nature-based solutions and re-Naturing cities'. European Commission, Brussels
33. ECHR (2016) Ünsped Paket Servisi SaN. Ve TiC. A.Ş. v. Bulgaria, no. 3503/08, 13 October 2015
34. ECHR (1991) Pine Valley developments ltd and others v. Ireland, 29 November 1991, Series A no 222
35. Mikša K, Kalinauskas M, Inácio M, Pereira P (2021) Implementation of the European Union floods directive—requirements and national transposition and practical application: Lithuanian case-study. Land Use Policy 100:104924. https://doi.org/10.1016/j.landusepol.2020.104924
36. Directive 2007/60/EC of the European Parliament and the Council of 23 October 2007 On the assessment and management of flood risks. Off. J. EC, L 288, 27–34
37. ECHR (1999) Iatridis v. Greece [GC], no. 31107/96, ECHR 1999-II
38. ECHR (2017) Vaskrsić v. Slovenia, no 31371/12, 25 April 2017
39. ECHR (2009) Guiso-Gallisay v. Italy (just satisfaction) [GC], no 58858/00, 22 December 2009
40. ECHR (2018) Osmanyan and Amiraghyan v. Armenia, no 71306/11, 11 October 2018
41. ECHR (2004) Gorraiz Lizarraga and others v. Spain, no. 62543/00, ECHR 2004-III
42. Anton DK, Shelton DL (2011) Environmental protection and human rights. Cambridge University Press, New York, p 986
43. van Wesenbeeck BK, IjFF S, Jongman B, Balog S, Kaupa S, Bosche L, Lange GM, Holm-Nielsen N, Nieboer H, Taishi Y, Kurukulasuriya P, Meliane I (2017) Implementing nature based flood protection: principles and implementation guidance (English). World Bank Group, Washington. http://documents.worldbank.org/curated/en/739421509427698706/Implementing-nature-based-flood-protection-principles-and-implementation-guidance. Accessed 12 Dec 2020
44. Wilkinson ME (2019) Commentary: Mr. Pitek's land from a perspective of managing hydrological extremes: challenges in Upscaling and transferring knowledge. In: Hartmann T, Slavíková L, McCarthy S (eds) Nature-based flood risk management on private land. Springer, Cham, pp 69–75. https://doi.org/10.1007/978-3-030-23842-1_7
45. Matczak P, Takács V, Goździk M (2019) Reversing the current: small scale retention programs in polish forests. In: Hartmann T, Slavíková L, McCarthy S (eds) Nature-based flood risk management on private land. Springer, Cham, pp 23–37. https://doi.org/10.1007/978-3-030-23842-1_3
46. Cook B, Forrester J, Bracken L, Spray C, Oughton E (2016) Competing paradigms of flood management in the Scottish/English borderlands. Disaster Prev Manag 25(3):314–328
47. Doberstein B, Fitzgibbons J, Mitchell C (2019) Protect, accommodate, retreat or avoid (PARA): Canadian community options for flood disaster risk reduction and flood resilience. Nat Hazards 98:31–50. https://doi.org/10.1007/s11069-018-3529-z

48. Löschner L, Seher W, Norbeck R, Kopf M (2019) Blauzone Rheintal: a regional planning instrument for future-oriented flood management in a dynamic risk environment. In: Hartmann T, Slavíková L, McCarthy S (eds) Nature-based flood risk management on private land. Springer, Cham, pp 141–154. https://doi.org/10.1007/978-3-030-23842-1_15
49. Crabbé A, Coppens T (2019) Swapping development rights in swampy land: strategic instruments to prevent floodplain development in Flanders. In: Hartmann T, Slavíková L, McCarthy S (eds) Nature-based flood risk management on private land. Springer, Cham, pp 69–75. https://doi.org/10.1007/978-3-030-23842-1_9
50. Jacobson T (2019) Too much water, not enough water: planning and property rights considerations for linking flood management and groundwater recharge. Water Int 44(5):588–606. https://doi.org/10.1080/02508060.2019.1619046
51. CES Consulting Engineers Salzgitter GmbH, Inros Lackner SE (2020) Upscaling nature-based flood protection in Mozambique's cities. World Bank, pp. 278. http://documents1.worldbank.org/curated/en/897311585301464586/pdf/Mozambique-Upscaling-Nature-Based-Flood-Protection-in-Mozambique-s-Cities-Urban-Flood-and-Erosion-Risk-Assessment-and-Potential-Nature-Based-Solutions-for-Nacala-and-Quelimane.pdf. Accessed 6 Dec 2020
52. Faisal IM, Kabir MR, Nishat A (1999) Non-structural flood mitigation measures for Dhaka City. Urban Water 1:145–153. https://doi.org/10.1016/S1462-0758(00)00004-2
53. Macháč J, Louda J (2019) UrbanWetlands restoration in floodplains: a case of the City of Pilsen, Czech Republic. In: Hartmann T, Slavíková L, McCarthy S (eds) Nature-based flood risk management on private land. Springer, Cham, pp 111–126. https://doi.org/10.1007/978-3-030-23842-1_12
54. Warner B, Damm C (2019) Relocation of dikes: governance challenges in the biosphere reserve "river landscape Elbe-Brandenburg". In: Hartmann T, Slavíková L, McCarthy S (eds) Nature-based flood risk management on private land. Springer, Cham, pp 171–180. https://doi.org/10.1007/978-3-030-23842-1_185
55. Collentine D, Futter MN (2018) Realising the potential of natural water retention measures in catchment flood management: trade-offs and matching interests. J Flood Risk Manag 11(1):76–84. https://doi.org/10.1111/jfr3.12269
56. Jack BK, Kousky C, Sims KRE (2008) Designing payments for ecosystem services: lessons from previous experience with incentive-based mechanisms. Proc Natl Acad Sci 105(28):9465–9470. https://doi.org/10.1073/pnas.0705503104
57. van Doorn-Hoekveld WJ, Goytia SB, Suykens C, Homewood S, Thuillier T, Manson C, Chmielewski P, Matczak P, van Rijswick HFMW (2016) Distributional effects of flood risk management – a cross-country comparison of preflood compensation. Ecol Soc 21(4):26–41. https://doi.org/10.5751/ES-08648-210426
58. Howarth W (2017) Integrated water resources management and reform of flood risk Management in England. J Environ Law 29:355–365. https://doi.org/10.1093/jel/eqx015
59. Monstadt J, Moss T (2008) Policy innovations in the aftermath of a disaster: contexts of floodplain restoration in Germany. In: Moss T, Monstadt J (eds) Restoring floodplains in Europe: policy contexts and project experiences. IWA Publishing, London, pp 63–87
60. Mansberger R, Seher W (2017) Land administration and land consolidation as part of Austrian land management. EU Agrarian Law 6(2):68–76. https://doi.org/10.1515/eual-2017-0010
61. Shahab S, Clinch JP, O'Neill E (2018) Timing and distributional aspects of transaction costs in transferable development rights programmes. Habitat Int 75:131–138. https://doi.org/10.1016/j.habitatint.2018.03.006
62. Kaplinsky ES (2018) A Canadian perspective on TDR: you call that a 'market'? In: Gerber J-D, Hartmann T, Hengstermann A (eds) Instruments of land policy dealing with scarcity of land. Routledge, London, pp 243–247
63. Dyca B, Muldoon-Smith K, Greenhalgh P (2020) Common value: transferring development rights to make room for water. Environ Sci Pol 114:312–320. https://doi.org/10.1016/j.envsci.2020.08.017

64. van Doorn-Hoekveld W (2014) Compensation in flood risk management with a focus on shifts in compensation regimes regarding prevention, mitigation and disaster management. Utrecht L Rev 10(2):216–238. https://doi.org/10.18352/ulr.279
65. Hartmann T (2009) Clumsy floodplains and the law: towards a responsive land policy for extreme floods. Built Environ 35(4):531–544. https://doi.org/10.2148/benv.35.4.531
66. Wahren A, Schwärzel K, Feger K-H (2012) Potentials and limitations of natural flood retention by forested land in headwater catchments: evidence from experimental and model studies. Flood Risk Manag 5:321–335. https://doi.org/10.1111/j.1753-318X.2012.01152.x
67. Wilkinson ME, Addy S, Quinn PF, Stutter M (2019) Natural flood management: small-scale progress and larger-scale challenges. Scott Geogr J 135(1–2):23–32. https://doi.org/10.1080/14702541.2019.1610571
68. Wunder S (2005) Payments for environmental services: some nuts and bolts. Occasional paper no. 42. Center for International Forestry Research, Nairobi
69. Kumar P, Kumar M, Garrett L (2014) Behavioural foundation of response policies for ecosystem management: what can we learn from Payments for Ecosystem Services (PES). Ecosyst Serv 10:128–136. https://doi.org/10.1016/j.ecoser.2014.10.005
70. Muñoz-Piña C, Guevara A, Torres JM, Braña J (2008) Paying for the hydrological services of Mexico's forests: analysis, negotiations and results. Ecol Econ 65(4):725–736. https://doi.org/10.1016/j.ecolecon.2007.07.031
71. Vedung E, van der Doelen FCJ (2010) The sermon: information programs in the public policy process choice, effects, and evaluation. In: Bemelmans-Videc ML, Rist RC, Vedung E (eds) Carrots, sticks and sermons: policy instruments and their evaluation. Comparative policy evaluation series, pp 103–128
72. Santoro S, Pluchinotta I, Pagano A, Pengal P, Cokan B, Giordano R (2019) Assessing stakeholders' risk perception to promote nature based solutions as flood protection strategies: the case of the Glinščica river (Slovenia). Sci Total Environ 655:188–201. https://doi.org/10.1016/j.scitotenv.2018.11.116
73. Biagini B, Miller A (2013) Engaging the private sector in adaptation to climate change in developing countries: importance, status, and challenges. Clim Dev 5(3):242–252. https://doi.org/10.1080/17565529.2013.821053
74. Kabisch N, Frantzeskaki N, Pauleit S, Naumann S, Davis M, Artmann M, Haase D, Knapp S, Korn H, Stadler J, Zaunberger K, Bonn A (2016) Nature-based solutions to climate change mitigation and adaptation in urban areas: perspectives on indicators, knowledge gaps, barriers, and opportunities for action. Ecol Soc 2(2):39. https://doi.org/10.5751/ES-08373-210239
75. van Hamm C, Klimmek H (2017) Partnerships for nature-based solutions in urban areas – showcasing successful examples. In: Kabisch N, Korn H, Stadler J, Bonn A (eds) Nature based solutions to climate change adaptation in urban areas, linkages between science, policy and practice. Springer, Cham, pp 275–289. https://doi.org/10.1007/978-3-319-56091-5_16

Socio-Economical Aspects of NBS

Aleksandra Figurek

Contents

Abstract The flood management needs to be undertaken in a more integrated manner. Incorporating the risk in flood management should be synchronized with the adequate measures which give their contribution to the reduction in the damage caused by a natural hazard. In this chapter, a multidisciplinary approach is used for presenting the socio-economic aspects of nature-based solutions (NBS). Implementation of NBS requires a more structured and comprehensive process that starts with the valuation of the services provided by the ecosystem. Several barriers are identified in the socio-economic area connected with the implementation of NBS and flood risks.

In the framework of the institutional setting, more actors or players are involved, with different resources, different values and preferences, and more views and perceptions. To select the most effective combination of measures, stakeholders required adequate analysis, with specific reference to the costs and benefits of the chosen actions.

Keywords Flood risk, Measures, Social innovation, Socio-economic analysis, Taxonomy

---

A. Figurek (✉)
Faculty of Agriculture, University of Banja Luka, Banja Luka, Bosnia and Herzegovina
e-mail: aleksandra.figurek@agro.unibl.org; aleksandrafigurek@yahoo.com

Carla S. S. Ferreira, Zahra Kalantari, Thomas Hartmann, and Paulo Pereira (eds.),
Nature-Based Solutions for Flood Mitigation: Environmental and Socio-Economic Aspects,
Hdb Env Chem (2022) 107: 489–504, DOI 10.1007/698_2021_764,

Published online: 11 April 2021

1 Introduction

European Commission defined NBS as "solutions that are inspired and supported by nature, which are cost-effective, simultaneously provide environmental, social and economic benefits and help build resilience" [1].

The multidisciplinary view of the socio-economic aspects in the connection of NBS underlines the establishment of a holistic approach, implementation of the adequate measures considering their synergies and potential trade-offs. This approach includes recognizing the environmental and socio-economic context of NBS design, implementation, and evaluation (calculation of the costs and benefits, coordinated for different stakeholders); formulation of the NBS in a direction which acts the multiple interconnected challenges; implementing NBS within various scales (stakeholders need to establish flexible management which will answer adequately to flood risks); monitoring of the realized activities [2, 3].

Adequate flood measures should be undertaken from the side of different stakeholders [4] including authorities such as urban planners, water resources engineers, disaster defense authorities, health and social services. Identification of the uncertainty in the direction what decision-actors which are involved in this process will undertake is not always known and make the prediction difficult – whether the decisions pay off or not [5]. This presents the key reason why quantitative risk analysis is often inadequate, and engagement of the different stakeholders is increasingly considered as an important factor in the implementation of the successful management measures [2, 3, 5, 6]. Flynn et al. [7] highlighted the need to account the social risk perception in risk management since the reality perceived affects stakeholders' decisions and could lead to failures in risk management actions [8–11].

Establishing the NBS is also seen as innovations that require engagement with multiple actors, providing co-benefits that bridge social and economic interests and thus can support new green economies and job benefits [12, 13]. Raymond et al. [2, 3] developed a seven-stage process for situating co-benefit assessment within policy and project implementation. They include identification of the problem or opportunity; selection and assess NBS and related actions; design NBS implementation processes; implementation of NBS; frequently engagement stakeholders and communicate co-benefits; transfer and upscale NBS; and monitor and evaluate co-benefits across all stages.

Implementation of NBS requires a more structured process with barriers identified in the socio-economic area.

2 Barriers in Establishing the NBS

The identification of the barriers which are present during the establishing of NBS is crucial for its successful implementation. The adequate implementation of NBS takes place in complex socio-ecological systems, in which the various elements

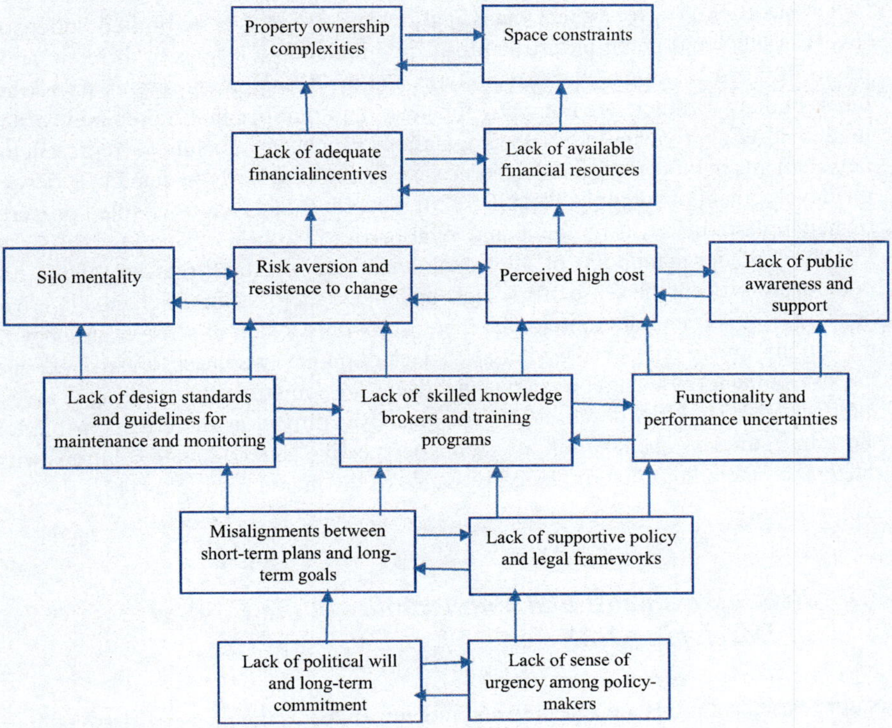

Fig. 1 ISM model, adapted from Sarabi et al. [16]

are highly intertwined [14]. Existing barriers in the NBS implementation are not mutually exclusive, which implies that timely and more effective policies should acknowledge the interdependencies among those solutions. The understanding of the important factors and their causalities can help in the reduction of the barriers and their overcoming [15].

The NBS barriers have also been searched and identified in the framework of several projects (e.g. [17]), including technical aspects of the NBS implementation, stakeholder activities, and various institutional factors. Sarabi et al. [16] defined a taxonomy of barriers by identifying the interdependencies among them. To provide a structural understanding and acknowledgement – how the barriers are related, the next interpretive structural model (ISM) presented in Fig. 1. is recommended.

The NBS barriers affect the uncertainties and election of the best ways to plan, implement, manage, and to monitor NBS activities. In this sense, it is of importance to design appropriate standards and guidelines for managing and timely monitoring of the NBS in different areas, having also in mind specific conditions that follow those areas. These guidelines are crucial to effectively respond to the specific challenges which arise from different resources and institutional characteristics [18].

Droste et al. [19] found that municipal revenue systems have limited options to invest in NBS and green infrastructures. The municipal revenues in different areas are mostly dedicated to social expenditures and the development of gray infrastructures. In that way, there are not many financial opportunities that should be invested in innovative approaches like NBS. Regulations and the law documents represent the key elements in this process and in some cases are not regularly updated in a view of the development of solutions like NBS [20]. Several studies have classified property ownership as a major institutional and legal barrier [21, 22].

From another standpoint of view, politicians are not always aware of the full potential of NBS in the sense of societal challenges, such as climate change in urban areas [23]. It is important to note that there is not only a lack of sense of urgency in the actions among policymakers but also a lack of public awareness toward NBS and its development [24]. Entrepreneurs and citizens are sometimes less willing to invest their resources in NBS because they see local government as more responsible for those investments. The solution for such a perspective is to co-create solutions with citizens by including them in the early stages of the planning process [17].

2.1 Risk Assessments and Vulnerability in the Context of Developing NBS

Nature-based solutions are accepted and recommended as the measures for enabling climate change mitigation and their adaptation to reduce flood risks and also for enhancing urban ecosystems [25, 26]. NBS combine technical, governance, finance, and social innovation, bringing together established ecosystem-based approaches, such as ecosystem services, natural capital, ecological engineering, green-blue infrastructure [27, 28]. The implementation of those measures should act in the direction of the flood difficulties and reduce risks to people and property as effectively as traditional gray infrastructures. Those measures should potentially offer many additional benefits, e.g. improving the natural habitat for wildlife, enhancing water and air quality, improving community sociocultural conditions [29]. The World Bank proposed comprehensive guidelines for the implementation of NBS to reduce flood risk [30]. The NBS framework and proposed guidelines can be implemented throughout NBS phases to assess vulnerability and risk and contribute to successful NBS management and monitoring the achievement of disaster risk reduction.

The risk assessment and vulnerability in the context of the implementation of NBS are interconnected. In most cases, they are used in the context of a wide range of risks to which households are exposed. There is a big concern for reducing welfare losses before they actually happen. The public policy identified flood risk management and vulnerability as central topics of all the broader approaches. Vulnerability is announced as the potential for a given receptor to experience harm during a flood event. It depends on the various elements as the susceptibility of a particular receptor to experience harm during the flood, then the ability of the receptor that has been harmed by the flood event to its recover. Flood risk components are based on the recognition that risks cannot be removed entirely, but only partially, and often at the expense of other societal goals, Sayers et al. [31].

Criteria for determining the risk of floods can be presented through the consequences of the harmful effects of floodwaters on human life and health, as well as on other material goods exposed to those harmful effects. These criteria are related to the identification of receptors or emitters of flood risk. Considering a large amount of data, their diversity, and the different effects of floods, this segment of analysis requires a multidisciplinary approach. In this sense, it is valuable to establish the SWOT (strengths, weaknesses, opportunities, threats)/PESTLE (political, economic, social, technical, legal, and environmental) analysis. With this analysis, there is the possibility to achieve multidisciplinarity to observe the problem of floods from all these aspects. PESTLE allows valorizing the impact of floods and also to identify the elements which are affected by floods.

The PESTLE framework [32] offers the chance to decompose all factors through SWOT analysis (factors on which the flood has a positive and negative impact). The mentioned factors can also be divided based on whether the impact is direct (i.e. whether it originates from the flood itself) or indirect. The identification of the positive influences from the system and environment are recognized as strengths and opportunities and negative are weaknesses and threats. Flood risk management is a policy with an excellent opportunity for establishing a synergy together with other aspects of water management. Risk reduction should be undertaken as part of an adaptive and dynamic decision-making process by which individuals and social factors have a chance to interact more intensively [33]. Perception of the stakeholders and understanding of natural disasters are socially constructed [34]. This is the reason why the differences in risk perception could lead to conflicting situations hampering the effectiveness of risk management measures [35]. Institutions should support decision-making processes in which multiple actors interact to deal with partially conflicting interests and also to resolve the social dilemmas [36]. The role of the institutions is recognized as key support for humans use to organize all forms of repetitive and structured interaction [37, 38].

NBS for reducing the flood risk show promising results in terms of risk reduction [39, 40]. The implementation of NBS proposed a set of general principles which have the focus on balancing ecosystem conservation as well as socio-economic benefits [41]. Timely monitoring is required to ensure that the implemented NBS continue to deliver the required risk reduction benefits in the long run.

2.2 Social Adaption Towards Understanding the Flood Risk

Understanding the flood risk and its factors has crucial social and political implications having in mind that the level of awareness of flood risk influences people's actions before and during a flood [42]. Such an attitude enables the creation of

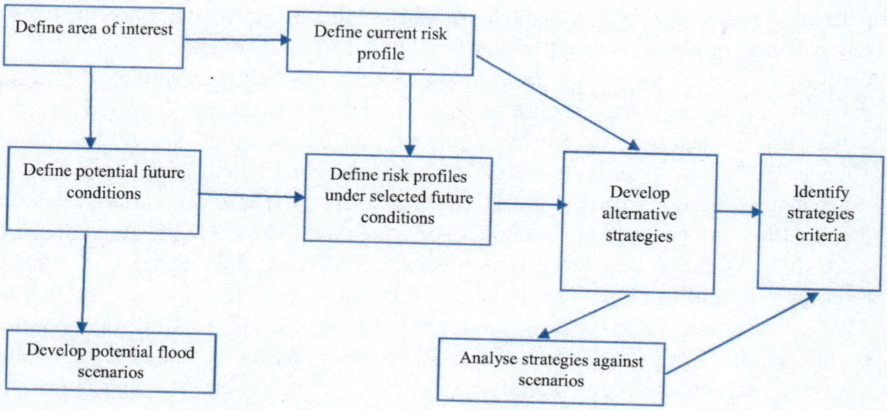

Fig. 2 System risk model (adapted from [31])

inclusive two-direction dialogue between the public and government about the main aspects of increasing the conditions for preparedness and effective responses for the flood. This presents the background for the policy-makers and institutional entities to establish the effective strategies which are in direction with the public expectations and present the way how to be accepted by the broader community (Fig. 2).

System risk models (both qualitative and quantitative) have an important role in the evaluation of the performance of a different range of measures towards future scenarios. The effectiveness of strategies in flood risk management is highly dependent on stakeholders' views and attitudes. This plays a critical role in terms of how individuals and institutions act to risk mitigation. As part of the decision-making process, competent authorities have the opportunity to adopt the transdisciplinary approach which includes scientific and cognitive-psychological knowledge to ensure that risk policy is neither purely science-based nor purely value-based [43]. Institutional entities need to find the best approach to integrate public views and societal values into the process which follows the risk analysis [44]. There is a need to facilitate communication between affected areas and communities with all stakeholders involved in flood relief efforts. On that road, the institutional entities need to adapt and sometimes to modify their strategies and continually reinforce risk perception and preparedness. Important aspects which need to be contained in this procedure are related to the media. Information provided by the mass media and communication channels as a source of indirect experience can influence risk perception [45–49].

Cologna et al. [50] established the framework for analyzing the effects of the media and its influence on risk perception, and how it affects adaptation policy, preparedness, and communication between stakeholders. Flood as a natural hazard creates a "window of opportunity" [51, 52] where the media has the potential and can make a significant contribution to risk education. This view is essential for the future activities of all connected stakeholders. The media and political entities also learn from existing experience and come to the knowledge of how risk perception communicates with the public about the projected impacts of flood risk.

The basic position of this concept presents the risk, and it is of great importance to understand its multiple dimensions. The main factors which influence the risk perception include previous experience of flood events; the crucial information provided by the mass media or communication channels; and trust in relevant institutions and authorities responsible for the flood defense measures [49].

3 Socio-Economic Analysis

The socio-economic analysis is often applied in the process of valorization of the disaster impact and their social and economic aspects (loss of human lives, destruction of housing and infrastructure, suspension of traffic and supply chains, etc.). Identification and measurement of areas under flooding risk are possible to realize through a socio-economic model, which allows comparison of the socio-economic vulnerability of different areas. This model consists of a set of indicators grouped into several categories and are presented in Fig. 3 [53].

The social aspect encompasses capacity between communities that affects their ability to adequately deal with natural hazards [54–57]. Social and demographic characteristics, such as population, housing, and land rights, present the important aspects of the protective measures [58, 59]. The proportion of house ownership [60] and access to land or land ownership [60, 61] are crucial indicators of social susceptibility and demonstrate the community's predisposition to experience damage to their homes or land due to natural hazards. The Human Development Index (HDI) as a composite social indicator is usually measured at the national level and represents overall social contexts [62].

Measures related to economic dimensions are more connected with the level of support, vitality, flexibility, and timely response of the community economy. Economic memory has often been given far more important than social memory, especially in disaster research. This is because of the easier accessibility and comparability of financial and economic information and the difficulty in the representation of subjective factors in the measurement framework [63, 64].

Economic indicators mostly include gross domestic product (GDP), poverty, and employment rate. In a certain respect, some of the economic indicators (e.g. GDP) may not be quantifiable at the local level, but they are adequate indicators for the planning and realization of measures that affect the economic strength of a community. Applied methods for assessing the co-benefits of NBS need to take into account the changing dynamics of the system at a variety of geographic and temporal scales [2, 3, 65–68].

Economic indicators that are directed towards resources, including return on investment and "profit" margins, require additional analysis. Indicators that quantify the cost–benefit of the future measures in addition to the involvement of stakeholders can enable a cost-added-value quantification [69].

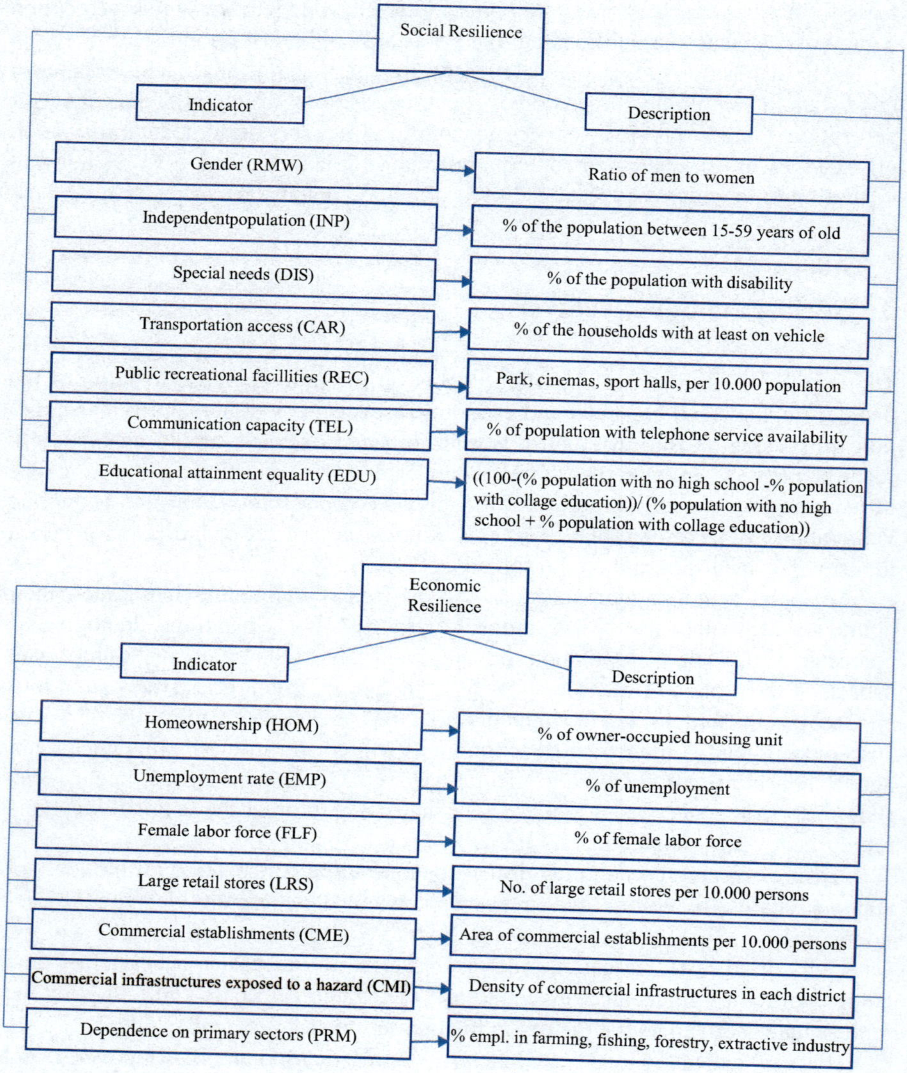

Fig. 3 Indicators for social and economic resilience dimensions, adapted from Moghadas et al. [53]

4 Integrated Assessment Models (IAMs)

Integrated assessment modeling (IAM) is a concept that is applied when categorizing and applying different approaches to the modeling process that aims to assess the impact of floods and the damage it causes. It is important to note the fact that integrated assessment implies and includes the development of methods for expanding the basic framework from several disciplines, with an emphasis on socio-economic and biophysical processes. Research and understanding of cause-and-effect relationships within several disciplines related to the perception of finding ways to overcome the negative effects are extremely important and the contribution of this concept [70–73]. The effects of natural hazards can have significant consequences on the economy and are included in the IAM. Due to climate change, it is necessary to continuously assess and present the disaster model, i.e. the risk assessment expected due to these changes [74].

Socio-economic and climate-related information should be strongly linked, to ensure a timely and adequate response from all stakeholders. In this sense, it is important to establish an interdisciplinary approach that can combine several different models. This procedure aims to develop an adequate methodology that will be directed towards better valorization and reduction of uncertainty that occurs due to variable climatic conditions. Modeling extreme weather events in flood-affected areas is of great importance for drawing up plans and institutional strategies. The economic scope of quantifying the consequences of harm includes models that aim to accurately assess and analyze all indicators that can contribute to a timely response and mitigation of future harm. It is also proposed to include other economic models, as well as partial and general equilibrium models [75].

The integrated climate assessment model, ICAM [76] includes different integrated modules: demographics and economics (the economy module); energy (the energy module); atmospheric composition and climate (the climate module); impacts of climate change (the damage module); and an intervention module. The economic dimension is mainly reflected through the adoption of economic models used to quantify the resources used, including management models.

The category that is an indispensable part of IAM is river flooding since it has a significant economic impact. Therefore, it is necessary to approach it with a sophisticated concept that will be based on global modeling and risk assessment of river floods with a detailed spatial solution [77]. Nordhaus [78, 79] gave his contribution to the development of IAMs and made attempts to improve these models to aid policymakers in making decisions about climate policy in the face of climate change uncertainty [80]. The accounting for river flood risk in a climate-economy IAM [81–83] is important as flood protection standards that reduce the probability of a river flooding. Anthoff and Tol [84] gave their contribution in creating the climate framework for uncertainty, negotiation, and distribution (FUND) model to estimate the river flood damage (account for flood protection scenarios to implement economically optimal protection). Ignjacevic et al. [85] made progress in estimations that are connected with the economic variability due to climate change. In addition, a CLIMRISK-RIVER flood risk model [85] is being developed and incorporated into the integrated climate economy (IAM) assessment model. With this model, it is possible to operate at the local level and with the preventive design of damage to river floods caused by climate change and different socio-economic scenarios of the flood adaption.

5 Integrated Early Warning Systems (EWS) Based on Micro-Network Frameworks

The establishment and functioning of an early warning system for water networks aim to reduce physical, social, and economic losses. Such early warning systems are reflected in an adequately developed technological infrastructure (micro-network) and include support for data processing and analysis using natural disaster forecasting (micro-network). They are a key factor in providing information that enables people and organizations to prepare in a timely and adequate manner for the launch of actions, which should prevent or mitigate disaster [86].

Water research modalities and its development are in direction of establishing significant solutions that primarily relate to water resources management, as well as solving problems related to unpredictable climate change. In this intention, the flood early warning systems (FEWS) were developed within the Urban Flood Project as a solution in monitoring the networks which are installed for flood defenses [87]. FEWS are based on the information and simulation results which are covered through an interactive decision support system. This helps dike managers, city authorities, and individuals to be informed in order to act timely in the case of emergency. FEWS provide a basis for stakeholders and local authorities to be able to strategically plan their activities and propose appropriate and optimal structural and non-structural measures for a given area, in terms of different climate change, urban development, and fire scenarios (Fig. 4).

6 Conclusions

Floods cause enormous harmful consequences, with the greatest damage being lost lives. Recurrence of floods, i.e. its frequency, indicates the need for detailed research. Future research should be more focused on the understanding of socio-economic dynamics in flood disasters since they represent a significant segment in the analysis of the impact of floods. The framework needs to be defined before floods and to include answers to questions related to the identification of hazards, risks, vulnerabilities, as well as measures to be followed to mitigate adverse effects. The plan should include segments relating to communities, the location of evacuation centers, evacuation routes, and the flood recovery process. Evacuation plans need to follow details about hydrological analyzes, to determine the degree of increase in water levels for different amounts of precipitation for a specific area. When used in conjunction with a flood warning system, those measures can help prevent loss of life and reduce flood damage. In this sense, it is necessary to organize presentations, workshops, written material, and leaflets so that residents are aware of flood risks and measures to reduce flood risks. Flood risk management should be undertaken in an organized and integrated manner. Integrated risk management should include all measures that contribute to reducing flood damage. To increase the chances of full

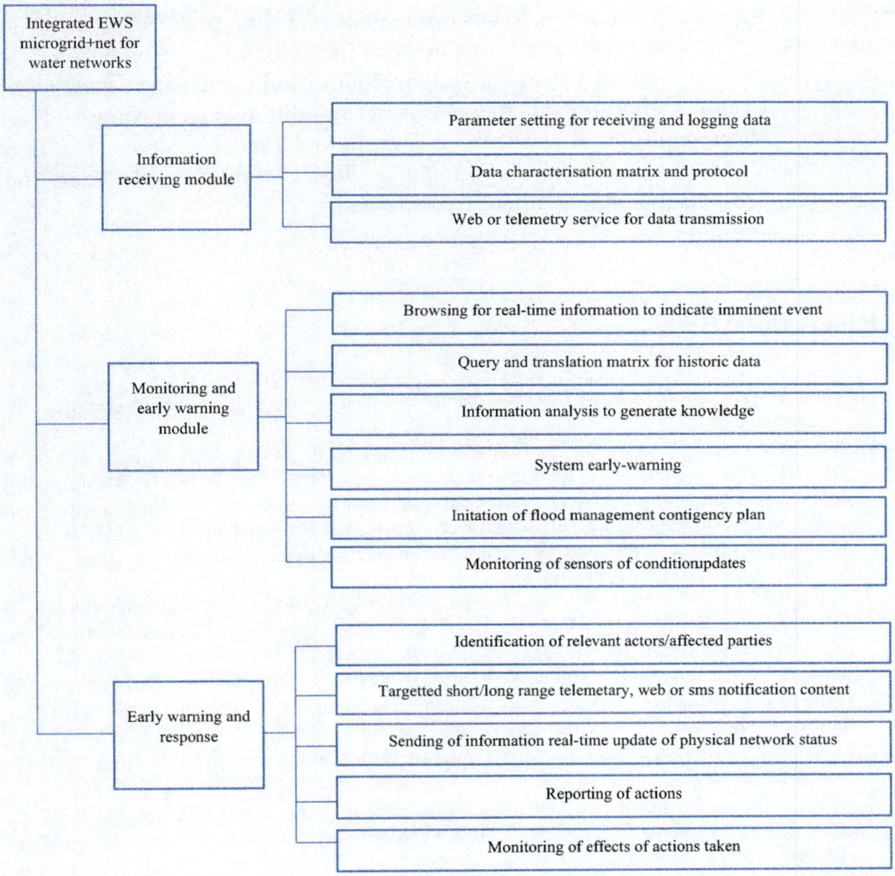

Fig. 4 Integrated EWS micro-network (adapted from [88])

implementation of the NBS and to maximize the sustainability and scope of results, a systematic approach to the capacity building should be applied. This includes also social and economic aspects, as well as people, organization, institutions, and a favorable environment for policies based on assessed needs. According to the perception of some stakeholders, the institutional issue is an important obstacle that can slow down the efficiency of the implementation of the NBS. Lack of risk awareness and low level of institutional cooperation must be addressed before the implementation of the NBS. Integrated socio-institutional and economic policies for risk awareness, training, and capacity building, institutional cooperation protocols should be established with stakeholders and decision-makers.

The existence of a comprehensive socio-economic analysis allows the consideration of significant indicators, which should identify key segments in terms of protection and action during floods. The result of the socio-economic analysis is a vulnerability assessment for each area with significant flood risk. Socio-economic indicators are grouped into several main categories concerning different social or economic aspects. The early warning system requires full coordination and cooperation between the authorities and relevant agencies at different administrative levels and units. A prerequisite for establishing such a system is the establishment of an adequate monitoring network, which requires clear command responsibility and division of responsibilities and effective coordination.

References

1. European Commission (2016) Topics: nature-based solutions. https://ec.europa.eu/research/environment/index.cfm?pg=nbs
2. Raymond CM, Berry P, Breil M, Nita MR, Kabisch N, de Bel M, Enzi V, Frantzeskaki N, Geneletti D, Cardinaletti M, Lovinger L, Basnou C, Monteiro A, Robrecht H, Sgrigna G, Muhari L, Calfapietra C (2017) An impact evaluation framework to support planning and evaluation of nature-based solutions projects. In: Report prepared by the EKLIPSE expert working group on nature-based solutions to promote climate resilience in urban areas. Centre for Ecology & Hydrology, Wallington, UK
3. Raymond CM, Frantzeskaki N, Kabisch N, Berry P, Breil M, Nita MR, Geneletti D, Calfapietra C (2017) A framework for assessing and implementing the co-benefits of nature-based solutions in urban areas. Environ Sci Pol 77:15–24. https://doi.org/10.1016/j.envsci.2017.07.008
4. Santoro S, Pluchinotta I, Pagano A, Pengal P, Cokan B, Giordano R (2019) Assessing stakeholders` risk perception to promote nature based solutions ad flood protection strategies: the case of the Glinscica river (Slovenia). Sci Total Environ 655:188–201
5. Rosenhead J, Mingers J (2005) Rational analysis for a problematic world revisited. 2nd edn. Wiley, Chichester
6. Renn O (1998) The role of risk perception for risk management. Reliab Eng Syst Saf 59:49–62
7. Flynn J, Slovic P, Mertz CK, Carlisle C (1999) Public support for earthquake risk mitigation in Portland, Oregon. Risk Anal 2:205–216
8. Bickerstaff K (2004) Risk perception research: socio-cultural perspectives on the public experience of air pollution. Environ Int 30(6):827–840. https://doi.org/10.1016/j.envint.2003.12.001
9. Figueiredo E, Valente S, Coelho C, Pinho L (2009) Coping with risk: analysis on the importance of integrating social perceptions on flood risk into management mechanisms– the case of the municipality of Aqueda, Portugal. J Risk Res 12(5):581–602
10. Harclerode MA, Lal P, Vedwan N, Wolde B, Miller ME (2016) Evaluation of the role of risk perception in stakeholder engagement to prevent lead exposure in an urban setting. J Environ Manag 184:132–142
11. Savadori L, Savio S, Nicotra E, Rumiati R, Finucane M, Slovic P (2004) Expert and public perception of risk from biotechnology. Risk Anal 24:1289–1299
12. Kabisch N, Korn H, Stadler J, Bonn A (2017) Nature-based solutions for societal goals under climate change in urban areas – synthesis and ways forward. In: Kabisch N, Korn H, Stadler J, Bonn A (eds) Nature-based solutions to climate change adaptation in urban areas – linkages between science, policy and practice. Springer, Berlin
13. Maes J, Jacobs S (2017) Nature-based solutions for Europe's sustainable development. Conserv Lett 10(1):121–124
14. McPhearson H, Kabisch G (2016) Advancing understanding of the complex nature of urban systems. Ecol Indic 70:566–573. https://doi.org/10.1016/J.ECOLIND.2016.03.054
15. Eisenack M, Hoffmann K, Oberlack P, Rotter T (2014) Explaining and overcOoming barriers to climate change adaptation. Nat Clim Chang 4:867. https://doi.org/10.1038/nclimate2350

16. Sarabi S, Han Q, Romme AG, de Vries B, Valkenburg R, den Ouden E (2020) Uptake and implementation of nature-based solutions: an analysis of barriers using interpretive structural modeling. J Environ Manag 270:110749. https://doi.org/10.1016/j.jenvman.2020.110749
17. Kabisch S, Korn B, Frantzeskaki P, Naumann D, Artmann H, Knapp K, Stadler, Zaunberger B (2016) Nature-based solutions to climate change mitigation and adaptation in urban areas. Ecol Soc 21(2). https://doi.org/10.5751/ES-08373-210239
18. Zuniga-Teran S, de Vito G, Ward S, Hart B (2019) Challenges of mainstreaming green infrastructure in built environment professions. J Environ Plann Manag 1:23. https://doi.org/10.1080/09640568.2019.1605890
19. Droste N, Schröter-Schlaack C, Hansjürgens B, Zimmermann H (2017) Implementing nature-based solutions in urban areas: financing and governance aspects. In: Kabisch K, Stadler B (eds) Nature-based solutions to climate change adaptation in urban areas: linkages between science, policy and practice. Springer, Berlin, pp 307–321. https://doi.org/10.1007/978-3-319-56091-5_18
20. Olorunkiya F, Wilkinson (2012) Risk: a fundamental barrier to the implementation of low impact design infrastructure for urban stormwater control. J Sustain Dev 5(9). https://doi.org/10.5539/jsd.v5n9p27
21. Liu J (2018) Green infrastructure for sustainable urban water management: practices of five forerunner cities. Cities 74:126–133. https://doi.org/10.1016/J.CITIES.2017.11.013
22. Pasquini C, Ziervogel (2013) Facing the heat: barriers to mainstreaming climate change adaptation in local government in the Western Cape
23. Hawxwell M, Maciulyt E, Sautter D (2019) Municipal governance for nature-based solutions. https://unalab.eu/system/files/2019-10/Municipal_Governance_for_Nature-based_Solutions_2019-10-24_1746.pdf
24. Wamsler W, Hanson AO, Stålhammar B, Falck G, Oskarsson S, Torffvit Z (2020) Environmental and climate policy integration: targeted strategies for overcoming barriers to nature-based solutions and climate change adaptation. J Clean Prod 247:119154. https://doi.org/10.1016/J.JCLEPRO.2019.119154
25. Cohen-Schacham E, Walters G, Janzen C, Maginnis S (2016) In: IUCN (ed) Naturebased solutions to address global societal challenges. IUCN, Gland, Switzerland. pp xiii+97
26. Denjean B, Altamirano MA, Graveline N, Giordano R, van der Keur P, Moncoulon D, Weinberg J, Máñez Costa M, Kozinc Z, Mulligan M, Pengal P, Matthews J, Van Cauwenbergh N, López Gunn E, Bresch DN (2017) Natural assurance scheme: a level playing field framework for Green-Grey infrastructure development. Environ Res 159:24–38. https://doi.org/10.1016/j.envres.2017.07.006
27. European Environment Agency (2017) Technical report no 14/2017. Green infrastructure and flood management. Promoting cost-efficient flood risk reduction via green infrastructure solutions
28. Nesshöver C, Assmuth T, Irvine KN, Rusch GM, Waylen KA, Delbaere B, Haase D, Jones-Walters L, Keune H, Kovacs E, Krauze K, Külvik M, Rey F, van Dijk J, Vistad OI, Wilkinson ME, Wittmer H (2016) The science, policy and practice of nature-based solutions: an interdisciplinary perspective. Sci Total Environ. https://doi.org/10.1016/j.scitotenv.2016.11.106
29. Dong X, Guo H, Zeng S (2017) Enhancing future resilience in urban drainage system: green versus grey infrastructure. Water Res 124:280–289
30. World Bank (2017) Implementing nature-based flood protection: principles and implementation guidance. World Bank, Washington
31. Sayers P, Yuanyuan L, Galloway G, Penning-Rowsell E, Fuxin S, Kang W, Yiwei C, Le Quesn T (2013) Flood risk management, a strategic approach. UNESCO, Paris
32. Stefanovic M, Gavrilovic Z, Bajcetic R (2014) Local community and torrential flood issues, handbook for local community and civil society organizations. Organization for European Security and Cooperation, Mission to Serbia
33. Slovic P, Finucane ML, Peters E, MacGregor DG (2004) Risk as analysis and risk as feelings: some thoughts about affect, reason, risk, and rationality. Risk Anal 24:31–322

34. Boholm Å (2003) The cultural nature of risk: can there be an anthropology of uncertainty? Ethnos 68(2):159–178
35. Giordano R, D'Agostino D, Apollonio C, Lamaddalena N, Vurro M (2013) Bayesian belief network to support conflict analysis for groundwater protection: the case of the Apulia region. J Environ Manag 115:136–146. https://doi.org/10.1016/j.jenvman.2012.11.011
36. Bisaro A, Hinkel J (2016) Governance of social dilemmas in climate change adaptation. Nat Clim Change 6(4):354
37. Ostrom E (2005) Understanding institutional diversity. Princeton University Press, Princeton
38. Van Loon-Steensma J, Slim PA (2012) The impact of erosion protection by stone dams on salt-marsh vegetation on two Wadden Sea barrier islands. J Coast Res 29(4):783–796
39. Bowler DE, Buyung-Ali L, Knight TM, Pullin AS (2010) Urban greening to cool towns and cities: a systematic review of the empirical evidence. Landsc Urban Plann 97(3):147–155
40. Rizvi AR (2014) Nature based solutions for human resilience: a mapping analysis of IUCN's ecosystem-based adaptation projects, IUCN, Geneva. https://portals.iucn.org/library/sites/library/files/documents/Rep-2014-008.pdf
41. Cohen-Shacham E, Walters G, Janzen C, Maginnis S (2016) Naturebased Solutions to Address Global Societal Challenges, xii, IUCN, Gland, Switzerland
42. Grothmann T, Reusswig F (2006) People at risk of flooding: why some residents take precautionary action while others do not. Nat Hazards 38:101–120
43. Renn O (2004) Perception of risks. The Geneva papers on risk and insurance. 29(1):102–114
44. Frewer L (2004) The public and effective risk communication. Toxicol Lett 149:391–397
45. Gavin NT, Leonard-Milsom L, Montgomery J (2011) Climate change, flooding and the media in Britain. Public Understand Sci 20(3):422–438
46. Kreibich H, Seifert I, Thieken AH et al (2011) Recent changes in flood preparedness of private households and business in Germany. Reg Environ Chang 11:59–71
47. Lamond JE, Proverbs DG (2009) Resilience to flooding: lessons from international comparison. Urban Des Planning 162(2):63–70
48. Thieken AH, Kreibich H, Müller M et al (2007) Coping with floods: preparedness, response and recovery of flood-affected residents in Germany in 2002. Hydrol Sci J 52(5):1016–1037
49. Wachinger G, Renn O, Begg C et al (2013) The risk perception paradox—implications for governance and communication of natural hazards. Risk Anal 33(6):1049–1065
50. Cologna V, Bark RH, Paavola J (2017) Flood risk perceptions and the UK media: moving beyond "once in a lifetime" to "be prepared" reporting. Clim Risk Manag 17(2017):1–10. https://doi.org/10.1016/j.crm.2017.04.005
51. Kasperson RE, Renn O, Slovic P et al (1988) The social amplification of risk: a conceptual framework. Risk Anal 8(2):177–187
52. Kingdon JW (1995) Agendas, alternatives, and public policies. Longman
53. Moghadas M, Asadzadeh A, Vafeidis A, Fekete A, Kottera T (2019) A multi-criteria approach for assessing urban flood resilience in Tehran, Iran. Int J Disaster Risk Reduct 35:101069. https://doi.org/10.1016/j.ijdrr.2019.101069
54. Burton CG (2012) The development of metrics for community resilience to natural disasters, University of South Carolina. https://doi.org/10.1007/s13398-014-0173-7.2
55. Cutter SL, Burton CG, Emrich CT (2010) Disaster resilience indicators for benchmarking baseline conditions. J Homel Secur Emerg Manag 7(14):1732. https://doi.org/10.2202/1547-7355
56. Cutter SL, Ash KD, Emrich CT (2014) The geographies of community disaster resilience. Glob Environ Chang 29:65–77. https://doi.org/10.1016/jgloenvcha.2014.08.005
57. Renschler CS, Frazier E, Arendt L, Cimellaro GP, Reinhorn M, Bruneau M (2010) Community resilience indices are integral of the geospatial - temporal functionality of components, or dimensions, of resilience population –Q. http://www.mceer.buffalo.edu/pdf/report/10-0006.pdf
58. Asare-Kyei D, Renaud FG, Kloos J, Walz Y, Rhyner J (2017) Development and validation of risk profiles of west African rural communities facing multiple natural hazards. PLoS One 12(3)

59. Satta A, Puddu M, Venturini S, Giupponi C (2017) Assessment of coastal risks to climate change related impacts at the regional scale: the case of the Mediterranean region. Int J Disast Risk Reduct 24:284–296
60. Sudmeier-Rieux K (2011) On landslide risk, resilience and vulnerability of mountain communities in central-eastern Nepal, PhD dissertation, University of Lausanne
61. Fedele G, Locatelli B, Djoudi H (2017) Mechanisms mediating the contribution of ecosystem services to human Well-being and resilience. Ecosyst Serv 28:43–54
62. Leal Filho W, Modesto F, Nagy GJ, Saroar M, Yannick Toamukum N, Ha'apio M (2018) Fostering coastal resilience to climate change vulnerability in Bangladesh, Brazil, Cameroon and Uruguay: a cross-country comparison, Mitig. Adapt Strategies Glob Change 23(4):579–602
63. Hewitt K (1997) Regions of risk: a geographical introduction to disasters. Longman, Harlow
64. Rashed T, Weeks J (2003) Assessing vulnerability to earthquake hazards through spatial multicriteria analysis of urban areas. Int J Geogr Inf Sci 17(6):547–576
65. Gari SR, Newton A, Icely JD (2015) A review of the application and evolution of the DPSIR framework with an emphasis on coastal social-ecological systems. Ocean Coast Manag 103:63–77. https://doi.org/10.1016/j.ocecoaman.2014.11.013
66. Menz MHM, Dixon KW, Hobbs RJ (2013) Hurdles and opportunities for landscape scale restoration. Science (80-) 339:526–527. https://doi.org/10.1126/science.1228334
67. Svarstad H, Petersen LK, Rothman D, Siepel H, Wätzold F (2008) Discursive biases of the environmental research framework DPSIR. Land Use Policy 25:116–125. https://doi.org/10.1016/j.landusepol.2007.03.005
68. Tscherning K, Helming K, Krippner B, Sieber S, Paloma SGY (2012) Does researchapplying the DPSIR framework support decision making? Land Use Policy 29:102–110. https://doi.org/10.1016/j.landusepol.2011.05.009
69. Connop S, Vandergert P, Eisenberg B, Collier MJ, Nash C, Clough J, Newport D (2016) Renaturing cities using a regionally-focused biodiversity-led multifunctional benefits approach to urban green infrastructure. Environ Sci Pol 62:1–13. https://doi.org/10.1016/j.envsci.2016.01.013
70. Easterling WE (1997) Why regional studies are needed in the development of fullscale integrated assessment modeling of global change precesses. Global Environ Change 7 (4):337–356
71. Holman IP, Loveland PJ, Nicholls RJ, Shackley S, Berry PM, Rounsevell MDA, Audsley E, Harrison PA, Wood R (2002) REGIS – regional climate change impact response studies in East Anglia and North West England. www.UKCIP.org.uk
72. Parson EA, Fisher Vanden K (1997) Integrated assessment models of global climate change. Annu Rev Energy Environ 22:589–628
73. Peirce M (1998) Computer-based models in integrated environmental assessment. A report produced for the European Environment Agency. Technical report no 14
74. Botzen WJW, Deschenes O, Sanders M (2019) The economic impacts of natural disasters: a review of models and empirical studies. Rev Environ Econ Pol. https://doi.org/10.1093/reep/rez004
75. Fernández FJ, Blanco M (2015) Modelling the economic impacts of climate change on global and European agriculture. Review of economic structural approaches. Economics 9 (2015–10):1–53
76. Dowlatabadi H (1998) Sensitivity of climate change mitigation estimates to assumptions about technical change. Energy Econ 20(5):473–493
77. Ward PJ et al (2015) Usefulness and limitations of global flood risk models. Nat Clim Chang 5 (8):712–715. https://doi.org/10.1038/nclimate2742
78. Nordhaus WD (1992) "The 'DICE' model: background and structure of a dynamic integrated climate-economy model of the economics of global warming
79. Nordhaus WD (2017) Revisiting the social cost of carbon. Proc Natl Acad Sci U S A 114 (7):1518–1523. https://doi.org/10.1073/pnas.1609244114

80. Tol RSJ (2018) The economic impacts of climate change. Rev Environ Econ Pol:4–25. https://doi.org/10.1093/reep/rex027
81. de Bruin KC, Dellink RB, Tol RSJ (2009) AD-DICE: an implementation of adaptation in the DICE model. Clim Chang 95(1–2):63–81. https://doi.org/10.1007/s10584-008-9535-5
82. Dumas P, Ha-Duong M (2013) Optimal growth with adaptation to climate change. Clim Chang 117(4):691–710. https://doi.org/10.1007/s10584-012-0601-7
83. Kuik O (2017) A simple river floods damage model for the fund model. Amsterdam
84. Anthoff D, Tol RSJ (2014) The climate framework for uncertainty, negotiation and distribution (FUND). technical description, version 3.9, 26. Www.Fund-Model.Org, pp 1–69. http://www.fund-model.org/versions
85. Ignjacevic P, Botzen WJ, Estrada F, Kuik O, Ward P, Tiggeloven T (2020) CLIMRISK-RIVER: accounting for local river flood risk in estimating the economic cost of climate change. Environ Model Softw 132:104784. https://doi.org/10.1016/j.envsoft.2020.104784
86. Horita FE, Albuquerque JP, Marchezini V, Mendiondo EM (2016) A qualitative analysis of the early warning process in disaster management. In: Proceedings of the 13th international conference on information systems for crisis response and management (ISCRAM), pp 1–9
87. Krzhizhanovskaya VV, Shirshov GS, Melnikova NB, Belleman RG, Rusadi FI, Broekhuijsen BJ et al (2011) Flood early warning system: design, implementation and computational modules. Proc Comput Sci 4:106–115
88. Adeyeye K, Bairi A, Emmitt S, Hyde K (2017) Socially-integrated resilience in building-level water networks using smart microgrid+net. In: 7th international conference on building resilience; using scientific knowledge to inform policy and practice in disaster risk reduction, ICBR2017, 27–29 November 2017, Bangkok, Thailand Procedia Engineering 212(2018): pp 39–46. https://doi.org/10.1016/j.proeng.2018.01.006

Part IV
Conclusion

Conclusions

Carla S. S. Ferreira, Zahra Kalantari, Thomas Hartmann, and Paulo Pereira

Contents

Abstract Mitigating and adapting to increasing flood risk driven by climate change and growing urbanization is still a challenge for humanity. Over the last decades, the Nature Based Solutions (NBS) approach has received increasing interest from governments, academia and society, but its implementation is still in its infancy. This volume presented an up-to-date compilation about the current knowledge on

C. S. S. Ferreira (✉)
Department of Physical Geography, Stockholm University, and Bolin Centre for Climate Research, Stockholm, Sweden

Navarino Environmental Observatory, Messinia, Greece
e-mail: carla.ferreira@natgeo.su.se

Z. Kalantari
Department of Physical Geography, Stockholm University, and Bolin Centre for Climate Research, Stockholm, Sweden

Navarino Environmental Observatory, Messinia, Greece

Department of Sustainable Development, Environmental Science and Engineering, KTH Royal Institute of Technology, Stockholm, Sweden
e-mail: zahra.kalantari@natgeo.su.se

T. Hartmann
Wageningen University, Wageningen, The Netherlands

J.E.P. University Ústí nad Labem, Usti nad Labem, Czech Republic
e-mail: thomas.hartmann@wur.nl

P. Pereira
Environmental Management Laboratory, Mykolas Romeris University, Vilnius, Lithuania

Carla S. S. Ferreira, Zahra Kalantari, Thomas Hartmann, and Paulo Pereira (eds.), *Nature-Based Solutions for Flood Mitigation: Environmental and Socio-Economic Aspects*, Hdb Env Chem (2022) 107: 507–514, DOI 10.1007/698_2021_777,

Published online: 3 July 2021

NBS regarding their potential to mitigate several types of floods recorded in different environments, the associated advantages and limitations, and the barriers affecting their wider implementation. This concluding chapter provides a synthesis of the main messages of the chapters included in this volume, and it highlights the relevant aspects that must be considered for an effective and broader application of NBS. In this regard, particular attention is given to the need to maximize their multifunctionality and to consider the transdisciplinary nature of the NBS approach.

Keywords Barriers, Flood mitigation, Multifunctionality, Nature-based solutions, Transdiciplinarity

Flooding is a complex problem affecting many civilizations worldwide. Despite all the efforts based on numerous measures and approaches developed and implemented over centuries, flood risks still represent a severe threat and a major societal challenge, expected to be aggravated in the future given the climate change projections for many geographical regions and the urbanization trends. The traditional flood protection and flood risk management approach can only provide limited answers to this problem. Nature-based solutions (NBS) promise a useful complement to grey infrastructure, but its implementation is still in its infancy. During the last decades, NBS for flood risk management have received increasing attention from policy and academia, with different projects being implemented at a wide range of scales in several areas all over the world. Experience in implementing NBS for flood mitigation is still scarce and the lack of evidence about their effectiveness remains an important barrier for a wider implementation of this approach [1]. By providing an up-to-date state of the art about NBS measures and their role on flood mitigation in rural, urban, and coastal areas, based on both literature review (Part I) and case studies from different parts of Europe (Part II), this book illustrated a wide variety of NBS already implemented and compiled the information available about their effectiveness on flood mitigation. The case studies, although adapted to local environmental and socio-economic contexts, provide practical examples and lessons that may be relevant to support NBS application in other areas.

Scale is an important aspect of NBS, playing an important role in their effectiveness to mitigate floods (*Kuriqi and Hysa*, Multi-Dimensional Aspects of Floods: Nature-Based Mitigation Measures from Basin to River Reach Scale). Some NBS measures proved relevant to mitigate local floods, e.g., in urban (*Pinto* et al., Assessment of NBS impact on pluvial flood regulation within urban areas: a case study in Coimbra, Portugal), mountain (*Nadal-Romero* et al., Impacts of land abandonment on flood mitigation in Mediterranean mountain areas), and coastal (*Kotsev* et al., Long-term impacts of land use change upon the natural flood storage reservoirs along the North Bulgarian Black Sea coast) areas, but effective flood risk mitigation requires a large-scale implementation of NBS as showcased by *Potočki* et al. (Hydrological aspects of nature-based solutions in flood mitigation in the Danube River Basin in Croatia – green vs. grey approach) and *Johnen* et al. (Modeling and evaluation of the effect of afforestation on the runoff generation within the Glinščica river catchment (Central Slovenia)). Some case studies presented have developed improvements on widely used methodologies by integrating well established concepts such as "sponge cities" (*Pavesi and Pezzagno*, From Sponge Cities to Sponge Landscapes with Nature-Based Solutions: a multidimensional approach to map suitable rural areas for flood mitigation and landscaping), surface runoff source areas (*Jakubínský* et al., Identification of surface runoff source areas as a tool for projections of NBS in water management) and "landscape connectivity" (*Kalantari* et al., Using landscape connectivity to identify suitable locations for nature-based solutions to reduce flood risk) to support the identification of the best sites to implement NBS at catchment scale.

One of the important advantages of NBS, and main reasons behind the increasing interest in this approach, is the multifunctionality of the solutions and the numerous environmental, social, and economic benefits provided by the associated ecosystem services [2]. Part I of this volume includes a synthesis of the ecosystem services provided by NBS, particularly those implemented in coastal (*Inácio* et al., Nature Based Solutions to mitigate coastal floods and associated socio-ecological impacts), urban (*Ferreira* et al., Nature-based solutions for flood mitigation and resilience in urban areas), and industrial (*Ilić* et al., The role of plants in water regulation and pollution control) areas, with a special focus on environmental chemical pollution (*Pereira* et al., Nature-based solutions impact on urban environment chemistry: Air, soil and water) and pathogen dispersion (*Bett* et al., The role of floods on pathogen dispersion) mitigation. Although empirical evidence on NBS co-benefits has been widely discussed, case studies with real assessments are still lacking. *Johnen* et al. (Modeling and evaluation of the effect of afforestation on the runoff generation within the Glinščica river catchment (Central Slovenia)) contributed to fulfil this knowledge gap, by performing cost–benefit analysis to assess the impact of afforestation on several co-benefits driven by the ecosystem services. By performing a literature review on current methodologies and frameworks used to perform environmental and socio-economic analysis of NBS, *Figurek* (Socio – Economical aspects of NBS) provides a relevant contribution on how to produce evidence on NBS relevant to support upscaling and a wider application of NBS.

As multifunctional measures and strategies, NBS approach brings together different disciplines. The implementation of NBS not only requires the collaboration between hydrologists, geographers, lawyers, social scientists, economists, ecologists, and engineers, but also involves, for example, spatial planning, governance issues, and public acceptance (Part III of the book). In this volume, *Bogdzevič* et al. (Legal implications of natural floods management – Lithuania case study) show the dispersion and discoordination between several legal instruments affecting flood risk management and the associated difficulties in implementing NBS, and *Kaufmann* et al. (Win-win for everyone? Reflecting on nature-based solutions for flood risk management from an environmental justice perspective) provide a deep reflection on NBS from an environmental justice perspective. To be effective, most NBS need to be implemented on private land which conveys additional challenges. In this volume, *Bogdzevič and Kalinauskas* (Sticks, carrots and sermons for implementing NBS on private property land) present different mechanisms to support NBS implementation on private land, discussing the experience in several European countries as well as some in North America, Asia, and Africa. Nevertheless, *Macháč* et al. (What nature-based flood protection solutions are best perceived by people? Lessons from field research in Czechia) demonstrated that stakeholders are more willing to implement some types of NBS in detriment of others. These chapters provide relevant contributions for better understanding the complexity in implementing NBS, but also present interesting ideas to overcome some of these problems.

The different disciplines involved in the implementation of NBS also raise communication challenges between different fields of knowledge and different groups of stakeholders involved in the implementation of these type of solutions. Although the need to improve communication between different experts and land-owners has been previously discussed in other books [3], this volume presents some interesting communication strategies to surpass the problem. For example, *Warachowska* et al. (A cooperative game for upstream-downstream river flooding risk prevention in four European river basins) present a game theory strategy to support communication and enhance collaboration between different stakeholders. The game theory approach promises to develop consensus and improve decision-making progress, which maximizes the interest of all the parts involved in NBS implementation.

So, this volume gathers a broad perspective on NBS by involving different disciplinary perspectives and in-depth knowledge of case studies. The collection of contributions allows some conclusions on the knowledge for understanding the value and limits of NBS and on the challenges of collaboration between multiple disciplines.

1 Enhanced Knowledge for Understanding the Value and Limits of Nature-Based Solutions

Few studies have monitored the impact of NBS, and the lack of knowledge is still a barrier to the widespread implementation of this approach. The existing barriers in the NBS implementation are not mutually exclusive, which implies that the implementation of NBS as a strategic complement to traditional flood risk management would require policies that are effective to realize the measures but at the same time also acknowledge the interdependencies among the different NBS. Such policy implementation is more complex than grey infrastructure realization. This complexity demands much attention for the preparation, planning, and governance of NBS. The understanding of the important technical but also socio-economic factors and their causalities contribute to the reduction of the potential barriers and their overcoming.

Overall, this volume presents novel aspects in NBS research such as the role of the plants in water regulation and water pollution control in urban and mining areas.

This is key to understand plant-based technologies for the purification and remediation of polluted water resources. Also, the application of game theory in the field of environmental chemistry was essential for understanding pollutants transport during flood events and flood risk mitigation based on NBS. This volume provides a novel and robust revision of the impacts of the air, soil, and water pollution on ecosystems, biodiversity, human health, and the NBS that can be used to minimize these impacts. For instance, the application of NBS for flood mitigation is discussed, for example, in the context of providing "space for water" and inclusion of natural retention areas as flexible and effective solutions to enhance the local biodiversity. Furthermore, new advances in the understanding of flood impacts on pathogen transmission are introduced and how this could be controlled looking at different drivers including climate change, land use change (e.g., urbanization and agricultural intensification), and changes in population are deliberated.

Important advances are made in the knowledge of the different environmental justice issues related with NBS, through recognition justice, procedural justice, and distributive justice by focusing on NBS projects trans-local consequences, whereas the current critical literature is focused particularly on urban NBS. In addition, this volume showcases pitfalls of incoherent legislation and formal institutions to implement flood mitigation, especially using NBS on the private property.

New insights are provided in the introduction of NBS large-scale approaches for appropriate placement and design by mapping and assessing the connectivity and identifying the major flow paths through the landscape. Also, new evidences are identified regarding the NBS cost-effective and long-term efficiency. Based on this principle, new examples are provided regarding *Sponge City* policies and identifying areas of potential flood retention in rural areas. This aspect is key to reduce the flood impacts on urban areas. For an effective flood management, public preferences and awareness are essential for the correct implementation of NBS. This aspect is explored in this volume and novel features on the use of agriculture areas to mitigate pluvial floods are assessed by exploring preferences of the residents in the river basins.

2 Nature-Based Solutions and Collaboration of Disciplines

Bringing together knowledge and experiences from NBS for flood mitigation can help to identify research gaps but also showcase the merits and shortcomings of NBS for flood mitigation. One of the aspects that this volume shows already in the outline is that NBS involve many different disciplines. While classical flood protection mainly focused on engineering solutions based on hydrological models, flood risk management already started to include spatial planning, ecology, and economics [4]. This is because flood risk management includes vulnerabilities and flood adaptation. NBS include flood mitigation outside the realm of traditional water management. This means that even more disciplines – each with its specific approaches – are involved. This means that the different disciplinary perspectives need to be incorporated – hydrology, engineering, law, planning, governance, ecology, soil science, agriculture, economics, and many more. The chapters included in this volume illustrate these various disciplinary perspectives and the frictions resulting from these approaches. Nature-based flood risk management is a challenge of multiple disciplines [4].

There are various ways in which disciplines collaborate for NBS design and implementation. Three main forms can be distinguished based on the degree of disciplinary integration [5]. Multidisciplinarity describes situations in which independent disciplines work in parallel on a specific project or challenge – each with its own disciplinary approaches and within its disciplinary boundaries [6]. Multidisciplinarity does not entail crossing disciplinary boundaries beyond sharing and comparing results and knowledge from the involved disciplines. Such disciplinary collaboration is useful for problems where multiple perspectives are needed but can be implemented separately or if the coordination of the expertise is institutionalized in one way or another. Interdisciplinarity goes beyond multidiciplinarity in terms of integration and cooperation of disciplines [7]. Interdisciplinarity entails integration of concepts, methods, and principles from independent disciplines [8]. This requires a common framework, understanding of terminology, and approaches that are fed from the different disciplines. Interdisciplinarity is suitable for complex problems that transcend disciplinary knowledge [6]. While multi- and interdisciplinarity can remain in the conceptual and analytical realm, transdisciplinarity is aiming at solving a common societal problem. Sometimes, transdisciplinarity is described as bridging academia and society or policy domains [8]. Transdisciplinarity is considered the most challenging form of disciplinary collaboration.

NBS seem to be largely at the stage of multidisciplinarity, where academics still try to figure out the mutual and different understandings of NBS and its implications. The complexity of the issue seems to point at the necessity to develop an interdisciplinary understanding. There are various hints for that: The multifunctional nature of NBS, potentially serving flood risk management, ecology, tourism and recreation, and agriculture at the same time, points at the need to integrate the knowledge and requirements from each discipline to benefit from the potential strength of NBS. This calls for interdisciplinary collaboration of disciplines. The spatial aspect of NBS asks for multiple disciplines to integrate even more: NBS require more land than traditional flood risk management. Traditional water management is fairly dominant in the spatial realm of rivers and water bodies – i.e., between embankments, but NBS entail encroaching on new spatial governance arenas which are barely served by certain disciplines. If water management wants to realize measures in the hinterland forests, on agricultural land, and on private land, the respective disciplines such as forest, agriculture and law need to contribute. This goes beyond academic collaboration – the need for more land in particular calls for a transdisciplinary approach to NBS.

While integration and inter- and transdisciplinarity sometimes seem to be treated as a panacea for all complex societal problems, this volume illustrates that the challenges but also the potentials of NBS are indeed crossing disciplinary boundaries, calling for disciplinary collaboration. By pointing out the potentials but also the frictions and tensions of the multiple disciplines involved in flood mitigation via NBS, this volume is a modest contribution towards a common conceptual framework and analytical methods. The volume, however, also shows that while inter- and transdisciplinarity for NBS are still in its infancy, it can be concluded from the different chapters that a common perspective on land can be the connecting bin that brings together essential disciplines for nature-based flood mitigation.

Acknowledgement This book was prepared in the context of the COST Action No. CA16209 Natural flood retention on private land, LAND4FLOOD (www.land4flood.eu), supported by COST (European Cooperation in Science and Technology) and can be considered an essential outcome of the project.

References

1. Kalantari Z, Ferreira CSS, Deal B, Destouni G (2019) Nature-based solutions for meeting environmental and socio-economic challenges in land management and development (editorial). Land Degrad Dev:1–4. https://doi.org/10.1002/ldr.3264
2. EC – European Commission (2015) Towards an EU research and innovation policy agenda for nature-based solutions & re-naturing cities. Final report of the Horizon 2020 Expert group on "Nature-Based Solutions & Re-Naturing Cities". European Commission, Brussels
3. Hartmann T, Slavíková L, McCarthy S (eds) (2019) Nature-based flood risk management on private land. Disciplinary perspectives on a multidisciplinary challenge. Springer, Heidelberg
4. Hartmann T, Juepner R (2017) The flood risk management plan between spatial planning and water engineering. J Flood Risk Manag 10(2):143–144
5. Slatin C et al (2004) Conducting interdisciplinary research to promote healthy and safe employment in health care: promises and pitfalls. Public Health Rep 119(1):60–72
6. Collin A (2009) Multidisciplinary, interdisciplinary, and transdisciplinary collaboration: implications for vocational psychology. Int J Educ Vocat Guid 9(2):101–110
7. Stock P, Burton RJ (2011) Defining terms for integrated (multi-inter-trans-disciplinary) sustainability research. Sustainability 3(8):1090–1113
8. Lawrence RJ (2010) Deciphering interdisciplinary and transdisciplinary contributions. Transdiscipl J Eng Sci 1(1):125–130